Lecture Notes in Computer Science 9828

Commenced Publication in 1973
Founding and Former Series Editors:
Gerhard Goos, Juris Hartmanis, and Jan van Leeuwen

More information about this series at http://www.springer.com/series/7409

Sven Hartmann · Hui Ma (Eds.)

Database and Expert Systems Applications

27th International Conference, DEXA 2016
Porto, Portugal, September 5–8, 2016
Proceedings, Part II

 Springer

Editors
Sven Hartmann
Clausthal University of Technology
Clausthal-Zellerfeld
Germany

Hui Ma
Victoria University of Wellington
Wellington
New Zealand

ISSN 0302-9743 ISSN 1611-3349 (electronic)
Lecture Notes in Computer Science
ISBN 978-3-319-44405-5 ISBN 978-3-319-44406-2 (eBook)
DOI 10.1007/978-3-319-44406-2

Library of Congress Control Number: 2016947400

LNCS Sublibrary: SL3 – Information Systems and Applications, incl. Internet/Web, and HCI

Printed on acid-free paper

This Springer imprint is published by Springer Nature
The registered company is Springer International Publishing AG Switzerland

Preface

This volume contains the papers presented at the 27th International Conference on Database and Expert Systems Applications (DEXA 2016), which was held in Porto, Portugal, during September 5–8, 2016. On behalf of the Program Committee, we commend these papers to you and hope you find them useful.

Database, information, and knowledge systems have always been a core subject of computer science. The ever-increasing need to distribute, exchange, and integrate data, information, and knowledge has added further importance to this subject. Advances in the field will help facilitate new avenues of communication, to proliferate interdisciplinary discovery, and to drive innovation and commercial opportunity.

DEXA is an international conference series which showcases state-of-the-art research activities in database, information, and knowledge systems. The conference and its associated workshops provide a premier annual forum to present original research results and to examine advanced applications in the field. The goal is to bring together developers, scientists, and users to extensively discuss requirements, challenges, and solutions in database, information, and knowledge systems.

DEXA 2016 solicited original contributions dealing with any aspect of database, information, and knowledge systems. Suggested topics included but were not limited to:

- Acquisition, Modeling, Management and Processing of Knowledge
- Authenticity, Privacy, Security, and Trust
- Availability, Reliability and Fault Tolerance
- Big Data Management and Analytics
- Consistency, Integrity, Quality of Data
- Constraint Modeling and Processing
- Cloud Computing and Database-as-a-Service
- Database Federation and Integration, Interoperability, Multi-Databases
- Data and Information Networks
- Data and Information Semantics
- Data Integration, Metadata Management, and Interoperability
- Data Structures and Data Management Algorithms
- Database and Information System Architecture and Performance
- Data Streams, and Sensor Data
- Data Warehousing
- Decision Support Systems and Their Applications
- Dependability, Reliability and Fault Tolerance
- Digital Libraries, and Multimedia Databases
- Distributed, Parallel, P2P, Grid, and Cloud Databases
- Graph Databases
- Incomplete and Uncertain Data
- Information Retrieval

- Information and Database Systems and Their Applications
- Mobile, Pervasive, and Ubiquitous Data
- Modeling, Automation and Optimization of Processes
- NoSQL and NewSQL Databases
- Object, Object-Relational, and Deductive Databases
- Provenance of Data and Information
- Semantic Web and Ontologies
- Social Networks, Social Web, Graph, and Personal Information Management
- Statistical and Scientific Databases
- Temporal, Spatial, and High-Dimensional Databases
- Query Processing and Transaction Management
- User Interfaces to Databases and Information Systems
- Visual Data Analytics, Data Mining, and Knowledge Discovery
- WWW and Databases, Web Services
- Workflow Management and Databases
- XML and Semi-structured Data

Following the call for papers, which yielded 137 submissions, there was a rigorous review process that saw each paper reviewed by three to five international experts. The 39 papers judged best by the Program Committee were accepted for long presentation. A further 29 papers were accepted for short presentation.

As is the tradition of DEXA, all accepted papers are published by Springer. Authors of selected papers presented at the conference were invited to submit extended versions of their papers for publication in the Springer journal *Transactions on Large-Scale Data- and Knowledge-Centered Systems (TLDKS)*.

We wish to thank all authors who submitted papers and all conference participants for the fruitful discussions. We are grateful to Bruno Buchberger and Gottfried Vossen, who accepted to present keynote talks at the conference.

The success of DEXA 2016 is a result of the collegial teamwork from many individuals. We like to thank the members of the Program Committee and external reviewers for their timely expertise in carefully reviewing the submissions. We are grateful to our general chairs, Abdelkader Hameurlain, Fernando Lopes, and Roland R. Wagner, to our publication chair, Vladimir Marik, and to our workshop chairs, A Min Tjoa, Zita Vale, and Roland R. Wagner.

We wish to express our deep appreciation to Gabriela Wagner of the DEXA conference organization office. Without her outstanding work and excellent support, this volume would not have seen the light of day.

Finally, we would like to thank GECAD (Research Group on Intelligent Engineering and Computing for Advanced Innovation and Development) at ISEP (Instituto Superior de Engenharia do Porto) for being our hosts for the wonderful days in Porto.

July 2016   Sven Hartmann
                                                                        Hui Ma

Organization

General Chairs

Abdelkader Hameurlain   IRIT, Paul Sabatier University Toulouse, France
Fernando Lopes   LNEG - National Research Institute, Portugal
Roland R. Wagner   Johannes Kepler University Linz, Austria

Program Committee Chairs

Hui Ma   Victoria University of Wellington, New Zealand
Sven Hartmann   Clausthal University of Technology, Germany

Publication Chair

Vladimir Marik   Czech Technical University, Czech Republic

Program Committee

Afsarmanesh, Hamideh	University of Amsterdam, The Netherlands
Albertoni, Riccardo	Italian National Council of Research, Italy
Anane, Rachid	Coventry University, UK
Appice, Annalisa	Università degli Studi di Bari, Italy
Atay, Mustafa	Winston-Salem State University, USA
Bakiras, Spiridon	Michigan Technological University, USA
Bao, Zhifeng	National University of Singapore, Singapore
Bellatreche, Ladjel	ENSMA, France
Bennani, Nadia	INSA Lyon, France
Benyoucef, Morad	University of Ottawa, Canada
Berrut, Catherine	Grenoble University, France
Biswas, Debmalya	Swisscom, Switzerland
Bouguettaya, Athman	RMIT, Australia
Boussaid, Omar	University of Lyon, France
Bressan, Stephane	National University of Singapore, Singapore
Camarinha-Matos, Luis M.	Universidade Nova de Lisboa, Portugal
Catania, Barbara	DISI, University of Genoa, Italy
Ceci, Michelangelo	University of Bari, Italy
Chen, Cindy	University of Massachusetts Lowell, USA
Chen, Phoebe	La Trobe University, Australia
Chen, Shu-Ching	Florida International University, USA
Chevalier, Max	IRIT - SIG, Université de Toulouse, France
Choi, Byron	Hong Kong Baptist University, Hong Kong, SAR China

Christiansen, Henning	Roskilde University, Denmark
Chun, Soon Ae	City University of New York, USA
Cuzzocrea, Alfredo	University of Trieste, Italy
Dahl, Deborah	Conversational Technologies, USA
Darmont, Jérôme	Université de Lyon (ERIC Lyon 2), France
de vrieze, cecilia	Bournemouth University, UK, Switzerland
Decker, Hendrik	Ludwig-Maximilians-Universität München, Spain
Deng, Zhi-Hong	Peking University, China
Deufemia, Vincenzo	Università degli Studi di Salerno, Italy
Dibie-Barthélemy, Juliette	AgroParisTech, France
Ding, Ying	Indiana University, USA
Dobbie, Gill	University of Auckland, New Zealand
Dou, Dejing	University of Oregon, USA
du Mouza, Cedric	CNAM, France
Eder, Johann	University of Klagenfurt, Austria
El-Beltagy, Samhaa	Nile University, Egypt
Embury, Suzanne	The University of Manchester, UK
Endres, Markus	University of Augsburg, Germany
Fazzinga, Bettina	ICAR-CNR, Italy
Fegaras, Leonidas	The University of Texas at Arlington, USA
Felea, Victor	Al. I. Cuza University of Iasi, Romania
Ferilli, Stefano	University of Bari, Italy
Ferrarotti, Flavio	Software Competence Center Hagenberg, Austria
Fomichov, Vladimir	National Research University Higher School of Economics, Moscow, Russian Federation
Frasincar, Flavius	Erasmus University Rotterdam, The Netherlands
Freudenthaler, Bernhard	Software Competence Center Hagenberg, Austria
Fukuda, Hiroaki	Shibaura Institute of Technology, Japan
Furnell, Steven	Plymouth University, UK
Garfield, Joy	University of Worcester, UK
Gergatsoulis, Manolis	Ionian University, Greece
Grabot, Bernard	LGP-ENIT, France
Grandi, Fabio	University of Bologna, Italy
Gravino, Carmine	University of Salerno, Italy
Groppe, Sven	Lübeck University, Germany
Grosky, William	University of Michigan, USA
Grzymala-Busse, Jerzy	University of Kansas, USA
Guerra, Francesco	Università degli Studi Di Modena e Reggio Emilia, Italy
Guzzo, Antonella	University of Calabria, Italy
Hameurlain, Abdelkader	Paul Sabatier University, France
Hamidah, Ibrahim	Universiti Putra Malaysia, Malaysia
Hara, Takahiro	Osaka University, Japan
Hartmann, Sven	TU Clausthal, Germany
Hsu, Wynne	National University of Singapore, Singapore
Hua, Yu	Huazhong University of Science and Technology, China
Huang, Jimmy	York University, Canada

Huptych, Michal   Czech Technical University in Prague, Czech Republic
Hwang, San-Yih   National Sun Yat-Sen University, Taiwan
Härder, Theo   TU Kaiserslautern, Germany
Iacob, Ionut Emil   Georgia Southern University, USA
Ilarri, Sergio   University of Zaragoza, Spain
Imine, Abdessamad   Inria Grand Nancy, France
Ishihara, Yasunori   Osaka University, Japan
Jin, Peiquan   University of Science and Technology of China, China
Kao, Anne   Boeing, USA
Karagiannis, Dimitris   University of Vienna, Austria
Katzenbeisser, Stefan   Technische Universität Darmstadt, Germany
Kim, Sang-Wook   Hanyang University, Republic of Korea
Kleiner, Carsten   University of Applied Sciences and Arts Hannover,
                               Germany
Koehler, Henning   Massey University, New Zealand
Kosch, Harald   University of Passau, Germany
Krátký, Michal   Technical University of Ostrava, Czech Republic
Kremen, Petr   Czech Technical University in Prague, Czech Republic
Küng, Josef   University of Linz, Austria
Lammari, Nadira   CNAM, France
Lamperti, Gianfranco   University of Brescia, Italy
Laurent, Anne   LIRMM, University of Montpellier 2, France
Léger, Alain   FT R&D Orange Labs Rennes, France
Lhotska, Lenka   Czech Technical University, Czech Republic
Liang, Wenxin   Dalian University of Technology, China
Ling, Tok Wang   National University of Singapore, Singapore
Link, Sebastian   The University of Auckland, New Zealand
Liu, Chuan-Ming   National Taipei University of Technology, Taiwan
Liu, Hong-Cheu   University of South Australia, Australia
Liu, Rui   HP Enterprise, USA
Lloret Gazo, Jorge   University of Zaragoza, Spain
Loucopoulos, Peri   Harokopio University of Athens, Greece
Lumini, Alessandra   University of Bologna, Italy
Ma, Hui   Victoria University of Wellington, New Zealand
Ma, Qiang   Kyoto University, Japan
Maag, Stephane   TELECOM SudParis, France
Masciari, Elio   ICAR-CNR, Università della Calabria, Italy
May, Norman   SAP SE, Germany
Medjahed, Brahim   University of Michigan - Dearborn, USA
Mishra, Harekrishna   Institute of Rural Management Anand, India
Moench, Lars   University of Hagen, Germany
Mokadem, Riad   IRIT, Paul Sabatier University, France
Moon, Yang-Sae   Kangwon National University, Republic of Korea
Morvan, Franck   IRIT, Paul Sabatier University, France
Munoz-Escoi, Francesc   Universitat Politecnica de Valencia, Spain
Navas-Delgado, Ismael   University of Málaga, Spain

Ng, Wilfred	Hong Kong University of Science and Technology, Hong Kong, SAR China
Ozsoyoglu, Gultekin	Case Western Reserve University, USA
Pallis, George	University of Cyprus, Cyprus
Paprzycki, Marcin	Polish Academy of Sciences, Warsaw Management Academy, Poland
Pastor Lopez, Oscar	Universidad Politecnica de Valencia, Spain
Pivert, Olivier	Ecole Nationale Supérieure des Sciences Appliquées et de Technologie, France
Pizzuti, Clara	ICAR-CNR, Italy
Poncelet, Pascal	LIRMM, France
Pourabbas, Elaheh	National Research Council, Italy
Qin, Jianbin	University of New South Wales, Australia
Rabitti, Fausto	ISTI, CNR Pisa, Italy
Raibulet, Claudia	Università degli Studi di Milano-Bicocca, Italy
Ramos, Isidro	Technical University of Valencia, Spain
Rao, Praveen	University of Missouri-Kansas City, USA
Resende, Rodolfo F.	Federal University of Minas Gerais, Brazil
Roncancio, Claudia	Grenoble University/LIG, France
Ruckhaus, Edna	Universidad Simon Bolivar, Venezuela
Ruffolo, Massimo	ICAR-CNR, Italy
Sacco, Giovanni Maria	University of Turin, Italy
Saltenis, Simonas	Aalborg University, Denmark
Sansone, Carlo	Università di Napoli Federico II, Italy
Sarda, N.L.	I.I.T. Bombay, India
Savonnet, Marinette	University of Burgundy, France
Sawczuk da Silva, Alexandre	Victoria University of Wellington, New Zealand
Scheuermann, Peter	Northwestern University, USA
Schewe, Klaus-Dieter	Software Competence Center Hagenberg, Austria
Schweighofer, Erich	University of Vienna, Austria
Sedes, Florence	IRIT, Paul Sabatier University, Toulouse, France
Selmaoui, Nazha	University of New Caledonia, New Caledonia
Siarry, Patrick	Université Paris 12 (LiSSi), France
Skaf-Molli, Hala	Nantes University, France
Srinivasan, Bala	Monash University, Australia
Sunderraman, Raj	Georgia State University, USA
Taniar, David	Monash University, Australia
Teisseire, Maguelonne	Irstea - TETIS, France
Tessaris, Sergio	Free University of Bozen-Bolzano, Italy
Teste, Olivier	IRIT, University of Toulouse, France
Teufel, Stephanie	University of Fribourg, Switzerland
Teuhola, Jukka	University of Turku, Finland
Thevenin, Jean-Marc	University of Toulouse 1 Capitole, France
Torra, Vicenc	University of Skövde, Sweden
Truta, Traian Marius	Northern Kentucky University, USA

Tzouramanis, Theodoros	University of the Aegean, Greece
Vaira, Lucia	University of Salento, Italy
Vidyasankar, Krishnamurthy	Memorial University of Newfoundland, Canada
Vieira, Marco	University of Coimbra, Portugal
Wang, Guangtao	NTU, Singapore
Wang, Junhu	Griffith University, Australia
Wang, Qing	The Australian National University, Australia
Wang, Wendy Hui	Stevens Institute of Technology, USA
Wijsen, Jef	Université de Mons, Belgium
Wu, Huayu	Institute for Infocomm Research, A*STAR, Singapore
Yang, Ming Hour	Chung Yuan Christian University, Taiwan
Yang, Xiaochun	Northeastern University, China
Yin, Hongzhi	The University of Queensland, Australia
Yokota, Haruo	Tokyo Institute of Technology, Japan
Zhao, Yanchang	RDataMining.com, Australia
Zhu, Qiang	The University of Michigan, USA
Zhu, Yan	Southwest Jiaotong University, China

External Reviewers

Liliana Ibanescu	UMR MIA-Paris, INRA, France
Paola Podestà	Italian National Council of Research, Italy
Luke Lake	Department of Human Services, Australia
Roberto Corizzo	University of Bari, Italy
Pasqua Fabiana Lanotte	University of Bari, Italy
Corrado Loglisci	University of Bari, Italy
Gianvito Pio	University of Bari, Italy
Weiqing Wang	The University of Queensland, Australia
Stephen Carden	Georgia Southern University, USA
Arpita Chatterjee	Georgia Southern University, USA
Tharanga Wickramarachchi	Georgia Southern University, USA
Hastimal Jangid	University of Missouri-Kansas City, USA
Loredana Caruccio	University of Salerno, Italy
Giuseppe Polese	University of Salerno, Italy
Valentina Indelli Pisano	University of Salerno, Italy
Virginie Thion	University of Rennes 1/IRISA, France
Grégory Smits	University of Rennes 1/IRISA, France
Hélène Jaudoin	University of Rennes 1/IRISA, France
Yves Denneulin	Grenoble INP, France
Ermelinda Oro	ICAR-CNR, Italy
Harekrishna Misra	Institute of Rural Management Anand, India
Vijay Ingalalli	LIRMM, France

Gang Qian	University of Central Oklahoma, USA
Lubomir Stanchev	California Polytechnic State University, USA
Xianying (Steven) Liu	IBM Almaden Research Center, USA
Alok Watve	Broadway Technology, USA
Xin Shuai	Thomson Reuters, USA
María del Carmen Rodríguez-Hernández	University of Zaragoza, Spain
Óscar Urra	University of Zaragoza, Spain
Samira Pouyanfar	Florida International University, USA
Hsin-Yu Ha	Florida International University, USA
Miroslav Blaško	Czech Technical University in Prague, Czech Republic
Bogdan Kostov	Czech Technical University in Prague, Czech Republic
Yosuke Watanabe	Nagoya University, Japan
Atsushi Keyaki	Tokyo Institute of Technology, Japan
Miika Hannula	The University of Auckland, New Zealand
Dominik Bork	University of Vienna, Austria
Michael Walch	University of Vienna, Austria
Nikolaos Tantouris	University of Vienna, Austria
Jingjie Ni	Hewlett-Packard Enterprise Company, USA
Prajwol Sangat	Monash University, Australia
Xiaotian Hao	HKUST, Hong Kong, SAR China
Ji Cheng	HKUST, Hong Kong, SAR China
Yiling Dai	Kyoto University, Japan
Arnaud Castelltort	University of Montpellier, France
Sabin Kafle	University of Oregon, USA
Shih-Wen George Ke	Chung Yuan Christian University, Taiwan
Yi-Hung Wu	Chung Yuan Christian University, Taiwan
Jorge Martinez-Gil	Software Competence Center Hagenberg, Austria
Loredana Tec	Software Competence Center Hagenberg, Austria
Senen Gonzalez	University of Chile, Chile
Nicolas Travers	CNAM, France
Fayçal Hamdi	CNAM, France
Camelia Constantin	University of Pierre et Marie Curie - Paris 6, France
Daichi Amagata	Osaka University, Japan
Masumi Shirakawa	Osaka University, Japan
Eleftherios Kalogeros	Ionian University, Greece
Stéphane Jean	LIAS/ISAE-ENSMA, France
Selma Khouri	LIAS/ISAE-ENSMA, France
Soumia Benkrid	ESI, Algiers, Algeria
Andrea Esuli	ISTI-CNR, Italy
Giuseppe Amato	ISTI-CNR, Italy
Imen Megdiche	IRIT, France
Fotini Michailidou	University of the Aegean, Greece
Christos Kalyvas	University of the Aegean, Greece

Eirini Molla   University of the Aegean, Greece
Sajib Mistry   RMIT University, Australia
Tooba Aamir   RMIT University, Australia
Azadeh Ghari Neiat   RMIT University, Australia
Rahma Jlassi   RMIT University, Australia

Keynotes

From Natural Language to Automated Reasoning

Bruno Buchberger

We outline the possible interaction between knowledge mining, natural language processing, sentiment analysis, data base systems, ontology technology, algorithm synthesis, and automated reasoning for enhancing the sophistication of web-based knowledge processing.

We focus, in particular, on the transition from parsed natural language texts to formal texts in the frame of logical systems and the potential impact of automating this transition on methods for finding hidden knowledge in big (or small) data and the automated composition of algorithms (cooperation plans for networks of application software).

Simple cooperation apps like IFTTT and the new version of SIRI demonstrate the power of (automatically) combining clusters of existing applications under the control of expressions of desires in natural language.

In the Theorema Working Group of the speaker quite powerful algorithm synthesis methods have been developed that can generate algorithms for relatively difficult mathematical problems. These methods are based on automated reasoning and start from formal problem specifications in the frame of predicate logic. We ask ourselves how the deep reasoning used in mathematical algorithm synthesis could be combined with recent advances in natural language processing for reaching a new level of intelligence in the communication between humans and the web for every-day and business applications.

The talk is expository and tries to draw a big picture of how we could and should proceed in this area but will also explain some technical details and demonstrate some surprising results in the formal reasoning aspect of the overall approach.

The Price of Data

Gottfried Vossen[1,2]

[1] ERCIS, University of Münster, Münster, Germany
vossen@wi.uni-muenster.de
[2] The University of Waikato Management School, Hamilton, New Zealand
vossen@waikato.ac.nz

Abstract. As data is becoming a commodity similar to electricity, as individuals become more and more transparent thanks to the comprehensive data traces they leave, and as data gets increasingly connected across company boundaries, the question arises of whether a price tag should be attached to data and, if so, what it should say. In this talk, the price of data is studied from a variety of angles and applications areas, including telecommunication, social networks, advertising, and automation; the issues discussed include aspects such as fair pricing, data quality, data ownership, and ethics. Special attention is paid to data market-places, where nowadays everybody can trade data, although the currency in which buyers are requested to pay may no longer be what they expect.

The term "Big Data" will always be remembered as *the* big buzzword of 2013 and, somewhat surprisingly, of several years thereafter. According to Bernard Marr[1], "the basic idea behind the phrase 'Big Data' is that everything we do is increasingly leaving a digital trace (or data), which we (and others) can use and analyze. Big Data therefore refers to that data being collected and our ability to make use of it." In earlier times, it was not unusual to leave analog traces, like purchase receipts from the grocery store, and neither was the idea to somehow monetize these traces. The owner of the grocery store would know his regular customers, and would try to keep old ones and attract new ones by offering them discount coupons or other incentives. With digital traces, business along such lines has exploded, become possible at a world-wide scale, and has reached nuances of everyday life that nobody would ever have thought of. So it is time to ask whether that data comes with a price tag and, if so, what it says.

This talk looks at the price of data from a variety of angles and application areas for which pricing is relevant. In telecommunication, for example, prices for making phone calls as well as for data (e.g., surfing the Web) have come down enormously over the last 20 years, due to increasingly cheaper technology as well as more and more competition. Search engines have made it popular to make money through advertising, where participants bid on keywords that may occur in search queries, and social networks generate revenue from letting companies have access to their user profiles and all the data that these contain. So what is the value of a user profile?

---

[1] http://www.datasciencecentral.com/profile/BernardMarr.

Data marketplaces [2, 4, 5, 9], on the other hand, are an emerging species of digital platform that revisits traditional marketplaces and their mechanisms. In a data marketplace, producers of data provide query answers to consumers in exchange for payment. In general, a data marketplace integrates public Web data with other data sources, and it allows for data extraction, data transformation and data loading, and it comprises meta data repositories describing data and algorithms. In addition, it consists of technology for 'uploading' and optimizing operators with user-defined-functionality, as well as trading and billing components. In return, the 'vendor' of this functionality receives a monetary contribution from a buyer. Essentially, everybody can trade data nowadays, and the roles of sellers and buyers may be swapped over time and be exchangeable. For a seller, the interesting issue is the question of how valuable some data may be for a customer (or what the competition is charging for the same or similar data); if that could be figured out, the seller could adapt the price he is asking accordingly.

From a more technical perspective, the pricing problem can be tackled from the point of view of data quality, and here it is possible to establish a notion of *fair pricing*. [6, 8] cast this problem into a universal-relation setting and study the impact of quantifiable data quality; they follow [1] who argue that relational *views* can be interpreted as versions of the 'information good' data and hence study the issue of pricing for competing data sources that provide essentially the same data but in different quality.

Fair pricing has been addressed in depth by [7], by demonstrating how the quality of relational data products can be adapted to match a buyer's willingness to pay by employing a *Name Your Own Price* (NYOP) model. Under that model, data providers can discriminate customers so that they realize the maximum price a customer is willing to pay, and data customers receive a product that is tailored to their own data quality needs and budgets. To balance customer preferences and vendor interests, a model is developed which translates fair pricing into a Multiple-Choice Knapsack optimization problem, thereby making it amenable to an algorithmic solution. The concept of trading data quality for a discount was previously suggested in [10, 11] and applied to both relational as well as XML data.

A final aspect to be mentioned in this context is that of data used in automation. Following [3], automation has become pervasive in recent years and has lead to the danger that people lose their specific abilities when supported or even replaced by machines, robots, or generally automated devices. Carr explains this, for example, with auto-pilots in airplanes: Often pilots are so reliant on an auto-pilot that they do not want to accept the fact the a decision the device has just made is wrong, and he gives examples where this has ended in disaster more than once. Hence the danger is that we overestimate the truth in data, that we trust it too much, so that, as a consequence, the quest for its price becomes obsolete.

References

[1] Balazinska, M., et al.: A discussion on pricing relational data. In: Tannen, V., et al. (eds) In Search of Elegance in the Theory and Practice of Computation. LNCS, vol. 8000, pp. 167–173. Springer, Heidelberg (2013)

[2] Balazinska, M., et al.: Data markets in the cloud: an opportunity for the database community. In: PVLDB 4.12, pp. 1482–1485 (2011)

[3] Carr, N.: The Glass Cage — Automation and Us. W.W. Norton & Company (2014)

[4] Muschalle, A., et al.: Pricing approaches for data markets. In: Proceedings of 6th BIRTE Workshop 2012. Istanbul, Turkey, pp. 129–144

[5] Schomm, F., et al.: Marketplaces for data: an initial survey. In: SIGMOD Record 42.1, pp. 15–26 (2013). http://doi.acm.org/10.1145/2481528.2481532

[6] Stahl, F., et al.: Fair knapsack pricing for data marketplaces. In: Proceedings of 20th East-European Conference on Advances in Databases and Information Systems (ADBIS). LNCS. Springer (2016)

[7] Stahl, F.: High-quality web information provisioning and quality-based data pricing. PhD thesis. University of Münster (2015)

[8] Stahl, F., et al.: Data quality scores for pricing on data marketplaces. In: Proceedings 8th ACIIDS Conference. Da Nang, Vietnam, pp. 214–225 (2016)

[9] Stahl, F., et al.: Data marketplaces: an emerging species. In: Haav, H., et al. (eds.) Databases and Information Systems VIII - Selected Papers from the Eleventh International Baltic Conference, DB&IS 2014, 8–11 June 2014, Tallinn, Estonia. Frontiers in Artificial Intelligence and Applications, vol. 270, pp. 145–158. IOS Press (2014). http://dx.doi.org/10.3233/978-1-61499-458-9-145

[10] Tang, R., et al.: Get a sample for a discount. In: Decker, H., et al. (eds.) Database and Expert Systems Applications. LNCS, vol. 8644, pp. 20–34. Springer International Publishing, Switzerland (2014)

[11] Tang, R., et al.: What you pay for is what you get. In: Decker, H., et al. (eds.) Database and Expert Systems Applications. LNCS, vol. 8056, pp. 395–409. Springer, Berlin (2013)

Contents – Part II

NoSQL, NewSQL

Multimedia Data

Personal Information Management

Semantic Web and Ontologies

Database and Information System Architectures

Query Answering and Optimization

Information Retrieval, and Keyword Search

Data Modelling, and Uncertainty

Contents – Part I

Data Clustering

Distributed and Big Data Processing

Decision Support Systems, and Learning

Data Streams

Data Integration, and Interoperability

Semantic Web, and Data Semantics

Social Networks, and Network Analysis

A Preference-Driven Database Approach to Reciprocal User Recommendations in Online Social Networks

Florian Wenzel[(✉)] and Werner Kießling

Institute for Computer Science, University of Augsburg, 86135 Augsburg, Germany
{wenzel,kiessling}@informatik.uni-augsburg.de

Abstract. Online Social Networks (OSN) are frequently used to find people with common interests, though such functionality is often based on mechanisms such as friends-of-friends that do not perform well for real life interactions. We demonstrate an integrated database-driven recommendation approach that determines reciprocal user matches, which is an important feature to reduce the risk of rejection. Similarity is computed in a data-adaptive way based on dimensions such as homophily, propinquity, and recommendation context. By representation of dimensions as unique preference database queries, user models can be created in an intuitive way and can be directly evaluated on datasets. Query results serve as input for a reciprocal recommendation process that handles various similarity measures. Performance benchmarks conducted with data of a commercial outdoor platform prove the applicability to real-life tasks.

Keywords: Reciprocal recommendations · Preference queries · OSN

1 Introduction

OSN are a prime medium to form new virtual and real-life connections, a behavior that is endorsed through user recommendation services. However, existing solutions neglect vast amounts of readily available user information and rather exploit the structural properties of the social graph [1]. While this approach is valid for some use cases, information-rich user models are favorable for scenarios that target real-life interactions such as finding companions for common activities. User models for this purpose should include aspects such as homophily or propinquity [4], concepts that govern the formation of social ties in real life. A corresponding user recommendation process should find partners in a reciprocal fashion, taking not only the preferences of the recommendation subject, but also those of the objects into account, to reduce the risk of rejection [5].

We present a recommendation approach that addresses these crucial points to provide semantically rich reciprocal recommendations. Activity-related, spatial, and social data together with friendship information is collected to create multidimensional user models. Each dimension is represented as unique preference

© Springer International Publishing Switzerland 2016
S. Hartmann and H. Ma (Eds.): DEXA 2016, Part II, LNCS 9828, pp. 3–10, 2016.
DOI: 10.1007/978-3-319-44406-2_1

database query, which guarantees fast and intuitive modeling and direct evaluation on corresponding datasets. We integrate the world's fastest in-memory analytical database *EXASolution*, which provides an efficient Skyline feature[1] based on our previous work. Query results of all users serve as input for a reciprocal recommendation process that determines similarity between the recommendation subject and each potential partner by applying similarity measures to each dimension, resulting in a similarity vector per comparison. The recommendation result is computed as Pareto-optimum of these vectors. A real-life scenario based on data of Europe's largest outdoor community *Outdooractive*[2] showcases our approach in action. Benchmarks illustrate the scalability of the process.

The remainder of this paper is organized as follows: Sect. 2 presents a motivating use case. Section 3 describes the basic framework for user preferences. Dimensions of information-rich user models are explained in Sect. 4, details of the recommendation process are presented in Sect. 5. Benchmark results are evaluated in Sect. 6, concluded by a summary and outlook in Sect. 7.

2   Use Case Scenario

To illustrate the recommendation approach, we follow a use case scenario based on anonymized data of over 125,000 members of the *Outdooractive* community. The individual stages of the recommendation process are depicted in Fig. 1.

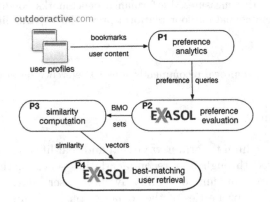

Fig. 1. Phases of the recommendation process

Given user Paul, we want to find users that join him on his next hiking adventure. Three dimensions of his user profile are of interest: preferences towards activities, current hometown, and demographic data. Most of this information can be either extracted directly or via preference elicitation in phase *P1*. Preferences considering activities include aspects such as difficulty, duration, or rating

---

[1] http://www.exasol.com/en/in-memory-database/overview/.

[2] http://www.outdooractive.com/en/.

of outdoor tours. The current hometown can be used to determine Points of Interest (POI) that are nearby. Demographic information helps to identify other users of similar age or gender. Since this information is also known for all other users, these preferences can be evaluated on specific database relations provided by the platform in phase *P2*. Resulting item sets can be used to compute user-to-user similarity, one dimension at a time, leading to similarity values for homophily (activities), propinquity (hometown), and social aspects (demographic information) in phase *P3*. Paul is looking for users of highest similarity in all these dimensions, but oftentimes there is no such perfect match. With the presented approach, Paul is able to retrieve best-matching users as Pareto-optimum of similarity vectors in phase *P4*. This way, he is guaranteed to get data-adaptive recommendations: the addition of new outdoor activities to the platform has a direct effect on the result of preference evaluation and might in turn lead to different recommendations.

3   Preference Framework

OSN profiles hold valuable information stored in numerical, categorical, and spatial attributes. To use it to full capacity, it has to be included into user models in phase *P1*. These in turn should be directly evaluable on datasets in phase *P2*. Towards this end, we follow the constructor-based framework of [2] which defines soft constraint preferences $P = (A, <_P)$ as strict partial orders on the domain of an attribute set A. Given a term $x <_P y$, which means "*I like y more than x*", the framework defines a *Best Matches Only* (BMO) query semantics by retrieving matches from an input relation R via a preference selection operator $\sigma[P](R)$:

$$\sigma[P](R) := \{t \in R \mid \neg\exists\ t' \in R : t[A] <_P t'[A]\} \tag{1}$$

The framework holds a taxonomy of base preference constructors operating on single attributes. All constructors are sub-constructors of a SCORE preference that minimizes a scoring function $f(x)$ so that $x <_P y \Leftrightarrow f(x) > f(y)$. To form classes of equivalent attribute values, a so-called d-parameter is applicable which extends $f(x)$ to $f_d(x) = \lceil \frac{f(x)}{d} \rceil$. Furthermore, complex constructors exist to combine base or complex preferences. Equal importance is expressed via *Pareto*, ordered importance via *Prioritization*.

To ensure scalability to large OSN datasets, we use the commercial *EXA-Solution* in-memory analytical database, based on a distributed and a parallel shared-nothing architecture. The system supports preferences via a *Skyline* feature, which is based on our previous work [2] and implements a distributed and parallel BNL-style algorithm [3]. The system provides a PREFERRING clause in addition to the SQL standard. A base preference is defined as numerical expression that has to be minimized or maximized as stated by keywords HIGH or LOW. Alternatively, Boolean expressions can be used for categorical domains. Complex preferences combine preference terms via keywords PLUS or PRIOR TO standing for *Pareto* or *Prioritization*. Preferences of single users such as Paul can now be modeled in the form of preference queries. These queries in turn can be evaluated to obtain preferred items of each dimension from specific datasets. Retrieved item sets finally serve as input for the recommendation process.

4   User Modeling

We focus on Paul whose profile contains an activity-related, a spatial, and a social dimension. First, each dimension d is expressed as preference P_d, next database relations R_d are assigned, and finally $\sigma[P_d](R_d)$ is evaluated to obtain a set of preferred items I_d. This covers phase $P1$ and $P2$ of Fig. 1. We assume w.l.o.g. that we obtain preferences either explicitly through user input or implicitly via elicitation or mining. As second step, we identify underlying datasets. *Outdooractive* as outdoor and tourism provider curates 3 database relations:

- *activity* (id INTEGER, category VARCHAR, tag VARCHAR, condition INTEGER, technique INTEGER, experience INTEGER, landscape INTEGER)
- *poi* (id INTEGER, category INTEGER, geom GEOMETRY)
- *user* (id INTEGER, age INTEGER, sex VARCHAR)

Preference selections $\sigma[P_d](R_d)$ are computed via *EXASolution* to return sets I_d in a data-adaptive fashion. This is a major advantage over static user comparisons. In case Paul favors difficult activities whereas a candidate prefers easy ones, a static comparison determines a low similarity. If the database only contains activities of medium difficulty then a data-adaptive approach could still detect high similarity. This also holds for user profiles without common attributes. If Paul prefers a high rating and a candidate certain tour tags, an edit distance would determine low similarity whereas item sets I_d might show a major overlap. Subsequent subsections describe preference queries for each user dimension.

4.1   Activity Dimension

Activity-related preferences are a strong motive for friendship [8]. The *activity* relation holds *category* and *tag* as categorical attributes. Paul's categorical preference are described by Boolean expressions. Values satisfying the IN condition are preferred. Numerical attributes range from 1 to the optimum 6. For these attributes, the syntax defines scoring functions that are minimized or maximized.

```
SELECT DISTINCT id
FROM activity
PREFERRING
  (category IN ('hiking','climbing')
  PLUS
  (LOW CASE WHEN landscape >4 THEN 0 ELSE
   ABS(landscape -5) END))
PRIOR TO
  (tag IN ('family -friendly','round tour'));
```

Paul prefers values for *landscape* that are 5 or above, else the distance to [5, 6] is minimized. The PLUS keyword indicates equal importance of the first two preferences, leading to Pareto evaluation. For intermediate results that are indifferent to this complex preference, the third base preference is evaluated as decisive factor as indicated by PRIOR TO. We denote the activity-related preference of a user u_a as $P_{u_a}^{act}$ which leads to preferred items $I_{u_a}^{act}$ as result of $\sigma[P_{u_a}^{act}](activity)$.

4.2   Spatial Dimension

Spatial preferences indicate preferred locations of a user, such as POI which are a base for propinquity [4]. *EXASolution* provides a spatial function ST_DISTANCE that computes the distance between the spatial attribute *geom* of a POI and Paul's hometown. The division by 5000 and CEIL function implement the d-parameter that forms equivalence classes of 5000 m. Within an equivalence class, locations with certain categories are preferred over others, again indicated by PRIOR TO and a Boolean expression. We denote the spatial preference of a user u_a as $P_{u_a}^{spat}$ which leads to preferred items $I_{u_a}^{spat}$ as result of $\sigma[P_{u_a}^{spat}](poi)$.

```
SELECT DISTINCT id
FROM poi
PREFERRING
(LOW (CEIL(ST_DISTANCE(geom, ST_SETSRID(
'POINT(695633.9 7104204.1)',3857))/5000)))
PRIOR TO
 (category IN ('hut','entertainment'));
```

4.3   Social Dimension

Demographic information is the base for status homophily, a concept predicting similarity for users with high overlap in dimensions such as age or sex [4]. Paul prefers users around his age and of opposite sex. The *age* preference is constructed with a CASE statement that assigns an optimal zero value to users holding the same age or an age that is up to 10 % lower. Equal importance is again expressed by PLUS. We denote the social preference of a user u_a as $P_{u_a}^{soc}$ which leads to preferred items $I_{u_a}^{soc}$ as result of preference selection $\sigma[P_{u_a}^{soc}](user)$.

```
SELECT DISTINCT id
FROM user
PREFERRING
(LOW CASE WHEN age>=22 AND age<=24
THEN 0 ELSE LEAST ((CEIL(ABS(22-age)/2)),
(CEIL(ABS(24-age)/2))) END)
PLUS (sex IN ('FEMALE'));
```

5   Recommendation Process

Given activity-related, spatial, and social preferences we obtain item sets $I_{u_a}^{act}$, $I_{u_a}^{spat}$, and $I_{u_a}^{soc}$ for a user u_a, together with the set $I_{u_a}^{fr}$ of friends of u_a. User models are compared one dimension at a time to determine the similarity of two users u_a and u_b, resulting in a vector s_{u_a,u_b} with functions f_i for each dimension:

$$s_{u_a,u_b} := (f_1(I_{u_a}^{act}, I_{u_b}^{act}), f_2(I_{u_a}^{spat}, I_{u_b}^{spat}), f_3(I_{u_a}^{soc}, I_{u_b}^{soc}), f_4(I_{u_a}^{fr}, I_{u_b}^{fr})) \qquad (2)$$

Similarity functions f_i are normalized to return values in the range $[0;1]$ with 1 as optimum. We refer to previous work published in [7] and select the *Ratio Model* of [6] as similarity measure. Starting with u_a, a vector s_{u_a,u_i} is calculated for each candidate u_i, resulting is a relation *similarity* (uid INTEGER, actsim DOUBLE, spatsim DOUBLE, socsim DOUBLE, friendsim DOUBLE) in phase *P3* of the overall process according to the following assignment:

$$actsim = s_{u_a,u_i}[1] \qquad spatsim = s_{u_a,u_i}[2]$$
$$socsim = s_{u_a,u_i}[3] \qquad friendsim = s_{u_a,u_i}[4]$$

For phase *P4* of Fig. 1, best-matching users for u_a are retrieved via a *Best Matching User* (BMU) query as skyline of dominating similarity vectors:

```
SELECT DISTINCT uid
FROM similarity
PREFERRING
  HIGH actsim PLUS HIGH spatsim PLUS
  HIGH socsim PLUS HIGH friendsim;
```

This process is inherently reciprocal since each dimension of the similarity vector holds a comparison of items of both the user being the subject of recommendation and the candidate user being the object. Both item sets are retrieved by preference queries. The process of Fig. 1 can be formalized as Algorithm 1 which in turn can be implemented as single User Defined Function (UDF).

Algorithm 1. Best-Matching User Algorithm (BMU)

---

input: set of users U, target user $u_t \in U$
output: set $R \subseteq U$ of best-matching users for u_t

BMU Algorithm(u_t, U)

    Phase 1: single user models
    for ($u_i \in U$): determine $P_{u_i}^{act}$, $P_{u_i}^{spat}$, $P_{u_i}^{soc}$;

    Phase 2: single user BMO-set calculation
    for ($u_i \in U$): calculate $I_{u_i}^{act} := \sigma[P_{u_i}^{act}]$(activity),
    $I_{u_i}^{spat} := \sigma[P_{u_i}^{spat}]$(poi), $I_{u_i}^{soc} := \sigma[P_{u_i}^{soc}]$(user);

    Phase 3: similarity vector calculation
    for ($u_i \in U \setminus u_t$):
    $s_{u_t,u_i} := (f_1(I_{u_t}^{act}, I_{u_i}^{act}), f_2(I_{u_t}^{spat}, I_{u_i}^{spat}), f_3(I_{u_t}^{soc}, I_{u_i}^{soc}), f_4(I_{u_t}^{fr}, I_{u_i}^{fr}))$;
    $S := S \cup s_{u_t,u_i}$

    Phase 4: BMU retrieval
    return $\sigma[P_{BMU}](S)$;

---

6 Benchmarks

An anonymized set of *Outdooractive* profiles is used to determine activity-related, spatial, and social preferences. Runtime is evaluated via an *EXASolution* cluster consisting of 8 nodes, each with 4 CPUs with 2 cores at 2 GHz and 16 GB of RAM. Corresponding database relations hold the following number of entries: *activity* holds 14,200,000 tuples, *poi* 592,000 tuples, and *user* 140,000 tuples.

▷ Activity-related preferences: Runtime is listed in Table 1. It grows linearly from 1,000 to 100,000 queries and increases with the number of nodes. The first observation indicates scalability. Increasing runtime with number of nodes occurs due to the distributed architecture of *EXASolution*. If preferences exhibit low selectivity then a significant communication overhead occurs. 10,000 preference queries can be computed within seconds, 100,000 queries in less than 5 min.

Table 1. Activity-related preferences

Queries	1 node (sec)	4 nodes (sec)	8 nodes (sec)
1,000	4.99	13.37	12.58
10,000	12.31	22.53	29.09
100,000	264.08	344.96	412.66

▷ Social preferences: Runtime is listed in Table 2. Scalability is given with increasing number of queries, however, runtime increases with number of nodes. 10,000 queries can be computed within seconds, 100,000 in under 10 min.

Table 2. Social preferences

Queries	1 node (sec)	4 nodes (sec)	8 nodes (sec)
1,000	19.51	21.38	22.34
10,000	19.21	26.96	34.66
100,000	593.17	675.85	804.74

▷ Spatial preferences: Runtime is listed in Table 3. For this query type, the addition of nodes does in fact have a positive effect. Since spatial queries include expensive distance calculations, runtime is higher compared to other dimensions.

Table 3. Spatial preferences

Queries	1 node (sec)	4 nodes (sec)	8 nodes (sec)
1,000	101.57	43.61	31.86
10,000	870.31	324.45	242.59
100,000	9385.35	3584.27	2707.40

Table 4 lists the evaluation of user models. A single model contains 3 preference queries. For this scenario, the addition of nodes has a major impact on query runtime. With 8 nodes, 100,000 user models can be computed in about 100 min, the same computation takes twice as long for a single node.

Table 4. Computation of user sets of different size

User	1 node (sec)	4 nodes (sec)	8 nodes (sec)
1,000	85.74	47.11	39.50
10,000	736.74	361.09	322.16
100,000	12468.10	6142.08	5954.21

Due to brevity, the total runtime of the recommendation process is going to be evaluated in subsequent publications with Algorithm 1 implemented as UDF.

7 Conclusion

We presented a preference-driven recommendation approach that permits a fast and intuitive creation of user models for a plurality of dimensions of OSN profiles. These information-rich models are vital for real-life interactions. As user models consist of preference queries, they can be directly evaluated on a database. This is the base for a data-adaptive and reciprocal recommendation process that incorporates preference dimensions of the recommendation subject and those of candidates. First benchmarks indicate scalability for large datasets. We are aware that this short paper left many interesting questions unanswered. How do we get from profiles to preference queries? How does the system perform against established recommendation techniques? The answers will require further research efforts and user studies and are part of an ongoing research agenda.

References

1. Chen, J., Geyer, W., Dugan, C., Muller, M., Guy, I.: Make new friends, but keep the old–recommending people on social networking sites. In: Conference on Human Factors in Computing Systems CHI, Boston, MA, USA, pp. 201–210 (2009)
2. Kießling, W., Endres, M., Wenzel, F.: The preference SQL system-an overview. IEEE Data Eng. Bull. **34**(3), 11–18 (2011)
3. Mandl, S., Kozachuk, O., Endres, M., Kießling, W.: Preference analytics in EXAS-olution. In: 16th GI Fachtagung BTW, Hamburg, Germany (2015)
4. McPherson, M., Smith-Lovin, L., Cook, J.M.: Birds of a feather: homophily in social networks. Ann. Rev. Sociol. **27**, 415–444 (2001)
5. Pizzato, L., Rej, T., Akehurst, J., Koprinska, I., Yacef, K., Kay, J.: Recommending people to people: the nature of reciprocal recommenders with a case study in online dating. User Model. User Adap. Inter. **23**(5), 447–488 (2013)
6. Tversky, A.: Features of similarity. Psychol. Rev. **84**(4), 327–352 (1977)
7. Wenzel, F., Kießling, W.: Aggregation and analysis of enriched spatial user models from location-based social networks. In: 1st International GeoRich Workshop in Conjunction with SIGMOD. Snowbird, UT, USA (2014)
8. Werner, C., Parmelee, P.: Similarity of activity preferences among friends: those who play together stay together. Soc. Psychol. Q. **42**(1), 62–66 (1979)

Community Detection in Multi-relational Bibliographic Networks

Soumaya Guesmi[✉], Chiraz Trabelsi, and Chiraz Latiri

LIPAH, Faculty of Sciences of Tunis, Université de Tunis El Manar, Tunis, Tunisia
soumaya.guesmi@fst.rnu.tn

Abstract. In this paper, we introduce a community detection approach from heterogeneous multi-relational network which incorporate the
multiple types of objects and relationships, derived from a bibliographic
networks. The proposed approach performs firstly by constructing the
relation context family (RCF) to represent the different objects and relations in the multi-relational bibliographic networks using the Relational
Concept Analysis (RCA) methods; and secondly by exploring such RCF
for community detection. Experiments performed on a dataset of academic publications from the Computer Science domain enhance the effectiveness of our proposal and open promising issues.

Keywords: Multi-relational bibliographic networks · Community
detection · RCA

1 Context and Motivation

The primary focus of this work is to extract emergent academic community structure from the bibliographic through the analysis of the different relationships
among the multi-relational bibliographic data. Although research attention on heterogeneous networks representation and efficient topological algorithm design, a
much more fundamental issue concerning the exploration of the heterogeneous
organization infrastructure and communities detection have not been skilfully
addressed. Indeed, A wide range of approaches have been proposed in the literature for communities detection in heterogeneous networks. However, they have
deeply focused on topological properties of these networks, ignoring the embedded
semantic information. To overcome this limitation, in recent years, Formal Concept Analysis (FCA) techniques are used for a conceptual clustering. Using FCA
aims to extract communities preserving knowledge shared in each community. In
such FCA based approaches, the inputs are bipartite graphs and the output is a
Galois hierarchy that reveals communities semantically defined with their shared
knowledge or common attributes. Vertices are designed as lattice extents and edges
are labeled by lattice intents (*i.e.*, shared knowledge). However, a Galois hierarchy
is not a satisfactory scheme since an exponential number of communities may be
obtained. Therefore, reduction methods should be introduced. In fact, only very
few researches have actually focused on this difficulty [4]. The authors in [5] used the

© Springer International Publishing Switzerland 2016
S. Hartmann and H. Ma (Eds.): DEXA 2016, Part II, LNCS 9828, pp. 11–18, 2016.
DOI: 10.1007/978-3-319-44406-2_2

iceberg method as well as the stability method as a Galois lattice reduction methods. Authors in [3] identify concepts with frequent intents above a set threshold. The main limit of this purpose, that some important concepts may be overlooked. Brandes et al. [1] combine both the iceberg and stability methods, it's argued that this approach yields good results for extracting pertinent communities based on concepts. As it's described in the survey conducted by Planti and Crampes [4], discovering communities based on FCA techniques is the most accurate, because it extracts communities using their precise semantics. Nonetheless, they fall short of giving simple and practical results. Therefore, a new research challenge consists on detecting communities from heterogeneous multi-relational networks. In order to discover communities with a well defined set of properties, we first need to extract the corresponding relations among multiple existing relations. In this paper, we introduce a query navigation approach based on the use of the RCA techniques [6] designed within a multiple academic databases for hidden relationships (or links) detection. This will have significant impact, it can help foster new collaborative teams, help with expertise discovery and in the long term, guide research teams reorganization consistency with collaboration patterns.

The paper is organized as follows. In the next section, we describe our community detection approach. Section 3 presents our experimental results, while Sect. 4 summarizes our contributions.

2   Proposed Community Detection Approach

In this section, we present our community detection approach which aims to model and to extract academic community structure from multi-relational bibliographic data. In order to achieve these goals, the proposed approach relies on two main stages: the multi-relational bibliographic hypergraph modelling stage; and the query navigation for communities discovering stage. We firstly proceed by describing the preliminary concepts of our proposal.

A. Preliminary Concepts

• **Formal context:** is a triplet $\mathcal{K} = (\mathcal{O}, \mathcal{A}, \mathcal{I})$, where $\mathcal{O}$ represents a finite set of objects, $\mathcal{A}$ is a finite set of items (or attributes) and $\mathcal{I}$ is a binary (incidence) relation (i.e., $\mathcal{I} \subseteq \mathcal{O} \times \mathcal{A}$). Each couple $(o, a) \in \mathcal{I}$ expresses that the object

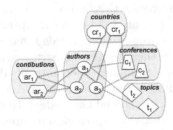

Fig. 1. An example of a multi-relational bibliographic hypergraph.

$o \in \mathcal{O}$ contains the item $a \in \mathcal{A}$. $\mathcal{O}$ is called one-valued context. A worth of interest link between the power-sets $\mathcal{P}(\mathcal{A})$ and $\mathcal{P}(\mathcal{O})$ associated respectively to the set of items $\mathcal{A}$ and the set of objects $\mathcal{O}$ [2].

• **Formal concept:** A pair $c = (O, A) \in \mathcal{O} \times \mathcal{A}$, of mutually corresponding subsets, *i.e.*, $O = \psi(A)$ and $A = \phi(O)$, is called a formal concept, where O is called *extent* of c and A is called its *intent*.

• **A partial order:** on formal concepts, *w.r.t.* set inclusion [2], is defined as: $\forall$ $c_1 = (O_1, A_1)$ and $c_2 = (O_2, A_2)$ two formal concepts, $c_1 \leq c_2$ if $O_2 \subseteq O_1$, or equivalently $A_1 \subseteq A_2$.

• **Galois concept lattice:** Given a context $\mathcal{K}$, the set of formal concepts $\mathcal{C}$ is a complete lattice $\mathcal{L}_{\mathcal{C}} = (\mathcal{C}, \leq)$, called *Galois (concept) lattice*, when $\mathcal{C}$ is considered with set inclusion between concepts intents (or extents) [2].

• **Relational Context Family (RCF):** is a pair (K, R) where $K = \{\mathcal{K}_i\}_{i=1,\ldots,n}$ is a set of (object-attribute) contexts $\mathcal{K}_i = (\mathcal{O}_i, \mathcal{A}_i, \mathcal{I}_i)$ and $\{r_{j,l}\}_{j,l \in \{1,\ldots,n\}}$ is a set of relational (object-object) contexts $r_{j,l} \subseteq \mathcal{O}_j \times \mathcal{O}_l$, where $\mathcal{O}_j$ (called the domain of $r_{j,l}$) and $\mathcal{O}_l$ (called the range of $r_{j,l}$) are the object sets of the contexts $\mathcal{K}_j$ and $\mathcal{K}_l$, respectively. $\mathcal{O}_j$ is called the domain of $r_{j,l}$ ($\mathrm{dom}(r_{j,l})$) and $\mathcal{O}_l$ is called the range of $r_{j,l}$ ($\mathrm{ran}(r_{j,l})$) [6].

A function *rel* is associated with a RCF which maps a context $\mathcal{K} = (\mathcal{O}, \mathcal{A}, \mathcal{I})$ $\in$ K to the set of all relations $r \in$ R starting at its object set $\mathcal{K} : rel(\mathcal{K}) = \{r \in$ R, where $\mathrm{dom}(r) = \mathcal{O}\}$. Hence, given a relation r and a quantifier f chosen within the set $F = \{\forall, \exists, \forall\exists, \geq, \geq_f, \leq, \leq_f\}$. k maps an object set from $\mathrm{ran}(r)$ to an object set from $\mathrm{dom}(r)$ as $k : F \times R \times \cup_{i=1,\ldots,n} \mathcal{P}(\mathcal{O}_i) \to \cup_{i=1,\ldots,n} \mathcal{P}(\mathcal{O}_i)$ [6]. Scaling a context along a relation consists in integrating the relation to the context in the form of one-valued attributes using a scaling operator. A context is scaled upon all the relevant relations originating from the context by augmenting $\mathcal{K}$ with all the resulting relational attributes. Thus, an object owns an attribute depending on the relationship between its link set and the extent of the concept, i.e., the instances of a relation r, say $r_k(o_i, o_j)$, where $o_i \in \mathcal{O}_i$ and $o_j \in \mathcal{O}_j$, are called links. The evolution of each context $\mathcal{K}_i \in$ K from the input RCF yields a sequence $\mathcal{K}_i^p$ whose zero member $\mathcal{K}_i^0 = (\mathcal{O}_i^p, \mathcal{A}_i^p, \mathcal{I}_i^p)$ is the input context $\mathcal{K}_i$ itself. From there on, each subsequent member is the complete relational expansion of the previous one upon the relations r from $rel(\mathcal{K}_i)$. This yields a global sequence of context sets $\mathcal{K}^p$ and the corresponding sequence of lattice sets, called the Concept Lattice Family (CLF). Thus, the concept lattice family is a set of lattices that correspond to the formal contexts, after enriching them with relational attributes.

In this work, we consider the exists scaling. Hence, let $r_{ij} \subseteq O_i \times O_j$ be a relational context. The exists scaled relation $r_{ij}^{\exists}$ is defined as $r_{ij}^{\exists} \subseteq O_i \times B(O_j, A, I)$, such that for an object o_i and a concept c:$(o_i, c) \in r_{ij}^{\exists} \Leftrightarrow \exists x, x \in o_i' \cap Extent(c)$.

B. Multi-relational Bibliographic Hypergraph Model

Three concepts are involved in our model: object context, relation context, and concept lattice family. As illustrated in Fig. 1, a set of authors $\{a_1, a_2, \ldots, a_n\}$, locates in a given country $\{cr_1, cr_2, \ldots, cr_p\}$, work closely with each other, under

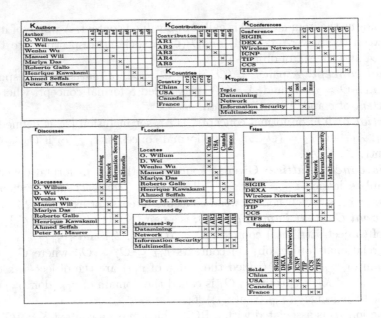

Fig. 2. Top. The objects contexts. **Bottom.** The relations contexts.

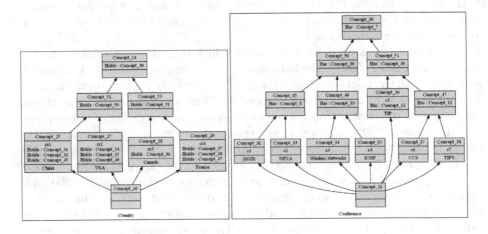

Fig. 3. Country and Conference lattices.

different topics $\{t_1, t_2, \ldots, t_k\}$; some of them share scientific contributions (contributions $\{ar_1, ar_2, \ldots, ar_m\}$ within a set of conferences $\{c_1, c_2, \ldots, c_l\}$). To generally describe such collaboration data, we define an object context as a set of objects or entities of the same type, *e.g.*, an author context is a set of authors and define a relation context as the interactions among objects contexts, *e.g.*, (author, topic) relation, (country, conference) relation., etc. We use a relational concept family to describe the relations contexts and the objects contexts constructed from a multi-relational bibliographic hypergraph. Figure 1 depicts the data schema of the handled multi-relational bibliographic hypergraph. The relational concept family is made of 5 objects contexts: $\mathcal{K}_{Authors}$, $\mathcal{K}_{Countries}$, $\mathcal{K}_{Conferences}$, $\mathcal{K}_{Topics}$, $\mathcal{K}_{Contributions}$; and 5 relations contexts: $r_{Locates}$, r_{Holds}, r_{Has}, $r_{Discusses}$ and $r_{Addressed-By}$. We report in Fig. 2 (Top) these 5 objects contexts and in Fig. 2 (Bottom) the 5 related relations contexts.

The overall process of RCA follows a multi-FCA method [6] which allows to build a set of lattices called Concept Lattice Family (*CLF*). It's an iterative process which generates at each step a set of concept lattices. First, the process constructs concept lattices using the objects contexts only. Then, in the following steps, it concatenates objects contexts with the relations contexts based on the existential scaling operator that produce scaled relations. Hence, the exists scaled relation translates the links between objects into conventional FCA attributes and extracts a collection of lattices whose concepts are linked by relations. Figure 3 depicts an example of Country and Conference lattices of the generated *CLF*.

C. Query Navigation for Communities Discovering

The second stage of the proposed approach aims to extract a set of academic communities by performing the following three steps:

• **Step 1: users' relational query submission:** the aim of this step is to transform the submitted user query to a Relational Query RQ which is composed of several Simple Queries (SQ). Hence, for a context $K = (A,O,I)$, a simple query denoted by $SQ = \{o_q\}$, is a set of objects satisfying the query (or the answer set) with $o_q \subset O$.

Definition 1 (Relational Query). A Relational Query $RQ = \{rq_0, rq_1, \ldots, rq_m\}$ on a relational context family(K,R) is a triplet $RQ = (q'_s, r_{st}, q'_t)$ with:
- q'_s and q'_t, source query and target query respectively, are a set of SQ.
- r_{st} is the relation between q'_s and q'_t. It leads one-to-one mapping between q'_s and q'_t.

• **Step 2: concept Lattice Family Exploration:** to explore the concept lattice family, we have to construct a query path QP which allows to know the path that we have to follow and specify the source and the target lattices.

Definition 2 (Query Path). Let $QP = \{qp_0, qp_1, \ldots, qp_n\}$ and qp_i is a pair $((q_s, L_s), (q_t, L_t))$ where L_s and $L_t \subset CLF$, the source and target lattices respectively. The Query Path QP is the inverse order of the relational query. It means $qp_0 = rq_m$ and $qp_n = rq_0$; with $q_{s0} = q'_{tm}$ and $q_{t0} = q'_{sm}$

• **Step 3: community detection:** in order to detect academic communities, we propose a new method called *Quering_Navigation* that leads to navigate between Galois Lattices based on the extracted query path QP. It takes as input the query path $QP = qp_k$ with $qp = ((qp_s, L_s),(qp_t, L_t))$ and outputs the identified community as an answer to the user query Q. *Query_Navigation* starts by handling all concepts C of the source lattice L_s, in order to extract the corresponding concepts (C_i) of the initial query path qp_0. *Query_Navigation* proceeds by identifying the concept extent of the lattice L_s and then extracts the concepts that contains an extent related to the query qp_s. The result of the initial phase is a set of concepts C_i that respond to the query qp_s. The second phase of *Query_Navigation*, consists in generating iteratively a set of concepts containing the set of concepts (C_i) extracted in the initial phase. It consists on handling the corresponding concept intent of the lattice L_t, for extracting the set of concepts (C_{i+1}) containing the C_i. If there is no more query path to be explored, *Query_Navigation* extracts the extent of the last selected concept (C_k). This set of C_k extent represents the set of individuals that constitutes the academic community returned to the user.

3   Experimental Evaluation

We collect data from two bibliographic databases. We use the well known database DBLP and we access on AMiner database for taking keywords, institutions and research topics in order to complete our conceptual hypergraph model. We keep only four research topics (Data Mining, Computer Network, Artificial Intelligence, Human Computer, Computer Graphics) and we pick only a few representative conferences for the five areas (11 conferences). The built dataset contains 914 contributions and 336 authors since 2010. The *Query_Navigation* algorithm is developed in JAVA and tested on a Windows 7 with Intel core i5 2.4 GHz and 8 GB of Ram.

Baseline Model: for enhancing the effectiveness of our approach, we have selected the most popular baseline communities structure which suggests communities as a set of authors belonging to the same affiliation. To carry out our experiments, we consider two simple queries (Q3 and Q4) and two relational queries (Q1 and Q2). We study whether our approach is able to capture the hidden relations between authors and if it can responds to different type of queries:

Q1: 4 entities, *i.e.*, Authors, Countries, Conferences and Topics; and 3 relations, *i.e.*, Locates, Holds and Has.
Q2: 3 entities, *i.e.*, Authors, Countries and Conferences; and 2 relations, *i.e.*, Locates and Holds.
Q3: 2 entities, *i.e.*, Authors and Countries; and 1 relation, *i.e.*, Locates.
Q4: 2 entities, *i.e.*, Authors and Topics; and 1 relation, *i.e.*, Discusses.

Furthermore, we consider two different ground truths [8]. The first ground truth $GT1$: each explicit authors' topic in the dataset is a ground truth community, it contains authors nodes which share the same topic. The second ground truth $GT2$: each explicit author conference is a ground truth community, it contains authors nodes which participate in the same conference.

Effectiveness of our approach: the performance is assessed by the measures of Recall, Precision and $F\beta$_measure, computed over all vertices [7]. These measures attempt to estimate whether the prediction of this vertices in the same community was correct. Given a set of algorithmic communities C and the ground truth communities S, precision indicates how many vertices are actually in the same ground truth community ($Precision = \frac{|T \cap S|}{|T|}$). The Recall indicates how many vertices are predicted to be in the same community in a retrieved community (Recall $= \frac{|T \cap S|}{|S|}$), and $F\beta$_measure is the harmonic mean of Precision and Recall ($F\beta$_ measure $= \beta \times \frac{Precision \times Recall}{Precision + Recall}$ where $\beta \in \{1,2\}$).

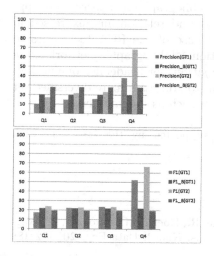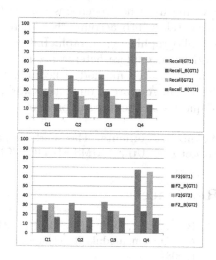

Fig. 4. Average score of the Precision, Recall, F1_measure and F2_measure of our approach *vs.* those of the baseline (B).

Thus, according to the sketched histograms in Fig. 4, we can point out that our approach outperforms the baseline. In fact, as expected, the Recall values of the baseline are much lower than those achieved by our approach among the two ground truths ($GT1$ and $GT2$). As we show, the average Recall achieves 83.87 % and 65,02 % comparing with the baseline which has 28,31 % and 14,58 % in term of Recall vs. an exceeding about 55.5 % and 50.4 % over the query $Q4$ among the two ground truths respectively. Indeed, in term of F2_measure our approach (67,61 %, 65,7 %) outperforms considerably the baseline (23,44 %, 16,5 %) over the query Q4 among GT1 and GT2 respectively, in this case we can say that the baseline have only a small number of communities detected fairly well and not many detected communities reflect to the ground truth communities.

However, the percentage of Precision for the baseline outperforms slightly our approach according to $Q1, Q2$ and $Q3$. Whereas, for $Q4$, our approach has an average of 68,57 % showing a drop of 28,31 % vs. an exceeding about 40.2 % against the baseline. A significant observation shows that the relational query

$Q1$ have better Recall (55,68 %) than that of the simple query $Q3$ and that of the relational query $Q2$ (44,85 %). We can conclude that the relational query improves the community structure and leads to extract relevant communities. Hence, considering four different queries, our approach outperforms the baseline in terms of Recall, F1_measure and F2_measure often by a large margin in the Recall score.

4 Conclusion

In this paper, we have presented a novel approach for academic communities discovering from heterogeneous multi-relational bibliographic networks. Our approach takes into account the different entities and relationships expressed in a bibliographic hypergraph. Indeed, we made use of the RCA techniques to model and explore heterogeneous multi-relational bibliographic network via a new introduced method, called: *Query_Navigation*, for academic communities detection. As part of our future work, we plan to address a more diversified set of queries by the integration of other quantifier such as ∀ quantifier.

Acknowledgements. This work is partially supported by the French-Tunisian project PHC-Utique RIMS-FD 14G 1404.

References

1. Brandes, U., Delling, D., Gaertler, M., Gorke, R., Hoefer, M., Nikoloski, Z., Wagner, D.: On modularity clustering. IEEE Trans. Knowl. Data Eng. **20**, 172–188 (2008)
2. Ganter, B., Wille, R.: Formal Concept Analysis: Mathematical Foundations. Springer, Berlin (1999)
3. Jay, N., Kohler, F., Napoli, A.: Analysis of social communities with iceberg and stability-based concept lattices. In: Medina, R., Obiedkov, S. (eds.) ICFCA 2008. LNCS (LNAI), vol. 4933, pp. 258–272. Springer, Heidelberg (2008)
4. Planti, M., Crampes, M.: Survey on social community detection. In: Social Media Retrieval, Computer Communications and Networks, pp. 65–85 (2013)
5. Roth, C., Obiedkov, S.A., Kourie, D.G.: On succinct representation of knowledge community taxonomies with formal concept analysis. Int. J. Found. Comput. Sci. **19**, 383–404 (2008)
6. Rouane-Hacene, M., Huchard, M., Napoli, A., Valtchev, P.: Relational concept analysis: mining concept lattices from multi-relational data. Ann. Math. Artif. Intell. **67**(1), 81–108 (2013)
7. Song, S., Cheng, H., Yu, J.X., Chen, L.: Repairing vertex labels under neighborhood constraints. PVLDB **7**, 987–998 (2014)
8. Yang, J., Leskovec, J.: Defining and evaluating network communities based on ground-truth. In: ICDM, pp. 745–754. IEEE Computer Society (2012)

Quality Prediction in Collaborative Platforms: A Generic Approach by Heterogeneous Graphs

Baptiste de La Robertie[✉], Yoann Pitarch, and Olivier Teste

Institut de Recherche en Informatique de Toulouse,
118 Route de Narbonne, 31071 Toulouse, France
{baptiste.delarobertie,yoann.pitarch,olivier.teste}@irit.fr

Abstract. As everyone can enrich or rather impoverish crowd-sourcing contents, it is a crucial need to continuously improve automatic quality contents assessment tools. Structural-based analysis methods developed for such quality prediction purposes generally handle a limited or manually fixed number of families of nodes and relations. This lack of genericity prevents existing algorithms for being adaptable to platforms evolutions. In this work, we propose a *generic* and *adaptable* algorithm, called *HSQ*, generalising various state-of-the-art models and allowing the consideration of graphs defined by an arbitrary number of nodes semantics. Evaluations performed over the two representative crowd-sourcing platforms *Wikipedia* and *Stack Exchange* state that the consideration of additional nodes semantics and relations improve the performances of state-of-the-art approaches.

Keywords: Link-analysis · Heterogeneous graphs · Quality

1 Introduction

Scientific literature has demonstrated strong correlations between users authority and contents quality on collaborative platforms [6,8,12,21]. Statistically, authoritative users are more likely to produce high quality content than others. Many state-of-the-art link analysis approaches exploit this *mutual reinforcement principle* between quality and authority for a quality assessment task [3,11,14,17,21]. However, most of them suffer from two major limitations. First, the lack of *genericity* of the formulations restricts them to a particular platform, making the solutions hardly transposable from one portal to another. Second, the lack of *adaptability* of the formulations prevents the algorithms from anticipating changes in the underlying graph. Thus, additional semantics of nodes or relations are most of the cases impossible to handle. These two limitations, shared by many structural-based algorithms, constitute the main motivations of our work. Our contributions are as follows:

- We propose a generic formulation of collaborative platforms using heterogeneous graphs and an unsupervised algorithm, *HSQ* (Heterogeneous Structural Quality), handling an unpredefined number of semantics of nodes and relations;

© Springer International Publishing Switzerland 2016
S. Hartmann and H. Ma (Eds.): DEXA 2016, Part II, LNCS 9828, pp. 19–26, 2016.
DOI: 10.1007/978-3-319-44406-2_3

- We demonstrate the genericity of the proposal by instanciating three different and recent state-of-the-art algorithms and show how to easily integrate new semantics of nodes and relations;
- We conduct empirical studies on two real data sets from the *Wikipedia* and *Stack Exchange* portals that demonstrate a significant interest of considering additional entities and relations for the quality assessment task in crowd-sourcing platforms.

2   Related Work

A first family of models for the quality assessment task on collaborative platforms exploit *contents signals*. Textual indices, numbers of citations or content length are some examples of content features used by *content-based* quality models. For example, on Wikipedia, it has been shown that the number of words per article [4] and the lifespan of the edits [1] are good quality predictors. However, content-based signals are too specific to a specific platform. Our work falls in the second family of approaches exploiting *structural signals* from the relations between the entities. Many works has empirically demonstrated correlations between users authority and contents quality, justifiying the wide range of PageRank [16] and HITS [13] based methods developped in the literature. On Wikipedia, a study of Dalip et al. [7] shows that structural features represent the most important family of predictors in a quality prediction task. More particularly, non considering such features leads to the greatest loss in terms of model quality. Hu et al. [10] propose to identify high quality articles on Wikipedia by exploiting this mutual dependency over a bipartite graph associating the articles to their contributors. Still on Wikipedia, a previous work [8] shows the interest of considering a co-edit relation between authors and reviewers to identify high quality articles. The study postulates that authoritative users get used to collaborate to produce high quality articles. Zhang et al. [20] apply the PageRank algorithm to on-line forums to identify authoritative users. Campbell et al. [5] and more recently Jurczyk et al. [12] make use of the HITS algorithm over a users-interaction graph to show a positive correlation between authority and quality. Recent analysis on Stack Overflow [15] and Quora [2,18] underlines the cyclic relation between content quality and producers authority.

   If this mutual reinforcement principle has been extensively exploited for simple graphs considering a few types of nodes and relations, it seems that no formulation has been proposed for more complex graphs and in particular for heterogeneous graphs.

3   Approach Description

Notations. Let $\mathcal{G} = (\mathcal{H}, \mathcal{V})$ be an heterogeneous graph defined over m families of nodes $\mathcal{H} = \{\mathcal{U}_i\}_{1 \leq i \leq m}$, and a set of binary relations $\mathcal{V} \subseteq \mathcal{H} \times \mathcal{H}$. We denote by n_i the number of entities in the family $\mathcal{U}_i$. Let $(\mathcal{U}_i, \mathcal{U}_j) \in \mathcal{V}$ be a pair of families. We note $\mathcal{V}_{ij}$ the relation defined over $\mathcal{U}_i \times \mathcal{U}_j$ and A_{ij} the associated adjacency matrix. We denote by $\mathbf{q}_i \in [0,1]^{n_i}$ the quality scores vector of the entities in the family $\mathcal{U}_i$.

Model. Firstly, for each pair of families $(\mathcal{U}_i, \mathcal{U}_j) \in \mathcal{V}$, we suppose a pair of *influence functions* (f_{ij}, g_{ji}) to model the *reinforcement principle*. Informally, the quality of the nodes in $\mathcal{U}_i$ influences the nodes quality in $\mathcal{U}_j$ and conversely, the nodes quality in $\mathcal{U}_j$ influences back the nodes quality in $\mathcal{U}_i$. This cyclic relation is illustrated in Fig. 1(a). More formally, we impose $\mathbf{x}_j = f_{ij}(\mathbf{y}_i)$ and $\mathbf{y}_i = g_{ij}(\mathbf{x}_j)$, with $\mathbf{x}_i \in [0,1]^{n_i}$ and $\mathbf{y}_i \in [0,1]^{n_i}$ being two vectors of *partial quality* scores. Note that if $(\mathcal{U}_i, \mathcal{U}_j) \notin \mathcal{V}$, we assume $f_{ij} = g_{ij} = 0$. Secondly, by considering linear aggregations of the different influences (see example in Fig. 1(b)), we have $\mathbf{x}_i = \sum_{j=1}^{m} f_{ji}(\mathbf{y}_j)$ and $\mathbf{y}_i = \sum_{j=1}^{m} g_{ij}(\mathbf{x}_j)$. In this work, we consider the case where influence functions are directly expressed by the adjacency matrices corresponding to each relation. Formally, $\forall \mathcal{V}_{ij} \in \mathcal{V}$, $f_{ij} = A_{ij}^T$ and $g_{ij} = A_{ij}$. Finally, by benoting $\mathbf{x}_i^{(t)}$ and $\mathbf{y}_i^{(t)}$ the partial quality scores at the tth iteration of a label propagation process, the proposed quality model is expressed as follow:

Fig. 1. (a) Reinforcement principle between two families $\mathcal{U}_i$ and $\mathcal{U}_j$ such that $(\mathcal{U}_i, \mathcal{U}_j) \in \mathcal{V}$. (b) Linear aggregation of incoming influence functions for family $\mathcal{U}_l$.

$$\mathbf{x}_i^{(t)} = \sum_{j=1}^{m} A_{ji}^T \sum_{k=1}^{m} A_{jk}\mathbf{x}_k^{(t-1)} \text{ and } \mathbf{y}_i^{(t)} = \sum_{j=1}^{m} A_{ij} \sum_{k=1}^{m} A_{kj}^T\mathbf{y}_k^{(t-1)} \quad (1)$$

The quality q_i for each family $\mathcal{U}_i \in \mathcal{V}$ is computed as an aggregation function $\mathcal{A}_i$ of the partial quality scores $\mathbf{x}_i$ and $\mathbf{y}_i$, formally $q_i = \mathcal{A}_i(\mathbf{x}_i, \mathbf{y}_i)$. In this work, $\mathcal{A}_i$ is the average function $\forall i \in \{1, ..., m\}$.

Computation. The proposed algorithm, *HSQ* (Heterogeneous Structural Quality), is an iterative label propagation procedure propagating the adjusted scores through the relations $\mathcal{V}_{ij}$. Main steps are the following. (1) **Initialization.** For each $\mathcal{U}_i \in \mathcal{H}$, set $\mathbf{x}_i^{(0)}$ and $\mathbf{y}_i^{(0)}$ to random vectors. (2) **Propagation.** For each $\mathcal{U}_i \in \mathcal{H}$, update scores $\mathbf{x}_i^{(t)}$ and $\mathbf{y}_i^{(t)}$ with Eq. (1). (3) **Normalization.** Set $||\mathbf{x}_i^{(t)}|| = 1$ and $||\mathbf{y}_i^{(t)}|| = 1$. (4) **Return** $\mathcal{A}_i(\mathbf{x}_i, \mathbf{y}_i)$.

Steps (2) and (3) are repeated until a convergence step is reached. Convergence of the algorithm for the trivial case $m = 1$ is demonstrated in [9]. For the general case $m \geq 1$, we stop the propagation when $\sum_{i=1}^{m} ||\mathbf{x}_i^{(t)} - \mathbf{x}_i^{(t-1)}||_2 + ||\mathbf{y}_i^{(t)} - \mathbf{y}_i^{(t-1)}||_2 \leq \epsilon$. The algorithm returns a vector of scores $q_i \in \mathbb{R}^{n_i}$ for each family of nodes $\mathcal{U}_i \in \mathcal{H}$. These scores should be ranked independently for each family in decreasing order of (predicted) quality.

Instances and Competitors. Wiki platforms are modelled with heterogeneous graphs using two families of nodes, the set of users and the set of articles (see Fig. 2(a)). Question and Answering websites are modelled with four families of nodes : users, answers, questions and comments (see Fig. 2(b)).

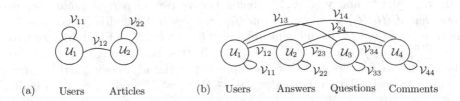

(a)   Users   Articles   (b)   Users   Answers   Questions   Comments

Fig. 2. (a) Wiki platform instance ($m = 2$). (b) Stack Exchange instance ($m = 4$).

On Wiki, the **Basic** model [10] constitutes a particular instance of the proposal, considering a bipartite graph ($m = 2$). Inter-user and inter-document relations are not considered, i.e., $\mathcal{V}_{11} = \mathcal{V}_{22} = \emptyset$. **HSQ** completes the previous model by considering collaborations $\mathcal{V}_{11}$ between users. Corresponding adjency matrix is such that $A_{11}(i, j)$ is the number of articles users i et j have co-edited. The degree of *collaboration* of the users is captured.

On Q&A websites, the **HITS** approach [11] and **NCR** model [21] are also particular instances of our model. In [11], a simple graph ($m = 1$) is considered, with $\mathcal{U}_1$ being the set of users. Authors assumes that $A_{11}(i, j) = 1$ if user j has answered at least once to a question formulated by user i. In [21], a graph with three families of nodes ($m = 3$) is considered, with $\mathcal{U}_1, \mathcal{U}_2$ and $\mathcal{U}_3$ being the set of users, answers and questions respectively. **HSQ** completes the NCR model by considering an additional set of entities $\mathcal{U}_4$ (the comments) and an inter-user relation $\mathcal{V}_{11}$. Adjency matrix associated to the inter-user relation is such that $A_{11}(i, j)$ is the number of answers i has provided *before* j to common questions. The *reactivity* of the users is captured.

4   Experiments

4.1   Datasets Description

Wikipedia.[1] A subset of roughly 23 000 articles was used. These articles were generated by 110 000 users and have been reviewed by the Editorial Team Assessment of the WikiProject. Each article is thus labelled according to the *WikiProject quality grading scheme* and belongs to one of the six class $FA \succ A \succ GA \succ B \succ C \succ S$. We assigned to each article i a numerical label y_i that respects

---

[1] https://en.wikipedia.org/wiki.

the user preferences. From $y_i = 0$ (class S, *Stub Articles*, i.e., very bad quality articles with no meaningful content) to $y_i = 5$ (class FA, *Featured Articles*, i.e., complete and professional articles). This scale is used as the ground truth in our evaluation. Recall we aim to rank articles by decreasing order of predicted quality. The repartition of the articles per class is summarized in Table 1.

Table 1. Statistics for the *Wikipedia* dataset.

Class	FA	A	GA	B	C	S
Label (y_i)	5	4	3	2	1	0
Number of articles	245	51	346	1 012	1 946	18 823

Stack Exchange.[2] The public dump of the *Stack Exchange* platform was used for evaluation. From October 2008 to September 2014, roughly 1 million of users, over 109 differents subplatforms, have generated more than 1.5 millions of questions, 2.5 millions of answers and 6.5 millions of comments. Numerical answers up votes, ranging from -65 to $2\,182$ for very popular answers are converted into integers. A first scale, noted b_s, is a binary scale where all negative answers, i.e., answers with score in $]-\infty, 0]$, constitute negative examples $(y_i = 0)$ while all answers with positive scores constitute positive examples $(y_i = 1)$. A second scale, used for ranking evaluation, noted r_s, is detailed in Table 2. Excepted for answers judged as bad quality (with negative scores), classes are balanced. Note that using b_s, we evaluate the capacity of the models to identify positive answers. Using r_s, the capacity of the models to rank answers in decreasing order of quality is evaluated.

Table 2. Answers scores discretization for the *Stack Exchange* dataset.

Class	A	B	C	D	E
Scores interval	$]-\infty, -1]$	$\{0\}$	$\{1\}$	$\{2,3\}$	$]3,\infty[$
Number of answers	52 540	542 562	629 443	629 443	651 825
Label y_i	0	1	2	3	4

4.2 Evaluation Metrics

The ranking over the articles and the answers is evaluated with the *Normalized Discount Cumulative Gain at k* (NDCG@k) [19]. Let σ be the permutation ordering the documents by decreasing order of predicted quality. The DCG@k is defined as $DCG(\sigma, k) = \sum_{i=1}^{k} \frac{2^{y_{\sigma(i)}} - 1}{\log(1+i)}$, where y_j is the label of document j. To compare different rankings, the normalized DCG is used,

---

[2] http://blog.stackoverflow.com/2009/06/stack-overflow-creative-commons-data-dump/.

$NDCG(\sigma, k) = \frac{DCG(\sigma,k)}{DCG(\sigma^*,k)}$, where σ^* is the optimal ranking. On Wikipedia, σ^* places all *Features Articles* on top, then all articles belonging to class A, and so on. On *Stack Exchange*, the degree of relevance of an answer is given by scale b_s or r_s. The average *NDCG@k* is reported over all the questions. We also evaluate the precision of the solutions. On *Wikipedia*, we report the fraction of positive predictions per class. On *Stack Exchange*, the average fraction of positive answers beyond the first k answers over all the questions is reported.

4.3   Experiment Results

Results on the *Wikipedia* and *Stack Exchange* datasets are summarized in Tables 3 and 4 respectively. In both cases, user parameter ϵ is fixed to 10^{-4}.

Table 3. Evaluations of the two solutions on the *Wikipedia* dataset.

	Model	FA	A	GA	B	C	S
NDCG	*Basic*	73.77	75.14	80.76	**81.87**	**84.11**	93.11
	HSQ	**74.39**	**75.75**	**81.54**	81.19	83.16	**93.80**
Prec.	*Basic*	62.45	0	8.67	**39.03**	**34.53**	**94.17**
	HSQ	**64.9**	0	**17.92**	29.55	30.27	93.16

Table 4. Evaluations of the three solutions on the *Stack Exchange* dataset using the *NDCG* metric on scales b_s and r_s and the *Precision* metric on scale b_s.

	Model		k=2	k=3	k=4	k=5	k=10	k=20
NDCG	*HITS*	b_s	88.38	88.89	90.13	92.47	95.01	95.39
		r_s	67.27	71.26	75.64	80.29	85.21	85.98
	NCR	b_s	89.22	89.64	90.81	93	95.37	95.72
		r_s	69.33	73.07	77.26	81.26	86.21	86.91
	HSQ	b_s	**89.38**	**89.92**	**91.15**	**93.27**	**95.5**	**95.82**
		r_s	**69.49**	**74.47**	**77.85**	**82.08**	**86.41**	**87.05**
Prec.	*HITS*		81.41	80.38	79.14	77.85	49.92	26.63
	NCR		82.52	81.28	79.89	78.31	50.00	26.65
	HSQ		**82.83**	**81.85**	**80.55**	**78.77**	**50.10**	**26.66**

On *Wikipedia*, regarding classes *FA* and *GA*, experiments are very conclusive. Proposed solution clearly outperforms *Basic* [10], suggesting a non-negligible benefit (+2 % and +9 % for *FA* and *GA* articles resp.) of considering the strength of collaborations to identify high quality articles. Interestingly, the co-edit relation integrated in *HSQ* is not helpful for discriminating mid or poor quality articles (classes *B*, *C*, and *S*. On *Stack Exchange*, the interest of the proposition is immediate. For both metrics, proposed solution outperforms competitors. We conclude that both users reactivity and users engagement bring discriminating informations to identify authoritative users and, therefore, high quality answers.

5 Conclusions

In the scientific litterature, structural-based analysis approaches for quality prediction purpose rely on graphs considering a few number of families of nodes and relations. Moreover, most of them suffer from a common lack of genericity and adaptability. To tackle these limitations, an unsupervised structural based algorithm, *HSQ*, was proposed. Base on a heterogeneous graph representation of the data, the proposal enables the reformulation of various state-of-the-art methods. By instanciating *HSQ* over the two major collaborative platforms *Wikipedia* and *Stack Exchange*, we have shown the genericity of the proposed solution. Experiment results have suggested that considering additional entities and interactions in the model was beneficial. In future work, we plan to study different influence functions in order to give different strengths for each family of entities.

References

1. Adler, B.T., de Alfaro, L.: A content-driven reputation system for the wikipedia. In: Proceedings of the 16th International Conference on World Wide Web, WWW 2007, pp. 261–270. ACM, New York (2007)
2. Agichtein, E., Castillo, C., Donato, D., Gionis, A., Mishne, G.: Finding high-quality content in social media. In: Proceedings of the International Conference on Web Search and Data Mining, WSDM 2008, pp. 183–194. ACM, New York (2008)
3. Bian, J., Liu, Y., Zhou, D., Agichtein, E., Zha, H.: Learning to recognize reliable users and content in social media with coupled mutual reinforcement. In: Proceedings of the 18th International Conference on World Wide Web, WWW 2009, pp. 51–60. ACM, New York (2009)
4. Blumenstock, J.E.: Size matters: word count as a measure of quality on wikipedia. In: Proceedings of the 17th International Conference on World Wide Web, WWW 2008, pp. 1095–1096. ACM, New York (2008)
5. Campbell, C.S., Maglio, P.P., Cozzi, A., Dom, B.: Expertise identification using email communications. In: Proceedings of the Twelfth International Conference on Information and Knowledge Management, CIKM 2003, pp. 528–531. ACM, New York (2003)
6. Chang, S., Pal, A.: Routing questions for collaborative answering in community question answering. In: Proceedings of the IEEE/ACM International Conference on Advances in Social Networks Analysis and Mining, ASONAM 2013, pp. 494–501. ACM, New York (2013)
7. Dalip, D.H., Gonçalves, M.A., Cristo, M., Calado, P.: Automatic assessment of document quality in web collaborative digital libraries. J. Data Inf. Qual. **2**(3), 14:1–14:30 (2011)
8. de La Robertie, B., Pitarch, Y., Teste, O.: Measuring article quality in wikipedia using the collaboration network. In: Proceedings of the IEEE/ACM International Conference on Advances in Social Networks Analysis and Mining, ASONAM 2015. ACM, New York (2015)
9. Golub, G.H., Van Loan, C.F.: Matrix Computations, 3rd edn. Johns Hopkins University Press, Baltimore (1996)

10. Hu, M., Lim, E.-P., Sun, A., Lauw, H.W., Vuong, B.-Q.: Measuring article quality in wikipedia: models and evaluation. In: Proceedings of the Sixteenth ACM Conference on Conference on Information and Knowledge Management, CIKM 2007, pp. 243–252. ACM, New York (2007)

11. Jurczyk, P., Agichtein, E.: Discovering authorities in question answer communities by using link analysis. In: Proceedings of the Sixteenth ACM Conference on Conference on Information and Knowledge Management, CIKM 2007, pp. 919–922. ACM, New York (2007)

12. Jurczyk, P., Agichtein, E.: Hits on question answer portals: exploration of link analysis for author ranking. In: Proceedings of the 30th Annual International ACM SIGIR Conference on Research and Development in Information Retrieval, SIGIR 2007, pp. 845–846. ACM, New York (2007)

13. Kleinberg, J.M.: Authoritative sources in a hyperlinked environment. J. ACM **46**(5), 604–632 (1999)

14. Li, B., Jin, T., Lyu, M.R., King, I., Mak, B.: Analyzing and predicting question quality in community question answering services. In: Proceedings of the 21st International Conference on World Wide Web, WWW 2012 Companion, pp. 775–782. ACM, New York (2012)

15. Movshovitz-Attias, D., Movshovitz-Attias, Y., Steenkiste, P., Faloutsos, C.: Analysis of the reputation system, user contributions on a question answering website: stackoverflow. In: Proceedings of the IEEE/ACM International Conference on Advances in Social Networks Analysis and Mining, ASONAM 2013, pp. 886–893. ACM, New York (2013)

16. Page, L., Brin, S., Motwani, R., Winograd, T.: The pagerank citation ranking: bringing order to the web. Technical report, Stanford Digital Library Technologies Project (1998)

17. Suryanto, M.A., Lim, E.P., Sun, A., Chiang, R.H.L.: Quality-aware collaborative question answering: methods and evaluation. In: Proceedings of the Second ACM International Conference on Web Search and Data Mining, WSDM 2009, pp. 142–151. ACM, New York (2009)

18. Wang, G., Gill, K., Mohanlal, M., Zheng, H., Zhao, B.Y.: Wisdom in the social crowd: an analysis of quora. In: Proceedings of the 22nd International Conference on World Wide Web, WWW 2013, pp. 1341–1352, Republic and Canton of Geneva, Switzerland, International World Wide Web Conferences Steering Committee (2013)

19. Yining, W., Liwei, W., Yuanzhi, L., Di, H., Wei, C., Tie-Yan, L.: A theoretical analysis of ndcg ranking measures. In: Proceedings of the 26th Annual Conference on Learning Theory (2013)

20. Zhang, J., Ackerman, M.S., Adamic, L.: Expertise networks in online communities: structure and algorithms. In: Proceedings of the 16th International Conference on World Wide Web, WWW 2007, pp. 221–230. ACM, New York (2007)

21. Zhang, J., Kong, X., Luo, R.J., Chang, Y., Ncr, P.: A scalable network-based approach to co-ranking in question-and-answer sites. In: Proceedings of the 23rd ACM International Conference on Conference on Information and Knowledge Management, CIKM 2014, pp. 709–718. ACM, New York (2014)

Analyzing Relationships of Listed Companies with Stock Prices and News Articles

Satoshi Baba[✉] and Qiang Ma

Kyoto University, Kyoto 606-8501, Japan
baba@db.soc.i.kyoto-u.ac.jp, qiang@i.kyoto-u.ac.jp

Abstract. To support decision making in our business and personal lives, we propose an integrated model of stock prices and a novel method based on it for analyzing the relationships of listed companies. In our integrated model, the stock price of a listed company consists of three factors: market-index, sector-index and pure-price. By utilizing this model, we can relax the affections of the market and sectors, and analyze the relationship between companies on the basis of their business performance by comparing their pure-prices. Experiments using a newly collected data set validated the proposed integrated model and methods.

Keywords: Relationship mining · Investment information analysis · Decision making support · Stock price analysis

1 Introduction

Modern companies connect and cooperate with each other, so analyzing relationships between organizations is an important and continuing topic for decision making support [8]. In this paper, we focus on relationships affecting company performance and propose a method for analyzing relationships between companies that uses stock prices and news articles. The stock price is one of the most important factors reflecting company performance. Generally, a rising stock price is correlated with good performance. We may identify the relationships between companies by checking similarities between the transitions in their stock prices.

Intuitively, it is possible to compare the stock prices of two companies to estimate their relationship in terms of business performance. If the trends of their stock prices are similar to each other, their business performance may be related. We calculate the relatedness of two companies on the basis of the sequences of their stock prices. There are two challenges when using stock prices to analyze the relationship between two companies.

1. Can we apply the raw data of stock prices to relationship mining? The answer is "no" because the stock price of a listed company may be affected by the market environment and other factors. The first challenge is to break the

This work was supported by a KAKENHI (No. 25700033) and SCAT Research Funding.

S. Hartmann and H. Ma (Eds.): DEXA 2016, Part II, LNCS 9828, pp. 27–34, 2016.
DOI: 10.1007/978-3-319-44406-2_4

stock price into usable components. We separate a company's stock price into three parts: market-index, sector-index, and pure-price (Sect. 3).

2. Should we use the stock price data for a company from the time when it became listed on a stock exchange? The answer is "no". One reason is that using much data requires complex calculation. Another reason is the relationships between companies are dynamic and change often. The second challenge is to select an appropriate data range of stock prices for further analysis (Sect. 4.4).

The major contributions of this paper can be summarized as follows.

- We propose an integrated model to extract the pure-price of a company by relaxing the affections of the market and the sector to which it belongs (Sect. 3).
- We propose a method to enable the relationships to be analyzed by comparing the value trends, not the absolute values (Sect. 4).
- We propose a method for selecting the sub-sequences of pure-prices for comparison by using news articles (Sect. 4.4).

2   Related Work

Various groups have studied companies' relationships. Jin et al. [5] proposed a method for discovering relationships between companies from the Web sites. They also proposed a method for building networks of companies. Only the text information on the Web sites is used to study lawsuit and partner relationships. We do not limit the kinds of relationships and study relationships on the basis of company business performance.

Michael et al. [3] showed that the result of a football game affects the stock prices of the football teams' sponsors. While they find the sponsor relationships manually, we find relationships automatically by utilizing the stock prices.

Many studies have investigated the relationship between stock prices and the text data such as that in news articles or SNS messages. Tetlock [7] studied how the pessimism from the Wall Street Journal columns affecting the value of the Dow Jones Industrial Average. Bollen et al. [2] investigated how the collective mood states reflect in Twitter postings correlated with the Dow Jones Industrial Average. These studies revealed that text data is an important information source for analyzing the stock price. In our study, we use the news articles as additional information to help identify appropriate data scopes of stock prices for comparison.

3   Integrated Model of Stock Prices

To avoid comparing two stock prices including affections of these companies' markets and the business sectors they belong to, we propose an integrated model of stock prices. We also propose a method that uses this model to relax the stock prices.

We separate the stock price of a company into three parts: market-index, sector-index, and pure-price. The market-index is the stock market index, the sector-index is the industry's stock index, and the pure-price reflects the company's business performance. The model integrating the three parts is based on seasonal adjustment [1]. The integrated model of stock prices is given by

$$X_t = C_t \times I_t \times M_t \tag{1}$$

where, X_t is a stock price, C_t is a pure-price, I_t is a sector-index and M_t is a market-index as of market closing on date t. I_t is the business sector stock index, and M_t is the market stock index (TOPIX, Tokyo Stock Exchange (TSE) 2nd stock index, etc.). The business sectors are determined by TSE. The pure-price is calculated using the integrated model.

4 Method for Analyzing Relationships Between Companies

4.1 Overview

Figure 1 shows the process flow of our method. We collect news articles and stock data in advance and store them in two databases. The user inputs either the target company name or company code for which the user wants to find related companies. The output is a ranked list of companies related to the target company.

To analyze the relationship between two given companies, at first, we normalize their pure-prices in order to estimate their relationship with change trends but not price values. We then automatically select the sub-sequences of normalized pure-prices by analyzing the news articles related to these companies. Next, we compare the sub-sequences to estimate whether the two companies related to each other.

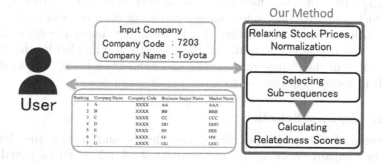

Fig. 1. Process flow

4.2 Data Collection

The stock price data download site[1] we use to obtain stock price data offers stock price indexes (TOPIX, business sector stock price index, etc.), and listed company's stock prices from 2007 to date. Our target companies were handled in Yahoo! Finance[2] and their stock price data did not have missing values. We used news articles published on or after September 14, 2010, so we used stock price data from 2010 to 2015.

The news source, ZAIKEI SHINBUN[3], is a Japanese online news site that offers categorized articles. We used news articles in "company and industry" category as they directly affecting stock prices. We used news articles from September 14, 2010 to December 30, 2015, the last day of the TSE in 2015. Companies for which there was not complete stock data for the selected period by news articles were excluded.

A news article was judged to be related to a company if the company's name was mentioned in it. We utilized the company names used in Yahoo! Finance.

4.3 Pure-Prices Normalization

The percentage change in stock price from one day to the next is an important index for analyzing stock prices. It is the percentage increase or decrease in price from the previous day's closing price to the current day's one. In this, we reveal the transition of stock prices from the previous day by change rates to investigate one from the dates of news articles.

Since calculation of the change rate requires stock price data for the previous day, the change rate for the day when the stock is first traded is 0. The equation for calculating the change rate is

$$\delta_d = \begin{cases} 0, & d = 1 \\ \dfrac{cp_d - cp_{d-1}}{cp_{d-1}} \times 100, & d > 1 \end{cases} \tag{2}$$

where, cp_{d-1} is the previous day's closing price, and cp_d the current day's one.

In some cases, the absolute value of δ_d is very large due to stock splits or reverse stock splits. We treat value changes following stock splits or reverse stock splits as outliers. We convert outliers to the average value of the price the day before and the day after the split. All change rates are divided by the highest absolute value of the change rate for the two companies being compared, and the result is normalized from -1 to 1.

4.4 Sub-sequence Selection

To overcome the huge computational complexity and handle the dynamic changes in company relationships, we compare stock prices with selected comparison ranges, i.e., we select sub-sequences of stock prices.

---

[1] http://k-db.com/.
[2] http://finance.yahoo.co.jp/.
[3] http://www.zaikei.co.jp/.

Company relationships change dynamically, and we should consider current relationships in decision making to look toward the future benefits. News articles are essential factors in determining whether stock prices increase or decrease [7], and it is highly possible that stock prices change a lot. Accordingly, we select the sub-sequence for comparison as the period from the date of the latest news article about an input company to December 30, 2015. Using this sub-sequence, we compare the pure-prices of the two companies being compared and calculate their relatedness score.

Our oldest news articles available were dated September 14, 2010. The current date is represented by t_c and the date of the news article being reported is represented by τ. If the comparison period $[\tau, t_c]$ is long enough (i.e., many pure-prices are available for comparison), the relatedness score is calculated by comparing the pure-prices in this period. If the period is short (i.e., few pure prices are available for comparison), the relatedness score is bipolar, i.e., either extremely large or extremely small. The relatedness score thus cannot be calculated if the sub-sequence is less than a certain length. In this study, we arbitrarily set the minimum length of the period $[\tau, t_c]$ is 30^4. In other words, the following equation holds:

$$(t_c - \tau) + 1 \geq 30 \tag{3}$$

4.5   Relatedness Score

Various methods are available for encoding time series data into strings and for evaluating correlation between time series data. We use three methods for estimating the relatedness of two companies by comparing their pure-prices in a certain duration, which is decided by considering news articles.

- Hamming distance: First, we encode the time series of pure-price into strings. k, the number of characters, is a parameter determined experimentally[5]. Then, we measure the difference between two strings. We decide the replacement cost on the basis of dictionary order. Relatedness score $r_h(A, B)$ is calculated using

$$r_h(A, B) = \frac{d_h(A, B)}{len(A)} \tag{4}$$

  where, $d_h(A, B)$ shows the Hamming distance between sub-sequences A, B and $len(X)$ shows the length of a string X. In this paper, we use $len(X)$ in the same definition. The closer $r_h(A, B)$ is to 0, The higher the relatedness score.

- SAX (Symbolic Aggregate approXimation) [6]: We use the length of a sub-sequence as w frames since the news event occurred and deal with daily pure-prices like data after piecewise aggregate approximation (PAA). Relatedness score $r_s(Q, C)$ is calculated using

$$r_s(Q, C) = \frac{d_m(Q, C)}{len(Q)} \tag{5}$$

---

[4] Future work includes determining a suitable minimum value.
[5] Future work includes considering a method for determining a suitable value of k.

where, $d_m(Q,C)$ shows the distance between sub-sequences Q, C by SAX. The closer $r_s(Q,C)$ is to 0, The higher the score.
- Partial correlation coefficient: We evaluate the correlation between two pure-prices after removing the effect of the time variable. Relatedness score $r_p(X,Y,T)$ is calculated using

$$r_p(X,Y,T) = \frac{pcor(X,Y,T)}{lendata(X)} \tag{6}$$

where, $pcor(X,Y,T)$ shows partial correlation coefficient between sub-sequences X, Y by removing the effect of the time data T. The closer $r_p(X,Y,T)$ is to 1, The higher the relatedness score. $lendata(X)$ represents the number of X elements.

5 Evaluation

We experimentally evaluated our integrated model of stock prices and our methods for analyzing relationships between companies. We calculated the value of nDCG(Normalized Discounted Cumulative Gain) [4] by comparing the relatedness ranking made by human participants with the one made using our methods. The four participants evaluated the relationships between each input company and the companies on the merged list using a five-grade mechanism, where 1 and 5 denote the minimally and highly related, respectively. We first compared the nDCG values of three methods for evaluating relatedness scores. Next we compared the nDCG value of the ranking made using the pure-prices with one made using the stock prices.

5.1 Data Sets

Ten companies we used were selected on the basis of their market capitalization on January 18, 2016 with up to two companies per business sector. We did not consider companies that did not have a complete set of stock prices from the date of the latest news article to December 30, 2015. The ten companies are JT(2914), Seven & i HOLDGS.(3382), Takeda Pharmaceutical Company Limited(4502), Toyota Motor Corporation(7203), Honda(7267), Canon(7751), Mitsubishi UFJ Financial Group(8411), Mizuho Financial Group(8411), Nippon Telegraph and Telephone Corporation(9432) and NTT docomo(9437). We used news articles from ZAIKEI SHINBUN from September 14, 2010 to December 30, 2015.

5.2 Parameter Tuning

We conducted an experiment for setting parameter k in the Hamming distance. In this experiment, we tested from $k = 10$ to $k = 10^2$ by 10 from the viewpoint of the computational complexity. As the input company, we used Toyota Motor Corporation. We calculated nDCG@30 for each k and selected the k with the highest nDCG@30, i.e., k = 80, to use in our evaluation.

5.3   Experimental Results

We first ran the three methods mentioned in Sect. 4.5 and obtain the top 15 companies ranked by each method. Then, we merged these result companies into one list and used them as the ranking targets for calculating the nDCG for each method.

Evaluation of Integrated Model. We compared the relatedness estimations with the pure-price and stock price for each company. The results are shown in Fig. 2. Figure 2 indicates that the partial correlation coefficient was more suitable than the two encoding methods in pure-price. This also indicates that pure-prices are more suitable than stock prices for estimating the relatedness between companies with large i.

Table 1 shows the nDCG@i($i = 5, 10, 15$) average values for the values plotted in Fig. 2. For each method, the nDCG value for the pure-prices was larger than that for the stock prices. These results demonstrate that our integrated model of stock prices is effective.

Evaluation for Different Comparison Ranges. We also compared the results when we used two different ranges of stock prices for relatedness estimation.

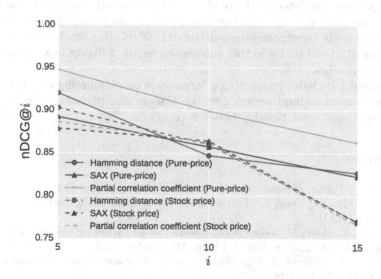

Fig. 2. Experimental results: nDCG@i

Table 1. Average values of nDCG@i($i = 5, 10, 15$)

	Hamming distance	SAX	Partial correlation coefficient
Pure-prices	0.865	0.857	0.903
Stock Prices	0.844	0.837	0.838

One comparison range was automatically determined by using news articles as described in Sect. 4.4. The other one was simply the range for all available data over the entail duration from September 14, 2010. We used the partial correlation coefficient method, which achieved the best nDCG score in the experiment described above, for relatedness estimation. The results are shown in Table 2. Table 2 shows that the nDCG@i values obtained by using news articles were larger than ones for the entire periods. These results show that using news articles not only reduces the computational complexity but also produces more accurate rankings of companies.

Table 2. nDCG@i (Two different comparison ranges)

	5	10	15
Period based on news article	0.947	0.899	0.862
Entire period	0.867	0.799	0.798

6 Conclusion

We have proposed an integrated model of stock prices and a novel method based on it for analyzing relationships between pairs of companies. An experimental evaluation using ten companies demonstrated that relaxing stock prices by using our integrated model and selecting sub-sequences by utilizing news articles are effective approaches.

Future work includes conducting a large-scale crowdsourcing experiment for further evaluation to improve our method. We are also planning to study ways of detecting implicit relationship between companies.

References

1. Arita, T.: A Primer of Seasonal Adjustment. Toyo Keizai Inc. (2012)
2. Bollen, J., Mao, H., Zeng, X.: Twitter mood predicts the stock market. J. Comput. Sci. **2**, 1–8 (2011)
3. Hanke, M., Kirchler, M.: Football championships and jersey sponsors' stock prices: an empirical investigation. Eur. J. Finan. **19**, 228–241 (2012)
4. Järvelin, K., Kekäläinen, J.: Cumulated gain-based evaluation of ir techniques. ACM TOIS **20**(4), 422–446 (2002)
5. Jin, Y., Matsuo, Y., Ishizuka, M.: Extracting inter-business relationship from world wide web. Trans. Japan. Soc. Artif. Intell. **22**, 48–57 (2007)
6. Lin, J., Keogh, E., Lonardi, S., Chiu, B.: A symbolic representation of time series, with implications for streaming algorithms. In: DMKD 2003, pp. 2–11 (2003)
7. Tetlock, P.C.: Giving content to investor sentiment: the role of media in the stock-market. J. Finan. **62**, 1139–1168 (2007)
8. Yamakura, K.: Inter-organization relationships and inter-organization relationships theory. Yokohama Bus. Adm. Soc. **16**(2), 166–178 (1995)

Linked Data

Approximate Semantic Matching over Linked Data Streams

Yongrui Qin[1]([⊠]), Lina Yao[2], and Quan Z. Sheng[3]

[1] University of Huddersfield, Huddersfield, UK
y.qin2@hud.ac.uk
[2] University of New South Wales, Sydney, Australia
lina.yao@unsw.edu.au
[3] The University of Adelaide, Adelaide, Australia
michael.sheng@adelaide.edu.au

Abstract. In the Internet of Things (IoT), data can be generated by all kinds of smart things. In such context, enabling machines to process and understand such data is critical. Semantic Web technologies, such as Linked Data, provide an effective and machine-understandable way to represent IoT data for further processing. It is a challenging issue to match Linked Data streams semantically based on text similarity as text similarity computation is time consuming. In this paper, we present a hashing-based approximate approach to efficiently match Linked Data streams with users' needs. We use the Resource Description Framework (RDF) to represent IoT data and adopt triple patterns as user queries to describe users' data needs. We then apply locality-sensitive hashing techniques to transform semantic data into numerical values to support efficient matching between data and user queries. We design a modified k nearest neighbors (kNN) algorithm to speedup the matching process. The experimentalresults show that our approach is up to five times faster than the traditional methods and can achieve high precisions and recalls.

Keywords: Internet of Things · Linked Data · Semantic matching · kNN classification

1 Introduction

The Semantic Web was first described by Berners-Lee et al. in 2001 [1]. It is considered as an evolution of the existing Web. Before Semantic Web, Web information was mainly produced for, and consumed by, humans. Most information on the World Wide Web was linked by hypertext. In this way information was presented in a convenient way for humans to access. Meanwhile, information available on the Web has been exploding as time goes on. People are creating photos, articles, videos, and many other kinds of information. Such information needs to be processed automatically. The Semantic Web was designed to make up for this situation.

© Springer International Publishing Switzerland 2016
S. Hartmann and H. Ma (Eds.): DEXA 2016, Part II, LNCS 9828, pp. 37–51, 2016.
DOI: 10.1007/978-3-319-44406-2_5

The Semantic Web stores information in a designed format so that the information is given well-defined meanings. However, the Semantic Web is not only about putting data on the Web. It is about links that make data easy for people/machines to explore and study [1]. Linked Data is such a technology that describes information, data and knowledge on the Semantic Web. The Resource Description Framework (RDF) is one of the most popular languages used to represent Linked Data.

In many domains, scientists have growing needs of integrating information and data. For example, computer science researchers would need integration of hardware knowledge and software knowledge in order to design systems. Environment scientists are looking for integration of hydrology, climatology, ecology and so on [2]. The Semantic Web is able to fulfill these needs as it provides "a common framework that allows data to be shared and reused across application, enterprise, and community boundaries" [3].

Furthermore, the Internet of Things (IoT) makes it possible to connect physical things to the Internet. Thus people are able to access remote sensing data and control the physical world from a distance [4]. Data that has been collected from IoT could be in various formats. IoT data could also be in large amount, which makes it difficult and costly for people to process manually [5]. This calls for the use of Semantic Web technologies to process data generated in the coming IoT era. One promising application scenario of Linked Data techniques is *smart city*. Figure 1 shows the structure of a smart city model based on Linked Data. In this system, data and information are collected via various kinds of devices, such as mobile phones, cars, cameras, sensors and so on. Sensing data is transformed to Linked Data streams in order to be processed automatically by machines. Then Linked Data streams are processed by the matching engine. Matching engine is the core component of the system. It combines different functionalities such as data processing, semantic query processing, matching algorithms, and so on. Further descripton of this scenario can be found in [6].

With the help of this smart city system, all the terminal devices are connected. Information about things and environments around the city, including temperature, humidity, traffic status, air pollution, and other information, is sent

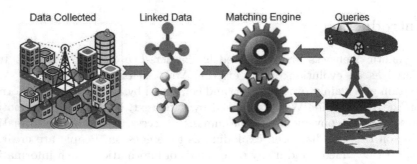

Fig. 1. Smart city model

to the matching engine in the format of Linked Data. In the meantime, queries coming from individuals, companies, devices or any other systems are sent to the matching engine as well. With a set of matching procedures, information that is best matched to the user queries will be returned to corresponding query senders.

However, in the Semantic Web, Linked Data in the RDF format cannot allow us to explore deeper into the semantic relations between different entries of data. The reason of this situation is that data in format of string does not support semantic matching efficiently. In IoT, we envision that data consumers are not likely to have complete knowledge and therefore supporting semantics-based matching is required in order to deliver relevant data to assorted consumers. In addition, semantic data is difficult to process due to the fact that different words might have similar underlying meanings. For instance, "master student" has a similar meaning with "PhD candidate" as they both refer to higher education students. However, they are completely different phrases in terms of texts. Machines could hardly find out their relationship efficiently based on the texts.

To address such problem, in this paper we adopt Locality Sensitive Hashing (LSH) techniques [7] to map semantic data into hashing values. LSH makes it possible to map different semantic data entries into a space based on their linguistic relations. In the same space, a word or phrase is closer to those that are more linguistic related to them. Using LSH, we are able to calculate semantic similarities of each pair of words/phrases based on their numerical values only. In other words, information can be semantically matched to specific queries based on their semantic hashing mappings. Specifically, in this work, we propose an approximate matching method, which modifies the naive k nearest neighbors (kNN) approach in order to make the matching process more efficient.

The main contributions of this paper are as follows. Firstly, we adapt the existing Locality Sensitive Hashing techniques to transform Linked Data streams and user queries into numerical values. We then develop a novel index construction approach for fast semantic matching based on the naive kNN classification approach. Finally, we conduct extensive experiments using a real-world dataset from DBPedia. The results show that our proposed system can disseminate Linked Data at a faster speed compared with the straightforward matching approach with thousands of registered queries.

The rest of this paper is organized as follows. In Sect. 2, we review the related work. We present some background knowledge, the framework and the technical details of our approach in Sect. 3. In Sect. 4, we report the results of our experimental study. Finally, we present some concluding remarks in Sect. 5.

2   Related Work

A large body of work has been done in the area of RDF based stream processing, such as Streaming SPARQL [8], Continuous Query Evaluation over Linked Streams [9], Sparkwave pattern [10], and EP-SPARQL language [11]. However, their focus is on *exact matching* over Linked Data streams, but not semantic matching. Further, they do not support large-scale query evaluation but focus on the evaluation of a single query or a small number of parallel queries over the streaming Linked Data.

Recent work on data summaries on Linked Data such as the work in [12] transforms RDF triples into a numerical space. Then data summaries are built upon numerical data instead of strings as summarizing numbers is more efficient than summarizing strings. In order to transform triples into numbers, hash functions are applied on the individual components (s, p, o) of triples. Thus a derived triple of numbers can be considered as a 3D point. Data summaries are designed mainly for indexing various Linked Data sources and used for identifying relevant sources for a given query. However, the data summaries approach does not support approximate matching. This is because in the data summaries approach, the hash functions are not locality sensitive. Other existing work introduced in [6,13] focuses on exact pattern matching, but not semantic matching.

The work in [7] presents an algorithm based on LSH to improve the performance of event detection system. It mainly focuses on first story detection (also known as new event detection). An algorithm based on LSH is developed to speed up the event detection process in order to efficiently detect new stories from Twitter posts. The challenge is that there are too many posts on Twitter which are not actual events. The focus in that work is processing Tweets, which is different from Linked Data and the Twitter event detection approach cannot be directly applied in matching over Linked Data streams.

3 Approximate Semantic Matching

In this section, we first briefly provide some necessary background knowledge on user queries and word vector representation. We then describe our approximate semantic matching approach in detail.

3.1 Preliminaries

User Queries. Similar to [14,15], triple patterns are adopted as the basic units of user queries in our system. A triple pattern is an expression of the form (s, p, o) where s and p are URIs or variables, and o is a URI, a literal or a variable. The eight possible triple patterns are: 1) (#s, #p, #o), 2) (?s, #p, #o), 3) (#s, ?p, #o), 4) (#s, #p, ?o), 5) (?s, ?p, #o), 6) (?s, #p, ?o), 7) (#s, ?p, ?o), and 8) (?s, ?p, ?o). Here, ? denotes a variable while # denotes a constant.

Words Vector Representation. Mikolov et al. proposed an efficient method to achieve vector representations for English words [16]. They proposed two new models for machine learning of word representations. More specific, they used numerical values (vectors) to represent words and compare semantic relations. The cosine similarity between two words can be approximated by the cosine similarity between their corresponding vectors. Such vector representations preserve the locality of words in the original text space and hence belong to the category of LSH techniques [7]. Based on the reported results, the accuracy of predicting semantic similarities between words based on vector representations could reach up to 70 % [17].

3.2 System Overview

Figure 2 shows the structure overview of the system. Linked Data collected from the real world will be sent to the system. Then the data will be hashed using LSH techniques. Meanwhile, users can send queries to the system. These queries are also hashed into numerical values. The core component of the system, Matching Engine, matches Linked Data streams against the queries and returns results to users.

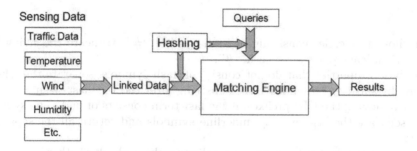

Fig. 2. System overview

3.3 Linked Data Processing

In the following, we focus on how to efficiently process Linked Data and support the semantic matching procedure.

Extract Last Terms. Each triple in the Linked Data streams contains either URI (like "http://example.org/example#John") or prefix (like "*xmlns : name*"). The prefix components are used to identify the resource, but they are not relevant to the major semantic meaning of the triple. In order to closely reflect the semantic meaning of the triple, we need to remove these prefix parts to get the last terms. Figure 3 shows an example of this procedure.

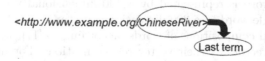

Fig. 3. Extract last terms of triples

In real world applications, to describe complex information, people need to deliver more information in a single triple. The triple in Fig. 3 is such an example. It has a phrase "ChineseRiver" as the last term. In this case, the last term is a composition of multiple words. The two words of the phrase in this example can be split up and the result is shown in Fig. 4. Below are some rules to extract and split the last terms:

Fig. 4. Split up complex last terms

- For those properties consisting of hash symbol "#", truncate the string by "#", then leave parts after hash "#".
- For those properties that do not consist of hash symbols, separate the whole string by slash "/", then leave the substring after the very last slash.
- After removing the URI prefixes, if the last term consists of underline symbol "_", separate the last term by underline symbols and return all the separate words.
- If the last term does not consist of underline symbols, check whether it contains capital letters. If so, separate each word starting with a capital letter.
- Apply any other known rules to split the last terms.

Hashing Semantic Data. Once we extract and split the last terms, we can hash these terms into numerical values using existing LSH techniques. Transforming Linked Data into numerical values has two main benefits:

- Numerical values can achieve faster speed in the comparing process than strings.
- Using numerical values to represent Linked Data provides convenience to compare the similarity between different words approximately and directly.

We choose the Google News dataset in the word2vec project [18] from Google as our LSH foundation. In this dataset, part of Google News data (about 100 billion words) [18] is selected and trained to build an LSH model for mappings between words and vectors. The final LSH model contains vectors for 3 million words, and each word is represented by a 300-dimensional vector. This means we can hash a single word to a 300-dimensional vector.

For phrases and compositions of words, according to [17], we simply use the addition of their vectors as their vector representations. For instance, we will have the vector for "ChineseRiver" to be the sum of two vectors of "Chinese" and "River":

$$\text{"V(ChineseRiver)} = \text{V(Chinese)} + \text{V(River)"}$$

An example of hashing triples is shown in Fig. 5. In this figure, the triple contains only the last terms without prefixes. Each word of the triple could be represented as a 300-dimensional vector, so finally the whole triple can be represented by a 900-dimensional vector.

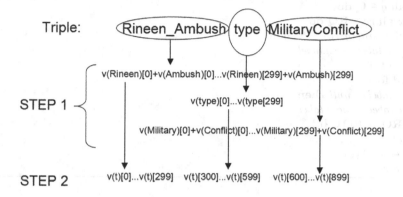

Fig. 5. Hashing example

3.4   Index Construction

Next, we build an index for user queries, which are triple patterns. Since a triple pattern also contains *subject, predicate, object*, matching a triple pattern to a query is actually comparing these three parts. In this work, all these parts have been transformed to numerical values. As in the Google word2vec project, where we obtain the Google News dataset, the measurement for testing similarity between two words is cosine similarity, we need to build the index based on cosine similarity.

Basically, the larger the cosine similarity is, the smaller the cosine distance is, and the two words are more related [19]. Here we are building a query index that is actually a kNN pre-trained data classification model for a given query set. To build the model, we need to classify all the data entries (queries) in this query set. The query index is built with a threshold θ, which defines the smallest value of cosine similarity that two queries in one classification should have, and a set of queries. Algorithm 1 shows the pseudocode of this step.

In order to improve the performance of our system, we select the representatives of queries in each class of queries. For each class, we simply take the first query as its representative. After processing with this algorithm, we successfully build an index of queries. In our system, this index contains vectors of the query patterns, class labels of all query patterns, and a representative query set.

3.5   Matching Data to Queries

Note that, the naive matching algorithm has a large timing cost since it has to compare the incoming triple with all user queries. If we use the naive matching

---

Algorithm 1. Pseudocode of Classifying Queries

Input: a set of queries **Q**, threshold of cosine similarity θ
Output: Classification result **U**, and representative queries **RQ**

   $\mathbf{U} \leftarrow \emptyset$
   $\mathbf{RQ} \leftarrow \emptyset$
   for all $q \in \mathbf{Q}$ **do**
      for all $rq \in \mathbf{RQ}$ **do**
         if $cosine(q, rq) > \theta$ **then**
            $q.label \leftarrow rq.label$
         end if
      end for
      if $q.label = null$ **then**
         $q.label \leftarrow new\ label$
         $\mathbf{RQ} \leftarrow \mathbf{RQ} \cup \{q\}$
      end if
      $\mathbf{U} \leftarrow \mathbf{U} \cup \{q\}$
   end for

---

method, we will find out all semantically matched results (under some given threshold θ) because the naive method will compare the triple against all the user queries in a brute force way. The problem is that the matching process is inefficient. To improve the performance of our system, we propose to adapt the kNN approach, which aims to find out the most semantically matched queries at a higher speed. The tradeoff is to sacrify some matching quality, such as with slightly lower recall and F1 scores (detailed definitions of these terms will be provided in Sect. 4).

In our adapted kNN approach, once we have built the kNN classification model (the query index), we are able to complete the "Matching Engine" shown in the system overview (Fig. 2) by implementing semantic matching logic on top of this model. The main idea is that, when we receive a newly incoming triple in the Linked Data stream, the system will identify k nearest classes to that triple. To obtain these k nearest classes, we first compute cosine similarity between the triple and each representative query in RQ, and then select k classes whose representative queries achieve top k cosine similarities. Then the triple will be matched against all the queries inside these k classes to find out all the queries that semantically match this triple. Since we only compare with k nearest classes of queries, not all queries in all classes, the matching process can be significantly accelerated and completed with high matching quality.

An example of this matching process is shown in Fig. 6. Q is the collection of the queries. Suppose in order to build the query index, these queries are classified into four classifications: $C1$, $C2$, $C3$ and $C4$. There is one representative query, drawn in yellow and circled, in each class.

When a triple t arrives at the system, the system computes the cosine similarities between t and all representative queries in RQ. Then we obtain top k (suppose $k = 2$) representative queries as the results. Assuming in this case, $C1$ and $C3$ are the two classes whose representative queries achieve best cosine

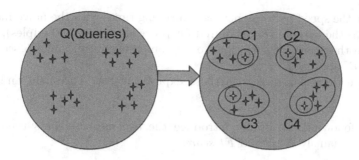

Fig. 6. Modified kNN classification method

similarity. We match triple t with all the queries in $C1$ and $C3$ by computing cosine similarity of each query and t. If the cosine similarity of a query q and t is greater than threshold θ, q will be a semantically matched query. After all queries in $C1$ and $C3$ are examined, we will obtain the final matching results. The core matching logic is shown in the following equation:

$$Result_{kNN} = \forall q \in C_1 \cup C_3 \wedge cosine(q, t) > \theta$$

To sum up, our approximate semantic matching consists of two main steps: *classification* and *matching*. The classification step has a time complexity of $O(d \times |RQ|^2)$, where $|RQ|$ is the number of classes and d is the number of dimensions of a word vector. Meanwhile, the matching step has a time complexity of $O(k \times d \times |Q|/|RQ|)$, where k is the parameter for matching and $|Q|/|RQ|$ is the average number of queries in a class.

4   Performance Evaluation

The experiments have been conducted using real-world data, which is a set of events extracted from DBpedia, provided by the authors of the work in [20]. We used these events in RDF format to form a Linked Data stream so as to simulate the sensing data streaming process in the smart city scenario. The event set contains resources of type `dbpedia-owl:Event`. Each event is a triple of the form `<eventURI, rdf:type, dbpedia-owl:Event>`. Examples of various event types that can be found in the event set are: "Football Match", "Race", "Music Festival", "Space Mission", "Election", "10th-Century BC Conflicts", "Academic Conferences", "Aviation Accidents and Incidents in 2001", etc. The experimental machine was running Windows 7, with Intel's Core i5 CPU and 8 GB RAM.

To the best of our knowledge, this is the first attempt to support semantic matching over Linked Data streams. To evaluate the performance of our system, a set of experiments were conducted to evaluate time, recall and F1 score by comparing with the naive matching approach:

- Evaluate the speed performance by comparing times with the naive matching approach (the average time required for processing every 300 triples).
- Evaluate the accuracy performance by comparing recall and F1 score with naive kNN approach.
- There are three parameters in these experiments: k, Threshold and Query Number.

In the following, we briefly introduce the two measurements used in our experiments, namely *recall* and *F1 score*:

- Recall is the percentage of the number of matched queries in our system divided by the number of all matched queries in the naive matching approach. Recall can be calculated using:

$$Recall = \frac{Number_{matched\_queries}}{Number_{naive\_matched\_queries}} \qquad (1)$$

- F1 score is also a measurement of matching accuracy, which can be calculated by using:

$$F_1 = 2 \cdot \frac{recall \cdot precision}{recall + precision} \qquad (2)$$

In this work, the precision is always 100% since our system matches triples and queries in the same way as the naive approach does (note that our system only selects queries that have $cosine(q, t) > \theta$ in the top k classes, and the naive method also returns all the queries that have $cosine(q, t) > \theta$). Therefore, any matched query of our results must be a matched query in the naive method's matching results as well.

In each experiment, we changed one parameter and kept the other two at their default values, so we had three group of experiments. Note that the time used in our system consists of two parts. The first part is the **classification time**, and the second part is the time used to find all matched queries during the matching process on top of the classification model, which we call the **matching time**.

4.1 Experimental Results—Parameter: k

The results with change of k are illustrated in Figs. 7 and 8. In this set of experiments, we set the number of queries as 1,500, and threshold as 0.6. We set the default threshold to 0.6 as this value can best balance matching speed and matching quality. We tested k in the range of $[1, 5]$. From the results we can observe that the classification time does not change too much while the matching time has an obvious increasing trend. Our approach is about 4 times faster than the naive approach when $k = 1$ and is about 3 times faster when $k = 5$. In terms of Recall and F1 score, both of them increase gradually when increasing the value of k. In most cases, Recall and F1 score are higher than 85%. This indicates that our approach can achieve high matching quality.

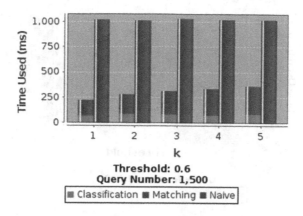

Fig. 7. Experiment: Time—k

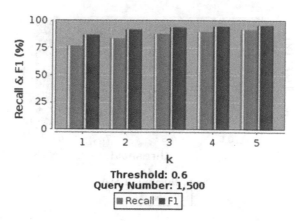

Fig. 8. Experiment: Recall & F1 Score — k

4.2   Experimental Results—Parameter: Threshold

The results with change of threshold are illustrated in Figs. 9 and 10. In this set of experiments, we set the number of queries as 1,500, and $k = 3$, because when the query number is 1,500, we can observe the normal performance gain that our approach can achieve and when $k = 3$, our approach shows a good balance between matching speed and matching quality. Meanwhile, the threshold increases from 0.5 to 0.8. From the results we can see that in terms of the time cost, our approach outperforms the naive approach by several times. When the threshold is 0.5, the matching time cost is high due to the formation of large query classes under low similarity threshold. This is also confirmed by the larger proportion of matching time cost obtained when threshold is 0.5 or 0.6. When threshold is larger, such as at 0.8, it is expected that the average size of each class is small. Therefore, we observe small matching cost compared with classification time. In terms of matching quality, both recall and F1 score are higher than

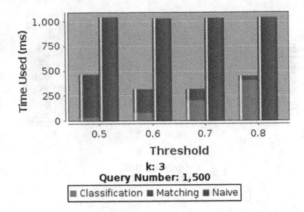

Fig. 9. Experiment: Time — Threshold

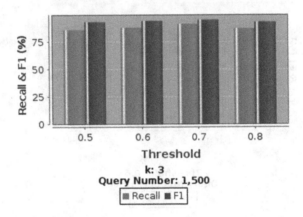

Fig. 10. Experiment: Recall & F1_ Score — Threshold

85 % in most cases. This demonstrates that our approach is very robust under different similarity thresholds.

4.3 Experimental Results—Parameter: Query Number

The results with change of query number are illustrated in Figs. 11 and 12. In this set of experiments we set preconditions as: $k = 3$, $Threshold = 0.6$. The query number is ranging from 500 to 3,000. The total matching time costs of both approaches are increasing approximately in a linear manner against the increasing number of queries to be matched. But the total time cost of the naive approach is observed to increase at a faster rate. Meanwhile, the matching quality is also improved with more queries. This should be because better classification results can be obtained with more queries. But after the number increases to and above 1,500, the matching quality stays quite stable.

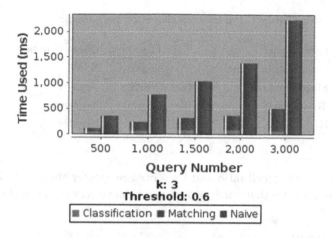

Fig. 11. Experiment: Time — Query Number

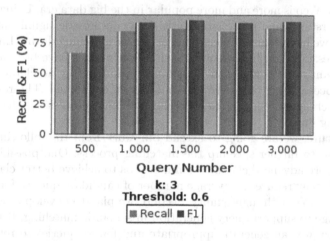

Fig. 12. Experiment: Recall & F1_ Score — Query Number

4.4 Discussion

By conducting the above three sets of experiments, we can summarize the effects that the three parameters have on the system performance. Table 1 shows the effects that each parameter has on the performance. In this table, "Positive" means it either accelerates the matching speed or improves the recall ratio and F1 score. "Negative" means the opposite way of "Positive". "N/A" means that this parameter does not affect the corresponding performance feature.

To sum up, our system has obvious advantage in the matching speed than the naive approach. Increasing the three parameters (i.e., k, threshold, query number) will normally cause higher matching time cost of the system. Meanwhile, increasing k has a positive effect on the recall ratio and F1 score. Through all the experiments, we demonstrate that our system has enhanced the matching speed significantly.

Table 1. ParameterŚ effects on performance

Performance	k	Threshold	Query Number
Classification Time	N/A	Negative	N/A
Matching Time	Negative	Positive	Negative
Total Time	Negative	Negative	Negative
Recall & F1 Score	Positive	N/A	Negative

In the meantime, the recall ratio and F1 score are greater than 85 % for most of the time. This indicates that our approach can achieve very high matching quality.

5 Conclusion

The Semantic Web is more and more popular in the big data era. Using machines to read, understand, and process semantic data can provide significant benefits. In this work, we have focused on enabling semantic matching during Linked Data streams processing. Locality-sensitive hashing techniques have been adapted to support semantic matching with high quality and better acceleration in the matching process. A set of experiments have been conducted. The results show that our matching system can speedup the matching process significantly with high matching quality.

In the future, we are going to extend our work from the following aspects. First, we plan to further speedup the matching process. One possible solution is to adopt more advanced classification methods to achieve better classification results, which may reduce the average number of candidate queries for matching a given RDF triple with high quality. Second, we plan to develop a new type of query language to support query generation in semantic matching. It is interesting to see how we can generate appropriate and fewest queries to reflect users' information needs possibly described in plain text in the semantic matching scenarios.

Acknowledgments. Authors would like to thank Zheng Jing for the implementation of the matching system and thank anonymous reviewers for their valuable comments.

References

1. Berners-Lee, T., Hendler, J., Lassila, O., et al.: The Semantic Web (2001)
2. Shadbolt, N., Hall, W., Berners-Lee, T.: The semantic web revisited. IEEE Intell. Syst. **21**(3), 96–101 (2006)
3. Koivunen, M.-R., Miller, E.: W3c semantic web activity. Semant. Web Kick-Off Finl. **63**, 27–44 (2001)
4. Hermann, K.: Internet of Things. Real-time Systems. Springer, Heidelberg (2011)

5. Qin, Y., Sheng, Q.Z., Falkner, N.J.G., Dustdar, S., Wang, H., Vasilakos, A.V.: When things matter: A survey on data-centric Internet of Things. J Netw. Comput. Appl. **64**, 137–153 (2016)
6. Qin, Y., Sheng, Q.Z., Curry, E.: Matching over linked data streams in the internet of things. IEEE Int. Comput. (Internet) **19**(3), 21–27 (2015)
7. Petrovic, S., Osborne, M., Lavrenko, V.: Streaming first story detection with application to twitter. In: Proceedings of Human Language Technologies: Conference of the North American Chapter of the Association of Computational Linguistics (HLT-NAACL), pp. 181–189 (2010)
8. Bolles, A., Grawunder, M., Jacobi, J.: Streaming SPARQL - extending SPARQL to process data streams. In: Proceedings of the 5th European Semantic Web Conference (ESWC), pp. 448–462 (2008)
9. Le-Phuoc, D., Dao-Tran, M., Xavier Parreira, J., Hauswirth, M.: A native and adaptive approach for unified processing of linked streams and linked data. In: Aroyo, L., Welty, C., Alani, H., Taylor, J., Bernstein, A., Kagal, L., Noy, N., Blomqvist, E. (eds.) ISWC 2011, Part I. LNCS, vol. 7031, pp. 370–388. Springer, Heidelberg (2011)
10. Komazec, S., Cerri, D., Fensel, D.: Sparkwave: Continuous schema-enhanced pattern matching over rfd data streams. In: Proceedings of the 6th ACM International Conference on Distributed Event-Based Systems (DEBS), pp. 58–68 (2012)
11. Anicic, D., Fodor, P., Rudolph, S., Stojanovic, N.: EP-SPARQL: a unified language for event processing and stream reasoning. In: Proceedings of the 20th International Conference on World Wide Web (WWW), pp. 635–644 (2011)
12. Harth, A., Hose, K., Karnstedt, M., Polleres, A., Sattler, K.-U., Umbrich, J.: Data summaries for on-demand queries over linked data. In: Proceedings of the 19th International Conference on World Wide Web (WWW), pp. 411–420 (2010)
13. Qin, Y., Sheng, Q.Z., Falkner, N.J.G., Shemshadi, A., Curry, E.: Towards efficient dissemination of linked data in the internet of things. In: Proceedings of the 23rd ACM Conference on Information and Knowledge Management (CIKM), pp. 1779–1782. Shanghai, China (2014)
14. Seaborne, A.: RDQL - a query language for RDF. In: W3C Member Submission (2001)
15. Liarou, E., Idreos, S., Koubarakis, M.: Evaluating conjunctive triple pattern queries over large structured overlay networks. In: Proceedings of the 5th International Semantic Web Conference (ISWC), pp. 399–413 (2006)
16. Mikolov, T., Chen, K., Corrado, G., Dean, J.: Efficient Estimation of Word Representations in Vector Space. CoRR abs/1301.3781 (2013)
17. Mikolov, T., Sutskever, I., Chen, K., Corrado, G.S., Dean, J.: Distributed representations of words and phrases and their compositionality. In: Proceedings of the 27th Annual Conference on Neural Information Processing Systems (NIPS), pp. 3111–3119 (2013)
18. Mikolov, T., et al.: The word2Vec Project. https://code.google.com/p/word2vec/, Retrieved December 2015
19. Leskovec, J., Rajaraman, A., Ullman, J.D.: Mining of Massive Datasets, 2nd edn. Cambridge University Press, Cambridge (2014)
20. Hasan, S., ÓRiain, S., Curry, E.: Towards unified and native enrichment in event processing systems. In: Proceedings of the 7th ACM International Conference on Distributed Event-Based Systems (DEBS), pp. 171–182 (2013)

A Mapping-Based Method to Query MongoDB Documents with SPARQL

Franck Michel[(✉)], Catherine Faron-Zucker, and Johan Montagnat

University of Nice Sophia Antipolis, CNRS, I3S (UMR 7271), Nice, France
{fmichel,faron,johan}@i3s.unice.fr

Abstract. Accessing legacy data as virtual RDF stores is a key issue in the building of the Web of Data. In recent years, the MongoDB database has become a popular actor in the NoSQL market, making it a significant potential contributor to the Web of Linked Data. Therefore, in this paper we address the question of how to access arbitrary MongoDB documents with SPARQL. We propose a two-step method to (i) translate a SPARQL query into a pivot abstract query under MongoDB-to-RDF mappings represented in the xR2RML language, then (ii) translate the pivot query into a concrete MongoDB query. We elaborate on the discrepancy between the expressiveness of SPARQL and the MongoDB query language, and we show that we can always come up with a rewriting that shall produce all correct answers.

Keywords: SPARQL access to legacy data · MongoDB · Virtual RDF store · Linked data · xR2RML

1 Introduction

The Web-scale data integration progressively becomes a reality, giving birth to the Web of Linked Data through the open publication and interlinking of data sets on the Web. It results from the extensive works achieved during the last years, aimed to expose legacy data as RDF and develop SPARQL interfaces to various types of databases.

At the same time, the success of NoSQL databases is no longer questioned today. Initially driven by major Web companies in a pragmatic effort to cope with large distributed data sets, they are now adopted in a variety of domains such as media, finance, transportation, biomedical research and many others[1]. Consequently, harnessing the data available from NoSQL databases to feed the Web of Data, and more generally achieving RDF-based data integration over NoSQL systems, are timely questions. In recent years, MongoDB[2] has become a very popular actor in the NoSQL market[3]. Beyond dealing with large distributed

---

[1] Informally attested by the manifold domains of customers claimed by major NoSQL actors.
[2] https://www.mongodb.org/.
[3] http://db-engines.com/en/system/MongoDB.

© Springer International Publishing Switzerland 2016
S. Hartmann and H. Ma (Eds.): DEXA 2016, Part II, LNCS 9828, pp. 52–67, 2016.
DOI: 10.1007/978-3-319-44406-2_6

data sets, its popularity suggests that it is also increasingly adopted as a general-purpose database. Arguably, it is likely that many MongoDB instances host valuable data about all sorts of topics, that could benefit a large community at the condition of being made accessible as Linked Open Data. Hence the research question we address herein: *How to access arbitrary MongoDB documents with SPARQL?*

Exposing legacy data as RDF has been the object of much research during the last years, usually following two approaches: either by materialization, *i.e.* translation of all legacy data into an RDF graph at once, or based on on-the-fly translation of SPARQL queries into the target query language. The materialization is often difficult in practice for big datasets, and costly when data freshness is at stake. Several methods have been proposed to achieve SPARQL access to relational data, either in the context of RDB-backed RDF stores [8,11,21] or using arbitrary relational schemas [4,17,18,23]. R2RML [9], the W3C RDB-to-RDF mapping language recommendation is now a well-accepted standard and several SPARQL-to-SQL rewriting approaches hinge upon it [17,19,23]. Other solutions intend to map XML [2,3] or CSV[4] data to RDF. RML [10] tackles the mapping of heterogeneous data formats such as CSV/TSV, XML and JSON. xR2RML [14] is an extension of R2RML and RML addressing the mapping of an extensible scope of databases to RDF. Regarding MongoDB specifically, Tomaszuk proposed a solution to use MongoDB as an RDF triple store [22]. The translation of SPARQL queries that he proposed is closely tied to the data schema and does not fit with arbitrary documents. MongoGraph[5] is an extension of the AllegroGraph triple store to query arbitrary MongoDB documents with SPARQL. Similarly to the Direct Mapping [1] the approach comes up with an ad-hoc ontology (e.g. each JSON field name is turned into a predicate) and hardly supports the reuse of existing ontologies. More in line with our work, Botoeva et al. recently proposed a generalization of the OBDA principles to MongoDB [6]. They describe a two-step rewriting process of SPARQL queries into the MongoDB aggregate query language. In the last section we analyse in further details the relationship between their approach and ours.

In this paper we propose a method to query arbitrary MongoDB documents using SPARQL. We rely on xR2RML for the mapping of MongoDB documents to RDF, allowing for the use of classes and predicates from existing (domain) ontologies. In Sect. 2 we shortly describe the xR2RML mapping language. Section 3 defines a database-independent abstract query language, and summarizes a generic method to rewrite SPARQL queries into this language under xR2RML mappings. Then Sect. 4 presents our method to translate abstract queries into MongoDB queries. Finally in Sect. 5 we conclude by emphasizing some technical issues and highlighting perspectives.

---

[4] http://www.w3.org/2013/csvw/wiki.

[5] http://franz.com/agraph/support/documentation/4.7/mongo-interface.html.

2   The xR2RML Mapping Language

The xR2RML mapping language [14] is designed to map an extensible scope of relational and non-relational databases to RDF. It is independent of any query language or data model. It is backward compatible with R2RML and it relies on RML for the handling of various data formats. It can translate data with mixed embedded formats and generate RDF collections and containers. Below we shortly describe the main xR2RML features and propose a running example.

An xR2RML mapping defines a logical source (`xrr:logicalSource`) as the result of executing a query (`xrr:query`) against an input database. An optional iterator (`rml:iterator`) can be applied to each query result. Data from the logical source is mapped to RDF terms (literal, IRI, blank node) by term maps. There exists four types of term maps: a subject map generates the subject of RDF triples, and multiple predicate and object maps produce the predicate and object terms. An optional graph map is used to name a target graph. Listing 1.2 depicts the `<#TmLeader>` xR2RML mapping.

Term maps extract data from query results by evaluating *xR2RML references*. The syntax of xR2RML references depends on the target database: a column name in case of a relational database, an XPath expression in case of a XML database, or a JSONPath[6] expression in case of NoSQL document stores like MongoDB or CouchDB. xR2RML references are used with property `xrr:reference` that contains a single xR2RML reference, and `rr:template` that may contain several references in a template string. In the running example below, the subject map uses a template to build IRI terms by concatenating http://example.org/project/ with the value of JSON field `"code"`. When the evaluation of an xR2RML reference produces several RDF terms, by default the xR2RML processor creates one triple for each term. Alternatively, it can group them in an RDF collection (`rdf:List`) or container (`rdf:Seq`, `rdf:Bag` and `rdf:Alt`) of terms optionally qualified with a language tag or data type.

Like R2RML, xR2RML allows to model cross-references by means of *referencing object maps*. A referencing object map uses values produced by the subject map of a mapping (the parent) as the objects of triples produced by another mapping (the child). Properties `rr:child` and `rr:parent` specify the join condition between documents of both mappings.

Running Example. To illustrate the description of our method, we define a running example that we shall use throughout this paper. This short example is specifically tailored to address the issues related to the SPARQL-to-MongoDB translation, it does not illustrate advanced xR2RML features, but more detailed use cases are provided in [7,14]. Let us consider a MongoDB database with one collection `"projects"` (Listing 1.1), that lists the projects held in a company. Each project is described by a name, a code and a set of teams. Each team is an array of members given by their name, and we assume that the last member is always the team leader. The xR2RML mapping graph in Listing 1.2 has one mapping: `<#TmLeader>`. The logical source is the MongoDB

---

[6] http://goessner.net/articles/JsonPath/.

```
{ "project":"Finance & Billing", "code":"fin",
  "teams":[
  [ {"name":"P. Russo"}, {"name":"F. Underwood"}],
  [ {"name":"R. Danton"}, {"name":"E. Meetchum"} ]] },
{ "project":"Customer Relation", "code":"crm",
  "teams":[
  [ {"name":"R. Posner"}, {"name":"H. Dunbar"}]] }
```

Listing 1.1. MongoDB collection "projects" containing two documents

```
<#TmLeader>
    xrr:logicalSource [xrr:query "db.projects.find({})"];
    rr:subjectMap [rr:template
      "http://example.org/project/{$.code}".];
    rr:predicateObjectMap [
      rr:predicate ex:teamLeader;
      rr:objectMap [ xrr:reference
        "$.teams[0,1][(@.length - 1)].name" ] ].
```

Listing 1.2. xR2RML example mapping graph

query "db.projects.find({})" that simply retrieves all documents from collection "projects". The mapping associates projects (subject) to team leaders (object) with predicate ex:teamLeader. This is done by means of a JSONPath expression that selects the last member of each team using the calculated array index "[(@.length - 1)]".

3 Translating SPARQL Queries into Abstract Queries Under xR2RML Mappings

Various methods have been defined to translate SPARQL queries into another query language, that are generally tailored to the expressiveness of the target query language. Notably, the rich expressiveness of SQL and XQuery makes it possible to define semantics-preserving SPARQL rewriting methods [2,8]. By contrast, NoSQL databases typically trade off expressiveness for scalability and fast retrieval of denormalised data. For instance, many of them hardly support joins. Therefore, to envisage the translation of SPARQL queries in the general case, we propose a two-step method. Firstly, a SPARQL query is rewritten into a pivot abstract query under xR2RML mappings, independently of any target database (illustrated by step 1 in Fig. 1). Secondly, the pivot query is translated into concrete database queries based on the specific target database capabilities and constraints. In this paper we focus on the application of the second step to the specific case of MongoDB. The rest of this section summarizes the first step to provide the reader with appropriate background. A complete description is provided in [16].

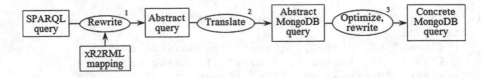

Fig. 1. Overview of the SPARQL-to-MongoDB query translation process

```
<AbsQuery> ::=
  <Query> | <Query> FILTER <filter> | <AtomicQuery>
<Query> ::=
  <AbsQuery> INNER JOIN <AbsQuery> ON {v₁,...vₙ} |
  <AbsQuery> AS child INNER JOIN <AbsQuery> AS parent
            ON child/<Ref> = parent/<Ref> |
  <AbsQuery> LEFT OUTER JOIN <AbsQuery> ON {v₁,...vₙ}|
  <AbsQuery> UNION <AbsQuery>
<AtomicQuery> ::= {From, Project, Where}
```

Listing 1.3. Grammar of the Abstract Pivot Query Language

The grammar of our pivot query language is depicted in Listing 1.3. Operators INNER JOIN ON, LEFT OUTER JOIN ON and UNION are entailed by the dependencies between graph patterns of the SPARQL query, and SPARQL filters involving variables shared by several triple patterns result in a FILTER operator. The computation of these operators shall be delegated to the target database if it supports them (i.e. if the target query language has equivalent operators like SQL), or to the query processing engine otherwise (e.g. MongoDB cannot process joins). Each SPARQL triple pattern *tp* is translated into a union of atomic abstract queries (<AtomicQuery>), under the set of xR2RML mappings likely to generate triples matching *tp*. Components of an atomic abstract query are as follows:

- *From* is the mapping's logical source, i.e. the database query string (xrr:query) and its optional iterator (rml:iterator).
- *Project* is the set of xR2RML references that must be projected, i.e. returned as part of the query results. In SQL, projecting an xR2RML reference simply means that the column name shall appear in the SELECT clause. As to MongoDB, this amounts to projecting the JSON fields mentioned in the JSON-Path reference.
- *Where* is a conjunction of abstract conditions entailed by matching each term of triple pattern *tp* with its corresponding term map in an xR2RML mapping: the subject of *tp* is matched with the subject map of the mapping, the predicate with the predicate map and the object with the object map. Three types of condition may be created:

(i) a SPARQL variable in the triple pattern is turned into a not-null condition on the xR2RML reference corresponding to that variable in the term map, denoted by *isNotNull(<xR2RML reference>)*;

(ii) A constant triple pattern term (IRI or literal) is turned into an equality condition on the xR2RML reference corresponding to that RDF term in the term map, denoted by *equals(<xR2RML reference>, value)*;

(iii) A SPARQL filter condition f about a SPARQL variable is turned into a filter condition, denoted by *sparqlFilter(<xR2RML reference>, f)*.

Finally, an abstract query is optimized using classical query optimization techniques such as the self-join elimination, self-union elimination or projection pushing. In [16] we show that, during the optimization phase, a new type of abstract condition may come up, *isNull(<xR2RML reference>)*, in addition to logical operators *Or()* and *And()* to combine conditions.

Running Example. We consider the following SPARQL query that aims to retrieve projects in which "*H. Dunbar*" is a team leader.

SELECT ?proj WHERE {?proj ex:teamLeader "H. Dunbar".}

The triple pattern, denoted by tp, is translated into the atomic abstract query {*From, Project, Where*}. *From* is the query in the logical source of mapping <#TmLeader>, i.e."db.projects.find({})". The detail of calculating *Project* is out of the scope of this paper; let us just note that, since the values of variable ?proj (the subject of tp) shall be retrieved, only the subject map reference is projected, i.e. the JSONPath expression "$.code". The *Where* part is calculated as follows:

– tp's subject, variable ?proj, is matched with <#TmLeader>'s subject map; this entails condition C_1: isNotNull($.code).
– tp's object, "H. Dunbar", is matched with <#TmLeader>'s object map; this entails condition C_2: equals($.teams[0,1][(@.length-1)].name, "H. Dunbar").

Thus, the SPARQL query is rewritten into the atomic abstract query below:

```
{ From:     {"db.projects.find({})"},
  Project:  {$.code AS ?proj},
  Where:    {isNotNull($.code),
     equals($.teams[0,1][(@.length-1)].name, "H. Dunbar") }}
```

The JSON documents needed to answer this abstract query shall verify condition $C_1 \wedge C_2$. In the next section, we elaborate on the method that allows to rewrite such conditions into concrete MongoDB queries.

4   Translating an Abstract Query into MongoDB Queries

In this section we briefly describe the MongoDB query language, then we define rules to transform an atomic abstract query into an abstract representation of a

```
AND(<exp₁>, <exp₂>, ...)       →  $and:[<exp₁>,<exp₂>,...]
OR(<exp₁>, <exp₂>, ...)        →  $or:[<exp₁>,<exp₂>,...]
WHERE(<JavaScript exp>)        →  $where:"<JavaScript exp>"
ELEMMATCH(<exp₁>,<exp₂>...)    →  $elemMatch:{<exp₁>,<exp₂>...}
FIELD(p₁) ... FIELD(pₙ)        →  "p₁. ... .pₙ":
SLICE(<exp>, <number>)         →  <exp>:{$slice:<number>}
COND(equals(v))                →  $eq:v
COND(isNotNull)                →  $exists:true, $ne:null
COND(isNull)                   →  $eq:null
NOT_EXISTS(<exp>)              →  <exp>:{$exists:false}
COMPARE(<exp>, <op>, <v>)      →  <exp>:{<op>:<v>}
NOT_SUPPORTED                  →  ∅
CONDJS(equals(v))              →  == v
CONDJS(isNotNull)              →  != null
```

Listing 1.4. Abstract MongoDB query representation and translation to a concrete query string

MongoDB query (step 2 in Fig. 1). Finally, we define additional rules to optimize and rewrite an abstract representation of a MongoDB query into a union of executable MongoDB queries (step 3 in Fig. 1).

4.1   The MongoDB Query Language

MongoDB provides a JSON-based declarative query language consisting of two major mechanisms. The *find* query retrieves documents matching a set of conditions. It takes a query and a projection parameters, and returns a cursor to the matching documents. Optional modifiers amend the query to impose limits and sort orders. Alternatively, the *aggregate* query allows for the definition of processing pipelines: each document of a collection passes through the stages of the pipeline, that allows for richer aggregate computations. As a first approach, this work considers the *find* query method, hereafter called the *MongoDB query language*. As an illustration let us consider the following query:

```
db.projects.find(
  {"teams.0":{$elemMatch:{"age":{$gt:30}}}}, {"code":1})
```

It retrieves documents from collection "projects", whose first team (array "teams" at index 0) has at least one member (operator $elemMatch) over 30 years old (operator $gt). The projection parameter, {"code":1}, states that only the "code" field of each matching document must be returned.

The MongoDB documentation[7] provides a rich description of the query language, that however lacks formal semantics. Recently, attempts were made to clarify this semantics while underlining some limitations and ambiguities: [5] focuses mainly on the *aggregate* query and ignores some of the operators we

---

[7] https://docs.mongodb.org/manual/tutorial/query-documents/.

use in our translation, such as $where, $elemMatch, $regex and $size. On the other hand, [13] describes the *find* query, yet some restrictions on the operator $where are not formalized. Hence, in [15] we specified the grammar of the subset of the query language that we consider. We also defined an abstract representation of MongoDB queries, that allows for handy manipulation during the query construction and optimization phases. Listing 1.4 details the constructs of this representation and their equivalent concrete query string. In the COMPARE clause definition, <op> stands for one of the MongoDB comparison operators: $eq, $ne, $lt, $lte, $gt, $gte, $size and $regex. The NOT_SUPPORTED clause helps keep track of parts of the abstract query that cannot be translated into an equivalent MongoDB query element; it shall be used when rewriting the abstract query into a concrete query (Sect. 4.3).

4.2   Query Translation Rules

Section 3 introduced a method that rewrites a SPARQL query into an abstract query in which operators INNER JOIN, LEFT OUTER JOIN and UNION relate atomic abstract queries of the form {*From, Project, Where*}. The latter are created by matching each triple pattern with candidate xR2RML mappings. The *Where* part consists of *isNotNull, equals* and *sparqlFilter* abstract conditions about xR2RML references (JSONPath expressions in the case of MongoDB).

MongoDB does not support joins, while unions and nested queries are supported under strong restrictions, and comparisons are limited (e.g. a JSON field can be compared to a literal but not to another field of the same document). Consequently, operators INNER JOIN, LEFT OUTER JOIN, and to some extend UNION and FILTER, shall be computed by the query processing engine. Conversely, the abstract conditions of atomic queries can be translated into MongoDB queries[8].

Given the subset of the MongoDB query language considered, the recursive function *trans* in Fig. 2 translates an abstract condition on a JSONPath expression into a MongoDB *find* query using the formalism defined in Listing 1.4. It consists of a set of rules applicable to a certain pattern. The JSONPath expression in argument is checked against each pattern in the order of the rules (0 to 9) until a match is found. We use the following notations:

- <JP>: denotes a possibly empty JSONPath expression.
- <JP:F>: denotes a non-empty JSONPath sequence of field names and array indexes, e.g. .p.q.r, .p[10]["r"].
- <bool_expr>: is a JavaScript expression that evaluates to a boolean.
- <num_expr>: is a JavaScript expression that evaluates to a positive integer.

Rule R0 is the entry point of the translation process (JSONPath expressions start with a '$' character). Rule R1 is the termination point: when the JSONPath expression has been fully parsed, the last created clause is the condition clause COND, producing e.g. "$eq:value" for an equality condition, or "$exists:true,

---

[8] In the current state of this work we do not consider SPARQL filter conditions.

R0	**trans**($, <cond>) → ∅
	trans($<JP>, <cond>) → **trans**(<JP>, <cond>)

R1 trans(∅, <cond>) → **COND**(<cond>)

R2 *Field alternative (a) or array index alternative (b)*
(a) **trans**(<JP:F>["p","q",...]<JP>, <cond>) →
    OR(**trans**(<JP:F>.p<JP>, <cond>), **trans**(<JP:F>.q<JP>, <cond>), ...)
(b) **trans**(<JP:F>[i,j,...]<JP>, <cond>) →
    OR(**trans**(<JP:F>.i<JP>, <cond>), **trans**(<JP:F>.j<JP>, <cond>), ...)

R3 *Heading field alternative (a) or heading array index alternative (b)*
(a) **trans**(["p","q",...]<JP>, <cond>) →
    OR(**trans**(.p<JP>, <cond>), **trans**(.q<JP>, <cond>), ...)
(b) **trans**([i,j,...]<JP>, <cond>) →
    OR(**trans**(.i<JP>, <cond>), **trans**(.j<JP>, <cond>), ...)

R4 *JavaScript filter on array elements, e.g.,* `$.p[?(@.q)].r`
    trans([?(<bool_expr>)]<JP>, <cond>) →
    ELEMMATCH(**trans**(<JP>, <cond>), **trans**JS(<bool_expr>))

R5 *Array slice: n last elements (a) or n first elements (b)*
(a) **trans**(<JP:F>[-<start>:]<JP>, <cond>) →
    trans(<JP:F>.*<JP>, <cond>) **SLICE**(<JP:F>, -<start>)
(b) **trans**(<JP:F>[:<end>]<JP>, <cond>) →
    trans(<JP:F>.*<JP>, <cond>) **SLICE**(<JP:F>, <end>)
    trans(<JP:F>[0:<end>]<JP>, <cond>) →
    trans(<JP:F>.*<JP>, <cond>) **SLICE**(<JP:F>, <end>)

R6 *Calculated array index, e.g.,* `$.p[(@.length - 1)].q`
(a) **trans**(JP_1>[(<num_expr>)]<JP_2>, <cond>) → **NOT_SUPPORTED**
    if <JP_1> contains a wildcard or a JavaScript filter expression
(b) **trans**(<JP:F>[(<num_expr>)], <cond>) → **AND**(
    EXISTS(<JP:F>),
    WHERE('this<JP:F>[replaceAt("this<JP:F>", <num_expr>)] **CONDJS**(<cond>')))
(c) **trans**(<JP:F_1>[(<num_expr>)]<JP:F_2>, <cond>) → **AND**(
    EXISTS(<JP:F_1>),
    WHERE('this<JP:F_1>[replaceAt("this<JP:F_1>", <num_expr>)]<JP:F_2>
    CONDJS(<cond>')))

R7 *Heading wildcard*
(a) **trans**(.*<JP>, <cond>) → **ELEMMATCH**(**trans**(<JP>, <cond>))
(b) **trans**([*]<JP>, <cond>) → **ELEMMATCH**(**trans**(<JP>, <cond>))

R8 *Heading field name or array index*
(a) **trans**(.p<JP>, <cond>) → **FIELD**(p) **trans**(<JP>, <cond>)
(b) **trans**(["p"]<JP>, <cond>) → **FIELD**(p) **trans**(<JP>, <cond>)
(c) **trans**([i]<JP>, <cond>) → **FIELD**(i) **trans**(<JP>, <cond>)

R9 *No other rule matched, expression <JP> is not supported*
    trans(<JP>, <cond>) → **NOT_SUPPORTED**

Fig. 2. Translation of a condition on a JSONPath expression into an abstract MongoDB query (function *trans*)

$ne:null" for a not-null condition. Rules R2 to R8 deal with the different types of JSONPath expressions. In case no rule matches, the translation fails and rule R9 creates the NOT_SUPPORTED clause, that shall be dealt with later on. Rule R4 deals with the translation of JavaScript filters on JSON arrays, where character '@' stands for each array element. It delegates their processing to function *transJS* (described in [15]). For instance, the filter "[?(@.age>30)]" is translated into the MongoDB sub-query "age":{$gt:30}.

Due to the space constraints, we do not go through the comprehensive justification of each rule in Fig. 2, however the interested reader is referred to [15].

Running Example. The *Where* part of the abstract query presented in Sect. 3 comprises two conditions:

C_1: isNotNull($.code), and
C_2: equals($.teams[0,1][(@.length - 1)].name, "H. Dunbar").
Here are the rules applied at each step of the translation of C_1 and C_2.
$M_1 \leftarrow$ trans(C_1) = trans($.code, isNotNull):
R0: $M_1 \leftarrow$ trans(.code, isNotNull)
R8 then R1: $M_1 \leftarrow$ FIELD(code) COND(isNotNull)
$M_2 \leftarrow$ trans(C_2) =
       trans($.teams[0,1][(@.length-1)].name, equals("H. Dunbar"))
R0: $M_2 \leftarrow$ trans(.teams[0,1][(@.length-1)].name, equals("H. Dunbar"))
R2 splits the alternative "[0,1]" into two members of an OR clause:
$M_2 \leftarrow$ OR(trans(teams.0.[(@.length-1)].name, equals("H. Dunbar")),
        trans(teams.1.[(@.length-1)].name, equals("H. Dunbar"))).
R6(c) processes the calculated array index "(@.length-1)" in each OR member:
$M_2 \leftarrow$ OR(AND(EXISTS(.teams.0),
   WHERE('this.teams[0][this.teams[0].length-1].name=="H. Dunbar"')),
   AND(EXISTS(.teams.1),
   WHERE('this.teams[1][this.teams[1].length-1].name=="H. Dunbar"')))

4.3 Rewriting of the Abstract MongoDB Query Representation into a Concrete MongoDB Query

Rules R0 to R9 translate a condition on a JSONPath expression into an abstract MongoDB query. Yet, several potential issues hinder the rewriting into a concrete query: (i) a NOT_SUPPORTED clause may indicate that a part of the JSON-Path expression could not be translated into an equivalent MongoDB operator; (ii) a WHERE clause may be nested beneath a sequence of AND and/or OR clauses although the MongoDB $where operator is valid only in the top-level query; (iii) unnecessary complexity such as nested ORs, nested ANDs, etc., may hamper performances. Those issues are addressed by two sets of rewriting rules, O1 to O5 and W1 to W6. They require the addition of the UNION clause to those in Listing 1.4. UNION is semantically equivalent to the OR clause but, whereas ORs are processed by the MongoDB database, UNIONs shall be computed by the query processing engine.

Query Optimization. Rules O1 to O5 in Fig. 3 perform several query optimizations. Rules O1 to O4 address issue (iii) by flattening nested OR, AND and UNION clauses, and merging sibling WHEREs. Rule O5 addresses issue (i) by removing the clauses of type NOT_SUPPORTED while still making sure that the query returns all the correct answers:

O1 *Flatten nested AND, OR and UNION clauses:*
$\quad$ AND($C_1,...\ C_n$, AND($D_1,...\ D_m$,)) → AND($C_1,...\ C_n$, $D_1,...\ D_m$)
$\quad$ OR($C_1,...\ C_n$, OR($D_1,...\ D_m$,)) → OR($C_1,...\ C_n$, $D_1,...\ D_m$)
$\quad$ UNION($C_1,...\ C_n$, UNION($D_1,...\ D_m$,)) → UNION($C_1,...\ C_n$, $D_1,...\ D_m$)
O2 *Merge ELEMMATCH with nested AND clauses:*
$\quad$ ELEMMATCH($C_1,...\ C_n$, AND($D_1,...\ D_m$,)) → ELEMMATCH($C_1,...\ C_n$, $D_1,...\ D_m$)
O3 *Group sibling WHERE clauses:*
$\quad$ OR(..., WHERE("W_1"), WHERE("W_2")) → OR(..., WHERE("(W_1) ‖ (W_2)"))
$\quad$ AND (..., WHERE("W_1"), WHERE("W_2")) → AND(..., WHERE("(W_1) && (W_2)"))
$\quad$ UNION(..., WHERE("W_1"), WHERE("W_2")) → UNION(..., WHERE("(W_1) ‖ (W_2)"))
O4 *Replace AND, OR or UNION clauses of one term with the term itself.*
O5 *Remove NOT_SUPPORTED clauses:*
(a) AND($C_1,...\ C_n$, NOT_SUPPORTED) → AND($C_1,...\ C_n$)
(b) ELEMMATCH($C_1,...\ C_n$, NOT_SUPPORTED) → ELEMMATCH($C_1,...\ C_n$)
(c) OR($C_1,...\ C_n$, NOT_SUPPORTED) → NOT_SUPPORTED
(d) UNION($C_1,...\ C_n$, NOT_SUPPORTED) → NOT_SUPPORTED
(e) FIELD(...)... FIELD(...) NOT_SUPPORTED → NOT_SUPPORTED

Fig. 3. Optimization of an abstract MongoDB query

- O5(a): If a NOT_SUPPORTED clause occurs in an AND clause, it is simply removed. Let $C_1,...C_n$ be any clauses and N be a NOT_SUPPORTED clause. Since $C_1\wedge...\wedge C_n \supseteq C_1\wedge...\wedge C_n\wedge N$, the rewriting widens the condition. Hence, all matching documents are returned. However, non-matching documents may be returned too, that shall be ruled out later on.
- O5(b): A logical AND implicitly applies to members of an ELEMMATCH clause. Therefore, removing the NOT_SUPPORTED has the same effect as in O5(a).
- O5(c) and (d): A NOT_SUPPORTED is managed differently in an OR or UNION clause. Since $C_1 \vee ... \vee C_n \subseteq C_1 \vee ... \vee C_n \vee N$, removing N would return a subset of the matching documents. Instead, we replace the whole OR or UNION clause with a NOT_SUPPORTED clause. This way, the NOT_SUPPORTED issue is raised up to the parent clause and shall be managed at the next iteration. Iteratively, the NOT_SUPPORTED clause is raised up until it is eventually removed (cases AND and ELEMMATCH above), or it ends up in the top-level query. The latter is the worst case in which the query shall retrieve all documents.
- O5(e): Similarly to O5(c), a sequence of fields followed by a NOT_SUPPORTED clause must be replaced with a NOT_SUPPORTED clause to raise up the issue to the parent clause.

Pulling Up WHERE Clauses. By construction, rule R6 ensures that WHERE clauses cannot be nested in an ELEMMATCH, but they may show in AND and OR clauses. Besides, rules O1 to O4 flatten nested OR and AND clauses, and merge sibling WHERE clauses. Therefore, a WHERE clause may be either in the top-level query (in this case the query is executable) or it may show in one of the following patterns (where W stands for a WHERE clause):

W1 OR($C_1,...C_n$, W) $\rightarrow$ UNION(OR($C_1,...C_n$), W)
W2 OR($C_1,...C_n$, AND($D_1,...D_m$, W)) $\rightarrow$ UNION(OR($C_1,...C_n$), AND($D_1,...D_m$, W))
W3 AND($C_1,...C_n$, W) $\rightarrow$ ($C_1,...C_n$, W)   if the AND clause is a top-level query object or under a UNION clause.
W4 AND($C_1,...C_n$, OR($D_1,...D_m$, W)) $\rightarrow$ UNION(AND($C_1,...C_n$, OR($D_1,...D_m$)), AND($C_1,...C_n$, W))
W5 AND($C_1,...C_n$, UNION($D_1,...D_m$)) $\rightarrow$ UNION(AND($C_1,...C_n$, D_1),... AND($C_1,...C_n$, D_m))
W6 OR($C_1,...C_n$, UNION($D_1,...D_m$)) $\rightarrow$ UNION(OR($C_1,...C_n$), D_1, ...D_m))

Fig. 4. Pulling up WHERE clauses to the top-level query

OR(...,W,...), AND(...,W,...), OR(..., AND(...,W,...), ...), AND(..., OR(...,W,...), ...).
In such patterns, rules W1 to W6 (Fig. 4) address issue (ii) by "pulling up"
WHERE clauses into the top-level query. Here is an insight into the approach:

- Since OR(C, W) is not a valid MongoDB query, it is replaced with query
  UNION(C, W) which has the same semantics: C and W are evaluated separately against the database, and the UNION is computed later on by the query
  processing engine.
- AND(C,OR(D,W)) is rewritten into OR(AND(C,D), AND(C,W)) and the OR is
  replaced with a UNION: UNION(AND(C,D), AND(C,W)). Since an logical AND
  implicitly applies to the top-level terms, we can finally rewrite the query into
  UNION((C,D), (C,W)) which is valid since W now shows in a top-level query.

Rewriting rules W1 to W6 are a generalization of these examples. They ensure
that a query containing a nested WHERE can always be rewritten into a union of
queries wherein the WHERE shows only in a top-level query. Hence we formulate
Theorem 1, for which a proof is provided in [15].

Theorem 1. *Let C be an equality or not-null condition on a JSONPath expression. Let $Q = (Q_1 ... Q_n)$ be the abstract MongoDB query produced by trans(C).*
Rewritability: *It is always possible to rewrite Q into a query $Q' = \text{UNION}(Q'_1,$
$.., Q'_m)$ such that $\forall i \in [1, m]$ Q'_i is a valid MongoDB query, i.e. Q'_i does not
contain any NOT_SUPPORTED clause, and a WHERE clause only shows at the
top-level of Q'_i.*
Completeness: *Q' retrieves all the documents matching condition C. If Q contains at least one NOT_SUPPORTED clause, then Q' may retrieve additional documents that do not match condition C.*

Running Example. For the sake of readability, below we denote the JavaScript
conditions in M_1 and M_2 as follows: *JScond$_0$* stands for

```
this.teams[0][this.teams[0].length-1].name=="H. Dunbar", and JScond₁ for
this.teams[1][this.teams[1].length-1].name=="H. Dunbar".
```

In Sect. 4.2 we have translated conditions C_1 and C_2 into abstract MongoDB queries M_1 and M_2. The MongoDB documents needed to answer the SPARQL query shall be retrieved by the query AND(M_1, M_2) =

AND(FIELD(code) COND(isNotNull), OR(
    AND(EXISTS(.teams.0), WHERE('JScond$_0$'))
    AND(EXISTS(.teams.1), WHERE('JScond$_1$'))))

Applying subsequently rules W2 and O4 replaces the inner OR with a UNION:

AND(FIELD(code) COND(isNotNull), UNION(
    AND(EXISTS(.teams.0), WHERE('JScond$_0$'))
    AND(EXISTS(.teams.1), WHERE('JScond$_1$'))))

Rule W5 pulls up the UNION clause:

UNION(
    AND(FIELD(code) COND(isNotNull), AND(EXISTS(.teams.0), WHERE('JScond$_0$'))),
    AND(FIELD(code) COND(isNotNull), AND(EXISTS(.teams.1), WHERE('JScond$_1$'))))

Finally, O1 merges the nested ANDs and W3 removes the resulting top-level AND:

UNION(
    (FIELD(code) COND(isNotNull), EXISTS(.teams.0), WHERE('JScond$_0$')),
    (FIELD(code) COND(isNotNull), EXISTS(.teams.1), WHERE('JScond$_1$')))

The abstract query can now be rewritten into a union of two valid queries:

{"code":{ $exists:true, $ne:null }, "teams.0":{ $exists:true },
    $where:'this.teams[0][this.teams[0].length-1].name == "H. Dunbar"'}
{"code":{ $exists:true, $ne:null }, "teams.1":{ $exists:true },
        $where:'this.teams[1][this.teams[1].length-1].name == "H. Dunbar"'}

The first query retrieves the document below, whereas the second query returns no document.

```
{ "project":"Customer Relation", "code":"crm",
  "teams":[ [ {"name":"R. Posner"}, {"name":"H. Dunbar"}]]}
```

Finally, the application of triples map <#TmLeader> to the query result produces one RDF triple that matches the triple pattern tp:

<http://example.org/project/crm> ex:teamLeader "H. Dunbar".

5 Discussion, Conclusion and Perspectives

In this document we proposed a method to access arbitrary MongoDB documents with SPARQL. This relies on custom mappings described in the xR2RML mapping language which allows for the reuse of existing domain ontologies. First, we introduced a method to rewrite a SPARQL query into a pivot abstract query independent of any target database, under xR2RML mappings. Then, we devised a set of rules to translate this pivot query into an abstract representation of a MongoDB query, and we showed that the latter can always be rewritten into a union of concrete MongoDB queries that shall return all the documents required to answer the SPARQL query.

Due to the limited expressiveness of the MongoDB *find* queries, some JSON-Path expressions cannot be translated into equivalent MongoDB queries. Consequently, the query translation method cannot guarantee that query semantics be preserved. Yet, we ensure that rewritten queries retrieve all matching documents, possibly with additional non-matching ones. The RDF triples thus extracted are subsequently filtered by evaluating the original SPARQL query. This preserves semantics at the cost of an extra SPARQL query evaluation.

In a recent work, Botoeva et al. proposed a generalization of the OBDA principles to support MongoDB [6]. Both approaches have similarities and discrepancies that we outline below. Botoeva et al. derive a set of type constraints (literal, object, array) from the mapping assertions, called the MongoDB database *schema*. Then, a relational view over the database is defined with respect to that schema, notably by flattening array fields. A SPARQL query is rewritten into a relational algebra (RA) query, and RA expressions over the relational view are translated into MongoDB *aggregate* queries. Similarly, we translate a SPARQL query into an abstract representation (that is not the relational algebra) under xR2RML mappings. The mappings are quite similar in both approaches although xR2RML is slightly more flexible: class names (in triples `?x rdf:type A`) and predicates can be built from database values whereas they are fixed in [6], and xR2RML allows to turn an array field into an RDF collection or container. To deal with the tree form of JSON documents we use JSONPath expressions. This avoids the definition of a relational view over the database, but this also comes with additional complexity in the translation process. Finally, [6] produces MongoDB *aggregate* queries, with the advantage that a SPARQL 1.0 query may be translated into a single target query, thus delegating all the processing to MongoDB. Yet, in practice, some aggregate queries may be very inefficient, hence the need to decompose RA queries into sub-queries, as underlined by the authors. Our approach produces *find* queries that are less expressive but whose performance is easier to anticipate, thus putting a higher burden on the query processing engine (joins, some unions and filtering). In the future, it would be interesting to characterise mappings with respect to the type of query that shall perform best (single vs. multiple separate queries, find vs. aggregate). A lead may be to involve query plan optimization logics such as the bind join [12] and the join reordering methods applied in the context of distributed SPARQL query engines [20].

More generally, the NoSQL trend pragmatically gave up on properties such as consistency and rich query features, as a trade-off to high throughput, high availability and horizontal elasticity. Therefore, it is likely that the hurdles we encountered with MongoDB shall occur with other NoSQL databases.

Implementation and Evaluation. To validate our approach we have developed a prototype implementation[9] available under the Apache 2 open source licence. Further developments on query optimization are on-going, and in the short-term we intend to run performance evaluations. Besides, we are working on two real-life use cases. Firstly, in the context of the Zoomathia research project[10], we proposed to represent a taxonomic reference, designed to support studies in Conservation Biology, as a SKOS thesaurus [7]. It is stored in a MongoDB database, and we are in the process of testing the SPARQL access to that thesaurus using our prototype. Secondly, we are having discussions with researchers in the fields of ecology and agronomy. They intend to explore the added value of Semantic Web technologies using a large MongoDB database of phenotype information. This context would be a significant and realistic use case of our method and prototype.

References

1. Arenas, M., Bertails, A., Prud'hommeaux, E., Sequeda, J.: A Direct Mapping of Relational Data to RDF (2012)
2. Bikakis, N., Tsinaraki, C., Stavrakantonakis, I., Gioldasis, N., Christodoulakis, S.: The SPARQL2XQuery interoperability framework. WWW **18**(2), 403–490 (2015)
3. Bischof, S., Decker, S., Krennwallner, T., Lopes, N., Polleres, A.: Mapping between RDF and XML with XSPARQL. J. Data Seman. **1**(3), 147–185 (2012)
4. Bizer, C., Cyganiak. R.: D2R server - publishing relational databases on the semantic web. In: ISWC (2006)
5. Botoeva, E., Calvanese, D., Cogrel, B., Rezk, M., Xiao, G.: A formal presentation of MongoDB (Extended version). Technical report (2016)
6. Botoeva, E., Calvanese, D., Cogrel, B., Rezk, M., Xiao, G.: OBDA beyond relational DBs: a study for MongoDB. In: International Workshop on Description Logics 2016, vol. 1577 (2016)
7. Callou, C., Michel, F., Faron-Zucker, C., Martin, C., Montagnat, J.: Towards a shared reference thesaurus for studies on history of zoology, archaeozoology and conservation biology. In: SW for Scientific Heritage, Workshop of ESWC (2015)
8. Chebotko, A., Lu, S., Fotouhi, F.: Semantics preserving SPARQL-to-SQL translation. Data Knowl. Eng. **68**(10), 973–1000 (2009)
9. Das, S., Sundara, S., Cyganiak, R.: R2RML: RDB to RDF mapping language (2012)
10. Dimou, A., Vander Sande, M., Colpaert, P., Verborgh, R., Mannens, E., Van de Walle, R.: RML: A generic language for integrated RDF mappings of heterogeneous data. In: LDOW (2014)
11. Elliott, B., Cheng, E., Thomas-Ogbuji, C., Ozsoyoglu, Z.M.: A complete translation from SPARQL into efficient SQL. In: IDEAS 2009, pp. 31–42. ACM (2009)
12. Haas, L., Kossmann, D., Wimmers, E., Yang, J.: Optimizing queries across diverse data sources. In: VLDB, pp. 276–285 (1997)

---

[9] https://github.com/frmichel/morph-xr2rml.
[10] http://www.cepam.cnrs.fr/zoomathia.

13. Husson, A.: Une sémantique statique pour MongoDB. In: 25th Journées Francoph-
    ones des Langages Applicatifs (JFLA), pp. 77–92 (2014)
14. Michel, F., Djimenou, L., Faron-Zucker, C., Montagnat, J.: Translation of relational
    and non-relational databases into RDF with xR2RML. In: WebIST, pp. 443–454
    (2015)
15. Michel, F., Faron-Zucker, C., Montagnat, J.: Mapping-based SPARQL access to a
    MongoDB database. Technical report, CNRS, 2015. https://hal.archives-ouvertes.
    fr/hal-01245883
16. Michel, F., Faron-Zucker, C., Montagnat, J.: A generic mapping-based query trans-
    lation from SPARQL to various target database query languages. In: WebIST
    (2016)
17. Priyatna, F., Corcho, O., Sequeda, J.: Formalisation and experiences of R2RML-
    based SPARQL to SQL query translation using Morph. In: WWW (2014)
18. Rodríguez-Muro, M., Kontchakov, R., Zakharyaschev, M.: Ontology-based data
    access: Ontop of databases. In: Alani, H., Kagal, L., Fokoue, A., Groth, P., Bie-
    mann, C., Parreira, J.X., Aroyo, L., Noy, N., Welty, C., Janowicz, K. (eds.) ISWC
    2013, Part I. LNCS, vol. 8218, pp. 558–573. Springer, Heidelberg (2013)
19. Rodríguez-Muro, M., Rezk, M.: Efficient SPARQL-to-SQL with R2RML mappings.
    J. Web Seman. **33**, 141–169 (2015)
20. Schwarte, A., Haase, P., Hose, K., Schenkel, R., Schmidt, M.: FedX: optimization
    techniques for federated query processing on linked data. In: Aroyo, L., Welty, C.,
    Alani, H., Taylor, J., Bernstein, A., Kagal, L., Noy, N., Blomqvist, E. (eds.) ISWC
    2011, Part I. LNCS, vol. 7031, pp. 601–616. Springer, Heidelberg (2011)
21. Sequeda, J.F., Miranker, D.P.: Ultrawrap: SPARQL execution on relational data.
    J. Web Seman. **22**, 19–39 (2013)
22. Tomaszuk, D.: Document-oriented triplestore based on RDF/JSON. In: Logic, phi-
    losophy and computer science, pp. 125–140. University of Bialystok (2010)
23. Unbehauen, J., Stadler, C., Auer, S.: Accessing relational data on the web with
    SparqlMap. In: Takeda, H., Qu, Y., Mizoguchi, R., Kitamura, Y. (eds.) JIST 2012.
    LNCS, vol. 7774, pp. 65–80. Springer, Heidelberg (2013)

Incremental Maintenance of Materialized SPARQL-Based Linkset Views

Elisa S. Menendez[1(✉)], Marco A. Casanova[1], Vânia M.P. Vidal[2],
Bernardo P. Nunes[1,3], Giseli Rabello Lopes[4],
and Luiz A.P. Paes Leme[5]

[1] Department of Informatics, PUC-Rio, Rio de Janeiro, RJ, Brazil
{emenendez, casanova}@inf.puc-rio.br,
Bernardo.nunes@uniriotec.br
[2] Computer Science Departament, UFC, Fortaleza, CE, Brazil
vvidal@lia.ufc.br
[3] Department of Applied Informatics, UNIRIO, Rio de Janeiro, Brazil
[4] Federal University of Rio de Janeiro, UFRJ, Rio de Janeiro, RJ, Brazil
giseli@dcc.ufrj.br
[5] Computer Science Institute, UFF, Niteroi, RJ, Brazil
lapaesleme@ic.uff.br

Abstract. In the Linked Data field, data publishers frequently materialize *linksets* between two different datasets using link discovery tools. To create a linkset, such tools typically execute *linkage rules* that retrieve data from the underlying datasets and apply matching predicates to create the links, in an often complex process. Also, such tools do not support linkset maintenance, when the datasets are updated. A simple, but costly strategy to maintain linksets up-to-date would be to fully re-materialize them from time to time. This paper presents an alternative strategy, called *incremental*, for maintaining linksets, based on idea that one should re-compute only the links that involve the updated resources. The paper discusses in detail the incremental strategy, outlines an implementation and describes an experiment to compare the performance of the incremental strategy with the full re-materialization of linksets.

Keywords: RDF views · Linksets · SPARQL update · Linked data

1 Introduction

The Linked Data initiative defines best practices for publishing and interlinking data on the Web using RDF triples to represent the data (Berners-Lee 2006). Briefly, a dataset is simply a set of RDF triples. A link is an RDF triple of the form (s,p,o), where s and o are resources defined in two distinct datasets. A linkset is a set of links. SPARQL is the standard query language used to query RDF datasets. A SPARQL-*based view* is a view defined by a SPARQL query.

Link discovery tools help create and materialize linksets by matching resources retrieved from two datasets. The first step to configure a link discovery tool typically defines what amounts to SPARQL-based views that specify sets of resources with useful

© Springer International Publishing Switzerland 2016
S. Hartmann and H. Ma (Eds.): DEXA 2016, Part II, LNCS 9828, pp. 68–83, 2016.
DOI: 10.1007/978-3-319-44406-2_7

properties. We refer to such views as *catalogue views* since they act as a catalogue of resources. The second step defines a set of *linkage rules* that specify conditions that resources must fulfill to be matched.

When a dataset is updated, the maintenance of a linkset requires attention since the resources may no longer meet the graph template used in the corresponding catalogue view. A trivial case is when a resource used in a link is removed from the original dataset; in this case, the link becomes invalid and must also be removed. The work in (Casanova *et al.* 2014) specified an incremental strategy to keep linksets updated, similar to the incremental strategies for relational view maintenance.

This paper extends the work reported in (Casanova *et al.* 2014) in three directions: (1) it presents in detail an incremental strategy to maintain materialized linksets; (2) it outlines an implementation of the proposed incremental strategy; (3) it describes experiments to measure the performance of the proposed incremental strategy and compare it with a re-materialization strategy.

The paper is organized as follows. Section 2 reviews related work. Section 3 contains basic definitions and a simple example. Section 4 details the incremental strategy to maintain linksets. Section 5 outlines a tool that implements the proposed incremental strategy and describes experiments conducted to assess the effectiveness of the incremental strategy. Finally, Sect. 6 contains the conclusions and discusses directions for future research.

2 Related Work

Several tools were developed to help solve the problem of finding links between different datasets. The LInk Discovery Framework for MEtric Spaces (LIMES) proposes algorithms that work efficiently with large knowledge bases (Ngomo and Auer 2011). The LIMES developers started with the idea of filtering obvious non-match instances to reduce the number of comparisons and improve matching time. The Silk Linking Framework (Volz *et al.* 2009b) offers a second example.

The *Web of Data – Link Maintenance Protocol* (WOD-LMP) (Volz *et al.* 2009a) is a protocol that helps link maintenance. It covers three use cases: (1) Link Transfer to Target – the source sends notifications to the target when a link is created or deleted; (2) Request of Target Change List – the source requests to the target a list of changes in a specified time range; (3) Subscription of Target Changes – the source sends the links notifications and the target stores this information to further notify the source about changes in the pointed resources.

DSNotify (Popitsch and Haslhofer 2011) is a general-purpose change detection framework that notifies linked datasets about events (create, remove, move, update) in their remote resources. To deal with these changes, DSNotify uses its own OWL Lite vocabulary, called DSNotify Eventset Vocabulary, which allows a detailed description (what, how, when and why) about the events.

We note that LIMES and Silk, although popular link discovery tools, do not support linkset maintenance, whereas WOD-LMP and DSNotify deal with change notification, but not with the actual linkset maintenance, as addressed in this paper.

The work reported in this paper is also related to strategies for materialized view maintenance. In the context of relational databases, a strategy for view maintenance is called *incremental* if only part of the view is modified to reflect the updates in the database (Gupta *et al.* 1993; Staud and Jarke 1996).

This strategy was adapted to maintain RDF views over relational databases (Vidal *et al.* 2013). In all such contexts, incremental view maintenance generally outperforms full view re-computation. However, we cannot directly adopt the familiar strategies proposed for incremental maintenance over relational datasets, since complex SPARQL updates pose new challenges, when compared with SQL updates.

The work reported in this paper is also closely related to strategies designed to maintain RDF mirrors[1], slices (Ibáñez *et al.* 2014) and views (Hung *et al.* 2004; Vidal *et al.* 2015; Endris *et al.* 2015) over RDF datasets, since the main part of our strategy is to compute the resources that affect the catalogue views used in the linksets. However, there is no work in the literature that deals with *complex* SPARQL-based views. Also, the proprietary systems that support the incremental maintenance of views, such as Oracle RDF Store, can only deal with small inserts.

Furthermore, we cannot consider that a linkset is a regular RDF view computed from two datasets, since they are materialized using complex linkage rules, which typically involve similarity measures that cannot be expressed with a SPARQL query. Hence, even if there was a solution for the maintenance of SPARQL-based views in the literature, we still would not be able to direct use it.

As already mentioned in the introduction, the work reported in this paper differs from previous work by the authors (Casanova *et al.* 2014) in three aspects. First, it presents in detail the incremental strategy to keep linksets updated, which includes a normalization process for views defined by SPARQL queries and a discussion on how to synthesize queries that compute sets of affected resources. Second, it briefly outlines an implementation of the proposed strategy. Lastly, based on the implementation, it describes experiments to measure the performance of the incremental strategy when compared with a full re-materialization strategy, a question neglected in the literature.

3   Catalogue Views and Linkset Views

3.1   Basic Definitions and Notation

To make the paper self-contained, we introduce an abstract notation to define *catalogue views* and *linkset views*, based on a minimum set of simple SPARQL 1.1 constructs (Harris and Seaborne 2013). The abstract notation is convenient since it highlights the aspects involved in the construction of materialized views and linksets.

Catalogue views and linkset views depend on the notion of a *simple construct query*, which intuitively defines the catalogue of resources. A SPARQL query F is a *simple construct query*, or a *simple query*, iff

---

[1] https://github.com/dbpedia/dbpedia-live-mirror.

- The CONSTRUCT clause of F has exactly one template of the form "$?x$ $rdf{:}type$ C" and a list of templates of the form "$?x$ P_k $?p_k$", where C is a class and P_k is a property, for $k = 1,...,n$. We say that $V_F = \{C,P_1,...,P_n\}$ is the *vocabulary* of F;
- F contains a single FROM clause, specifying the dataset used to evaluate F;
- The WHERE clause of F contains the pattern of the values that will be mapped to the resources and properties of the CONSTRUCT clause; the WHERE clause is such that it does not contain negations or the MINUS operator (Sect. 4.2 will discuss the reasons for restricting the WHERE clause).

A *catalogue view definition* is a pair $v = (V,F)$, where F is a simple construct query, called the *view mapping*, and V is the vocabulary of F, called the *view vocabulary*. Whenever possible, we will simply refer to F as the view definition.

Assume that a dataset contains a single set of RDF triples. Let T be the dataset specified in the FROM clause of F and $\sigma_T(t)$ be the state of T at time t. When evaluated against $\sigma_T(t)$, the simple query F returns a set of triples, which we denote $F[\sigma_T(t)]$.

A *materialization* of F at time t is the process of computing $F[\sigma_T(t)]$ and storing it as part of a dataset. We could naturally expand the abstract notation for a simple view to indicate the dataset and provide a name for the materialization of the view.

A *linkset view definition* is a quintuple $l = (p,F,G,\pi,\mu)$, where

- p is the link property
- F and G are simple queries whose vocabularies have the same cardinality n and whose FROM clauses specify the datasets over which l is evaluated
- π is a permutation of $(1,...,n)$, called the *alignment* of l
- μ is a $2n$-relation, called the *match predicate* of l

Let $V_F = \{C,P_1,...,P_n\}$ and $V_G = \{D,Q_1,...,Q_n\}$ be the vocabularies of F and G, respectively. Intuitively, π indicates that, for each $k = 1,...,n$, the match predicate will compare values of P_k with values of Q_m, where $m = \pi(k)$. The notion of alignment could be generalized to permit more sophisticated alignments and mappings.

Let T be the dataset specified in the FROM clause of F and U be the dataset specified in the FROM clause of G. We say that l is *evaluated over T and U* and that l is *from T to U*. Let $\sigma_T(t)$ and $\sigma_U(t)$ be the states of T and U at time t. The linkset view definition l induces a set of triples, denoted $l[\sigma_T(t),\sigma_U(t)]$, as follows:

$(s,p,o) \in l[\sigma_T(t),\sigma_U(t)]$ iff there are triples
$(s,\ rdf{:}type,\ C),\ (s,\ P_1,\ s_1),\ ...,\ (s,\ P_n,\ s_n) \in F[T]$ and
$(o,\ rdf{:}type,\ D),\ (o,\ Q_1,\ o_1),\ ...,(o,\ Q_n,\ o_n) \in G[U]$ such that
$(s_1,\ ...,\ s_n,\ o_{m1},\ ...,\ o_{mn}) \in \mu$, where $\pi\ (k)$, for each $k = 1,...,n$

Again, a *materialization* of l is the process of computing the set $l[\sigma_T(t),\sigma_U(t)]$ and storing it as part of a dataset. Also, we could expand the abstract notation to indicate the dataset and provide a name for the materialization of a linkset view definition.

3.2   Running Example

To illustrate catalogue views and linkset views, consider the dataset called Internet Movie Database (IMDb), which contains triples about movies, actors, etc. Suppose that

IMDb has a fictitious endpoint <http://imdb.org/sparql>, with default graph <http://imdb.org/data>, and uses the ontology in Fig. 1. Also consider the DBpedia dataset, which contains triples extracted from Wikipedia pages. It uses the endpoint <http://dbpedia.org/sparql> with default graph <http://dbpedia.org> and the ontology partially presented in Fig. 1.

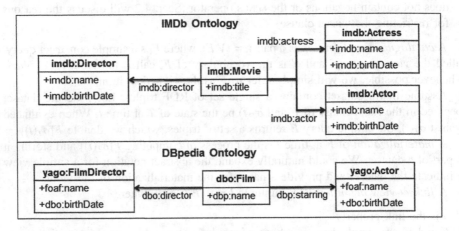

Fig. 1. Simplified fragments of IMDb and DBpedia Ontologies.

Suppose that a user wants to link the directors in *IMDb* with those in DBpedia by comparing their names and birth dates. For that purpose, s/he uses two catalogue views, *M* and *D*, respectively over IMDb and DBpedia. Suppose that *M* is:

```
CONSTRUCT { ?x rdf:type yago:FilmDirector.
            ?x foaf:name ?nm . ?x dbo:birthDate ?bt }
FROM <http://imdb.org/data>
WHERE { ?x rdf:type imdb:Director .
        ?x imdb:name ?nm . ?x imdb:birthDate ?bt }
```

and that *D* is:

```
CONSTRUCT { ?x rdf:type yago:FilmDirector.
            ?x foaf:name ?nm . ?x dbo:birthDate ?bt }
FROM <http://dbpedia.org>
WHERE { ?x rdf:type yago:FilmDirector.
        ?x foaf:name ?nm . ?x dbo:birthDate ?bt }
```

Lastly, the user creates the linkset view definition $f = (owl:sameAs,M,D,\pi,\mu)$ to materialize *owl:sameAs* links indicating that a director in *M* and a director in *D* are the same real-world object. As the match predicate, the user may choose the Levenshtein distance (Levenshtein 1966) with threshold <2 to compare the names of the directors and assume that the birthdates match only if they are equal. Then, for example, if views M and D respectively have two resources u and v with names "Tim Burton" and "Tim Button" and the same birthdate, then the linkset will have a triple $(u, owl:sameAs, v)$, since the names have a Levenshtein distance of 1 (replace "t" by "r"), which satisfies the accepted threshold.

4 Incremental Maintenance of Linkset Views

4.1 Overview

Consider the *materialized linkset maintenance problem*, defined as follows: "Given two datasets, T and U, and a materialized linkset L from T to U, maintain L when updates on T or U occur".

A possible solution is to incrementally maintain L, that is, update L based on the updates on T or U. However, L does not contain the triples capturing the property values that generated the links. Hence, it is obviously impossible to detect when an update u on T or U affects L by just looking at the links in L. Thus, to incrementally maintain L, we propose a strategy that overcomes this lack of information by capturing the changes that must be applied to L using the information about the updates and the mappings of the catologue views adopted to define L.

Let V be a collection of catalogue views over T. Let u be an update on T and $\sigma_T(t_0)$ and $\sigma_T(t_1)$ be the states of T before and after u (the discussion is symmetric for updates on U). In the first process required by our incremental strategy, we need to capture the changes that affect each view v in V following four main steps:

(1) Compute the set R^+ of resources in v affected by the inserted triples of u.
(2) Compute the set R^- of resources in v that are affected by the deleted triples of u.
(3) For s in $R^- \cup R^+$, retrieve the (new) property values of s from $\sigma_T(t_1)$, denoted P.
(4) Associate R^- and P with the update timestamp t_u.

Let F be a catalogue view and $F \in V$. Let $F[R^-(t_u)]$ be a collection of deleted resources of F associated with a given update timestamp t_u. Let $F[P(t_u)]$ be a collection of new property values of F associated with a given update timestamp t_u. Let t_1 be the current timestamp. Let $R^-[t_0,t_1]$ be the set of *accumulated deleted resources*, where $r \in R^-[t_0,t_1]$ iff $r \in F[R^-(t_u)]$ and $t_0 < r(t_u) < t_1$. Let $P[t_0,t_1]$ be the set of *accumulated property values*, where $p \in P[t_i,t_j]$ iff $p \in F[P(t_u)]$ and $t_i < p(t_u) < t_j$.

Suppose that L is a materialized linkset specified by the linkset view definition $l = (p,F,G,\pi,\mu)$, where G is a catalogue view over U and $G[\sigma_U(t)]$ denote the set of triples that G returns when execute over state $\sigma_U(t)$ of U. In the second process of the incremental strategy, we incrementally update L following two main steps:

(1) Delete from L all links whose subject or object occurs in $R^-[t_0,t_1]$.
(2) Try to match $P[t_0,t_1]$ with the property values of a resource in $\sigma_U(t_1)$; if a match is found, add a link to L.

4.2   Normalization of View Mappings

Recall from Sect. 3.1 that the WHERE clause of a simple query F does not contain negations or the MINUS operator. In this section, we show how to transform the *triple patterns* of the WHERE clause of F into a *normalized form*, which simplifies the discussion in Sect. 4.3.

Table 1 summarizes the allowed types of property paths in triple patterns (column 2) and the corresponding normalized form (column 3). Briefly, the *Normalization Process* iteratively runs through the triple patterns of the WHERE clause and replaces their complex property paths by simpler ones until all paths are *predicate paths*, that is, paths of length one. Note that a property path generates one or more simpler triple patterns in a single *group graph pattern*, in the case of *Inverse Paths*, *Sequence Paths*, *Fixed Length Path* and *One or More Path* expressions. But a property path generates two simpler triple patterns in different *group graph patterns* with an UNION clause, in the case of *Alternative Path*, *Zero or More Path* and *Zero or One Path*. Additionally, a triple pattern marked with "($\sqrt{}$)" needs **no further processing** to avoid a loop in the process. The original triple patterns are replaced by the normalized triple patterns in F. The output of the *Normalization Process* is a normalized view F' and a list of *predicate triple patterns*, denoted L_P, that is, triple patterns with predicate paths or predicate variables. Table 2 illustrates how the normalization works.

Table 1. Property path normalization.

Property Path	Original Triple Pattern	Normalized Triple Pattern
Predicate Path	?x *iri* ?y	do nothing
Inverse Path	?x ^*elt* ?y	?y *elt* ?x
Sequence Path	?x *elt1/elt2* ?y	?x *elt1* ?o1 . ?o1 *elt2* ?y
Alternative Path	?x *elt1|elt2* ?y	{ { ?x *elt1* ?y } UNION { ?x *elt2* ?y } }
Fixed Length Path ($n > 0$) (*)	?x *elt{n}* ?y	?x *elt* ?o$_1$. ?o$_1$ *elt* ?o$_2$. … ?o$_n$ *elt* ?y
One or More Path	?x *elt+* ?y	?x *elt** ?o1 . ($\sqrt{}$) ?o1 *elt* ?o2 . ?o2 *elt** ?y . ($\sqrt{}$)
Zero or More Path	?x *elt** ?y	{ { ?x *elt** ?o1 . ($\sqrt{}$) ?o1 *elt* ?o2 . ?o2 *elt** ?y } ($\sqrt{}$) UNION { ?x *elt{0}* ?y } } ($\sqrt{}$)
Zero or One Path	?x elt? ?y	{ { ?x *elt* ?y } UNION { ?x *elt{0}* ?y } } ($\sqrt{}$)

(*) This syntactical form is not included in the specification of SPARQL 1.1, but it is supported by several triplestore systems.

Table 2. Example of the normalization process.

View Mapping
`CONSTRUCT { ?x imdb:workedWith ?a }` `WHERE { ?x ^imdb:director/(imdb:actress|imdb:actor) ?a }`
Normalized Form

`CONSTRUCT { ?x :workedWith ?a }`
`WHERE { ?o imdb:director ?x.`
`        {{ ?o imdb:actress ?a } UNION { ?o imdb:actor ?a }} }`

L_P **- List of Predicate Triple Patterns**
`[?o imdb:director ?x, ?o imdb:actress ?a, ?o imdb:actor ?a ]`

4.3 Computing Affected Resources and New Property Values

We first summarize the notation to be used in what follows:

- T and U are datasets and u is an update on T
- $\sigma_T(t_i)$ is the state of T at time t_i, $i = 0,1$
- F and G are catalogue view definitions over T and U, respectively
- $\sigma_F(t_i) = F[\sigma_T(t_i)]$ is the state of the view defined by view F at time t_i, $i = 0,1$
- $l = (p,F,G,\pi,\mu)$ is a linkset view definition over v and w
- $\sigma_l(t_i) = l[\sigma_T(t_i),\sigma_U(t_i)]$ is the state of l at t_i, $i = 0,1$
- $\Delta^- x(t_0,t_1) = x(t_0) - x(t_1)$ and $\Delta^+ x(t_0,t_1) = x(t_1) - x(t_0)$

where X is either T, U, F, G or l

A *deletion resources set query* of F for u, denoted F_u^-, is any query that computes a set of resources that contains the set of resources visible through F and affected by the deletions in u. Likewise, an *insertion resources set query* of F for u, denoted F_u^+, is any query that computes a set of resources that contains the set of resources visible through F and affected by the insertions in u. More precisely, we define:

- A *deletion resources set query of F for u*, denoted F_u^-, is a SPARQL query such that, for any states $\sigma_T(t_0)$ and $\sigma_T(t_1)$ of T such that $\sigma_T(t_0)$ and $\sigma_T(t_1)$ are the states before and after u, we have $\{r \ / \ \exists p \exists o((r,p,o) \in \Delta_v^-(t_0,t_1)) \} \subseteq F_u^-[(\sigma_T(t_0)]$
- An *insertion resources set query of F for u*, denoted F_u^+, is a SPARQL query such that, for any states $\sigma_T(t_0)$ and $\sigma_T(t_1)$ of T such that $\sigma_T(t_0)$ and $\sigma_T(t_1)$ are the states before and after u, we have $\{r \ / \ \exists p \exists o((r,p,o) \in \Delta_v^+(t_0,t_1)) \} \subseteq F_u^+[(\sigma_T(t_0)]$

Recall from Sect. 3.1 that we restrict view mappings to use the types of property paths in the second column of Table 1 and not to contain negations or the MINUS operator. This restriction has one important consequence, stated as follows.

A view mapping F over a dataset T is *monotonic* iff, for any two states $\sigma_T(t)$ and $\sigma_T(u)$, if $\sigma_T(t) \subseteq \sigma_T(u)$ then $F[\sigma_T(t)] \subseteq F[\sigma_T(u)]$.

Proposition 1: Assume that F is a view mapping whose WHERE clause uses the types of property paths listed in Table 1 and does not contain negations or the MINUS operator. Then, F is monotonic.

Monotonicity permits us to consider only deletions when constructing deletion resources set queries of F for u and, likewise, only insertions when constructing insertion resources set queries. Intuitively, if F were not monotonic, an insertion into $\sigma_T(t_0)$ might propagate to a deletion from $F[\sigma_T(t_0)]$ and, likewise, a deletion from $\sigma_T(t_0)$ might propagate to an insertion into $F[\sigma_T(t_0)]$.

Canonical Deletion and Insertion Resources Set. Let W_F be the WHERE clause and g_F be the graph in the FROM clause of F. Assume that F has already been normalized and let L_F be the set of predicate triple patterns that occur in W_F. Suppose that we materialize the set of deleted triples specified in the update u in state $\sigma_T(t_0)$ into a named graph g^-.

Assume that the predicate triple patterns in L_P are "$a_k\ b_k\ c_k$", for $k = 1,...,n$. Table 3 shows the template that generates the *canonical deletion resources set query* for F and u, denoted CF_u^-. Recall that the variable ?x identifies the resource of the catalogue view as defined in Sect. 3.1. Note that the results of CF_u^- are inserted into another named graph, denoted R^-, in which each resource is associated with the view identification and the timestamp of the update, denoted t_u. The template for the *canonical insertion resources set query* for F and u, denoted CF_u^+, is similarly defined, except that g^- is replaced by g^+, a named graph for the set of inserted triples u^+, and R^- is replaced by R^+, a named graph with the results of CF_u^+. We again resort to an example to illustrate the process of constructing CF_u^-. Table 3 recalls the definition of view M from Sect. 3.2, shows an update example, the query to populate g^- represented by the named graph <http://imdb.org/deletions> and finally the synthesized query CM_u^-.

Note that, if CM_u^- is executed before u is applied, the graph <http://imdb.org/data> has the necessary data to match the triple

(:Tim_Burton imdb:name 'Tim Burton').

Indeed, returning to the general discussion, let $\sigma_T(t_0)$ and $\sigma_T(t_1)$ be the states of T before and after an update u is applied. We say that $CF_u^-[\sigma_T(t_0)]$, the result of executing CF_u^- in state $\sigma_T(t_0)$, is *the set of affected resources computed by CF_u^- in state $\sigma_T(t_0)$*. Likewise, we say that $CF_u^+[\sigma_T(t_1)]$, the result of executing CF_u^+ in state $\sigma_T(t_1)$, is *the set of affected resources computed by CF_u^+ in state $\sigma_T(t_1)$*. After CF_u^- $[\sigma_T(t_0)]$ is computed, u can actually be applied and the triples in the named graph g^- can be cleared. However, CF_u^+ has to be executed after u is applied, otherwise the state of T would not have the necessary data to match the triples in g^+.

To summarize, the process of computing the affected resources R^- and R^+ follows four main steps: (1) intercept u and populate g^+ and g; (2) execute F_u^-, populating R^-; (3) execute u; (4) execute F_u^+, populating R^+.

We stress that CF_u^- and CF_u^+ are just a possible solution. Note that they can be synthesized at design time, right after the normalization, since the template will not

Table 3. Example of the computation of affected resources.

Template of CF_u^-

`INSERT { GRAPH <R⁻> { ?x :view F . ?x :timestamp t_u } }`
`WHERE { { GRAPH <g⁻> { a₁ b₁ c₁ } . GRAPH <g_F> { W_F } }`
`  UNION { GRAPH <g⁻> { a₂ b₂ c₂ } . GRAPH <g_F> { W_F } }`
`... UNION { GRAPH <g⁻> { aₙ bₙ cₙ } . GRAPH <g_F> { W_F } } }`

M – the view mapping

`CONSTRUCT { ?x rdf:type yago:FilmDirector.`
`            ?x foaf:name ?nm . ?x dbo:birthDate ?bt }`
`WHERE { ?x rdf:type imdb:Director.`
`          ?x imdb:name ?nm. ?x imdb:birthDate ?bt}`

u – the update	**Query to populate g^-**

`WITH <http://imdb.org/data>`
`DELETE DATA {:Tim_Burton`
`      imdb:name 'Tim Burton'}`
	`INSERT DATA {`
	`GRAPH <http://imdb.org/deletions> {`
	`:Tim_Burton imdb:name 'Tim`
	`Burton'}}`

L_P -`?x rdf:type imdb:Director, ?x imdb:name ?nm, ?x imd:birthDate ?bt`
CM_u^- – canonical deletion resources set query

`INSERT { GRAPH <http://imdb.org/deletionsResources>`
`            { ?x :view M . ?x :timestamp "1" } }`
`WHERE{{GRAPH <http://imdb.org/deletions>{ ?x rdf:type imdb:Director}`
`      GRAPH <http://imdb.org/data> { ?x rdf:type imdb:Director .`
`          ?x imdb:name ?nm .?x imdb:birthDate ?bt }`
`  UNION{GRAPH <http://imdb.org/deletions> { ?x imdb:name ?nm }`
`      GRAPH <http://imdb.org/data> { ?x rdf:type imdb:Director .`
`          ?x imdb:name ?nm .?x imdb:birthDate ?bt } }`
`  UNION{ GRAPH <http://imdb.org/deletions> { ?x imdb:birthDate ?bt }`
`      GRAPH <http://imdb.org/data> { ?x rdf:type imdb:Director .`
`          ?x imdb:name ?nm .?x imdb:birthDate ?bt } } }`

change, only the data deleted or inserted. Furthermore, CF_u^- is correct in the following sense (a similar result holds for CF_u^+).

Proposition 2: Let F be a view mapping and u be an update over a dataset T. Let CF_u^- be the canonical deletion resources set query for F. Then, for any states $\sigma_T(t_0)$ and $\sigma_T(t_1)$ of T such that $\sigma_T(t_0)$ and $\sigma_T(t_1)$ are the states before and after u,

$$\{ r \,/\, \exists p \exists o(\ (r,p,o) \in \Delta_v^-(t_0, t_1)\ ) \} \subseteq CF_u^-[(\sigma_T(t_0)].$$

New Property Values. After computing the graphs of the affected resources, we proceed to compute the named graph with the new property values, denoted P. Let W_F be the WHERE clause, C_F be the CONSTRUCT clause and g_F be the graph in the FROM clause of F. Table 4 shows the template and an example of the query to compute P.

Table 4. Computing the new property values.

```
Template:
INSERT { GRAPH <P> { C_F . ?x :view F . ?x :timestamp t_u } }
WHERE
{ { { SELECT DISTINCT ?d WHERE { GRAPH <R⁻> { ?d ?p ?o } }
    UNION
    { SELECT DISTINCT ?i WHERE { GRAPH <R⁺> { ?i ?p ?o } } }
    GRAPH <g_F> { W_F }   FILTER ( ( ?x = ?i ) || ( ?x = ?d ) ) }
```

```
Example:
INSERT { GRAPH <http://imdb.org/newProperties>
         { ?x rdf:type yago:FilmDirector. ?x foaf:name ?nm .
           ?x dbo:birthDate ?dt . ?x :view M . ?x :timestamp "1" } }
WHERE
{ { { SELECT DISTINCT ?d
    WHERE{ GRAPH <http://imdb.org/deletedResources> { ?d ?p ?o } }
    UNION
    { SELECT DISTINCT ?i
      WHERE{ GRAPH <http://imdb.org/insertedResources> { ?i ?p ?o } } }
    GRAPH <http://imdb.org/data> { ?x rdf:type imdb:Director .
                          ?x imdb:name ?nm . ?x imdb:birthDate ?bt }
    FILTER ( ( ?x = ?i ) || ( ?x = ?d ) ) }
```

Table 5. Example of the linkset update process.

Template	
DELETE{ GRAPH <L> { ?s ?p ?o } } WHERE { GRAPH <R⁻> { ?s :view F. ?s :timestamp ?t. FILTER (?t > t_0) } GRAPH <L> { ?s ?p ?o } }	

$R^-[t_0, t_1]$	$R^+[t_0, t_1]$
imdb:Tim_Burton :view "M" . imdb:Tim_Burton :timestamp "1"	imdb:Ridley_Scott :view "M" . imdb:Ridley_Scott :timestamp "1"

P	
imdb:Tim_Burton rdf:type yago:FilmDirector . imdb:Tim_Burton foaf:name "Tim Burton"; dbo:birthDate "1958-08-25" . imdb:Tim_Burton :view "M" ; :timestamp "1" . imdb:Ridley_Scott rdf:type yago:FilmDirector . imdb:Ridley_Scott foaf:name "Ridley Scott";dbo:birthDate "1937-11-30". imdb:Ridley_Scott :view "M" ; :timestamp "1"	

Updating the Linkset	

DELETE{ GRAPH <http://linkset/directors> { ?s ?p ?o } }
 WHERE { GRAPH <http://imdb.org/deletionsResources>
       { ?s :view "M" . ?s :timestamp ?t . FILTER (?t > "0") }
       GRAPH <http://linkset/directors> { ?s ?p ?o } }	

Old state of linkset *l*	New state of linkset *l*
imdb:Tim_Burton owl:sameAs dbr:Tim_Burton	imdb:Ridley_Scott owl:sameAs dbr:Ridley_Scott

4.4   Updating a Materialized Linkset

Finally, the linkset can be updated according to the set of resources that were affected after the last timestamp maintenance. Let L be a materialized linkset, we first need to delete all links involving a resource in $R^-[t_0,t_1]$ according to the template of Table 5. Additionally, Table 5 shows an example of the linkset update process supposing that L was materialized in a named graph <http://linkset/directors>.

Then, the matching process is re-executed, using the triples in $P[t_0,t_1]$, instead of the whole view, and the new links are finally added to the materialized linkset.

We conclude this section with an observation about how the canonical queries are synthesized. We note that P also considers the resources in the deleted set R^- when computing the new property values. This is necessary since CF_u^- computes a superset R^- of the set of resources affected by deletions. That is, there might be a resource $r \in R^-$ that forced the deletion of a link of the form (r,p,o) from L, but r might not actually be affected by the deletions. Therefore, the algorithm has to recompute all such links. However, the problem of detecting the exact set of resources affected by deletions (or insertions) is NP-Complete, which is proved by a transformation from Subgraph Isomorphism. Thus, synthesizing a deletion resources set query that returns the exact set of resources affected by a set of deletions is infeasible, unless P = NP.

5   Implementation and Evaluation

5.1   Architecture

The *Linkset Maintainer* tool implements the strategy detailed in Sect. 4. The tool was developed in the Java 7 programming language, using the Eclipse Luna IDE, JBoss Application Server 7 and ARQ API as the SPARQL Processor.

Figure 2 summarizes the architecture of the tool. At initialization time, for each view F over a dataset T, the *View Controller* normalizes F and, at run time, it computes the set of affected resources and new property values of F with respect to updates submitted to T, as already discussed in Sects. 4.2 and 4.3. At initialization time, the *Linkset Controller* for a linkset l over views F and G registers itself with the *View Controllers* for F and G and computes the initial state of l. At run time, it retrieves the sets of deleted resources and the sets of new property values of F and G, computes the accumulated set of deleted resources and the accumulated set of property values and incrementally maintains the linkset according to timestamp of the last maintenance, as discussed in Sect. 4.4.

The current implementation of the *Linkset Controller* uses Silk as the link discovery tool, since it provides an API that enables the matching process to be executed programmatically. The user only specifies the linkage rules and the tool automatically does the rest. The user may adopt other discovery tool, but s/he will have to manually manage the tool.

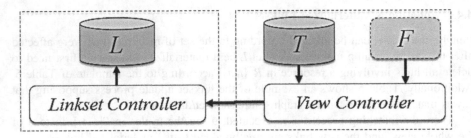

Fig. 2. Linkset view maintainer architecture.

5.2   Evaluation Setup

In order to compare the performance of the incremental strategy with the full re-computation of linksets, we selected two datasets: an *IMDb* dump, with 44,855,096 triples, and the DBpedia endpoint, which at the time of the experiments had 883,644,235 triples. All experiments were executed on a computer with an Intel Core i5 1,7 GHz processor and 4 GB RAM, running OS X Yosemite 10.10.2.

We defined views about movie directors for each dataset. The view "IMDb_Director" has 41,929 resources and "DBpedia_Director" has 9,937 resources. Then, we materialized an owl:sameAs linkset of directors, using these views, by comparing their names and birth dates. The resulting linkset had 4,565 links. We performed updates on the *IMDb* dataset that affected view "IMDb_Director". All updates were similar to the following, except that they differ on the LIMIT clause to get an exact number of affected resources:

```
DELETE WHERE { ?s rdf:type imdb:Director } LIMIT 1
```

5.3   Experiments

We first compared the full rematerialization and the incremental strategy for the directors linkset in the presence of the updates described in Sect. 5.2. Figure 3 shows the runtime of the updates, varying the number of affected resources. For each update, the runtime of the incremental strategy includes the time to compute the deleted resources and new property values, execute the update, and update the linkset. Likewise, the runtime of the full rematerialization includes the time to execute the update and rematerialize the linkset.

Since the queries to compute the new property values depend on the affected resources, after some point, it may become disadvantageous to use the incremental strategy. In the case of the directors linkset, this point was around 32 K resources, which is 78 % of the total number of resources of view "IMDb_Director", which has 41 K resources. However, if the number of affected resources is small, incremental maintenance is far better than full rematerialization, as expected.

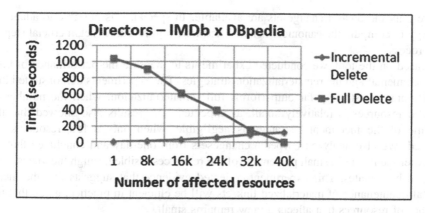

Fig. 3. Maintenance performance of linkset directors.

We run a second experiment to assess, in a real-world situation, the percentage of resources, visible through views, that are affected by updates on a dataset. We based the experiments on the *DBpedia changesets*, that is, sets of changed triples extracted from Wikipedia, which are organized by year, month, day and hour and separated by the type of the update (added, removed, reinserted and cleared). For this second experiment, we defined two views over DBpedia about actors (49,308 resources) and actresses (7,309 resources), in addition to the directors' view. We then analyzed the number of view resources affected by the changesets from an entire day (April 28, 2015). We computed the number of updated resources by changeset and how many of these resources were visible through any of the views. Table 6 summarizes the results.

Considering each changeset as a single update, Table 6 shows an average of 99 resources per changeset, of which only 2 % were visible through any of the views. Furthermore, the max number of affected resources in a single changeset was only 44. Therefore, this second experiment provides evidence that the number of affected resources tends to be small in real-world situations.

Table 6. Analysis of DBpedia changesets.

	Total	Sets	Avg	Max
Updated resources	551,236	5,568	99	975
View resources	13,199	5,568	2	44

6 Conclusions

We first detailed an incremental strategy to keep linksets updated. We focused on how to compute the sets of affected resources that are visible through a view. Then, we showed how to keep the linksets updated based on such sets.

We presented the Linkset Maintainer, a tool that implements the incremental strategy. The tool was designed for an environment where it is possible to intercept the updates submitted to a dataset. However, the tool can be adapted to an environment where only the dataset changesets are available. In special, it is possible to adapt the strategy to compute the canonical deletion and insertion resources set, a crucial step of the process.

Based on the tool, we conducted experiments to measure the performance of both the incremental and the rematerialization strategies. The experiments demonstrated that the incremental strategy far outperforms full rematerialization, when the number of affected resources is relatively small, as expected. The results also showed that the runtime of the incremental strategy is negligible, when only a few resources are affected. We also analyzed DBpedia changesets from one day and concluded that, in the experiments, just a small percentage of the resources visible through the views were affected by updates. This experiment collected evidence that suggests that the incremental maintenance of materialized linksets will be efficient in practice, given that the number of resources that affects a view remains small.

As future work, we plan to continue the development of the tool to improve performance and to provide a better user interface to help the definition of views and linksets. Finally, we plan to make the tool freely available.

Acknowledgments. This work was partly funded by CNPq under grants 153908/2015-7, 557128/2009-9, 444976/2014-0, 303332/2013-1, 442338/2014-7 and 248743/2013-9 and by FAPERJ under grants e E-26-170028/2008 and E-26/201.337/2014.

References

Berners-Lee, T.: Linked Data (2006). http://www.w3.org/DesignIssues/LinkedData.html

Bouza, A.: The Movie Ontology (2010). http://www.movieontology.org

Casanova, M.A., Vidal, V.M.P., Lopes, G.R., Leme, L.A.P.P., Ruback, L.: On Materialized sameAs linksets. In: Decker, H., Lhotská, L., Link, S., Spies, M., Wagner, R.R. (eds.) DEXA 2014, Part I. LNCS, vol. 8644, pp. 377–384. Springer, Heidelberg (2014)

Endris, K.M., Faisal, S., Orlandi, F., Auer, S., Scerri, S., et al.: Interest-based RDF update propagation. In: Arenas, M., et al. (eds.) ISWC 2015. LNCS, vol. 9366, pp. 513–529. Springer, Heidelberg (2015)

Gupta, A., Mumick, I., Subrahmanian, V.: Maintaining views incrementally. In: ACM SIGMOD Record, pp. 157–166 (1993)

Harris, S., Seaborne, A.: SPARQL 1.1 Query Language (2013). http://www.w3.org/TR/sparql11-query/

Hung, E., Deng, Y., Subrahmanian, V.: Maintaining RDF views. In: Technical report CS-TR-4612 (UMIACS-TR-2004-54), University of Maryland (2004)

Ibáñez, L.-D., Skaf-Molli, H., Molli, P., Corby, O.: Col-graph: towards writable and scalable linked open data. In: Mika, P., et al. (eds.) ISWC 2014, Part I. LNCS, vol. 8796, pp. 325–340. Springer, Heidelberg (2014)

Levenshtein, V.: Binary codes capable of correcting deletions, insertions, and reversals. In: Soviet Physics Doklady, vol. 10 (1966)

Ngomo, A., Auer, S.: LIMES - a time-efficient approach for large-scale link discovery on the web of data. In: Proceedings of the IJCAI 2011, pp. 2312–2317 (2011)

Popitsch, N., Haslhofer, B.: DSNotify – a solution for event detection and link maintenance in dynamic triplesets. J. Web Semant. **9**(3), 266–283 (2011)

Staudt, M., Jarke, M.: Incremental maintenance of externally materialized views. In: Proceedings of the VLDB 1996, pp. 75–86 (1996)

Vidal, V.M.P., Casanova, M.A., Cardoso, D.S.: Incremental maintenance of RDF views of relational data. In: Meersman, R., Panetto, H., Dillon, T., Eder, J., Bellahsene, Z., Ritter, N., De Leenheer, P., Dou, D. (eds.) ODBASE 2013. LNCS, vol. 8185, pp. 572–587. Springer, Heidelberg (2013)

Vidal, V.M., Casanova, M.A., Arruda, N., Roberval, M., Leme, L.P., Lopes, G.R., Renso, C.: Specification and incremental maintenance of linked data mashup views. In: Zdravkovic, J., Kirikova, M., Johannesson, P. (eds.) CAiSE 2015. LNCS, vol. 9097, pp. 214–229. Springer, Heidelberg (2015)

Volz, J., Bizer, C., Gaedke, M.: Web of Data Link Maintenance Protocol - Maintaining Links Between Changing Linked Data Sources (2009a). http://www4.wiwiss.fu-berlin.de/bizer/silk/wodlmp

Volz, J., Bizer, C., Gaedke, M., Kobilarov, G.: Discovering and maintaining links on the web of data. In: Bernstein, A., Karger, D.R., Heath, T., Feigenbaum, L., Maynard, D., Motta, E., Thirunarayan, K. (eds.) ISWC 2009. LNCS, vol. 5823, pp. 650–665. Springer, Heidelberg (2009b)

Data Analysis

Aggregate Reverse Rank Queries

Yuyang Dong[✉], Hanxiong Chen, Kazutaka Furuse, and Hiroyuki Kitagawa

Department of Computer Science, University of Tsukuba, Tsukuba, Ibaraki, Japan
tou@dblab.is.tsukuba.ac.jp, {chx,furuse,kitagawa}@cs.tsukuba.ac.jp

Abstract. Recently, reverse rank queries have attracted significant research interest. They have real-life applicability, such as in marketing analysis and product placement. Reverse k-ranks queries return users (preferences) who favor a given product more than other people. This helps manufacturers find potential buyers even for an unpopular product. Similar to the cable television industry, which often bundles channels, manufacturers are also willing to offer several products for sale as one combined product for marketing purposes.

Unfortunately, current reverse rank queries, including Reverse k-ranks queries, only consider one product. To address this limitation, we propose the *aggregate reverse rank queries* to find matching user preferences for a set of products. To resolve this query more efficiently, we propose the concept of pre-processing the preference set and determining its upper and lower bounds. Combining these bounds with the query set, we proposed and implemented the tree pruning method (TPM) and double-tree method (DTM). The theoretical analysis and experimental results demonstrated the efficacy of the proposed methods.

Keywords: Similarity search · Aggregate reverse rank queries · Tree-based method

1 Introduction

Top-k and reverse k-rank queries are two different kinds of view-models. The top-k query is a user view-model that helps consumers by obtaining the best k products that match a user's preference. On the other hand, the reverse k-rank query [18] supports manufacturers by discovering potential consumers through retrieving the most appropriate user preferences. Therefore, it is a manufacturer view-model and can be used as a tool for identifying customers and estimating product marketing.

Figure 1 shows an example of a reverse 1-rank query. Five different cell phones (p_1–p_5) are scored on "smart" and "ratings" in a table (Fig. 1(a)). The preferences of two users Tom and Jerry are in another table (Fig. 1(b)) and consist of the weights for all attributes. The score of a cell phone based on user preference is determined from the inner product of the cell phone attributes vector and user preference vector. Without loss of generality, we assumed that minimum values are preferable. The results of the reverse 1-rank query are given in the last cells

© Springer International Publishing Switzerland 2016
S. Hartmann and H. Ma (Eds.): DEXA 2016, Part II, LNCS 9828, pp. 87–101, 2016.
DOI: 10.1007/978-3-319-44406-2_8

(a) User preferences and Ranks

user	w[smart]	w[rating]	Ranks
Tom	0.8	0.2	p3,p2,p1,p4,p5
Jerry	0.3	0.7	p2,p5,p3,p4,p1

(b) Cell phone ranks and R-1Rank

	p[smart]	p[rating]	Score on Tom	Score on Jerry	Rank in Tom	Rank in Jerry	R-1Rank
p1	6	7	6.2	6.7	3rd	5th	Tom
p2	2	3	2.2	2.7	2nd	1st	Jerry
p3	1	6	2.0	4.5	1st	3rd	Tom
p4	7	5	6.6	5.6	4th	4th	Tom
p5	8	2	6.8	3.8	5th	2nd	Jerry

Fig. 1. The example of reverse 1-rank queries.

of Fig. 1(b). For example, Tom believes that p_1 is the third-best phone, while Jerry thinks that p_1 is the fifth-best. To manufacturers, Tom is more likely to buy p_1 than Jerry; hence, the reverse 1-rank query returns Tom as the result.

Motivation. Manufacturers use "product bundling" for marketing purposes. Product bundling is offering several products for sale as one combined product. It is a common feature in many imperfectly competitive product markets. For example, Microsoft Co., Ltd. includes a word processor, spreadsheet, presentation program, and other useful software into a single Office Suite. The cable television industry often bundles various channels into a single tier to expand the channel market. Manufacturers of video games are also willing to group a popular game with other games of the same theme in the hope of obtaining more benefits by selling them together.

Because product bundling is an important business approach, helping manufacturers target buyers for their bundled products is important. Unfortunately, the reverse k-rank query and other kinds of reverse ranking queries are all designed for just one product. To address this limitation, we propose a new query definition that finds k customers with the smallest aggregate rank values, where the rank of a product set is defined as the sum of each product's rank. We call this approach *aggregate reverse rank queries* (*AR-k queries*).

Group IQI = 2	sum Rank in Tom	sum Rank in Jerry	AR-1Rank
p1,p2	5 (3 + 2)	6 (5 + 1)	Tom
p2,p3	3 (2 + 1)	4 (1 + 3)	Tom
p4,p5	9 (4 + 5)	6 (4 + 2)	Jerry

Fig. 2. The example of aggregate reverse 1-rank queries.

Figure 2 shows an example of an AR-1 query. There are three groups of bundled products: $\{p_1, p_2\}$, $\{p_2, p_3\}$, and $\{p_4, p_5\}$. The aggregate rank of $\{p_1, p_2\}$ is 5 according to Tom's preferences and 6 according to Jerry's. Thus, the AR-1 query returns Tom as the result because Tom prefers this bundle the most.

Contribution. This paper makes the following contributions:

- To the best of our knowledge, we are the first to address the "one product" limitation of reverse k-rank queries. We propose a new AR-k query that returns the k user preferences that best match a set of products.
- We propose the concept of pre-processing preferences to determine possible upper and lower bounds. This process can be done before the AR-k query is issued to enhance its efficiency and is implemented with the proposed tree-pruning method (TPM) and double-tree method (DTM).
- Along with the theoretical analysis, we also performed experiments on both real and synthetic data. The experimental results validated the efficiency of the proposed methods.

The rest of this paper is organized as follows: Sect. 2 summarizes related work. Section 3 states the definitions. In Sect. 4, we present the method of bounding the query set. Sections 5 and 6 propose two solutions (TPM and DTM) of AR-k. Experimental results are shown in Sect. 7 and Sect. 8 concludes the paper.

2   Related Work

Ranking is an important property for evaluating the position of a product. Many variants of rank-aware queries have been widely researched.

Ranking Query (Top-k Query). The most basic approach is the top-k query. When given a user preference, the top-k query returns k products with minimal ranking scores found by a score function. One possible approach to the top-k problem is the onion technique [1]. This algorithm pre-computes and stores convex hulls of data points in layers like an onion. [4] is an important investigation that describes and classifies top-k query processing techniques in relational databases.

Reverse Rank Query (RRQ). Reverse top-k queries [10,12] have been proposed to evaluate the impact of a potential product on the market based on the preferences of users who treat it as a top-k product. For an efficient reverse top-k process, Vlachou et al. [13] proposed a branch-and-bound algorithm (BBR) using boundary-based registration and a tree base. Vlachou et al. [11,14] have reported various applications of reverse top-k queries. However, in order to answer the reverse query for some less-popular objects, [18] proposed the reverse k-rank query to find the top-k user preferences with the highest rank for a given object among all users.

Other Reverse Queries. Other related research on reverse queries is listed below. Given a data point, queries are performed to find result sets containing this data point. In contrast to the nearest-neighbor search, Korn and

Muthukrishnan [5] proposed the reverse nearest-neighbour (RNN) query. Besides the nearest neighbor, Yao et al. [17] proposed the reverse furthest neighbor (RFN) query to find points where the query point is deemed as the furthest neighbor. For reverse k nearest neighbor (RKNN), Yang et al. [15] analyzed and compared notable algorithms from [2,7–9,16]. RKNN differs from RRQ because it evaluates the relative L_p distance between two points in one Euclidean space. However, RRQ focuses on the absolute ranking among all objects, and scores are found via the inner product function. In addition, RKNN treat the user preference and product as the same kind of point in the same space, while RRQ has two data sets of different data spaces. The reverse skyline query uses the advantages of products to find potential customers based on the dominance of competitors' products [3,6]. The preference of each user is described as a data point representing the desirable product. But in RRQ, the preference is described as a weight vector.

3   Problem Statement

The assumption of the product database, preference database and the score function between them are same with the related research [10,13,18]. Let there be a product data set P and preference data set W. Each $p \in P$ is a d-dimensional vector that contains d non-negative scoring attributes. p is represented as a point $p = (p[1], p[2], ..., p[d])$, where $p[i]$ is the attribute value of p in the ith dimension. The preference $w \in W$ is also a d-dimensional weighting vector, and $w[i]$ is a non-negative weight that evaluates $p[i]$, where $\sum_{i=1}^{d} w[i] = 1$. The score is defined as the inner product of p and w, which is expressed by $f(w, p) = \sum_{i=1}^{d} w[i] \cdot p[i]$. Given a query q, which is in the same space as, but not necessarily an element of P, the reverse k-rank query [18] is defined as follows.

Definition 1 ($rank(w, q)$). *Given a point set P, weighting vector w, and query q, the rank of q by w is $rank(w, q) = |S|$, where $S \subseteq P$ and $\forall p_i \in S, f(w, p_i) < f(w, q) \wedge \forall p_j \in (P - S), f(w, p_j) \geq f(w, q)$.*

Definition 2 (*reverse k-ranks query*). *Given a point set P, weighting vector set W, positive integer k, and query q, the reverse k-rank query returns the set S, $S \subseteq W$, $|S| = k$, such that $\forall w_i \in S, \forall w_j \in (W - S), rank(w_i, q) \leq rank(w_j, q)$ holds.*

To deal with a query having more than one query point, we propose the AR-k query, which is formally defined as follows.

Definition 3 (*aggregate reverse rank query, AR-k*). *Given a point set P, weighting vector set W, positive integer k, and query point set Q, the AR-k query returns the set S, $S \subseteq W$, $|S| = k$, such that $\forall w_i \in S, \forall w_j \in (W - S)$, $ARank(w_i, Q) \leq ARank(w_j, Q)$ holds.*

Three aggregate evaluation functions were considered for ARank:
- **Sum:** $ARank(w, Q) = \sum_{q_i \in Q} rank(w, q_i)$.

- **Maximum:** $ARank(w, Q) = Max_{q_i \in Q}\{rank(w, q_i)\}$.
- **Minimum:** $ARank(w, Q) = Min_{q_i \in Q}\{rank(w, q_i)\}$.

There are many other possible definitions for ARank(w,Q). We considered the above because they are the most likely to be used in real applications. Suppose that there is a set of products offered by a manufacturer and we want to help them find the most potential buyers. Then, the above three evaluating functions correspond to the following requests:

Sum: find buyers who more strongly believe that this product set is better than other people. **Maximum/Minimum:** find buyers who more strongly believe that the best/worst product in this set is better than other people.

The rest of this paper only focuses on **Sum AR-k** because **Maximum** and **Minimum** can be solved simply by using the technique of the existing reverse k-rank query. From a technical point of view, for maximum score, let q' be the query of Q such that $f(w, q') = \max_{q_i \in Q}\{f(w, q_i)\}$ with respect to w, then the rank of q', $rank(w, q')$, is also equal to $Max_{q_i \in Q}\{rank(w, q_i)\}$. Thus, we can process Maximum AR-k simply by applying the reverse k-rank query to q'. **Minimum** can be solved in a similar manner.

4 Bounding the Query Set in Advance

A naive solution to an AR-k query is to sum up the ranks for $q \in Q$ one by one against each $w \in W$ and $p \in P$. This is inefficient, especially when Q is large. Our idea is to bound the query set Q with respect to W. In this section, we introduce a sophisticated method of bounding Q with two points $Q.up$ and $Q.low$ from a subset of W. Denoted by $W_t = \{w_t^{(i)}\}_1^d$, this subset is the set of *top-weighting vectors* for all dimensions, as defined in the following,

Definition 4 *(top-weighting vector). Given a set of weighting vector W, let e_i be the direction vector for dimension i such that $e_i[i] = 1$ and $e_i[j] = 0, i \neq j$ and let $cos(a, b) = a \cdot b/(|a||b|)$ be the cosine similarity between vectors a and b. The top-weighting vector for dimension i is defined by $w_t^{(i)}$ where $w_t^{(i)} \in W$ and $\forall w \in W, cos(w_t^{(i)}, e_i) \geq cos(w, e_i)$.*

W_t can be found before the query set Q is issued, so it can be considered as cost less in terms of query processing. Because W_t contains the border of the weighting vector in all dimensions, we can use it to find the upper border and lower border points set of Q.

Definition 5 *(upper and lower border query sets Q_u and Q_l). Given a d-dimensional query points set Q,*

$Q_u = \{q_i | q_i \in Q \wedge \forall q_j \in Q, \exists w_t^{(i)} \in W_t, f(w_t^{(i)}, q_i) \geq f(w_t^{(i)}, q_j)\}$ *and*
$Q_l = \{q_i | q_i \in Q \wedge \forall q_j \in Q, \exists w_t^{(i)} \in W_t, f(w_t^{(i)}, q_i) \leq f(w_t^{(i)}, q_j)\}$.

By the definition, for each $w_t^{(i)}$ there is a corresponding $q_i \in Q_u$ (Q_l) such that q_i's score with respect to $w_t^{(i)}$ is the largest (smallest) among Q. Different $w_t^{(i)}$ may correspond to a same q_i and vice versa. Generally, it is easy to find the minimum bounding rectangle (MBR) of a point set X, and let its upper-right and lower-left corners be MBR(X).up and MBR(X).low, respectively. We show below that $Q.up = $ MBR(Q_u).up and $Q.low = $ MBR(Q_l).low bound the query set Q for the AR-k query.

Figure 3 shows the geometric view for the example of $Q.low$ and $Q.up$ where $Q = \{q_1, q_2, q_3\}$. $w_t^{(1)} = w_5$ and $w_t^{(2)} = w_1$ are the top-weighting vectors in dimensions 1 and 2, respectively. Each $w_t^{(i)}$ is also a normal vector of the hyper-planes $H(w_t^{(i)})$. For $Q.up$, in 2-dimensional space, the hyper-planes $H(w_t^{(1)})$ are the dashed lines l_1 which are perpendicular to $w_t^{(1)}$. By sweeping l_1 parallelly from far infinity toward the original point $(0,0)$, q_1 is the first point that is touched. Hence, q_1's score with respect to w is equal to $max_{q \in Q} f(w_t^{(1)}, q)$, and q_1 is included in Q_u. In the same manner, l_2 touches q_3 first, so $q_3 \in Q_u$. $Q.up = $ MBR(Q_u).up upper-bounds the scores for Q_u. Similarly, sweeping the perpendicular dashed lines l_3 and l_4 from $(0,0)$ toward infinity both touch q_2, hence $Q_l = \{q_2\}$ and $Q.low = q_2$.

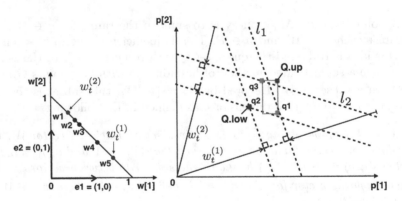

Fig. 3. A 2-dimensional example. $w_t^{(1)} = w_5$, $w_t^{(2)} = w_1$ and $Q_u = \{q_1, q_3\}$, $Q_l = \{q_2\}$, $Q.low = MBR(Q_l).low = q_2$, $Q.up = MBR(Q_u)$.up

Theorem 1 *(Correctness of Q.up and Q.low). Given top-weighting vectors set W_t, the d-dimensional query point set Q, $Q.up$ and $Q.low$. For $w \in W$ and $q \in Q$, $f(w, Q.low) \leq f(w, q) \leq f(w, Q.up)$ always holds.*

Proof. By contradiction. For $Q.up$, assume that $\exists q \in Q, q \notin Q_u$ holds so that $f(w, q) \geq f(w, Q.up)$. Therefore, $\exists q[i] > Q.up[i], i \in [1, d]$, so there must exist a $w_t^{(j)} \in W_t, j \in [1, d]$ that makes $f(w_t^{(j)}, q)$ the maximum value, and q should in Q_u. This leads to the contradiction.[1] A similar contradiction occurs with $Q.low$.

---

[1] The geometric view is that there exists a hyper-plane $H(w_t^{(j)})$ that first touches q rather than others.

We can use the rank of $Q.low$ to infer the bounds of the aggregate rank of Q.

Lemma 1 *(Aggregate rank bounds of Q for w): Given a set of query points Q and a weighting vector w, the lower bound of $ARank(w,Q)$ is $|Q| \times rank(w, Q.low)$, and the upper bound of $ARank(w, Q)$ is $|Q| \times rank(w, Q.up)$.*

Proof. $\forall q_i \in Q$, $\forall w[i] \geq 0$, it holds that $f(w, q_i) \geq f(w, Q.low)$ hence $rank(w, q_i) \geq rank(w, Q.low)$. By definition, $ARank(w, Q) = \sum_{q_i \in Q} rank(w, q_i) \geq |Q| \times rank(w, Q.low)$. Similarly, $|Q| \times rank(w, Q.up)$ is the upper bound of $ARank(w, Q)$.

Having W_t, the time cost of finding $Q.low$ and $Q.up$ is reduced from $O(|Q| \times |W|)$ to only $O(|Q| \times d)$, where d is the dimension of data. Considering that $|Q| \times d$ is much smaller than the size of the data set, the overhead of finding $Q.low$ and $Q.up$ is very small.

5   Tree-Pruning Method (TPM)

To enhance efficiency, our first approach, which is the *tree pruning method (TPM)*, indexes the data set P with the R-tree to group similar points and uses the bounds of MBRs (i.e., the R-tree entries) to reduce computing costs.

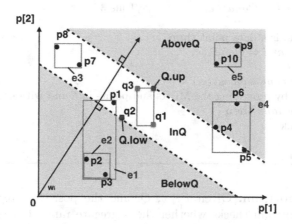

Fig. 4. The partitioned space of BelowQ, InQ and AboveQ based on $Q.low$ and $Q.up$ with a single w_i in 2d space of data set P.

First, we introduce how TPM filters P with $Q.low$ and $Q.up$. Figure 4 shows the geometric view for an example of two-dimensional data. The two dashed lines cross the boundaries ($Q.low$ and $Q.up$), and they are perpendicular to the weighting vector w_i. The space is partitioned into three parts, which are marked as $BelowQ$, InQ, and $AboveQ$ in Fig. 4. For example, e_2 is in $BelowQ$ and e_5 is in $AboveQ$. MBRs in $BelowQ$ and $AboveQ$ can be filtered by checking the upper and lower boundaries. Formally, the pruning rules are as follows.

- *Rule* 1. (MBR in *BelowQ*) If $f(w, e_p.up) < f(w, Q.low)$, count the number of points in e_p because $\forall p \in e_p, \forall q \in Q, f(w, q) > f(w, p)$ holds.
- *Rule* 2. (MBR in *AboveQ*) If $f(w, e_p.low) > f(w, Q.up)$, then discard e_p because $\forall p \in e_p, \forall q \in Q, f(w, q) < f(w, p)$ holds.
- *Rule* 3. (MBR in *InQ*) If $f(w, e_i.low) > f(w, Q.low)$ and $f(w, e_i.up) < f(w, Q.up)$, then add e_i to candidate for further examination.

Algorithm 1. ARank-P

Input: $P, w, Q, minRank$
Output: include: rnk; discard: -1;
1: $rnk \Leftarrow 0, Cand \Leftarrow \emptyset$
2: $heapP.enqueue(RtreeP.Root())$
3: **while** $heapP.isNotEmpty()$ **do**
4:    $e_p \Leftarrow heapP.dequeue()$
5:    **for** each $e_i \in e_p$ **do**
6:       **if** $f(w, e_i.low) < f(w, Q.up)$ **then**
7:          **if** e_i in $BelowQ$ **then**
8:             $rnk \Leftarrow rnk + e_i.size() \times |Q|$ //Rule 1
9:             **if** $rnk \geq minRank$ **then**
10:                **return** -1
11:          **else if** e_i in InQ **then**
12:             $Cand \Leftarrow Cand \cup e_i$ //Rule 3
13:          **else**
14:             **if** e_i is a data point **then**
15:                $Cand \Leftarrow Cand \cup e_i$
16:             **else**
17:                $heapP.enqueue(e_i)$
18: Refine $Cand$ by processing the MBRs and points in $Cand$ with each q.
19: **if** $rnk \leq minRank$ **then**
20:    **return** rnk
21: **else**
22:    **return** -1

ARank-P Algorithm. Given P, w, Q, and the positive integer $minRank$, the *ARank* algorithm checks whether the aggregate rank of Q is smaller than the given $minRank$. It also returns the value of the aggregate rank when $ARank(w, Q) < minRank$. As shown by Algorithm 1, *ARank* uses the R-tree to prune similar points in a group (MBR). In this algorithm, the counter rnk is used to count the aggregate rank of Q (Line 1). Then, the algorithm recursively checks the MBRs in the R-tree of P from the root (Line 2). If e_i belongs to $BelowQ$, the counter rnk is increased by $e_i.size() \times |Q|$ (Lines 7–8) based on Lemma 1. When rnk becomes greater than $minRank$, the algorithm returns -1 to terminate (Lines 9–10). If e_i in InQ, we add e_i into the candidate set $Cand$ for refinement (Lines 11–12). In other situations, when a leaf node of entries is encountered, the point is added into $Cand$ for refinement (Lines 14–15). Otherwise, e_i is added to the queue (Line 17). After traversal of RtreeP, refinement

---

Algorithm 2. Tree-Pruning Method (**TPM**)

Input: P, W, Q

Output: result set *heap*

1: initialize *heap* with first k weighting vectors and aggregate ranks of $|Q|$
2: $minRank \Leftarrow heap$'s last rank.
3: **for** each $w \in W-$ {first k element in W} **do**
4:    $rnk \Leftarrow$ ARank-P$(P, w, Q, minRank)$
5:    **if** $rnk \neq -1$ **then**
6:       $heap.insert(w, rnk)$
7:       $minRank \Leftarrow$ last rank of *heap*.
8: **return** *heap*

---

is performed where the $Cand$ set is checked for each $q \in Q$ and rnk is updated (Line 18). Note that $Cand$ contains both the MBR and single p in the space part of InQ. The refinement also considers the upper and lower bounds of the MBR to filter each q. Finally, rnk is the aggregate rank if $rnk < minRank$ or -1 is returned, which indicates that the current w is not a result.

TPM Algorithm. The TPM algorithm first initializes *heap* with the first k weighting vectors and their aggregate ranks of Q (Line 1). Then, for the other weighting vectors, the ARank-P Algorithm is called to check the aggregate rank of the query set Q (Line 4). If the current w can make the rank of Q better than the last rank in *heap*, this w is inserted into *heap* with its rank. Then, *heap* automatically updates itself by removing the last element and inserting a new w and aggregate rank while keeping the sorted order of rank (Line 6). Then, $minRank$ is updated by the last rank in the updated *heap* (Line 7). Eventually, the algorithm returns *heap* as the result of the aggregate reverse k-rank query.

6 Double-Tree Method (DTM)

TPM uses an R-tree to manage similar p and avoid computing with MBRs. However, TPM is limited in that it evaluates each w one by one, and its efficiency declines when the W set is large. This limitation inspired us to remove redundant computing by grouping similar w. We propose the double-tree method (**DTM**), which also indexes W set in an R-tree. The R-trees for P and W are denoted as *RtreeP* and *RtreeW*, respectively. Figure 5 shows the three parts of $BelowQ$, InQ and $AboveQ$, which are separated by the bounds of the MBR e_w in *RtreeW* and $Q.up$ (Q.low). Based on the MBR features in *RtreeP* and *RtreeW*, we can obtain the score bounds of a single data point on the MBR e_w of *RtreeW*.

Lemma 2 (*Score bound of p*): *Given an MBR with the weighting vector e_w in RtreeW and $p \in P$, the score $f(w, p)$ is lower-bounded by $f(e_w.low, p)$ and upper-bounded by $f(e_w.up, p)$.*

Proof. For $w \in e_w$, $\forall w[i] \geq e_w.low[i]$ holds, so $\sum_{i=1}^{d} e_w.low[i] \cdot p[i] \leq \sum_{i=1}^{d} w[i] \cdot p[i]$ hence $f(w, p) \geq f(e_w.low, p)$. Similarly, $f(w, p) \leq f(e_w.up, p)$.

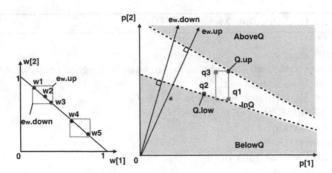

Fig. 5. The space part of BelowQ, InQ and AboveQ based on $Q.low$ and $Q.up$ with a MBR e_w in 2d space of data set P.

The score bounds of the MBR e_p of $RtreeP$ based on e_w of $RtreeW$ can also be inferred from the following lemma.

Lemma 3 (*Score bound of MBR*): *Given the MBR e_w of RtreeW and MBR e_p of RtreeP, the score of every $p \in e_p$ is lower-bounded by $f(e_w.low, e_p.low)$ and upper-bounded by $f(e_w.up, e_p.up)$.*

Proof. For $p \in e_p$, $\forall i, p[i] \leq e_p.low[i]$ holds based on the proof in Lemma 2, so $\sum_{i=1}^{d} e_w.low[i] \cdot e_p.low[i] \leq \sum_{i=1}^{d} e_w[i].low \cdot p[i] \leq \sum_{i=1}^{d} w[i] \cdot p[i]$. Hence, $f(w, p) \geq f(e_w.low, e_p.low)$. Similarly, $f(w, p) \leq f(e_w.up, e_p.up)$ holds.

Based on the above lemmas, we can build the bounds of the aggregate rank for Q on the MBR e_w.

Theorem 2 (*Aggregate rank bounds of Q for e_w*): *Given the set of query points Q and the MBR of the weighting vector e_w, the lower bound of rank for every $w \in e_w$ is $|Q| \times rank(e_w.low, Q.low)$, and the upper bound of $ARank(w, Q)$ is $|Q| \times rank(e_w.up, Q.up)$.*

Proof. This is similar to the proof for Lemma 1.

The ARank-P algorithm checks the rank of Q with the single w. This time, we propose using ARank-WP to check a group of w, e_w. For e_w, Algorithm 3 helps check these $w \in e_w$ with Q and $minRank$. The algorithm returns 1 if all $w \in e_w$ make the Q rank in $minRank$ and returns -1 if none of $w \in e_w$ makes Q rank better than $minRank$. The algorithm returns 0 if it needs to check the child entries of e_w.

Unlike the TPM algorithm in Sect. 5, DTM uses two R-trees to index the P and W. Hence, it can prune both the weighting vectors and points. Algorithm 4 starts from the root of $RtreeW$ and calls Algorithm 3 to check the aggregate rank of Q on e_w (Line 9). If $flag$ is 0, all child MBRs are added to $heapW$ for the next loop (Lines 10–11). If $flag$ is 1, this means that every w in e_w makes Q rank better than $minRank$. Thus, we can call Algorithm 1 to compute the

---

Algorithm 3. ARank-WP

Input: $P, e_w, Q, minRank$
Output: include: 1; discard: -1; uncertain : 0;
1: $rnk \Leftarrow 0, Cand \Leftarrow \emptyset$
2: $heapP.enqueue(RtreeP.root())$
3: **while** $heapP.isNotEmpty()$ **do**
4:    $e_p \Leftarrow heapP.dequeue()$
5:    **for each** $e_i \in e_p$ **do**
6:       **if** $f(e_w.low, e_i.low) < f(e_w.up, Q.up)$ **then**
7:          **if** e_i in $BelowQ$ **then**
8:             $rnk \Leftarrow rnk + e_i.size() \times |Q|$
9:             **if** $rnk \geq minRank$ **then**
10:                **return** -1
11:          **else if** e_i in InQ **then**
12:             $Cand \Leftarrow Cand \cup e_i$
13:          **else**
14:             **if** e_i is a data point **then**
15:                $Cand \Leftarrow Cand \cup e_i$
16:             **else**
17:                $heapP.enqueue(e_i)$
18: Refine $Cand$ and process the MBRs and points in $Cand$ with each q.
19: **if** $rnk \leq minRank$ **then**
20:    **return** 1
21: **else**
22:    **return** 0

Algorithm 4. Double-tree method (**DTM**)

Input: P, W, Q
Output: result set $heap$
1: initialize $heap$ with the first k weighting vectors and the aggregate ranks of $|Q|$
2: $minRank \Leftarrow heap$'s last rank.
3: $heapW.enqueue(RtreeW.root())$
4: **while** $heapW.isNotEmpty()$ **do**
5:    $e_w \Leftarrow heapW.dequeue()$
6:    **if** e_w is a single weighting vector **then**
7:       call the function ARank-P and update $minRank$.
8:    **else**
9:       $flag \Leftarrow$ ARank-WP$(P, e_w, Q, minRank)$
10:       **if** $flag = 0$ **then**
11:          $heapW.enqueue($all subMBR $\in e_w)$
12:       **else**
13:          **if** $flag = 1$ **then**
14:             **for each** $w \in e_w$ **do**
15:                call the function ARank-P and update $minRank$.
16: **return** $heap$

---

rank of each w in e_w and update *heap* and *minRank* (Lines 14–15). When the leaf node of a single w is being checked, Algorithm 1 is called just like in TPM (Lines 6–7). When the algorithm terminates, *heap* is returned as the result of the aggregate reverse rank query.

Table 1 summarizes the comparison of space and time complexities for NA (naive) and the proposed TPM and DTM. NA has the highest cost in terms of time complexity because $O(|P| \cdot |W|)$. However, it requires no extra index and only needs $O(k)$ space complexity. The proposed TPM and DTM algorithms need space to store the R-tree but have lower computation costs.

Table 1. Time complexity, space complexity for algorithms NA, TPM and DTM.

Algorithm	Index	Time complexity	Space complexity
NA	None	$O(|P| \cdot |W|)$	$O(k)$
TPM	RtreeP	$O(|W| \cdot \log |P|)$	$O(\log |P|)$
DTM	RtreeP, RtreeW	$O(\log |W| \cdot \log |P|)$	$O(\log |P| + \log |W|)$

7   Experiment

We present the experimental evaluation of the naive, TPM, and DTM algorithms for AR-k. All algorithms were implemented in C++, and the experiments were run on a Mac with 2.6 GHz Intel Core i7, 16 GB RAM. The page size was 4K.

Data Set. Both synthetic and real data were employed for the data set P. The synthetic data sets were uniform (UN), clustered (CL) and anti-correlated (AC) with an attribute value range of $[0, 1)$ that were generated as in [13, 18]. We also performed comparison experiments on two real data sets: HOUSE and NBA[2]. HOUSE contains 201760 six-dimensional tuples and represents the annual payments of American families (gas, electricity, water, heating, insurance, and property tax) in 2013. NBA is a 20960-tuple data set of box scores of players in the NBA from 1949 to 2009. We extracted the NBA statistics for points, rebounds, assists, blocks, and steals to form a 5-d vector that represents a player. For data set W, we also had the UN and CL data sets, which were generated in the same manner as the data sets of P. We generated Q by using clustered data.

Experimental Results for Synthetic Data. Figure 6 shows the experimental results for the synthetic data sets (UN, CL, AC) with varying dimensions d (2–5), where both data sets P and W contained 100 K tuples. Q had five query points, and we wanted to find the five best preferences ($k = 5$) for this Q. Figures 6a–c show that TPM and DTM were at least 10 times faster than NA in terms of CPU time. DTM performed the best because it skipped checking each p and w and was stable for all dimensional cases. Tree-based methods perform less querying for CL data than other data distributions because it is easier to index clustered data with the R-tree. Figures 6d–f show that DTM had less I/O usage

---

[2] NBA: http://www.databasebasketball.com/; HOUSE: https://usa.ipums.org/usa/.

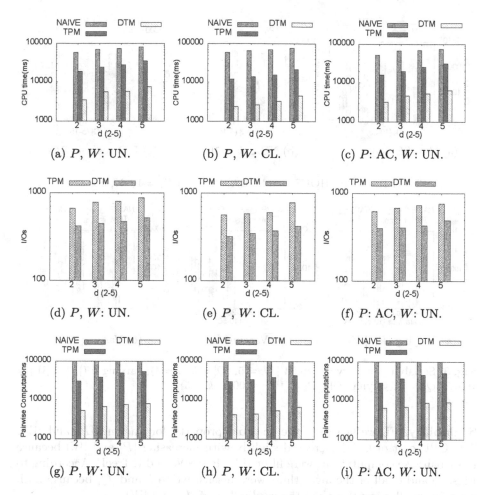

Fig. 6. Comparison results of CPU time (a, b and c), I/O cost (d, e, f) and Pairwise computations (g, h and i) on synthetic data, $|P| = |W| = 100K$, all with $|Q| = 5, k = 10$.

than TPM for all kinds of data. Figures 6g–i show pairwise computations with p and w for calculating the scores. DTM needed fewer computations because it can prune both points and weighting vectors with double R-trees.

Experimental Results for Real Data. Figure 7a shows the performance with the HOUSE data set and different k (10–50). DTM again performed the best. We found that the major dimensions of HOUSE were similar to an exponential distribution. The NBA data set was used to solve another practical query: who likes a team more than others? We selected five, ten, and fifteen players from the same team as Q and then generated the data set W as various user preferences. As expected, DTM found the answer the fastest. Figure 7c shows the I/O cost of the two proposed tree-base algorithms (TPM and DTM). DTM required less I/O usage.

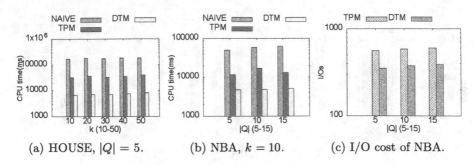

(a) HOUSE, $|Q| = 5$.   (b) NBA, $k = 10$.   (c) I/O cost of NBA.

Fig. 7. Real data, HOUSE and NBA, W:UN, $|W| = 100K$, $k = 10$.

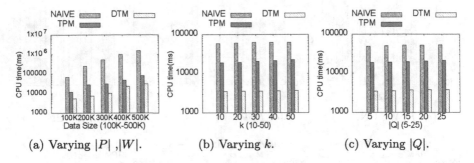

(a) Varying $|P|$,$|W|$.   (b) Varying k.   (c) Varying $|Q|$.

Fig. 8. Scalability on varying $|P|$,$|W|$ ($k = 5$, $|Q| = 5$, $d = 4$); varying k ($|P| = |W| = 100K$, $|Q| = 5$); varying $|Q|$ ($|P| = |W| = 100K$, $k = 5$).

Scalability. Figure 8a shows the scalable property for varying $|P|$ and $|W|$. The CPU cost of DTM increased slightly with increasing $|P|$ and $|W|$ because most pairwise computations were filtered by R-treeW and R-treeP. According to Figs. 8b and c, all of the algorithms were insensitive to k and $|Q|$ because both were far smaller in value than the cardinality of $|P|$ and $|W|$.

8   Conclusion

Reverse rank queries have become important tools in marketing analyzing. However, related research on reverse rank queries has only focused on one product. We propose the aggregate reverse rank query to address the situation of multiple query products for applying to the product bundling. We devised the TPM and DTM methods for efficient querying. TPM is a tree-based pruning method that prunes unnecessary products with the help of an R-tree. DTM uses two R-trees to manage products and user preferences and prune both of them. We compared the methods through experiments on both synthetic data and real data and the results show that DTM is the most efficient one.

As future work, we first plan to investigate approaches for other ARank functions, such as evaluating the aggregate rank by the harmonic average of each rank. We also want to consider approximate solutions for AR-k queries.

References

1. Chang, Y.-C., Bergman, L.D., Castelli, V., Li, C.-S., Lo, M.-L., Smith, J.R.: The onion technique: Indexing for linear optimization queries. In: SIGMOD Conference, pp. 391–402 (2000)
2. Cheema, M.A., Lin, X., Zhang, W., Zhang, Y.: Influence zone: efficiently processing reverse k nearest neighbors queries. In: Proceedings of the 27th ICDE 2011, pp. 577–588 (2011)
3. Dellis, E., Seeger, B.: Efficient computation of reverse skyline queries. In: Proceedings of the 33rd International Conference on VLDB, pp. 291–302 (2007)
4. Ilyas, I.F., Beskales, G., Soliman, M.A.: A survey of top-k query processing techniques in relational database systems. ACM Comput. Surv. **40**(4), 11:1–11:58 (2008)
5. Korn, F., Muthukrishnan, S.: Influence sets based on reverse nearest neighbor queries. In: Proceedings of the ACM SIGMOD, pp. 201–212 (2000)
6. Lian, X., Chen, L.: Monochromatic and bichromatic reverse skyline search over uncertain databases. In: Proceedings of the ACM SIGMOD, pp. 213–226 (2008)
7. Stanoi, I., Agrawal, D., El Abbadi, A.: Reverse nearest neighbor queries for dynamic databases. In: ACM SIGMOD Workshop, pp. 44–53 (2000)
8. Tao, Y., Papadias, D., Lian, X.: Reverse knn search in arbitrary dimensionality. In: Proceedings of the 13th International Conference on VLDB, pp. 744–755 (2004)
9. Tao, Y., Papadias, D., Lian, X., Xiao, X.: Multidimensional reverse k NN search. VLDB J. **16**(3), 293–316 (2007)
10. Vlachou, A., Doulkeridis, C., Kotidis, Y., et al.: Reverse top-k queries. In: ICDE, pp. 365–376 (2010)
11. Vlachou, A., Doulkeridis, C., Kotidis, Y.: Identifying the most influential data objects with reverse top-k queries, pp. 364–372 (2010)
12. Vlachou, A., Doulkeridis, C. , Kotidis, Y., et al.: Monochromatic and bichromatic reverse top-k queries, pp. 1215–1229 (2011)
13. Vlachou, A., Doulkeridis, C., Kotidis, Y., et al.: Branch-and-bound algorithm for reverse top-k queries. In: SIGMOD Conference, pp. 481–492 (2013)
14. Vlachou, A., Doulkeridis, C., et al.: Monitoring reverse top-k queries over mobile devices. In: MobiDE, pp. 17–24 (2011)
15. Yang, S., Cheema, M.A., Lin, X., Wang, W.: Reverse k nearest neighbors query processing: experiments and analysis. Proc. VLDB Endowment **8**, 605–616 (2015)
16. Yang, S., Cheema, M.A., Lin, X., Zhang, Y.: SLICE: reviving regions-based pruning for reverse k nearest neighbors queries. In: IEEE 30th International Conference on Data Engineering, ICDE, pp. 760–771 (2014)
17. Yao, B., Li, F., Kumar, P.: Reverse furthest neighbors in spatial databases. In: Proceedings of the 25th ICDE, pp. 664–675 (2009)
18. Zhang, Z., Jin, C., Kang, Q.: Reverse k-ranks query. PVLDB **7**(10), 785–796 (2014)

Abstract-Concrete Relationship Analysis of News Events Based on a 5W Representation Model

Shintaro Horie[1]([✉]), Keisuke Kiritoshi[2], and Qiang Ma[1]

[1] Kyoto University, Kyoto 606-8501, Japan
`horie@db.soc.i.kyoto-u.ac.jp`, `qiang@i.kyoto-u.ac.jp`
[2] NTT Communications, Atsugi, Japan
`k.kiritoshi@ntt.com`

Abstract. In a follow-up news article, description of previous news events may be abbreviated or summarized. This feature makes news article difficult to understand if the reader has no knowledge about the previous events. In such a case, providing concrete and detailed descriptions is helpful. In this paper, we propose a five element, who, what, whom, when, and where (5W) model and extraction method with completion functionality. With this model, a news event is represented using these 5Ws. To discover abstract and concrete descriptions of a given event, we propose the novel concept of abstractiveness based on this model. The abstractiveness of an event description is defined based on the difficulty of imagining and identifying that event. Currently, we estimate the abstractiveness of an event by considering the abstract levels and comprehensivity of its 5Ws to identify that event. We also propose a method for estimating the abstractiveness of an event and analyzing the abstract-concrete relationships between news events based on the 5W model. The experimental results indicate that our model, concept, and method are effective for extracting a concrete event description.

Keywords: News event analysis · Abstractiveness · Understanding support · Relationship analysis

1 Introduction

News is an important information source for personal and business activities. Supporting readers' news understanding is an important challenge. Many methods for supporting such understanding, for example, suggesting a related article, have been proposed [1–3].

In some news articles that are follow-up reports of a certain topic, the background and previous information may be briefly introduced in a short description. In such a case, it is not easy to understand if readers have any knowledge about that topic. An example is shown as follows.

© Springer International Publishing Switzerland 2016
S. Hartmann and H. Ma (Eds.): DEXA 2016, Part II, LNCS 9828, pp. 102–117, 2016.
DOI: 10.1007/978-3-319-44406-2_9

"The Park Geun-hye administration is drawing flak for its poor response to the Middle East respiratory syndrome outbreak, despite rising public concerns of the surging number of confirmed or suspected patients."[1]

This sentence states that President Park's handling of MERS is a "poor response". If readers have not read about this topic previously, they would not understand what "poor response" means. In addition, since there is no concrete and detailed information about the handling of MERS, it is not easy to know the actions of President Park regarding MERS and why the reporters claimed these actions are "poor responses". It is difficult to understand what actually happened.

We define an event as a certain entity's action and represent that event by using five elements: *Who, What, Whom, When,* and *Where* (5Ws). We propose a 5W model and extraction method based on these elements to support readers' understanding by clarifying what happened and investigated ways to discover abstract and concrete descriptions of a given event. We thus propose the concept of abstractiveness. The abstractiveness of an event description is defined based on the difficulty of imagining and identifying that event. We also propose a method for estimating the abstractiveness level of an event and analyzing the abstract-concrete relationship between events.

The major contributions of this paper are summarized as follows.

1. We propose the 5W model to represent news event. We model an event into five elements of *Who, What, Whom, When,* and *Where*. We also propose an element-extraction method based on dependency parsing and completing each element. We complement *Who* and *Whom* elements by applying co-reference resolution and *Where* and *When* elements by applying clustering methods based on latent Dirichlet allocation (LDA) (Sect. 3).
2. The novel concept of *abstractiveness* and its estimating method are proposed. The method is used to analyze the abstract-concrete relationship between two events. With this method, we first estimate whether events denote the same concrete event by comparing the 5Ws of their descriptions. Then, we compare their abstractivenss levels to analyze their relationship. The experimental results show that this method is effective for estimating the abstractiveness of an event (Sect. 4).

2 Related Work

Many systems and methods have been proposed for supporting readers' news understanding. A typical method is recommending news article to support understanding. NewsCube analyzes and provides multiple aspects of news events to support news understanding [1]. Kiritoshi et al. [3,4] proposed a method for supporting news understanding by gathering diverse information for a news event.

---

[1] http://www.koreaherald.com/view.php?ud=20150603001128.

Mihalcea et al. [5] presented an automatic keyword annotation system called Wikify!, which automatically extracts keywords and links them to relevant pages in Wikipedia or another encyclopedia. Wikify helps readers understand the entities mentioned in news articles. Like Wikify!, NewsStand answers the question, "Where did a certain event happen?" or "What is happening at a certain location?" by associating news articles with a particular location [2]. Their research is similar to ours in that the extraction target is not an article but an event and involves estimation of geolocation by using words in article as clues. Event extraction is part of an information extraction task on the Web and widely adapted for various domains [6]. Radinsky et al. [7] analyzed the causal relationship among events by using event extraction to clarify what caused an event. They expanded property exemplification of events theory [10], which represents an event with objects such as actors, instruments, actions, time, and location. We assume our concept of event abstractiveness is represented by elements constructing an event. Therefore, we represent an event with our 5W model to estimate the abstract-concrete relationships among events.

Tanaka et al. [8] proposed a method for estimating the concrete level of a Web page by estimating the term concreteness. Concreteness levels of terms are estimated using the Medical Research Council Psycolinguistic Database² on the basis of concreteness and image-ability defined by Allan et al. [11]. Whereas they try to estimate document concreteness by using the concreteness levels of terms, our purpose was to estimate event abstractiveness.

3   5W Model

We represent an event description by using the 5Ws: *Who, Whom, What, When,* and *Where* for analyzing the abstractiveness of an event description. Our 5W model is represented on the basis of our assumption that an event is represented by an entity conducting a certain action, the action, target entity of the action, and spatiotemporal information.

Given an occurrence as a sentence, we decompose the sentence into the associate subject, object, verb, time, and location as *Who, Whom, What, When,* and *Where* elements, respectively. Basically, we decompose these elements by using dependency parsing, i.e., we obtain a word dependency relation through predicates such as the subject (*nsubj*) or direct object (*dobj*). For example, the sentence "Fumio arranging Seoul visit to settle 'comfort women' row." is parsed as ⟨*Fumio, nsubj, arranging*⟩ and ⟨*arrangin, dobj, visit*⟩. We use this relation to obtain the subject and object relationship through the verb.

The element What is represented using the verb of the target sentence and is not abbreviated. On the other hand, the other elements (Who, Whom, Where, and When) may be abbreviated because we may infer them with the context and surrounding text. Hereafter, we explain how to complete these elements when they are abbreviated.

---

² http://www.psych.rl.ac.uk.

Completion of Elements Who and Whom: Elements Who and Whom are completed by applying co-reference resolution methods. Currently, we use the Standford Core NLP [12] tool. Additionally, we use Accurate Online Disambiguation of Entities (AIDA) [9] to disambiguate.

Completion of Elements When and Where: When temporal (When) and/or location (Where) information are abbreviated, based on the assumption that descriptions with the same context will share the same spatiotemporal information, we complete the elements When and Where with a clustering method on the basis of the topic distribution obtained from LDA [13].

Please note that if the time and location information have been mentioned before and there is no change, writers do not describe them again and again and readers can infer them from the context. However, sometimes, the spatiotemopral information is not mentioned just because it is an abstract description. In our 5W model, to represent an event, if the spatiotemporal information is not mentioned, we first try to use the context to complete it. If we cannot detect the spatiotemporal information, no values will be assigned to When and Where elements.

The details of our method for extracting and completing the elements When and Where are described in Algorithm 1. With element extraction, if a sentence has temporal expression and location terms, we extract them as When and Where elements (line 5). If we cannot find sptaiotemporal information of the event from the sentence, we try to infer them by using the context information (lines 6–18). Context means a cluster containing sentences that describe the same topics. The idea is simple, the events described in the same cluster (context) share sptatiotemporal information (line 3 in Algorithm 1). For clustering, we apply LDA to sentences as documents then cluster sentences by using the obtained topic distribution. By estimating the number of topics for LDA, we calculate *perplexity* [13] with 5-fold cross-validation and use the lowest topic number.

Given the topic distributions of each sentence, sentences are clustered by *spectral clustering* [15] based on topic distribution. The reason we use spectral clustering is that we use the Jensen-Shannon divergence for calculating probabilistic distribution distance. Since we have to determine the number of clusters for using spectral clustering, we use the number of topics as that of clusters calculated before.

However, if we cannot find explicit spatiotemporal information from the cluster, we try to represent the elements When and Where with ranges.

In lines 19–21, we infer geolocation as including all geolocations in an article. To find the common place to include each location, we use Geonames[3] to obtain administrative districts (e.g. city and country) and create a tree structure. For including all locations in an article, we find the common place by using the lowest common ancestor (LCA) [16], described in line 20, of all location nodes. For example, if we obtain two locations such as New York City and Albany in an article, we obtain New York State as the LCA.

---

[3] http://www.geonames.org.

Algorithm 1. Estimating When and Where element

---

1 **Input:** Sentence set S, published date t_p, sentence cluster label set L
2 **Output:** Event set E
3 $C \leftarrow \texttt{Clustering}(S)$
4 **for** *sentence cluster* $S_c \in C$ **do**
5      $E_c \leftarrow \{e_i | e_i \leftarrow EventExtraction(s_i), s_i \in S_c\}$
6   **for** $e_i \in E_c$ **do**
7   **if** e_i^{Where} *is None or* e_i^{When} *is null* **then**
8   **for** $e_j \in E_c$ **do**
9   **if** e_i^{Where} *is null* **and** e_j^{Where} *is not null* **then**
10                      $e_i^{Where} \leftarrow e_j^{Where}$
11   **if** e_i^{When} *is null* **and** e_j^{When} **then**
12                      $e_i^{When} \leftarrow e_j^{When}$
13   **if** e_i^{Where} *is not null* **and** e_i^{When} *is not null* **then**
14   break
15   **if** e_i^{Where} *is null* **then**
16              $e_i^{Where} \leftarrow \texttt{LowestCommonAncestor}(\{location | location \in s_k, s_k \in S\})$
17   **if** e_i^{When} *is null* **then**
18              $begin \leftarrow min\{time | time \in s_k, s_k \in S\}, end \leftarrow t_p$
19              $e_i^{When} \leftarrow (begin, end)$

On the other hand, by inferring the largest period as possible, the beginning of the period is determined as the oldest date written in the article, and the published date of the article is adopted for the end of the period (line 23).

The distance between sentences s_1 and s_2 is calculated by normalizing each sentence topic distribution $\boldsymbol{\theta}_{s_1}$ and $\boldsymbol{\theta}_{s_2}$ and using the Jensen-Shannon divergence as follows.

$$Dist(s_1, s_2) = D_{JS}(\boldsymbol{\theta}_{s_1} || \boldsymbol{\theta}_{s_2}) \tag{1}$$

$$D_{JS}(\boldsymbol{\theta}_{s_1} || \boldsymbol{\theta}_{s_2}) = \frac{1}{2} D_{KL}(\boldsymbol{\theta}_{s_1} || M) + \frac{1}{2} D_{KL}(\boldsymbol{\theta}_{s_2} || M)$$

$$M = \frac{1}{2}(\boldsymbol{\theta}_{s_1} + \boldsymbol{\theta}_{s_2})$$

$$D_{KL}(\boldsymbol{\theta}_{s_1} || \boldsymbol{\theta}_{s_2}) = \sum_{t \in Topic} \boldsymbol{\theta}_{s_1}(t) log \frac{\boldsymbol{\theta}_{s_1}(t)}{\boldsymbol{\theta}_{s_2}(t)}$$

4   Analysis of Relationship Among Events

This section describes the methods of abstractiveness estimation and relationship classification to analyze abstractiveness-concreteness relationship.

4.1   Estimate of Abstractiveness

We define the abstractiveness of an event description as the difficulties to imagine and identify that event. Currently, we infer the abstractiveness of an event description by following two aspects, (1) ambiguity of the 5W elements and (2) sufficiency level of the 5W elements to identify.

(1) Ambiguity of 5W elements: The ambiguity of an element is estimated based on the range of meaning. In other words, estimating how many meanings each element have.

For example, for the *When* element, the following sentence mentions events from the 1990s to the present.

"Congress has not approved major gun-control legislation since the 1990s."[4]

Hence, the more widely a period is described for the event, the higher abstractiveness the element has.

(2) Sufficiency level of 5W elements to identify: The sufficiency level of the 5W elements denotes whether there are a sufficient amount of elements to identify that event. Note that the 5Ws have strong co-relationships with each other, and we do not need all elements to identify an event in many cases. For example, if we know the subject (*Who*) and exact time (*When*), it is possible to infer the location (*Where*) in many cases.

In other words, the 5Ws are dependent on each other, and we can refer to one element by using the others. Hence, we calculate the abstractiveness of an event by estimating how many events have elements and how much abstractiveness those elements have.

Abstractiveness of Who and Whom

– "Japan Foreign Minister arranging Seoul visit to settle 'comfort women' row."
– "Japan Foreign Minister Kishida says arranging visit to South Korea."[5]

The above sentences describe the same event. However, the subjects are different, i.e., "Japan Foreign Minister" and "Japan Foreign Minister Kishida". If only focusing on the subject, referencing the name is more concrete.

We consider the ambiguity of the *Who* or *Whom* from the above example. There have been many "Japan Foreign Ministers" in Japan, but "Japan Foreign Minister Kishida" is attached to only one person. We use a semantic class for representing with such relationships. For example, "Kishida" is considered an instance of the "Japan Foreign Minister" class. We estimate the abstractiveness of the subject and object by using this "instance-of" relationship obtained from ontology.

To obtain the "instance-of" relationship, we use the knowledge-base system called YAGO [14] developed from online knowledge resources (e.g. Wikipedia

---

[4] http://www.reuters.com/article/us-usa-obama-guns-idUSKBN0UM0AU20160108.
[5] http://uk.reuters.com/article/uk-japan-southkorea-idUKKBN0U801M20151225.

and WordNet). YAGO contains knowledge of more than 10 million entities and contains more than 120 million facts that represent relationships between entities. Moreover, classes have a hierarchical structure (e.g. the highest class of all named entities is "Entity".) We assume all subjects and objects are named entities and regard the element (Who or Whom) at the "Entity" class as the highest abstractiveness level (level 1). For example, let c be the element Who; thus, $abst_{Who}(c) = 1$ if and only if $class(c)$ is "Entity"

The function $abst_{Depth}$ to estimate abstractiveness is non-linearly decreased by the depth of elements. For example, the abstractiveness between "Government minister" and "Shinzo Abe" is different from that between "Person" and "Government minister", even if they have the same hierarchy in YAGO. We define the abstractiveness of the depth function as non-linear by introducing λ, which is the constant term for decreasing regularization.

The difference in depth between "Government minister" and "Shinzo Abe" is the same as that between "Person" and "Government minister". However, the difference in the abstractiveness of the latter pair should be greater than that of the former.

$$abst_{Depth}(c, \lambda) = \exp\left(-\frac{depth(c) - 1}{\lambda}\right), \tag{2}$$

where the depth function returns the depth of element c in a hierarchical structure. The abstractiveness of the Who and Whom elements is represented using this function as follows:

$$abst_{Who}(c) = abst_{Depth}(c, \lambda_{Who}) \tag{3}$$
$$abst_{Whom}(c) = abst_{Depth}(c, \lambda_{Whom}) \tag{4}$$

The constant term λ is defined respectively as λ_{Who} and λ_{Whom} because we assume that each element has a different rate of decrease. We assign a value to both terms as $abst = 0.5$ when the subject or object is the country class, as mentioned above.

Abstractiveness of What Element. The abstractiveness of the What element is calculated using the hierarchical structure of the semantics frame and ambiguity of frames to which the verb belongs.

A semantic frame, composed of the lexical database called Framenet[6], represents the semantic role in sentences with entities as participants. In Framenet, the concept of the verb is called *frame*, the entity evoking the frame is called *frame entity*, and the word belonging to a concept is called *lexical unit*. For example, the lexical unit "cook" has the "Cooking creation" or "Heating Instrument" concept, and this is called a frame. Usually, a lexical unit belongs to several

---

[6] https://framenet.icsi.berkeley.edu/fndrupal/.

frames. For example, "Cooking creation" inherits from the "Intentionally create" frame. There is a hierarchical relationship among certain frames; therefore, we use this hierarchical structure for calculating the abstractiveness level of a verb. Figure 1 shows the hierarchical structure of a frame.

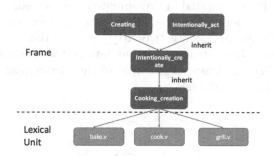

Fig. 1. Example of frame hierarchical structure

We define the abstractiveness of the What element with the depth of the hierarchical structure and number of frames as

$$Depth_{F_v} = \frac{1}{|F_v|} \sum_{f_i \in F_v} abst_{Depth}(depth(f_i), \lambda_{What}),$$

where v is the target verb for calculating abstractiveness, F_v is the frame set including v as a lexical unit, f_i is a member of F_v, $depth(f_i)$ represents the function that returns the depth of f_i, and $abst_{Depth}$ is the abstractiveness of the depth function defined in Eq. (2). This formula is used to calculate the mean abstractiveness as the depth of frames in F_v.

To consider semantic ambiguity, we also append expression $1 - \exp(-\frac{|F_v|-1}{\beta})$ to the abstractiveness of What $abst_{What}(v)$ to ensure that the abstractiveness is 0 when v belongs only to one frame and 1 when v belongs to the largest number of frames. β is a constant term adjusting as abstractiveness is 0.5 when v belongs to the average number of frames. The abstractiveness of the What element is defined by the depth of hierarchical structure and the semantic ambiguity as follows.

$$abst_{What}(v) = \alpha(1 - \exp(-\frac{|F_v| - 1}{\beta})) + (1 - \alpha)Depth_{F_v}, \tag{5}$$

where α is a constant term. For simplicity, we set $\alpha = 0.5$ to be the average among hierarchical structures and depth abstractiveness.

Abstractiveness of Where Element. We estimate the abstractiveness of Where by using GeoNames, which contains geo-information. It has over one million place names and has an administrative district class such as city or country.

We also construct a hierarchical structure by obtaining hierarchical relationships (e.g. "White House" located in "Washington") from GeoNames. We design the abstractiveness of location to be; the deeper the location depth, the less abstractiveness. For example, since "Country" is lower than "Earth" or "Continent" in this hierarchical structure, it has less abstractiveness than "Earth" or "Continent". Figure 2 shows an example of the hierarchical structure for a location obtained from GeoNames.

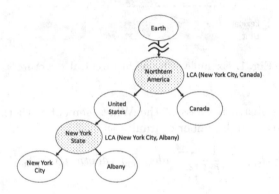

Fig. 2. Example of hierarchical structure in GeoNames

As mentioned in Sect. 4.1, there is a complemented location. The location or date information obtained by this complement should be higher in abstractiveness than those from explicit descriptions.

Therefore, the longer the distance $Dist(s_i, s_j)$, the higher the abstractiveness we assign. The $Dist(s_i, s_j)$ is the distance between sentences s_i and s_j from which events e_i and e_j are extracted, respectively, e_j is in the same cluster as e_i and e_j, referenced by e_i to estimate the *Where* element (Sect. 4.1).

The abstractiveness of Where $abst_{Where}$ is defined as follows.

$$abst_{Where}(s_1, s_2, l) = Dist(s_1, s_2)(abst_{Depth}(l, \lambda_{Where}) - 1) - abst_{Depth}(l, \lambda_{Where}), \tag{6}$$

where λ_{where} is defined as a constant term. We assign a value to constraint λ_{Where} as $abst_{Where} = 0.5$ when l is at the prefecture depth.

Abstractiveness of When Element. The abstractiveness of the *When* element is calculated by analyzing time duration. For example, "Mar, 2016" has much more abstractiveness than "on Mar 01, 2016". Currently, we count the number of days. Let t be the time span, then we define the abstractiveness of When $abst_{When}$ as

$$abst_{When}(s_1, s_2, t, \lambda_{When}) = \exp\left(-\frac{t-1}{\lambda_{When}}\right)(1 + Dist(s_1, s_2)) - 1, \quad (7)$$

where $Dist(s_i, s_j)$ is the distance between sentences s_i and s_j from which event e_i and e_j are extracted, respectively, e_j is the same cluster as e_i and e_j referenced by e_i to estimate the *When* element (Sect. 4.1). Moreover, λ_{When} is a constant term that needs to adjust the increase in abstractiveness. We assign a value to λ_{When} as $abst_{When} = 0.5$ when t is 1 month.

Abstractiveness of Event. As mentioned above, the 5Ws are closely related. Hence, if a certain element cannot be obtained, that element can be estimated from other elements if they have sufficiently low abstractiveness. We also have to consider the difference in the abstractiveness of every element for identifying what event occurred.

To simplify representation, we assume that; the higher abstractiveness of an event, the higher each included element is by formulation as the linear sum of each element. Therefore, we define the abstractiveness of an event as follows.

$$abst_{Event}(e_i) = \frac{1}{|E|} \sum_{c \in 5W} w_c abst_c(e_{i,c}). \quad (8)$$

The weights represent each element's contribution to the abstractiveness of the event, denoted as $w = (w_{Who}, w_{What}, w_{Whom}, w_{Where}, w_{When})$. The weights are assigned a value to represent the importance of each element to estimate abstractiveness. We assign them a higher value in the order of Who, When, Where, Whom, and What. Let the 5W set be $5W = \{Who, What, Whom, Where, When\}$, and an event e_i represented by the 5Ws as $e_i = (e_i^{Who}, e_i^{What}, e_i^{Whom}, e_i^{Where}, e_i^{When})$.

4.2   Relationship Analysis

Two event descriptions having an abstract-concrete relationship should share the same context and represent the same event. It is meaningless to estimate the relationship of descriptions on different events.

As mentioned above, we calculate the abstractiveness of events independently. However, the range of event abstractiveness strongly depends on the topic domain. Therefore, we compare two events based on abstractiveness only if they share context and denote the same concrete event.

Two descriptions sharing a context and denoting the same concrete event are classified using a support vector machine (SVM) through feature extraction to find event context.

The relationship is represented by the hierarchical structure of elements. For example, if an element e_i^c in event e_i is an ancestor of an element e_j^c in event e_j on the hierarchical structure, we consider that e_j^c is semantically a subset of e_i^c. Then we represent this relationship as $e_j^c \subseteq e_i^c$ and construct a binary set $R_{i,j} = \{\delta_{i,j}^c | \delta_{i,j}^c = \{0,1\}, c \in 5W\}$ to represent features of the relationship between events e_i and e_j. $\delta_{i,j}^c$ is defined as follows.

$$\delta_{i,j}^c = \begin{cases} 1, & \text{if } e_i^c \subseteq e_j^c \quad or \quad e_j^c \subseteq e_i^c \\ 0, & \text{otherwise.} \end{cases} \tag{9}$$

For the *When* element, we assume a time span t_i is a subset of t_j if t_i is completely included in t_j (e.g. "Mar 1, 2015" is included in "Mar, 2015"). Furthermore, for the *Who* and *Whom* elements, even though the hierarchical structure based on "instance-of" relationship represents only the generalization of an entity, this structure cannot represent the relation between named entities (e.g. "belong to" relationship such as President Obama belongs to the Democratic Party). For this reason, we use the linking relationship obtained from YAGO and assume the subject or object entity c_i is a subset of c_j, represented as $c_i \subseteq c_j$ and $\delta_{c_i,c_j} = 1$, if c_1 is linked from c_2.

In addition, we estimate the relationship between events by considering the distance. The distance is represented as sentence and document distances, which are sources of each event and calculated as $Dist(d_1, d_2)$, $Dist(s_1, s_2)$ by LDA and the Jensen-Shannon divergence.

5   Experiments

5.1   Evaluation of Events Classification

In this section, we discuss the evaluation of what features are effective for classification and whether it is feasible to estimate the abstractiveness of an event description.

A data set was constructed from three topics for evaluation. We selected each topic composed randomly of ten articles. We chose the news topics continuously reported as follows.

- Volkswagen scandal about CO2 emissions
- US Presidential election 2016
- Political topics about Shinzo Abe

All the articles were scraped from only The New York Times[7] to circumvent the effect of different forms of writing.

We created event pairs by selecting 10 query events and 30 relationship candidate events for each query. As a result, we obtained a total of 300 event pairs. Table 1 lists the events and source article topics. The details of this procedure are as follows.

---

[7] http://www.nytimes.com.

Table 1. Example of events

Event	News topics	Who	What	Whom	Where	When
1	Volkswagen scandal	Volkswagen	halt	sales	U.S	2015-09-20 - 2015-09-20
2	US Presidential Election	Donald Trump	created	pool	Iowa	2015-06-01 - 2015-09-30
3	Japan Political Topic	Shinzo Abe	told	Parliament	Japan	2015-01-19 - 2015-01-25

1. Obtain ten articles per topic.
2. Extract events from each article.
3. Choose ten query events from all extracted events
4. Randomly choose a candidate event having a containing a relationship between the subject and query event to form a pair between query and candidate events.

We labeled all 300 event pairs manually. As a result, there were 72 out of the 300 pairs having abstract-concrete relationships. We define a query event and candidate event as having an abstract-concrete relation if one event is a concrete example of the other or very similar to the other. Table 2 lists the average Area Under the ROC Curves (AUCs) of a feature combinations by 5-fold cross validation using these labeled data. In Table 2, d-dist and s-dist respectively represent the distance of documents and sentences calculated by the Jensen-Shannon distance of topic distribution. 5Ws represent the binary relationship feature of each element. From the results in Table 2, the combination that has the highest AUC is (*Who, Whom, When, document distance*).

The Receiver Operating Characteristic (ROC) curves are illustrated in Fig. 3, which were calculated using 70 % of the data set as a training set and rest as the test set. The figure illustrates that this combination's true positive rate is significantly higher than those of other combinations around the false positive rate of 0.1, indicated with red dashed line.

Table 2. Relation Classification result

Feature combination	AUC	Standard deviation
Who, Whom, When, d-dist	0.680430	0.072135
What, When, d-dist	0.663564	0.069450
Who, What, When, d-dist	0.660500	0.062955
What, Whom, When, d-dist	0.660500	0.062955
Who, What, Whom, When, d-dist	0.653965	0.076205
When, s-dist	0.652335	0.107128
Whom, When, s-dist	0.634689	0.122228
Whom, s-dist	0.626186	0.088462
Who, s-dist	0.626186	0.088462
Who, When, s-dist	0.625353	0.131786

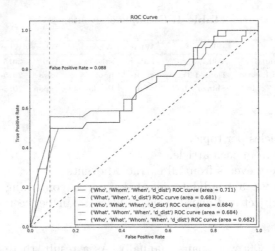

Fig. 3. ROC curves of top 5 feature combinations (Color figure online)

The results of the top 5 combinations show that they mutually include *When* and document-distance features, suggesting that these two features are very effective for classification.

Next, we investigated a misclassified event pair. During the investigation, the false positive rate was higher than the false negative rate. This was due to the use of the element-complete functionality of our estimation method. The following sentences were misclassified.

1. "And when he indulged in the pandering to Iowa institutions that is typical of political supplicants here, he did so in his exaggerated, almost comic style – as if he were playing the role of presidential candidate."
2. "But other said they think Trump should be focusing on the next contests."

They were extracted from the news article, "Trump Calls for Iowa Election Do-over", and "Ted Cruz Wins Republican Caucuses in Iowa". The misclassified events extracted from each sentence are listed in Table 3.

Table 3. Misclassified events

Event	Who	What	Whom	Where	When
1	Donald Trump	did	None	Iowa	2016-02-01 - 2016-02-01
2	Donald Trump	focusing	None	Iowa	2016-02-01 - 2016-02-01

As shown in Table 3, our event-extraction method yielded results in which these two events had the same Who, Whom, Where, and When elements. Hence, SVM-classified the two events as having an abstract-concrete relationship.

The subject of the two events was Donald Trump, but the subject of the source sentence started with "They Think". Therefore, these events were labeled as having no abstract-concrete relationship.

5.2   Evaluation of Events Abstractiveness

We constructed an event data set for evaluating abstractiveness. We treating 31 events including a query event as one set in the data set mentioned in Sect. 5.1. This event data set was scored on a 6-point scale, where 0 and 5 denote the lowest and highest abstractiveness level, respectively.

For the event-abstractiveness evaluation, we used the *word count score* method as the baseline. The word count score is the number of words in a sentence from an extracted event; the smaller words, the lower the event abstractiveness. The ranking results are listed in Table 4. nDCG (Normalized Discounted Cumulative Gain) [17] is the evaluation method. We calculated the average of nDCG for the top k and denoted this measure as nDCG@k. Our proposed method exhibited higher values of nDCG@10 and nDCG@15. However, the word count method was better at nDCG@5, due to, we believe, the difference among news topics. Table 5 lists the results of each topic. Our proposed method exhibited the highest nDCG for the news topic of the US presidential election. On the other hand, it exhibited the lowest nDCG for Volkswagen scandal.

Table 4. Experimental results of abstractiveness ranking

Data set	nDCG@ 5		nDCG@10		nDCG@15	
	Baseline	Proposed	Baseline	Proposed	Baseline	Proposed
0	0.777416	0.670639	0.749571	0.645485	0.764630	0.722483
1	0.361862	0.543290	0.353669	0.555646	0.393572	0.591406
2	0.500233	0.804516	0.556167	0.795289	0.602773	0.737972
3	0.540533	0.448647	0.558714	0.538196	0.541825	0.637689
4	0.431555	0.411875	0.475255	0.510520	0.538099	0.552380
5	0.851829	0.486527	0.763066	0.582364	0.783946	0.635375
6	0.658663	0.661857	0.704356	0.625359	0.763631	0.618023
7	0.460658	0.519607	0.577883	0.615285	0.658203	0.664889
8	0.451236	0.445189	0.539381	0.500821	0.552930	0.567032
9	0.473012	0.412977	0.662608	0.566966	0.685796	0.583734
Ave	**0.5507**	0.546512	0.594067	**0.599211**	0.628541	**0.635072**

The news topics Volkswagen scandal and political topics about Shinzo Abe mention many past events. In many cases, our method failed to estimate the *When* element. On the other hand, for the topic of US president election, these events were reported in real time and it was easy to estimate the time information.

Table 5. Evaluation of abstractiveness ranking for each news topic

Topic	nDCG@ 5		nDCG@10		nDCG@15	
	Baseline	Proposed	Baseline	Proposed	Baseline	Proposed
US Presidental Election	0.297349	**0.663529**	0.444528	**0.666567**	0.506401	**0.639824**
Political topics about Shinzo Abe	**0.797380**	0.591886	**0.677343**	0.581203	**0.626547**	0.569199
Volkswagen scandal	0.159954	**0.317679**	0.308995	**0.319780**	**0.473534**	0.358939

6   Conclusion

We proposed the 5W model for news events and an event-extraction method with completion functionality. We also proposed the concept of abstractiveness and an abstract/concrete relationship analysis method on the basis of the 5W model. The experimental results indicate that the element *When* and document distance features are more effective for analyzing abstract-concrete relationships. Compared with base line methods, our proposed method exhibited better nDCG values for ranking event descriptions based on their abstractiveness. In the near future, we plan to carry out further experiments to improve our method. We also plan to develop an application system to support news understanding.

Acknowledgment. This work is partly supported by KAKENHI (No. 25700033) and SCAT Research Funding.

References

1. Park, S., Kang, S., Chung, S., Song, J.: NewsCube: delivering multiple aspects of news to mitigate media bias. In: SIGCHI, pp. 443–452 (2009)
2. Teitler, B.E., Lieberman, M.D., Panozzo, D., Sankaranarayanan, J., Samet, H., Sperling, J.: NewsStand: a new view on news. In: SIGSPATIAL, p. 18 (2008)
3. Kiritoshi, K., Ma, Q.: A diversity-seeking mobile news app based on difference analysis of news articles. In: Chen, Q., Hameurlain, A., Toumani, F., Wagner, R., Decker, H. (eds.) DEXA 2015. LNCS, vol. 9262, pp. 73–81. Springer, Heidelberg (2015)
4. Kiritoshi, K., Ma, Q.: Named entity oriented difference analysis of news articles and its application. IEICE Trans. Inf. Syst. **E99–D**(4), 906–917 (2016)
5. Mihalcea, R., Csomai, A.: Wikify!: linking documents to encyclopedic knowledge. In: CIKM, pp. 233–242 (2007)
6. Liao, S., Grishman, R.: Using document level cross-event inference to improve event extraction. In: ACL, pp. 909–918 (2010)
7. Radinsky, K., Davidovich, S., Markovitch, S.: Learning causality for news events prediction. In: WWW, pp. 475–484 (2012)
8. Tanaka, S., Jatowt, A., Kato, M.P., Tanaka, K.: Estimating content concreteness for finding comprehensible documents. In: WSDM, pp. 782–792 (2013)
9. Hoffart, J., Yosef, M.A., Bordino, I., Frstenau, H., Pinkal, M., Spaniol, M., Taneva, B., Thater, S., Weikum, G.: Robust disambiguation of named entities in text. In: EMNLP, pp. 782–792 (2011)
10. Kim, J.: Supervenience and Mind: Selected Philosophical Essays. Cambridge University Press, Cambridge (1993)

11. Allan, P., Yuille, J.C., Madigan, S.A.: Concreteness, imagery, and meaningfulness: values for 925 nouns. American Psychological Association (1968)
12. Manning, C.D., Surdeanu, M., Bauer, J., Finkel, J.R., Bethard, S., McClosky, D.: The Stanford CoreNLP natural language processing toolkit. In: ACL, pp. 55–60 (2014)
13. Hoffman, M.D., Bach, F.R., Blei, D.M.: Online learning for latent dirichlet allocation. In: Advances in Neural Information Processing Systems, pp. 856–864 (2010)
14. Suchanek, F.M., Kasneci, G., Weikum, G.: Yago: a large ontology from wikipedia and wordnet. Web Semant. Sci. Serv. Agents World Wide Web 6(3), 203–217 (2008)
15. Von Luxburg, U.: A tutorial on spectral clustering. Stat. Comput. 17(4), 395–416 (2007)
16. Schieber, B., Vishkin, U.: On finding lowest common ancestors: simplification and parallelization. SIAM J. Comput. 17(6), 81–93 (1988)
17. Jrvelin, K., Keklinen, J.: Cumulated gain-based evaluation of IR techniques. ACM TOIS 20(4), 422–446 (2002)
18. National Center for Biotechnology Information. http://www.ncbi.nlm.nih.gov

Detecting Maximum Inclusion Dependencies without Candidate Generation

Nuhad Shaabani$^{(\boxtimes)}$ and Christoph Meinel

Hasso-Plattner-Institut, University of Potsdam,
Prof.-Dr.-Helmert-Str. 2-3, 14482 Potsdam, Germany
{nuhad.shaabani,christoph.meinel}@hpi.de

Abstract. Inclusion dependencies (INDs) within and across databases are an important relationship for many applications in data integration, schema (re-)design, integrity checking, or query optimization. Existing techniques for detecting all INDs need to generate IND candidates and test their validity in the given data instance. However, the major disadvantage of this approach is the exponentially growing number of data accesses in terms of the number of SQL queries as well as I/O operations. We introduce MIND2, a new approach for detecting n-ary INDs ($n > 1$) without any candidate generation. MIND2 implements a new characterization of the maximum INDs we developed in this paper. This characterization is based on set operations defined on certain metadata that MIND2 generates by accessing the database only 2 × the number of valid unary INDs. Thus, MIND2 eliminates the exponential number of data accesses needed by existing approaches. Furthermore, the experiments show that MIND2 is significantly more scalable than hypergraph-based approaches.

Keywords: Mind2 · Inclusion dependency · Data integration · Data profiling

1 Introduction

Inclusion dependencies (INDs) present an important part of metadata about relationships between attributes in relational datasets [2]. An IND states that all tuples of some attribute-combination in one relation are contained in the tuples of some other attribute-combination in the same or a different relation. This makes INDs important for many tasks, such as data integration [17], integrity checking [3], query optimization [4], or schema (re-)design [10].

However, in many real-life databases knowledge about INDs is often unknown, or is lost, or does not correspond any more to the dataset structure. Furthermore, a lot of production databases are constantly changing over time so that metadata quickly become out-of-date. Thus, there is a high demand for effective and scalable approaches for mining valid INDs from a given dataset.

The problem of n-ary IND discovery ($n > 1$) is NP-hard [5]. Existing algorithms in related work for exhaustively discovering all INDs in a dataset can

© Springer International Publishing Switzerland 2016
S. Hartmann and H. Ma (Eds.): DEXA 2016, Part II, LNCS 9828, pp. 118–133, 2016.
DOI: 10.1007/978-3-319-44406-2_10

be divided into two approaches: levelwise-based approaches such as MIND [13] and hypergraph-based approaches as FIND$_2$ [6,7] and ZIGZAG [14]. But what all these algorithms have in common is that they apply the projection invariance of INDs [2,11]: a n-ary valid IND implies sets of k-ary valid INDs ($1 \leq k \leq n$). Thus, the number of all valid INDs implied by a n-ary valid IND is 2^n.

For discovering a single valid IND σ of size n, the levelwise approach [12] has to discover $2^n - 1$ implied INDs before even considering σ. This means for MIND that it has to execute 2^n SQL queries for validation. Experiments conducted by [6,14] show that levelwise algorithms do not scale beyond a maximum IND size of 8.

Attempting to reduce the exponential number of database accesses needed by the Apriori-based approach, FIND$_2$ and ZIGZAG transform the IND discovery problem into a discovery problem in a hypergraph whose nodes are all valid unary INDs, respectively. FIND$_2$ maps the IND discovery problem to the hyperclique discovery problem while ZIGZAG maps it to the minimal traversal discovery problem. Both problems are polynomial in the number of edges, and therefore exponential in the number of nodes in the hypergraph because the number of edges in a hypergraph of n nodes is bounded by 2^n. In principle, both algorithms first discover unary and binary INDs by enumeration and validation. Then optimistically assume that all high-arity INDs constructed from validated unary and binary INDs (or in general, from validated INDs in the previous iteration) are likely to be valid. That assumption makes both algorithms extremely sensitive to an overestimation of valid unary and binary INDs. A high number of such small INDs can cause many invalid larger IND candidates to be generated and validated against the database. Furthermore, hypergraph-based algorithms have high complexity, and are scalable only for sparse hypergraphs [6,8].

Research Question. The research question we address in this paper is how we can find all n-ary valid INDs ($n > 1$) between two relations without generating candidates and testing them against the database.

Contributions. We answer the research question by developing MIND$_2$ (short for **M**aximum **IN**clusion **D**ependency **D**iscovery), a novel approach for mining all maximum INDs without any candidate generation, where a maximum IND is a valid IND that is not implied by any other valid IND.

Having the set of all valid unary INDs, denoted by $\mathcal{I}_u$, discovered, MIND$_2$ computes the unary IND coordinates C_u for every valid unary IND $u \in \mathcal{I}_u$. Unary IND coordinates is a new concept we introduce in this paper (see Definition 4). To compute all unary IND coordinates MIND$_2$ executes only $2 \times |\mathcal{I}_u|$ simple SQL select queries with an order by clause. After computing all unary IND coordinates MIND$_2$ does not access the database any more because MIND$_2$ computes the set of all maximum INDs, denoted by $\mathcal{I}_M$, by only applying set operations on the unary IND coordinates (see Sect. 3). We compare the performance of MIND$_2$ with that of FIND$_2$ using real and synthetic datasets. They experiments show that MIND$_2$ is much more faster than FIND$_2$. Furthermore, they show that MIND$_2$'s scalability, on contrast to FIND2's scalability, is not influenced by a high number of small valid INDs.

2   Preliminaries

Let $\mathcal{A}$ be a finite set of attributes. For $A_1, \ldots, A_n \in \mathcal{A}$ and for a symbol R, $R[A_1, \ldots, A_n]$ is called a relational schema over $A_1, \ldots, A_n$ and R is the relation name. An instance of R, identified by r, is a finite set of tuples over R. A tuple over R is an element from $dom(A_1) \times \cdots \times dom(A_n)$, where $dom(A_i)$ defines the set of the possible values of attribute A_i $(1 \leq i \leq n)$. The number of attributes in R is $|R|$ and the number of tuples in r is $|r|$. We refer to a tuple in r as r_i, where i $(1 \leq i \leq |r|)$ is the tuple-ID in r. ID_R indicates the set of all tuple-IDs in r. For an attribute sequence $X = [A_{i_1}, \ldots, A_{i_m}]$, we define $\pi_X(R)$ as the projection of R on X. Accordingly, $r_i[X] = \pi_X(r_i)$ indicates the projection of the tuple r_i on X. Furthermore, we identify the selection of a tuple r_i from r with $\sigma_{ID_R=i}(R)$. That is, $\{r_i\}$ is the result of $\sigma_{ID_R=i}(R)$. Accordingly, $\sigma_{ID_R<i}(R)$ identifies the set of all tuples in r with an ID less than i. Thus, $\sigma_{ID_R<i}(R) = \{r_k \in r \mid k < i\}$.

Definition 1. *(IND). Let $R[A_1, \ldots, A_{|R|}]$ and $S[B_1, \ldots, B_{|S|}]$ be two relational schemata. For $n \geq 1$, let X be a sequence of n attributes from R and Y a sequence of n attributes from S. An **inclusion dependency** (IND) over R and S is an assertion of the form $R[X] \subseteq S[Y]$ where n is the size of the IND. For $n = 1$ the IND is called a unary IND (uIND).*

*Let r and s be instances of R and S, respectively. An IND $R[X] \subseteq S[Y]$ is **valid** according to r and s if and only if $\forall r_i \in r, \exists s_j \in s$ such that $r_i[X] = s_j[Y]$.*

In particular, INDs are a prerequisite for foreign keys, which are a necessity for suggesting join paths, data linkage, and data normalization.

3   Principles of Mind2

We consider two relational schemata $R[A_1, \ldots, A_{|R|}]$ and $S[B_1, \ldots, B_{|S|}]$ with corresponding instances r and s. To formulate the basic ideas of detecting all maximum INDs between R and S, we identify the set of all unary INDs with Σ_u and the set of all INDs with Σ. Furthermore, we introduce the following sets.

The set of all valid unary INDs between R and S according to r and s

$$\mathcal{I}_u = \{u \in \Sigma_u \mid u \text{ is valid according to } r \text{ and } s\}$$

The set of all valid INDs between R and S according to r and s

$$\mathcal{I} = \{I \in \Sigma \mid I \text{ is valid according to } r \text{ and } s\}$$

We represent every IND $\sigma = R[X] \subseteq S[Y]$ with $X = [A_{i_1}, \ldots, A_{i_n}]$ and $Y = [B_{i_1}, \ldots, B_{i_n}]$ as a set of all unary INDs implied by it. In other words, we present σ as the set $\{A_{i_1} \subseteq B_{i_1}, \ldots, A_{i_n} \subseteq B_{i_n}\}$. Furthermore, we identify the set of all attributes occurring on the left hand side of σ with $LHS(\sigma)$ and the set of all attributes occurring on the right hand side of σ with $RHS(\sigma)$. Thus, we have $LHS(\sigma) = \{A_{i_1}, \ldots, A_{i_n}\}$ and $RHS(\sigma) = \{B_{i_1}, \ldots, B_{i_n}\}$. Representing an IND as a set allows us to characterize the computation of the set of all maximum INDs $\mathcal{I}_M$ as set operations.

Based on the set presentation, we introduce the concept of a maximum IND.

Definition 2 *(Maximum IND). Let $I \in \mathcal{I}$ be a valid IND between R and S. I is a maximum IND if and only if there is no $I' \in \mathcal{I}$ such that $I \subset I'$ holds. We denote the set of all maximum INDs between R and S with $\mathcal{I}_M$.*

Having $\mathcal{I}_M$ discovered, we can derive the set of all valid INDs $\mathcal{I}$ as

$$\mathcal{I} = \{I \mid \exists M \in \mathcal{I}_M : I \subseteq M\}$$

The set $\mathcal{I}_M$ can be considered as a concise representation of the set $\mathcal{I}$. Thus, our goal in this work is to directly compute $\mathcal{I}_M$ without any intermediate IND sets.

Table 1. Running Example

R

ID_R	A_1	A_2	A_3	A_4	A_5
1	a	b	c	d	e
2	f	g	i	j	k

S

ID_S	B_1	B_2	B_3	B_4	B_5
1	a	b	c	d	$\perp$
2	$\perp$	$\perp$	c	d	$\perp$
3	$\perp$	$\perp$	c	d	e
4	f	g	i	$\perp$	$\perp$
5	f	g	$\perp$	j	k

Example 1. According to the two relations presented in Table 1, we have

$\mathcal{I}_u = \{u_i = A_i \subseteq B_i \mid 1 \le i \le 5\}$
$\mathcal{I} = \{\{u_1, u_2\}, \{u_1, u_3\}, \{u_2, u_3\}, \{u_1, u_2, u_3\}, \{u_1, u_4\}, \{u_2, u_4\}, \{u_1, u_2, u_4\}, \{u_4, u_5\}\} \cup \mathcal{I}_u$
$\mathcal{I}_M = \{\{u_1, u_2, u_3\}, \{u_1, u_2, u_4\}, \{u_4, u_5\}\}$
E.g. $\sigma = \{u_1, u_5\} \notin \mathcal{I}$ (i.e. not valid) because $r_1[LHS(\sigma)] = r_1[\{A_1, A_5\}] = \{(a, e)\} \not\subseteq \pi_{RHS(\sigma)}(S)$.

The first principle of computing $\mathcal{I}_M$ is formulated as follows.

Principle 1. For every tuple pair $r_i \in r$ and $s_i \in s$, we compute M^{ij}, the maximum IND between $\sigma_{ID_R=i}(R)$ and $\sigma_{ID_S=j}(S)$ according to r_i and s_j ($1 \le i \le |r|$ and $1 \le j \le |s|$). To characterize the set M^{ij} we introduce two new concepts: attribute value-positions and unary valid IND coordinates.

Definition 3 *(Attribute Value-Positions). The value positions of an attribute $A \in U$, $U \in \{R, S\}$, is the set $P_A = \pi_{\{ID_U, A\}}(U)$*

Definition 4 *(Unary IND Coordinates). The coordinates of a valid unary IND $u \in \mathcal{I}_u$ is the set $C_u = \{(i,j) \mid \exists(i,v) \in P_{LHS(u)}$ and $\exists(j,v') \in P_{RHS(u)} : v = v'\}$*

The coordinates of a valid unary IND $u \in \mathcal{I}_u$ is the set of all tuple-ID pairs (i,j) where the value of attribute $LHS(u)$ in the tuple $r_i \in r$ is identical with the value of attribute $RHS(u)$ in the tuple $s_i \in s$. In other words, $(i,j) \in C_u$ if and only if $r_i[LHS(u)] = s_j[RHS(u)]$.

Having the coordinates of all unary INDs generated, we can compute the maximum IND M^{ij} between any tuple pair (r_i, s_j) without any database access based on the following lemma.

Lemma 1. *M^{ij} consists of all unary INDs $u \in \mathcal{I}_u$ with $(i,j) \in C_u$. In other words, $M^{ij} = \{u \in \mathcal{I}_u \mid (i,j) \in C_u\}$.*

Proof. Let $M^{ij} = \{u_1, \ldots, u_n\}$ be the set of all valid uINDs with $(i,j) \in C_{u_k}$ where $1 \leq k \leq n$. Based on Definition 4, there is $(i, v_k) \in P_{LHS(u_k)}$ and $(j, v'_k) \in P_{RHS(u_k)}$ with $v_k = v'_k$ for every $k \in \{1, \ldots, n\}$. This means that $(v_1, \ldots, v_n) = (v'_1, \ldots, v'_n)$. In other words, $r_i[LHS(M^{ij})] = s_j[RHS(M^{ij})]$. Based on Definition 1, M^{ij} is a valid IND between $\sigma_{ID_R=i}(R)$ and $\sigma_{ID_S=j}(S)$ according to r_i and s_j.

We now have to show that M^{ij} is maximum. We assume that M^{ij} is not maximum. This means based on Definition 2 that there is a valid IND M_1^{ij} with $M^{ij} \subset M_1^{ij}$. Thus, M_1^{ij} contains some $u' \in \mathcal{I}_u$ with $(i,j) \notin C_{u'}$. This means that the value of attribute $LHS(u')$ in r_i is different from the value of attribute $RHS(u')$ in s_j. Thus, $r_i[LHS(M_1^{ij})] \neq s_j[RHS(M_1^{ij})]$ which means that M_1^{ij} is not valid. Thus, our assumption is wrong. □

Table 2. The coordinates of all valid uINDs between R and S in Table 1

i	P_{A_i}	P_{B_i}	$C_{A_i \subseteq B_i}$
1	$\{(1,a),(2,f)\}$	$\{(1,a),(2,\perp),(3,\perp),(4,f),(5,f)\}$	$\{(1,1),(2,4),(2,5)\}$
2	$\{(1,b),(2,g)\}$	$\{(1,b),(2,\perp),(3,\perp),(4,g),(5,g)\}$	$\{(1,1),(2,4),(2,5)\}$
3	$\{(1,c),(2,i)\}$	$\{(1,c),(2,c),(3,c),(4,i),(5,\perp)\}$	$\{(1,1),(1,2),(1,3),(2,4)\}$
4	$\{(1,d),(2,j)\}$	$\{(1,d),(2,d),(3,d),(4,\perp),(5,j)\}$	$\{(1,1),(1,2),(1,3),(2,5)\}$
5	$\{(1,e),(2,k)\}$	$\{(1,\perp),(2,\perp),(3,e),(4,\perp),(5,k)\}$	$\{(1,3),(2,5)\}$

Example 2. Based on our running example, the second column in Table 2 lists the value positions P_{A_i} of R's attributes while the value positions P_{B_i} of S's attributes are listed in the third column. The last column in this table shows the coordinates of all valid unary INDs between R and S (see Example 1). E.g. for $A_5 \subseteq B_5$, we have $(1,e) \in P_{A_5}$ and $(3,e) \in P_{B_5}$. Therefore, $C_{A_5 \subseteq B_5}$ contains the pair $(1,3)$. Also, $(2,5) \in C_{A_5 \subseteq B_5}$ because $(2,k) \in P_{A_5}$ and $(5,k) \in P_{B_5}$.

The maximum INDs M^{ij} between r_i and s_j ($1 \le i \le 2$ and $1 \le j \le 5$) are

$$M^{1,1} = \{u_1, u_2, u_3, u_4\}, \quad M^{1,2} = \{u_3, u_4\}, \quad M^{1,3} = \{u_3, u_4, u_5\}, \quad M^{1,4} = M^{1,5} = \emptyset$$
$$M^{2,1} = M^{2,2} = M^{2,3} = \emptyset, \quad M^{2,4} = \{u_1, u_2, u_3\}, \quad M^{2,5} = \{u_1, u_2, u_4, u_5\}$$

E.g. let us explain the content of the maximum IND $M^{1,2}$ between r_1 and s_2. We have $(1,2) \in C_{u_3}$. Therefore, $u_3 \in M^{1,2}$. Also, $u_4 \in M^{1,2}$ because $(1,2) \in C_{u_4}$. But $u_1, u_2, u_5 \notin M^{1,2}$ because $(1,2) \notin C_{u_1}$, $(1,2) \notin C_{u_2}$, and $(1,2) \notin C_{u_5}$.

In the next step, we compute the set of all maximum INDs between every tuple $r_i \in r$ and the relation s based on the following principle, respectively.

Principle 2. For every tuple $r_i \in r$, we compute $\mathcal{I}_M^i$, the set of all maximum INDs between $\sigma_{IDR=i}(R)$ and S according to r_i and s. To characterize the set $\mathcal{I}_M^i$, we introduce the following operator.

Definition 5 (ϕ-operator). $\phi : 2^\Sigma \to 2^\Sigma$, $\phi(\mathcal{S}) = \{\sigma \mid \nexists \sigma' \in \mathcal{S} : \sigma \subset \sigma'\}$

Operator ϕ takes a set of INDs and returns each IND that is not included in any other IND in this set. Thus, we conclude: $\phi(\mathcal{S}) \subseteq \mathcal{S}$ for any $\mathcal{S} \in 2^\Sigma$.

Lemma 2. $\mathcal{I}_M^i = \phi(\mathcal{I}^i)$, where $\mathcal{I}^i$ is the set of all non-empty M^{ij} ($1 \le j \le |s|$).

Proof. Every $M^{ij} \in \mathcal{I}^i$ is a valid (but not necessary a maximum) IND between $\sigma_{IDR=i}(R)$ and S. But what we want to have is all maximum INDs from $\mathcal{I}^i$. Based on Definition 5, ϕ-operator solves this task. Thus, $\mathcal{I}_M^i = \phi(\mathcal{I}^i)$ is the set of all maximum INDs between $\sigma_{IDR=i}(R)$ and S. □

Example 3. Based on Example 2, we have
$$\mathcal{I}^1 = \{M^{1,1}, M^{1,2}, M^{1,3}\}, \quad \mathcal{I}_M^1 = \phi(\mathcal{I}^1) = \{M^{1,1}, M^{1,3}\}$$
$$\mathcal{I}^2 = \{M^{2,4}, M^{2,5}\}, \quad \mathcal{I}_M^2 = \phi(\mathcal{I}^2) = \{M^{2,4}, M^{2,5}\}$$

We can now compute $\mathcal{I}_M$, the set of all maximum INDs between R and S, from the sets $\mathcal{I}_M^i$ ($1 \le i \le |r|$) based on Principle 3.

Principle 3. To explain the main idea behind Principle 3, let us consider the two relations in Table 1. What are the maximum INDs between them if we know $\mathcal{I}_M^1$ and $\mathcal{I}_M^2$ computed in Example 3? First, the intersection between any two INDs $M_1 \in \mathcal{I}_M^1$ and $M_2 \in \mathcal{I}_M^2$ is a valid IND between R and S. E.g., $M^{1,1} \cap M^{2,4} = \{u_1, u_2, u_3\}$ is a valid IND between R and S. Second, after computing the intersection between each pair $(M_1, M_2) \in \mathcal{I}_M^1 \times \mathcal{I}_M^2$, taking all maximum sets from the result gives us the set of all maximum INDs (see Example 4). We generalize these two ideas as follows.

Definition 6 (ψ-operator). $\psi : 2^\Sigma \times 2^\Sigma \to 2^\Sigma$, $\psi(\mathcal{S}_1, \mathcal{S}_2) = \{\sigma \mid \exists(\sigma_1, \sigma_2) \in \mathcal{S}_1 \times \mathcal{S}_2 : \sigma = \sigma_1 \cap \sigma_2 \text{ and } \sigma \ne \emptyset\}$

In words, for two sets $\mathcal{S}_1$ and $\mathcal{S}_2$ of INDs the ψ-operator takes every tuple (σ_1, σ_2) from $\mathcal{S}_1 \times \mathcal{S}_2$ and computes the intersection between σ_1 and σ_2. To characterize the computation of the set $\mathcal{I}_M$, we define the ρ-operator.

Definition 7 (*ρ-operator*). *Let $\mathcal{I}_M$ be the set of all $\mathcal{I}_M^i$ ($1 \leq i \leq |r|$).*

$$\rho(\mathcal{I}_M) = \begin{cases} \mathcal{S} & \text{if } |\mathcal{I}_M| = 1 \text{ and } \mathcal{S} \in \mathcal{I}_M \\ \phi(\psi(\mathcal{S}, \rho(\mathcal{I}_M \setminus \{\mathcal{S}\}))) & \text{if } |\mathcal{I}_M| > 1 \text{ and } \mathcal{S} \in \mathcal{I}_M \end{cases}$$

Now, we can compute $\mathcal{I}_M$ as follows.

Lemma 3. $\mathcal{I}_M = \rho(\mathcal{I}_M)$

Proof. We prove the lemma by induction on the number of tuples i in r.

Basis Step: For $i = 1$, we have $\mathcal{I}_M = \{\mathcal{I}_M^1\}$. Thus, $\rho(\{\mathcal{I}_M^1\}) = \mathcal{I}_M^1 = \mathcal{I}_M$ based on the construction of the set $\mathcal{I}_M^1$.

Induction Assumption: For $1 \leq i < |r|$, let $\mathcal{I}'_M$ be the set of all $\mathcal{I}_M^i$ and $\mathcal{I}'_M$ be the set of all maximum INDs between $\sigma_{ID_R < |r|}(R)$ and S. We assume

$$\mathcal{I}'_M = \rho(\mathcal{I}'_M) \tag{1}$$

Inductive Step: Let $\mathcal{I}_M^{|r|}$ be the set of all maximum INDs between $\sigma_{ID_R = |r|}(R)$ and S. Thus, $\mathcal{I}_M = \mathcal{I}'_M \cup \mathcal{I}_M^{|r|}$. Based on assumption (1), we have to show

$$\mathcal{I}_M = \rho(\mathcal{I}_M) = \phi(\psi(\mathcal{I}_M^{|r|}, \rho(\mathcal{I}'_M))) = \phi(\psi(\mathcal{I}_M^{|r|}, \mathcal{I}'_M))$$

Every set in $\psi(\mathcal{I}_M^{|r|}, \mathcal{I}'_M)$ is a valid IND between R and S because the intersection of two valid INDs is a valid IND. We assume that there is a valid IND I with

$$I \notin \psi(\mathcal{I}_M^{|r|}, \mathcal{I}'_M) \tag{2}$$

Because I is a valid IND, there is $I_1 \in \mathcal{I}_M^{|r|}$ with $I \subseteq I_1$ and $I_2 \in \mathcal{I}'_M$ with $I \subseteq I_2$. Thus, $I \subseteq I_1 \cap I_2$, but $I_1 \cap I_2 \in \psi(\mathcal{I}_M^{|r|}, \mathcal{I}'_M)$. This means that assumption (2) is wrong. Consequently, $\psi(\mathcal{I}_M^{|r|}, \mathcal{I}'_M)$ contains all valid INDs between S and R. Based on Definition 5, $\phi(\psi(\mathcal{I}_M^{|r|}, \mathcal{I}'_M))$ is the set of all maximum INDs in $\psi(\mathcal{I}_M^{|r|}, \mathcal{I}'_M)$. Thus, $\mathcal{I}_M = \phi(\psi(\mathcal{I}_M^{|r|}, \mathcal{I}'_M))$. □

Example 4. Based on Example 3, we have $\mathcal{I}_M = \{\mathcal{I}_M^1, \mathcal{I}_M^2\}$. Accordingly,
  $\psi(\mathcal{I}_M) = \{M^{1,1} \cap M^{2,4}, M^{1,1} \cap M^{2,5}, M^{1,3} \cap M^{2,4}, M^{1,3} \cap M^{2,5}\}$
  $\psi(\mathcal{I}_M) = \{\{u_1, u_2, u_3\}, \{u_1, u_2, u_4\}, \{u_3\}, \{u_4, u_5\}\}$
  $\mathcal{I}_M = \rho(\mathcal{I}_M) = \phi(\psi(\mathcal{I}_M)) = \{\{u_1, u_2, u_3\}, \{u_1, u_2, u_4\}, \{u_4, u_5\}\}$
(compare with Example 1).

In the following section, we formulate MIND2 algorithmically. We also present its data structures. This formulation is the basis of our implementation of MIND2.

4   Mind$_2$

Overall Mind$_2$. MIND$_2$ consists of three major components. Algorithm 1 as the first component, is responsible for computing the unary IND coordinates C_u of each valid unary IND $u \in \mathcal{I}_u$ based on Definition 4. It also stores each generated set C_u in a separate file in an external repository *Repo* on a hard drive.

Then Algorithm 2 reads the generated coordinates at once and computes the set of all maximum INDs $\mathcal{I}_M$ incrementally according to the ascending order of the tuple-IDs $i \in ID_R$ in the left relation r. In other words, it computes the ρ-operator (see Definition 7) iteratively. Before the iteration in which the set $\mathcal{I}_M^i$ (the set of all maximum INDs between $\sigma_{ID_R=i}(R)$ and S) can be generated, Algorithm 2 computes all maximum INDs between $\sigma_{ID_R<i}(R)$ and S and stores them in $\mathcal{I}_M$. In other words, before the computation of $\mathcal{I}_M^i$ starts, the set $\mathcal{I}_M$ contains the maximum INDs between the tuples $\{r_k \in r \mid 1 \le k < i\}$ and s. Having $\mathcal{I}_M^i$ generated, Algorithm 2 replaces the current content of the set $\mathcal{I}_M$ with the result of the composite operation $\phi(\psi(\mathcal{I}_M, \mathcal{I}_M^i))$. This procedure continues until all tuple-IDs $i \in ID_R$ have been processed. At the end and based on Lemma 3, the set $\mathcal{I}_M$ contains all maximum INDs between R and S. At the beginning, we initialize $\mathcal{I}_M$ with $\{\mathcal{I}_u\}$ because $\{\{\mathcal{I}_u\}\}$ is an upper bound of $\mathcal{I}_M$.

The third component of MIND$_2$ is Algorithm 3 called by Algorithm 2 to compute the sets $\mathcal{I}_M^i$ ($1 \le i \le |r|$). It computes them based on Lemmas 1 and 2. Below, we explain these components in details.

```
Input      : I_u, Repo
Output     : C_u for every u ∈ I_u
1  foreach u ∈ I_u do
2  |    i2jsMap ← createMap(Int, Set)
3  |    A ← LHS(u); B ← RHS(u)
4  |    Cur_A ← createCursor(A)
5  |    Cur_B ← createCursor(B)
6  |    (i, v) ← Cur_A.next()
7  |    (j, v') ← Cur_B.next()
8  |    while Cur_A.hasNext() and Cur_B.hasNext()
   |    do
9  |    |    if v = v' then
10 |    |    |    ID_A ← {}; ID_B ← {}
11 |    |    |    (k, w) ← Cur_A.current()
12 |    |    |    while v = w do
13 |    |    |    |_  ID_A = ID_A ∪ {k};
   |    |    |        (k, w) ← Cur_A.next()
14 |    |    |    (k, w) ← Cur_B.current()
15 |    |    |    while v = w do
16 |    |    |    |_  ID_B = ID_B ∪ {k};
   |    |    |        (k, w) ← Cur_B.next()
17 |    |    |    if ID_A ≠ ∅ and ID_B ≠ ∅ then
18 |    |    |    |    foreach i ∈ ID_A do
19 |    |    |    |_   |_ i2jsMap.put(i, ID_B)
20 |    |    else if v > v' then
21 |    |    |_  (j, v') ← Cur_B.next()
22 |    |    else
23 |    |    |_  (i, v) ← Cur_A.next()
24 |    writer ← createWriter(u, Repo)
25 |    ID_A ← i2jsMap.keys(); sort(ID_A)
26 |    foreach i ∈ ID_A do
27 |    |    P_B ← i2jsMap.get(i); sort(ID_B)
28 |    |_   writer.write(u, i, ID_B)
```

Algorithm 1. genCoordinates

$A_1 \subseteq B_1$	1, [1]	
	2, [4, 5]	
$A_2 \subseteq B_2$	1, [1]	
	2, [4, 5]	
$A_3 \subseteq B_3$	1, [1, 2, 3]	
	2, [4]	
$A_4 \subseteq B_4$	1, [1, 2, 3]	
	2, [5]	
$A_5 \subseteq B_5$	1, [3]	
	2, [5]	

Fig. 1. The output of Algorithm 1 for the set of all valid unary INDs between R and S in the running example

Generating unary IND Coordinates. To compute the unary IND coordinates of a $u \in \mathcal{I}_u$, Algorithm 1 opens two cursors at once (lines 3–5): one for reading the sorted value positions of the attribute $A = LHS(u)$ and the other for reading the sorted value positions of the attribute $B = RHS(u)$ (see Definition 3 for the value positions of an attribute). The value positions of every attribute are sorted according to its values in the corresponding relation. In other words, for any $(i_1, v_1), (i_2, v_2) \in P_A$ $(\in P_B)$: the tuple (i_1, v_1) will be read by the corresponding cursor before the tuple (i_2, v_2) if the value v_1 occurs before the value v_2 in the sort sequence. Otherwise, (i_2, v_2) will be read before (i_1, v_1).

In the main *while*-loop (lines 8–23), Algorithm 1 moves the two cursors in such a way so that it can associate every tuple-ID $i \in ID_R$ with the set of all tuple-IDs $j \in ID_S$ for which both attributes A and B have the same value. In other words, the tuple-ID i is associated with the set $\{j \mid \exists(j, v) \in P_B : (i, v) \in P_A\}$. It saves this association temporary in the hash map *i2jsMap* (lines 17–19).

After finishing the reading of value positions of P_A and P_B, respectively, Algorithm 1 creates a file for the current unary IND u in the *for*-loop (lines 1–28) and saves every pair $(i, \{j \mid \exists(j, v) \in P_B : (i, v) \in P_A\})$ in a line in this file. The lines (records) are sorted in ascending order by the left tuple-IDs $i \in ID_R$ and in every line the IDs $j \in \{j \mid \exists(j, v) \in P_B : (i, v) \in P_A\}$ are also sorted in ascending order (lines 24–28). This policy of organizing the value positions is required by Algorithm 2.

MIND_2 needs only $2 \times |\mathcal{I}_u|$ database accesses because every cursor needs a simple *SQL select* statement with an *order by* clause for reading the value position of an attribute.

Example 5. Based on the attribute value positions listed in Table 2, Fig. 1 illustrates the output of the Algorithm refalgo:coordinatesGen. Every row in this figure represents a file containing the coordinates of an unary IND.

Generating Maximum INDs between R and S. Algorithm 2, as implementation of Principle 3 (see Sect. 3), generates the set of all maximum INDs $\mathcal{I}_M$ by computing the ρ-operator (see Definition 7) incrementally. It opens all files of the unary INDs coordinates generated by Algorithm 1 and reads them at once (lines 3–4). Every $u \in \mathcal{I}_u$ is associated with a sequential file reader for reading its coordinates C_u. The file readers are managed by a priority queue. For any two readers fr, fr', reader fr has a higher priority than fr' in the queue if and only if the tuple-ID i in the file entry (u, i, L) is smaller than the tuple-ID i' in (u', i', L') where (u, i, L) is the entry that fr can currently read and (u', i', L') is the entry that fr' can currently read. Managing the readers in this way allows Algorithm 2 to collect all unary INDs $u \in \mathcal{I}_u$ that have the same tuple-ID i $(i \in ID_R)$ in their coordinates (lines 7–18).

In every pass through the main *while*-loop (lines 6–29) the algorithm collects the elements (u, L) in the set $\mathcal{L}$ where all unary INDs u in these elements have the same tuple-ID $i \in ID_R$. Every list L in (u, L) is (based on its construction by Algorithm 1) the list of all tuple-IDs $j \in ID_S$, where the values of attribute $RHS(u)$ in these tuples and the value of $LHS(u)$ in tuple i are identical.

```
Input   : I_u, Repo
Output  : I_M

1  Queue ←createPriorityQueue()
2  foreach u ∈ I_u do
3      fr ← createFileReader(u, Repo)
4      Queue.add(fr)

5  I_M ← {I_u}
6  while Queue ≠ ∅ do
7      L ← ∅; Readers ← ∅
8      fr ← Queue.pull()
9      Readers ← Readers ∪ {fr}
       (u, i, L) ← fr.current()
10     L ← L ∪ {(u, L)}
11     while Queue ≠ ∅ do
12         fr' ← Queue.peek()
13         (u', i', L') ← fr'.current()
14         if i ≠ i' then break
15         fr ← Queue.pull()
16         Readers ← Readers ∪ {fr}
17         (u, i, L) ← fr.current()
18         L ← L ∪ {(u, L)}

19     I*_M ← genSubMaxINDs(L, I_M)
20     I_M ← φ(ψ(I_M, I*_M))
21     foreach u ∈ I_u : {u} ∈ I_M do
22         I_M ← I_M \ {{u}}

23     if I_M = ∅ then
24         I_M ← I_u; break

25     activeU ← ∪_{M ∈ I_M} M
26     foreach fr ∈ Readers do
27         if fr.hasNext() and
28            fr.u ∈ activeU then
29            fr.next();
              Queue.add(fr)
```

Algorithm 2. genMaxINDs

```
Input   : L, I*^{-1}_M
Output  : I*_M

1  Queue ←createPriorityQueue()
2  foreach (u, L) ∈ L do
3      lr ← createListReader(u, L)
4      Queue.add(lr)

5  UB ← ∅
6  while Queue ≠ ∅ do
7      Readers ← ∅
8      lr ← Queue.pull()
9      Readers ← Readers ∪ {lr}
       (u, j) ← lr.current()
10     M*^j ← {u}
11     while Queue ≠ ∅ do
12         lr' ← Queue.peek()
13         (u', j') ← lr'.current()
14         if j ≠ j' then break
15         lr ← Queue.pull()
16         Readers ← Readers ∪ {lr}
           (u, j) ← lr.current()
17         M*^j ← M*^j ∪ {u}

18     if ∃M ∈ I*^{-1}_M : M ⊆ M*^j then
19         UB ← UB ∪ {M}

20     if UB = I*^{-1}_M then
21         I*_M ← I*^{-1}_M; break

22     I*_M ← I*_M ∪ {M*^j}
23     foreach lr ∈ Readers do
24         if lr.hasNext() then
25            lr.next(); Queue.add(lr)

26  I*_M ← φ(I*_M)
```

Algorithm   3.   genSub-
MaxINDs

After creating the set $\mathcal{L}$ in the current pass of the main *while*-loop for a certain i, Algorithm 2 calls Algorithm 3 to compute the maximum INDs between $\sigma_{ID_R=i}(R)$ and S (line 19). We donate this set with $\mathcal{I}^*_M$ where the symbol $*$ is a placeholder for any $i \in ID_R$.

After computing maximum INDs $\mathcal{I}^*_M$ between $\sigma_{ID_R=i}(R)$ and S, the set of all maximum INDs $\mathcal{I}_M$ will be updated by applying the composite operation $\phi(\psi(\mathcal{I}_M, \mathcal{I}^*_M))$ in line 20 (see Definition 5 for ϕ-operator and Definition 6 for ψ-operator). The set $\mathcal{I}_M$ is initialized with the set $\{\mathcal{I}_u\}$ (line 5). If the updated set $\mathcal{I}_M$ contains only the unary INDs, the algorithm breaks the main *while*-loop and returns the set of all unary INDs as the maximum INDs (line 23–24). Otherwise, Algorithm 2 will update the queue only with readers of those unary INDs u which are contained at least in one set of $\mathcal{I}_M$ (lines 25–29).

Generating maximum INDs between $\sigma_{ID_R=i}(R)$ and S. Based on Principle 1 and Principle 2, Algorithm 3 computes the set of all maximum INDs between $\sigma_{ID_R=i}(R)$ and S from the set $\mathcal{L}$ while it exploits the set $\mathcal{I}^{*-1}_M$ to improve the performance. The set $\mathcal{L}$ generated by Algorithm 2 (lines 7–18) contains the elements (u, L): all unary INDs in these elements have the same left tuple-ID i in their coordinates while every list L in (u, L) is the sorted list of all tuple-IDs $j \in ID_S$ in the coordinates $(i, j) \in C_u$. The algorithm reads all the lists in the set $\mathcal{L}$ at once and uses a priority queue to manage the list readers in the same way in which Algorithm 2 manages the file readers of the unary INDs coordinates.

In the main *while*-loop we collect all unary INDs u in the set M^{*j} that have the same tuple-ID j in their coordinates (lines 7–17). The symbol $*$ in M^{*j} is a placeholder for the corresponding i. Thus, based on the properties of the elements (u, L) of the set $\mathcal{L}$, the set M^{*j} contains all unary INDs u that have (i, j) in their coordinates C_u. This means, according to Lemma 1, M^{*j} is the maximum IND between $\sigma_{ID_R=i}(R)$ and $\sigma_{ID_S=j}(S)$.

Every computed set M^{*j} is collected in the set $\mathcal{I}_M^*$ (line 22). This means, updating $\mathcal{I}_M^*$ by applying the ϕ-operator on it gives us, according to Lemma 2, the maximum INDs between $\sigma_{ID_R=i}(R)$ and S (line 26).

The objective of the input set $\mathcal{I}_M^{*-1}$ is to improve the performance of computing $\mathcal{I}_M^*$. The set $\mathcal{I}_M^{*-1}$ is the set of all maximum INDs between $\sigma_{ID_R<i}(R)$ and S. For every generated set M^{*j} Algorithm 3 checks if there is a set M in $\mathcal{I}_M^{*-1}$ such that M is a subset of M^{*j} (lines 18–19). If such a set exists, it is added to the set UB. If the set UB contains all sets from $\mathcal{I}_M^{*-1}$, then the algorithm breaks the execution and returns $\mathcal{I}_M^{*-1}$ as the maximum INDs between $\sigma_{ID_R=i}(R)$ and S (lines 20–21). This rule does not have any affect on the correctness of Algorithm 2. This is because the result of the composite operation $\phi(\psi(\mathcal{I}_M, \mathcal{I}_M^*))$ in Algorithm 2 is the set $\mathcal{I}_M$ itself if every set in $\mathcal{I}_M^*$ is a superset of a set in $\mathcal{I}_M$.

5   Experiments

The main aim of our experiments is to compare the scalability of MIND$_2$ with that of FIND$_2$. This is our focus because FIND$_2$ is developed to reduce the number of IND candidates required by Apriori-based approaches. Although ZIGZAG is also designed to handle long INDs, we limited our experiments to FIND$_2$. That is because, as discussed in Sects. 1 and 6, FIND$_2$ and ZIGZAG approach the IND discovery problem from similar directions and have many properties in common.

The number of rows varies between 500,000 and 16,000,000 rows in these experiments. The other important variable that has a big impact on the scalability of discovering the n-ray INDs between two relations is the number of the unary INDs. The number of unary INDs in the experiments varies between 8 and 19 unary INDs in the corresponding table pairs.

Experimental Conditions. We performed the experiments on Windows 7 Enterprise system with an Intel Core i5-3470 (Quad Core, 3.20 GHz CPU) and 8 GB RAM. We used Oracle 11g as the database server installed on the same machine. We implemented both algorithms in 64-bit Java 7. We implemented FIND$_2$ based on [6]. For MIND$_2$, we set the minimum Java heap size to 4 GB and the maximum to 6 GB. While for FIND$_2$, we set the Java stack size to 4 GB. FIND$_2$ validates IND candidates by applying the SQL query proposed in [15].

Experiment Groups 1. The purpose of these experiments is to compare the scalability of MIND$_2$ with that of FIND$_2$ by using a real-word dataset called

MUSICBRAINZ[1]. MUSICBRAINZ is an open music encyclopedia that collects music metadata and makes them available to the public. MUSICBRAINZ contains 27 GB of data. It contains 206 tables (relations) with 1,165 non-empty columns (attributes). We found a total of 24,881 valid unary INDs by applying S-INDD [18]. We detected pairs of tables where there is at least one valid n-ary IND with size greater than 2 between the tables of every pair. The number of tuples varies between 500,000 and 1,000,500 tuples. The results of these experiments are presented in Table 3. The acronym "TP." stands for table pair. The left part of Table 3 shows some statistics about detected INDs: the number of valid unary INDs ($|\mathcal{I}_u|$), the number of detected maximum INDs ($|\mathcal{I}_M|$), the size of the longest maximum INDs (n_{max}) accompanied by their number ((x Nr.)), and the size of shortest maximum INDs (n_{min}) accompanied by their number ((x Nr.)). The right part of Table 3 shows the needed time (in minutes) by MIND2 and FIND2 for detecting the valid INDs, respectively. The acronym "o.o.M." refers to out of memory exception. In most of these experiments, MIND2 outperforms FIND2 significantly. Furthermore, they show that MIND2's scalability, on the contrary to that of FIND2, is robust and not sensitive to the high number of small valid INDs. The reason why FIND2 terminates with an *out of memory* exception is the complexity of hypergraphs created by FIND2. If one of these hypergraphs is not sparse (irreducible), then the hyperclique-finding subroutine presented in [6] attempts to simplify the corresponding hypergraph by removing hyperedges from it. The removing of hyperedges performed by this subroutine of FIND2 is not defined deterministically. This behavior causes a lot of recursive calls and consumes a huge amount of memory. FIND2 needed less time than MIND2 only for the table pair 5 and 7, respectively. This is because the created hypergraphs for these table pairs are sparse, respectively.

Table 3. Comparing MIND2's runtime with FIND2's runtime using MUSICBRAINZ database (o.o.M. = out of memory, m = minutes)

TP.	$\|\mathcal{I}_u\|$	$\|\mathcal{I}_M\|$	n_{max} (x Nr.)	n_{min} (x Nr.)
1	19	75	5 (x 2)	2 (x 4)
2	17	25	3 (x 13)	2 (x 12)
3	15	28	3 (x 17)	2 (x 11)
4	15	56	3 (x 56)	-
5	14	28	3 (x 20)	2 (x 8)
6	13	23	3 (x 6)	2 (x 17)
7	12	26	3 (x 19)	2 (x 7)
8	12	11	3 (x 11)	-

TP.	MIND2	FIND2
1	184 m	o.o.M. after 250 m
2	4 m	o.o.M. after 40 m
3	2 m	o.o.M. after 33 m
4	1.5 m	o.o.M. after 322 m
5	15 m	4 m
6	15 m	o.o.M. after 33 m
7	22 m	6 m
8	11 m	30 m

Experiment Groups 2. The purpose of these experiments is to compare MIND2's performance with the performance of the best case for FIND2. The best case for FIND2 (also for ZIGZAG) is when FIND2 needs to build only the

---

[1] https://musicbrainz.org.

Table 4. Results of the experiments in groups 2 and 3 (# = number of, m = minutes)

DB	$\mathcal{I}_M$	n_{max} (x Nr.)
1	1	9 (x 1)
2	1	10 (x 1)
3	8	8 (x 8)
4	9	9 (x 9)

	#DB-Accesses		Runtime	
DB	FIND_2	MIND_2	FIND_2	MIND_2
1	37	18	57 m	11 m
2	46	20	100 m	12 m
3	509	18	263 m	9.5 m
4	1021	20	906 m	11.5 m

2-hypergraph and then finds only one clique representing a valid IND. This happens for example when the database contains only one valid IND σ of size $n > 2$. In this case, FIND_2 needs $n \times (n-1)/2$ database access to enumerate the valid binary INDs and one access to validate the clique. To demonstrate this case, we generated two synthetic databases DB 1 and DB 2. Both databases contain 16,000,000 tuples. DB 1 contains one valid maximum IND in size 9. While DB 2 contains one valid maximum IND in size 10. The results of these experiments are presented in Table 4 (rows 1 and 2 in each part of Table 4). FIND_2's runtime is dominated by the runtime of the required *SQL* queries for enumerating the valid binary INDs. Therefore, MIND_2 is up to 8x faster than FIND_2.

Experiment Groups 3. The purpose of these experiments is to show that in some cases FIND_2 needs the same exponential number of database accesses as needed by the Apriori approach. Let $\sigma = \{u_1, \ldots, u_n\}$ be an invalid n-ary IND with the property that every $(n-1)$-ary IND contained in σ is a valid IND. In this case, FIND_2 builds $n-2$ k-hypergraphs ($2 \leq k \leq n-1$) where every k-hypergraph has $\binom{n}{k}$ edges and contains only the same clique, namely $\{u_1, \ldots, u_n\}$. Thus, FIND_2 needs $\binom{n}{2} + \cdots + \binom{n}{n-1} + (n-1) = 2^n - 3$ *SQL* queries to discover the n valid $(n-1)$-ary INDs contained in σ. To illustrate this case, we also generated two synthetic databases DB 3 and DB 4, where every database has 10,000,000 tuples in average. DB 3 contains 8 valid INDs in size 8. While DB 4 contains 9 valid INDs in size 9. Table 4 (rows 3 and 4 in each part of this table) presents the results of these experiments. The FIND_2's runtime is dominated by the exponential number of the database accesses needed for the validation of the IND candidates. Therefore, MIND_2 is much more (up to 82x) faster than FIND_2.

6   Related Work

Kantola et al. [5] give an upper bound for the complexity of the IND-detecting problem and proof of its NP-completeness. Casanova et al. [3] formulate the simple axiomatization for INDs and prove that the decision problem for INDs is PSPACE-complete. Köhler and Link [9] investigated INDs and NOT NULL constraints under simple and partial semantics from theoretical point of view.

N-ary INDs. FIND_2 proposed by Koeller and Rundensteiner [6,7] begins by exhaustively validating unary and binary INDs, forming a 2-uniform hypergraph using unary INDs as nodes and binary INDs as edges. Then the algorithm proceeds in stages enumerated by a $k = 2, 3, \ldots$. In every stage k, all hypercliques are detected by HYPERCLIQUE algorithm [6] in the k-hypergraph, where every hyperclique represents an IND candidate. Then IND candidates are checked for validity in the database. Each invalid IND corresponding to hyperclique in the k-hypergraph is broken into $(k + 1)$-ary INDs contained in it. Then the $(k + 1)$-ary INDs form the edges of a $(k + 1)$-hypergraph. Edges corresponding to invalid $(k + 1)$-ary INDs are removed from the $(k + 1)$-hypergraph. The process is repeated for increasing k until no new cliques are found. DeMarchi and Petit [14] developed ZIGZAG algorithm based on borders of theories [12]. Initially and for a k specified by the user, ZIGZAG initializes the positive border and the negative border by applying an adaptation of the level-wise algorithm MIND until the level k is reached. Furthermore, ZIGZAG introduces the optimistic positive border computed by finding minimal hypergraph traversals in a hypergraph generated from the negative border. The algorithm iteratively updates the three borders as long as the optimistic positive border contains INDs that are not contained in the positive border. Every updating process combines a pessimistic bottom-up with an optimistic top-down search. In the bottom-up search ZIGZAG validates IND candidates against the database. While in the top-down approach it estimates the distance between invalid INDs and the positive border by counting the number of tuples that do not satisfy these invalid INDs. MIND proposed by Marchi et al. [13] applies the level-wise approach to generate IND candidates. MIND generates all $(k + 1)$-IND candidates from the valid k-INDs and the valid unary INDs. It is based on the view that the validity of $\sigma_1 = R[A_1, \ldots, A_k] \subseteq S[B_1, \ldots, B_k]$ and the validity of $\sigma_2 = R[A_{k+1}] \subseteq S[B_{k+1}]$ are necessary but not sufficient conditions for $\sigma = R[A_1, \ldots, A_k, A_{k+1}] \subseteq S[B_1, \ldots, B_k, B_{k+1}]$ to be valid. That is, if σ_1 or σ_2 is invalid, then it is impossible for σ to be valid. In this case, σ is pruned and no testing for its validity is necessary. In the other case, if both of σ_1 and σ_2 are valid, then σ has a chance to be valid and therefore becomes a candidate of size $k + 1$. This candidate is then validated against the database. After all the $(k + 1)$-ary IND candidates are generated and tested, the algorithm generates and tests $(k + 2)$-ary IND candidates.

Unary INDs. Shaabani and Meinel developed S-INDD [18], a scalable algorithm for discovering unary INDs in large datasets. S-INDD introduces the concept of attribute clustering. Deriving unary INDs from the attribute clustering eliminates the redundant intersection operations resulting from deriving them from the inverted index applied in [13]. Furthermore, Shaabani and Meinel have shown that SPIDER [1] is a special case of S-INDD and that S-INDD is much more scalable than SPIDER. SPIDER [1] is presented by Bauckmann et al. The algorithm first sorts the distinct values in all columns and then uses a parallel merge-sort like algorithm to compute all unary INDs simultaneously. Papenbrock et al. presented BINDER [19]. BINDER applies a divide and conquer technique for discovering unary INDs. The main goal of BINDER's approach was to improve SPIDER's performance. BINDER takes a further step to generate all n-ary INDs by applying string concatenations and the same Apriori strategy applied by MIND [13]. This approach results in an exponential number of I/O-operations and exponentially increases the original data size.

Foreign Key Discovery. Zhang et al. [20] applied approximation techniques for profiling foreign keys. Memari et al. [16] proposed algorithms for profiling foreign keys under the different semantics for NULL markers of the SQL Standard.

7   Conclusion and Future Work

We developed MIND2, a new approach for mining maximum inclusion dependency between two relations. MIND2 is based on a new characterization of maximum INDs. We achieved this characterization by only defining set operations on unary IND coordinates, a new concept we also introduced in this paper. Applying these set operations on unary IND coordinates enables discovering maximum INDs without any candidate generation, which has a big impact on a scalable discovery of long n-ary INDs. This work is the main milestone for our further works: as MIND2's performance is quadratically bounded by the number of tuples, we work in a distributed version of MIND2 in order to parallelize both the computation of unary IND coordinates and the computation of maximum INDs.

References

1. Bauckmann, J., Leser, U., Naumann, F.: Efficiently computing inclusion dependencies for schema discovery. In: ICDE Workshops (2006)
2. Casanova, M.A., Fagin, R., Papadimitriou, C.H.: Inclusion dependencies and their interaction with functional dependencies. In: PODS (1982)
3. Casanova, M.A., Tucherman, L., Furtado, A.L.: Enforcing inclusion dependencies and referential integrity. In: VLDB (1988)
4. Gryz, J.: Query folding with inclusion dependencies. In: Proceedings of the 14th IEEE Internation Conference on Data Engineering (ICDE 1998), pp. 126–133 (1998)
5. Kantola, M., Mannila, H., Räihä, K.J., Siirtola, H.: Discovering functional and inclusion dependencies in relational databases. JIIS **7**(7), 591–607 (1992)
6. Koeller, A., Rundensteiner, E.: Discovery of high-dimensional inclusion dependencies. Technical Reports WPI-CS-TR-02-15, Worcester Polytechnic Institute (2002)
7. Koeller, A., Rundensteiner, E.: Discovery of high-dimensional inclusion dependencies. In: ICDE, pp. 683–685 (2003)
8. Koeller, A., Rundensteiner, E.A.: Heuristic strategies for inclusion dependency discovery. In: Meersman, R. (ed.) OTM 2004. LNCS, vol. 3291, pp. 891–908. Springer, Heidelberg (2004)
9. Köhler, H., Link, S.: Inclusion dependencies reloaded. In: CIKM 2015 (2015)
10. Levene, M., Vincent, M.W.: Justification for inclusion dependency normal form. IEEE Trans. Knowl. Data Eng. **12**, 2000 (2000)
11. Liu, J., Li, J., Liu, C., Chen, Y.: Discover dependencies from data- a review. IEEE Trans. Knowl. Data Eng. **24**(2), 251–264 (2012)

12. Mannila, H., Toivonen, H.: Levelwise search and borders of theories in knowledgediscovery. Data Min. Knowl. Discov. **1**(3), 241–258 (1997)
13. Marchi, F.D., Lopes, S., Petit, J.M.: Efficient algorithms for mining inclusion dependencies. In: EDBT 2002, pp. 464–476 (2002)
14. Marchi, F.D., Petit, J.M.: Zigzag: A new algorithm for mining large inclusion dependencies in databases. In: ICDM (2003)
15. Marchi, F., Lopes, S., Petit, J.M.: Unary and n-ary inclusion dependency discovery in relational databases. JIIS **32**, 53–73 (2009)
16. Memari, M., Link, S., Dobbie, G.: SQL Data Profiling of Foreign Keys. In: Johannesson, P., Lee, M.L., Liddle, S.W., Opdahl, A.L., López, O.P. (eds.) Conceptual Modeling. LNCS, vol. 9381, pp. 229–243. Springer, Heidelberg (2015)
17. Miller, R.J., Hernández, M.A., Haas, L.M., Yan, L., Howard Ho, C.T., Fagin, R., Popa, L.: The clio project: Managing heterogeneity. SIGMOD Rec. **30**, 78–83 (2001)
18. Shaabani, N., Meinel, C.: Scalable inclusion dependency discovery. In: Renz, M., Shahabi, C., Zhou, X., Cheema, M.A. (eds.) DASFAA 2015. LNCS, vol. 9049, pp. 425–440. Springer, Heidelberg (2015)
19. Papenbrock, T., Sebastian Kruse, J.: Divide & conquer-based inclusion dependency discovery. VLDB **8**, 774–785 (2015)
20. Zhang, M., Hadjieleftheriou, M., Ooi, B.C., Procopiuc, C.M., Srivastava, D.: On multi-column foreign key discovery. VLDB **3**, 805–814 (2010)

NoSQL, NewSQL

Footprint Reduction and Uniqueness Enforcement with Hash Indices in SAP HANA

Martin Faust[1], Martin Boissier[1]([✉]), Marvin Keller[1], David Schwalb[1],
Holger Bischoff[2], Katrin Eisenreich[2], Franz Färber[2], and Hasso Plattner[1]

[1] Hasso Plattner Institute, Potsdam, Germany
{martin.faust,martin.boissier,marvin.keller,
david.schwalb,hasso.plattner}@hpi.de
[2] SAP SE, Walldorf, Germany
{holger.bischoff,katrin.eisenreich,franz.faerber}@sap.com

Abstract. Databases commonly use multi-column indices for composite keys that concatenate attribute values for fast entity retrieval. For real-world applications, such concatenated composite keys contribute significantly to the overall space consumption, which is particularly expensive for main memory-resident databases. We present an integer-based hash representation of the actual values for the purpose of reducing the overall memory footprint of a system while maintaining the level of performance. We analyzed the performance impact as well as the memory footprint reduction of hash-based indices in SAP HANA in a real-world enterprise database setting. For a live production SAP ERP system, the introduction of hash-based primary key indices alone reduces the entire memory footprint by 10 % with comparable performance.

Keywords: In-memory databases · Hash indices · Footprint reduction · Enterprise systems

1 Composite Keys in Enterprise Applications

Today's trends in hardware development render in-memory databases as a viable platform for enterprise applications. In-memory databases use compression techniques for the purpose of reducing the required main memory. We analyzed the primary keys of a large enterprise resource planning (ERP) installation of a Global 2000[1] company. We found that most tables' primary keys contain multiple columns as shown in Fig. 1(a). To achieve fast data retrieval on these tables, multi-column indices are used. Looking at the memory breakdown shown in Fig. 1(b), we see that composite keys account for nearly 30 % of the entire memory footprint.

In SAP HANA, these multi-column indices are stored as a simple concatenation of the primary key values (hereafter called *value-based indices*). Although various forms of compression are applied to these indices, they introduce additional data stored in DRAM and therefore further add to the memory footprint.

---

[1] Global 2000: http://www.forbes.com/global2000/.

S. Hartmann and H. Ma (Eds.): DEXA 2016, Part II, LNCS 9828, pp. 137–151, 2016.
DOI: 10.1007/978-3-319-44406-2_11

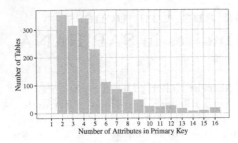
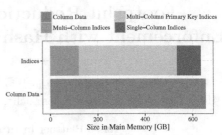
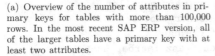

(a) Overview of the number of attributes in primary keys for tables with more than 100,000 rows. In the most recent SAP ERP version, all of the larger tables have a primary key with at least two attributes.

(b) Memory consumption: index structures consume almost as much main memory as the actual data, whereby multi-column primary key indices alone are responsible for over 400 GB.

Fig. 1. Statistics for a live production SAP ERP system of a Global 2000 company: (a) overview of primary key lengths and (b) break down of memory consumption.

In this paper, we evaluate whether we can reduce the size of composite keys by storing a hash-based integer representation of the composite values instead of the actual values concatenated while maintaining the same level of performance.

2 Production Enterprise System: SAP ERP and Columnar In-Memory Databases

An Enterprise Resource Planning (ERP) application is the central management software for large companies. We had the opportunity to analyze a live production system of an SAP ERP system of a Global 2000 company. This system stores over 10 billion records in 23,886 tables with a total main memory footprint of about 1.3 TB. 90 % of these tables have multi-column primary keys, emphasizing the impact a change of the primary key type could have. While analyzing a single instance does not cover the whole ERP market, we consider this system representative since SAP ERP systems have a share of 25 % of the global ERP market and are used by more than half of the Fortune 500 companies.

The ERP system runs on a columnar in-memory database optimized for OLxP workloads: SAP HANA. In-memory database systems like SAP HANA [5] and HYRISE [9] use a main/delta architecture to store database relations. Inserts and updates are handled by a comparatively small and write-optimized partition, called *delta partition*. The delta partition is frequently merged with the *main partition* [5,9]. The main partition is read-only, compressed, and read-optimized towards analytical workloads. This allows for fast analytical queries while still supporting sufficient transactional performance.

Each column is dictionary-encoded consisting of an attribute vector and a dictionary. The dictionary stores all distinct values in a sorted manner while the attribute vector contains bit-packed valueIDs for each record. These valueIDs reference the actual, uncompressed values stored in the dictionary by their offset.

3   Related Work

Database index structures have been optimized in many ways. The general goal is to increase lookup performance while minimizing the additional storage needed for these indices. But with changing trends in hardware, there is the need to further optimize these index structures for their target systems. In-memory databases require new and optimized in-memory indexing structures since traditional indexing strategies become inefficient on modern hardware [1].

Tree-Based Indices. Leis et al. [12] introduced the DRAM-optimized *adaptive radix trees* (ART). By adaptively choosing efficient data structures used in ART, they were able to achieve high space efficiency while surpassing the performance of traditional tree-based index structures.

Athanassoulis and Ailamaki [3] introduced a method of reducing memory requirements of tree-based index structures by employing probabilistic data structures (Bloom filters). By trading accuracy for size, they were able to reduce the footprint of tree-based indices by up to 4× for real-world scenarios while keeping the performance on par with traditional tree indexing. Their motivation was the trend of solid-state disks emerging as a viable alternative to traditional hard disk drives.

Hash-Based Indices. An alternative to tree-based structures is a hash-based data structure that is typically employed in two types: (1) hash tables with fixed size and no reorganization of data and (2) hash tables with variable size and dynamic reorganization. An example for the former is *chained bucket hashing* [8]. Dynamic structures include *extensible hashing* [4], *linear hashing* [10,13] and *modified linear hashing* [11]. Ross presented a method of hash probing for typical database workloads using SIMD instructions [16].

With the usage of in-memory databases with column stores, the problem of efficiently accessing disk blocks is replaced by accessing the main memory and therefore the problem of minimizing cache misses [14]. Sidirourgos and Kersten introduced column imprints as a cache conscious secondary indexing structure for column stores [17]. For each column, a histogram of a few equal-height bins is created. For every cache line of data, a bit vector is created with each bit corresponding to a bin of the histogram. A bit is set if the cache line contains at least one value in the corresponding bin. The authors have shown significantly improved query speed with a storage overhead of only 12 %.

Composite Keys. Faust et al. [6] introduced the composite group-key as an alternative indexing method for composite keys. They utilize the existing dictionary compression by concatenating the compressed values (i.e., valueIDs) of the primary key column's dictionaries and storing them in an additional data structure named key-identifier list. This structure contains integer values with 8, 16, 32, or 64 bits per key and is stored alongside a bit-packed position list to retrieve the record's position. Because they are storing an integer representation of the primary key attributes, the size of this index is significantly smaller than the size of the previously introduced index types that store the key attributes in an uncompressed format. Tests have shown that the composite group-key's performance is on par with established indexing methods while decreasing the storage requirements.

The default method of storing composite keys in SAP HANA is adding an additional column to the table that contains the concatenated values of all primary key columns. For each entry, a compressed (using *Golomb* or *Simple9* compression [2]) position list is stored for fast record retrieval. This allows database operations to only use a single column instead of having to scan every column of the composite primary key, but adds significantly to the overall size of the table, because the primary key values are basically stored twice. The additional key column consists of a sorted dictionary containing the key values, the attribute vector and the position list. For key lookups, the primary key values are concatenated into a single search string that is used in a binary search on the composite key column's dictionary. The position list is used to find the records' positions in constant time. Because this is the current default method of indexing composite primary keys in HANA, we compare the performance and storage requirements of this index type with hash-based indices introduced in the next section.

4   Hash-Based Unique Index

Hash-based indices hash the attributes of a composite-key to obtain a single, fixed-length representation.

4.1   Index Structure

The index is modeled as a dictionary-compressed column, and therefore contains a main and a delta partition. For the main partition, the sorted dictionary D_M stores hashed keys and is extended with an inverted index I_M to provide a mapping to row identifier. For this work, we assume that the inverted index establishes a one-to-one mapping of dictionary entries to position lists, hence, no additional logic is needed to support variable length position lists. Per definition, storing primary key values means there are 100 % unique values in the dictionary what makes traditional dictionary encoding pointless. To reduce the dictionary size, delta encoding is used. The inverted index has the same length like D_M and is bit-packed. The attribute vector is a bit-packed list of offsets in D_{Hash}. The hash index dictionary of the delta partition is unsorted and again each entry is extended by a position list. Figure 2 shows the schematic process to create the inverted hash-based index.

4.2   Lookup Algorithm

The index allows efficient point queries, i.e. the lookup of a key. For a primary key lookup, the predicate has to be translated into its hash representation for comparison with the values in the hash dictionary. This is achieved by concatenating the values of the primary key columns and applying the hash function to

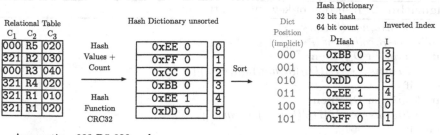

Assumption: 000 R5 020 and
321 R1 010 hash to the same key.

Fig. 2. Schematic overview of the hash-based multi-attribute index on the main partition (delta encoding not shown).

Algorithm 1. Lookup of key with n attributes

$h \leftarrow crc32(concat(k_0, ..., k_{n-1}))$
$match_M \leftarrow I_M[D_M[(h, min)..(h, max)]]$
$match_D \leftarrow I_D[D_D[(h, min)..(h, max)]]$
$MVCCverify(match_M), MVCCverify(match_D)$
$results \leftarrow []$
for P in (M, D) **do**
    for $rowID$ in $match_P$ **do**
        $equal \leftarrow True$
        for $i \leftarrow 0...(n-1)$ **do**
            $equal \leftarrow equal$ & $D_{Pi}[AV_i[rowID]] == k_i$
        end for
        if $equal$ **then**
            $results \leftarrow [results.rowID]$
        end if
    end for
end for
return results

---

it. The resulting hash value is used in a binary search to find matching hashes in the hash dictionaries of the main partition as well as the delta partition. The position of the matching rows is extracted from the inverted index. Because of possible hash collisions, the actual values of all matching tuples have to be compared to find the tuples matching all predicates. The lookup algorithm, including the handling of collisions is shown in Algorithm 1.

4.3 Insert Algorithm

A frequent operation accessing the index is the lookup of a non-existing key for uniqueness constraints, when a new tuple is about to be inserted. The lookup has to be performed first to find rows that would potentially cause uniqueness violations (see Sect. 4.2). For every matching hash that was found, the actual

Algorithm 2. Insertion of key with n attributes

$h \leftarrow crc32(concat(k_0, ..., k_{n-1}))$
$collisions_D, collisions_M \leftarrow []$
if (h, min) in D_M **then**
    $collisions_M \leftarrow I_M[D_M[(h, min)..(h, max)]]$
end if
if (h, min) in D_D **then**
    $collisions_D \leftarrow I_D[D_D[(h, min)..(h, max)]]$
end if
MVCCverify($collisions_M$)
MVCCverify($collisions_D$)
for P in (M, D) **do**
    for c in $collisions_P$ **do**
        $equal \leftarrow True$
        for $i \leftarrow 0...(n-1)$ **do**
            $equal \leftarrow equal$ & & $D_Pi[AV_i[c]] == k_i$
        end for
        if $equal$ **then**
            abort: unique violation
        end if
    end for
end for
$count \leftarrow |collisions_M| + |collisions_D|$ #all collisions refer to different keys
$InsertDelta(h, count)$

attribute values are compared to ensure the uniqueness constraint. If the actual values are different, the hash is inserted into the hash dictionary with an 8-byte collision counter. If the values match, the new record will not be inserted, because of a violation of uniqueness for the primary key. The insert algorithm with verification is detailed in Algorithm 2.

4.4 Limitations

Because the used hash is an integer representation of the whole primary key and does not store the actual attribute values, it is not possible to use hash-based indices for range queries or partial key lookups. SAP HANA automatically creates single-column indices on all attributes of the primary key that are hence used to answer non-full primary key selects. For the value-based index in contrast, range selects and partial key lookups can often be executed directly on the index via binary substring searches (depending on the selected attributes). Consequently, depending on the query filters on multiple columns have to be evaluated for the case of a hash-based primary key while a single access to the value-based index is sufficient for many typical OLTP queries. Further, for partitioned tables hash-based primary keys are only beneficial if the complete key is included in the partitioning criteria.

4.5   Hash Function

The hash function for the index ought to be fast and provide well-distributed hashes for continuously ascending keys. We use CRC32(C) as the hashing function for several reasons. First, cryptographic properties are not needed. Also, other hashing alternatives yield fewer collisions, but the number of expected collisions is limited anyway by SAP HANA's partition size limit of 2^{31} rows per partition. Second, recent Intel CPUs implement the CRC32 instruction in hardware (see Sect. 4.6) with a latency of only three CPU cycles.

Cyclic Redundancy Check (CRC) is a code commonly used for error-detection in digital networks or storage devices to detect unintentional changes in data. A message is encoded by appending a fixed-length check value. The check value is the remainder of the division of a given message by a specified polynomial. The receiver of a message can check its integrity by performing the same division and comparing the check values. The length of the remainder determines the name of the CRC. A CRC with a check value of n bits is called an n-bit CRC or CRCn. We use CRC32, i.e., the remainder has a length of 32 bits. For hash-based indices, we do not use CRC-32 to check data integrity. We use the check value as a shorter (32 bits) integer representation of the primary key values.

The message used as dividend in the polynomial division is the concatenation of the primary key values. To concatenate the key, we create a prefix-free encoding, by prefixing each key attribute with its length. The concatenated string follows the form ``<len(key1)>,key1;<len(key2)>,key2;". Since single partitions do not grow larger than two billion records, a hash length of 32 bits is sufficient.

4.6   CRC32: Hardware-Assisted Hashing

With the SSE4.2 instruction set, Intel added support for hardware-assisted CRC32C to their processors. Traditional CRC32, used for example in ZIP and Ethernet, uses the polynomial $0x04C11DB7$ as divisor while CRC32C, which is supported by SSE4.2, uses the Castagnoli polynomial $0x1EDC6F41$. The SSE4.2 instruction uses a precalculated, built-in lookup table for the Castagnoli polynomial and is therefore limited to this specific polynomial while software implementations can choose the polynomial best suited for their use case. Using different polynomials results in different checksums, i.e. different hashes for the same key.

The CRC32 instruction expects two parameters: a destination operand and a source operand. It uses the fixed polynomial (Castagnoli) to accumulate the CRC32 value for the source operand (i.e., the concatenated key values) and stores the result in the destination operand. The source operand can be a register or a memory location while the destination operand must be a register. This instruction can operate on a maximum data size of 64 bits and is implemented with a latency of three CPU cycles and a throughput of a single CPU cycle. To incrementally accumulate a CRC32 value, the result of the previous CRC32 operation is used to execute the CRC32 instruction again with new input.

The hardware implementation is 2–3× faster than highly optimized software implementations and its performance can be further increased by parallelizing the CRC computations [7]. These capabilities emphasize the viability of CRC32 for the use case of hash-based indices.

4.7   Collision Handling

By definition, any function that maps an unlimited range to a fixed range is prone to collisions. Collisions occur when a hash function creates the same hash for different values. Using CRC32, there are 2^{32} possible hash values. Although this is sufficient for the SAP HANA's maximum of two billion records that can be stored per partition, hash collisions are inevitable and have to be dealt with.

SAP HANA appends an 8-byte counter to the hashes before adding them to the dictionary. The value of this counter is unique and thereby ensures that all values in the dictionary are unique even if hashes for different values match. If a collision occurs while inserting a new record, the insert algorithm compares the actual values to enforce uniqueness, as described in Sect. 4.3. Collisions also have to be expected during key queries. As a consequence, lookups need to verify the actual key components against the predicate, as outlined in Sect. 4.2.

4.8   Column Merge

When merging the content of the delta partition into the main partition, a new main dictionary for the primary key is created. This dictionary contains all distinct hashed key values from the delta dictionary as well as from the old main dictionary. Since any insert into the table has to check for uniqueness in the main partition as well as in the delta partition, primary keys are ensured to be unique and thus the dictionaries can be directly merged. When a hash value of the delta partition already exists in the main partition, the collision counters are simply added and the inverted position list is updated.

4.9   Memory Footprint

Per dictionary entry, a 4-byte hash value is stored along with an 8-byte collision counter to resolve hash collisions. As mentioned earlier, the dictionary containing the hash values is compressed using delta encoding. Since there are only unique values stored in the dictionary, traditional dictionary encoding would have a negative effect on compression. Instead of storing the full values or compressing single values, delta encoding stores only the difference of consecutive values. As a rule of thumb, after compressing the hash-index's dictionary, the average size per entry is 8–10 bytes.

5   Evaluation

We evaluate the potential memory footprint reductions of the hash-based index both on tables of the analyzed live production enterprise system and on a synthetic table of the TPC-C benchmark.

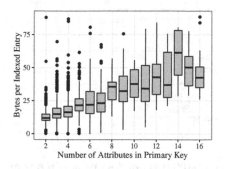
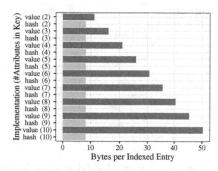

(a) ERP system of a Global 2000 company: space consumption for value-based primary keys (tables with over 100,000 entries).

(b) Bytes per indexed item for a table with 1M integer values.

Fig. 3. Space consumption of composite keys.

We evaluated three tables to cover a broad range of use cases for hash-based indices (an overview of the primary keys is shown in Table 1). The tables BSEG and SKA1 are both table copies of the live production SAP ERP system. BSEG is a transactional table storing accounting documents and is the central part of the financial module. SKA1 is a master data table storing the chart of accounts of the general ledger module. Since it is a master data table, it is considerably smaller than the BSEG table. The third table is TPC-C's largest transactional table ORDERLINE, which we created with a scaling factor of 2,000.

The benchmarks have been executed on the same system with a varying number of benchmark processes. Each benchmark process runs 16 threads (8 for the insert benchmarks) that share the same database connection. SELECT queries solely project the first attribute of the primary key in order to exclude time required for tuple materialization. The benchmark system was a four-socket server equipped with Intel Xeon E7-4880 v2 CPUs and 2 TB of DRAM running SAP HANA SPS 11, revision 111. Error bars denote the standard error.

5.1   Main Memory Footprint

We measured the space consumption of all multi-column indices of the analyzed live production enterprise system. Figure 3(a) shows a box plot of the bytes per indexed entry. For the 1,736 tables with more than 100,000 entries, the average size of an indexed key is about 24 bytes.

Table 1. Overview of the primary keys and their characteristics of benchmarked tables.

	Primary key attributes				
BSEG 70 M tuples	MANDT varchar(3)	BUKRS varchar(4)	BELNR varchar(10)	GJAHR varchar(4)	BUZEI varchar(3)
	Distinct values:	Distinct values:	Distinct values:	Distinct values:	Distinct values:
	1	476	7,777,105	31	999
ORDERLINE 600 M tuples	OL_W_ID integer	OL_D_ID integer	OL_O_ID integer	OL_NUMBER integer	-
	Distinct values:	Distinct values:	Distinct values:	Distinct values:	
	2,000	10	3,000	15	
SKA1 67,618 tuples	MANDT varchar(3)	KTOPL varchar(4)	SAKNR varchar(10)	-	-
	Distinct values:	Distinct values:	Distinct values:		
	1	54	53,598		

The size of value-based indices in large (>100,000 entries) tables in our analyzed system amounts to 386 GB. If we conservatively assume a size of 10 bytes per entry for the hash-based index (see Sect. 4.9), the memory footprint of all composite primary key indices can be reduced by up to 36 % (or 148 GB).

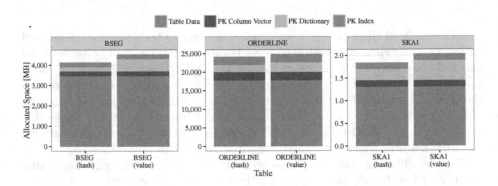

Fig. 4. Break down of memory consumptions for benchmarked tables.

The potential space savings depend on the characteristics of the primary key. Larger keys (i.e., longer concatenations of attribute values) result in larger savings when compressed to 10-byte hashes than smaller keys. Further, with increasing share of primary key columns compared to the total number of columns, the potential space savings of the whole table increase as well. Figure 3(b) illustrates the high impact, the size of the primary key has on memory savings.

Figure 4 shows a breakdown of the memory used by the three benchmarked tables. As discussed, the total memory savings by using the hash-based index depend on both the number of attributes of the primary key and on the data types of the attributes. For the BSEG table with five **varchar** attributes, the footprint reduction for the whole table is around ~10 % due to a 3× smaller dictionary. The reduction of the ORDERLINE table with four integer attributes is smaller with ~1 % for the whole table. To assess hash-based indices for SAP ERP systems it is important to know that the majority of primary key attributes are of type varchar. After analyzing the number of primary key attributes in all large tables of our ERP system (depicted in Fig. 1(a)), we saw that most of these large tables have primary keys with four or more attributes. By using the hash index instead of the value-based index we estimate a footprint reduction of the whole ERP system by 10 %.

5.2   Lookup Performance

We analyzed the latency for three kinds of select queries, all of which are typical for OLTP workloads. We discard OLxP and OLAP queries, because they are usually not accessing primary key indices.

Full Primary Key Selects. A full primary key select describes a lookup query that filters on all attributes of the composite primary key and therefore returns a single record or an empty result set. Our benchmark script executed 10,000 queries per thread and measured the end-to-end latency from sending the query till receiving the data records. The results are shown in Fig. 5. For full primary key selects, we saw a latency increase between 5–15 % for hash-based indices.

Partial Key Selects. Partial key queries describe SELECT statements that select on a true subset of the primary key attributes. These queries are very common in real-world applications and in particular in ERP systems. We modified the full primary key queries to not select on the last attribute of the primary key (e.g., ORDERLINE.OL_NUMBER).

As mentioned in Sect. 4.4, hash-based indices are not accessed for queries selecting anything but the complete primary key. For those queries, the single-column indices created on each primary key attribute are accessed instead. The query latencies are shown in Fig. 5. Depending on the size of the table, the hash-based index in on par with the value-based index (SKA1 table) or is clearly outperformed by up to two orders of magnitude (ORDERLINE table).

Range Queries. As a third reading access pattern, we evaluated range queries. Similar to partial key selects, range queries select on a subset of the primary key attributes but additionally execute a range selection (e.g., ORDERLINE.OL_NUMBER > 10 AND ORDERLINE.OL_NUMBER < 20). We select all rows with BSEG.BELNR and ORDERLINE.OL_NUMBER in a specified range. The size of the ranges was set to return 100 tuples on average.

As mentioned before, hash-based indices cannot be used for range queries. That means, that we are again testing the performance of the additional single columns indices compared to direct binary searches on the dictionary of the

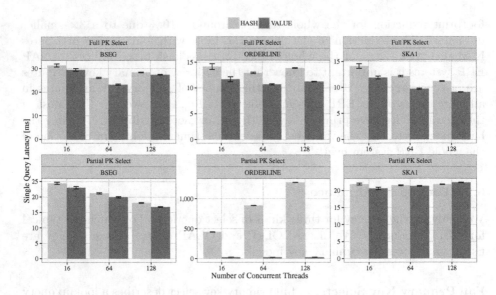

Fig. 5. Latency comparison of full and partial primary key selections.

value-based index. Similar to the partial select, the performance is depending on the size of the table with decreasing performance for increasingly large tables (see Fig. 6).

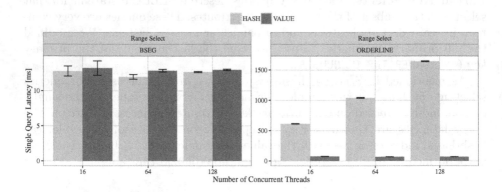

Fig. 6. Latency comparison of range selections.

5.3   Insert Performance

We measured the insert latency on the following three synthetic tables.

SYNTH3: a table with three attributes (varchar and two integers), all are part of the primary key.

SYNTH8: a table with eight attributes (three varchars and 5 integers), all are part of the primary key.

SYNTH100: a table with 100 attributes (30 integers, remainder varchars) of which eight are primary key columns (similar to SYNTH8).

All three tables contain 100 M tuples at the beginning of each test run. The results are shown in Fig. 7. The graphs show that the hash-based index is on par performance-wise with the value-based index for a variety of insert scenarios.

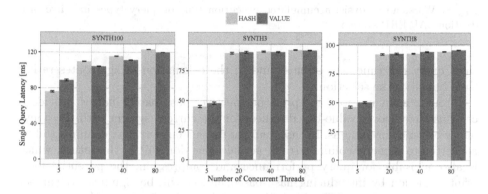

Fig. 7. Comparison for INSERT operations on synthetic tables (100 M tuples) with varying widths.

5.4   Applicability on Enterprise Workloads

The analysis of enterprise system workload by Krueger et al. [9] has shown a trend towards read-dominated workloads. Contrary to benchmarks like TCP-C, OLAP as well as OLTP workloads in modern enterprise applications consist of mostly read queries. Further, applications optimized for a column-based architecture and without materialized aggregates as in SAP's *simplified Financials* (sFIN) applications emphasize that trend [15]. The analysis of the sFIN workload, which is illustrated in Fig. 8, has shown that over 98 % of the application's total execution time is spent on read queries. 14 % of the total time are spent on primary key selects while the remaining 84 % are more complex select queries like joins and aggregations. Insert statements only account for 1.3 % of the total execution time.

We estimate the overall impact of hash indices based on the analyses in Sect. 5 to be rather low from a performance perspective. With the exception of range queries, the hash-based indices perform on par with the value-based indices for OLTP-like queries. Since the share of range queries on the (partial) primary key is rather low, the performance drop is neglectable. With an increasing share of complex and computation-intensive OLxP and OLAP queries in future systems, the performance of the primary key will have a decreasing impact. Especially

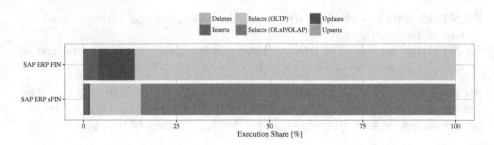

Fig. 8. Workload analysis: accumulated execution time of query types in a live production SAP ERP system.

since query run times are often bound to the calculation of aggregates rather than bound to the selection.

From a main memory footprint perspective, hash-based indices provide a clear advantage over value-based indices for the current system with the dominance of `varchar` columns. In case many of the current `varchar` columns will be converted to numeric columns (their actual value domain) in the future, which is also advisable for query performance and compression, the potential footprint reduction by introducing hash-based indices will be significantly smaller (compare table `ORDERLINE` with integer attributes in Fig. 4). In that case, the composite group-key is a viable alternative (see Sect. 3).

6   Conclusion

Hash-based primary key indices can be used to reduce the main memory footprint of an enterprise application while maintaining the level of performance for typical OLTP query patterns. We saw that footprint reductions and performance of hash-based indices depend on the characteristics of the tables they are applied to and the workload of the application. For recent enterprise systems optimized for column-based architectures, we expect a comparable performance when using hash-based indices over value-based indices while decreasing the entire main memory footprint by 10 %.

References

1. Ailamaki, A., et al.: DBMSs on a modern processor: where does time go? In: VLDB 1999, Proceedings of 25th International Conference on Very Large Data Bases, pp. 266–277 (1999)
2. Anh, V.N., Moffat, A.: Inverted index compression using word-aligned binary codes. Inf. Retr. **8**(1), 151–166 (2005)
3. Athanassoulis, M., Ailamaki, A.: BF-Tree: approximate tree indexing. Proc. VLDB Endowment **7**, 1881–1892 (2014)

4. Fagin, R., Nievergelt, J., Pippenger, N., Raymond Strong, H.: Extendible hashing a fast access method for dynamic files. ACM Trans. Database Syst. (TODS) **4**(3), 315–344 (1979)
5. Färber, F., et al.: SAP HANA database: data management for modern business applications. ACM Sigmod Rec. **40**(4), 45–51 (2012)
6. Faust, M., Schwalb, D., Plattner, H.: Composite group-keys. In: Jagatheesan, A., Levandoski, J., Neumann, T., Pavlo, A. (eds.) IMDM 2013/2014. LNCS, vol. 8921, pp. 139–150. Springer, Heidelberg (2015)
7. Gopal, V., et al.: Fast CRC computation for iSCSI Polynomial using CRC32 instruction. Technical report, Intel Corporation (2011)
8. Knuth, D.E.: The Art of Computer Programming: Sorting and Searching, vol. 3. Pearson Education, USA (1998)
9. Krueger, J., et al.: Fast updates on read-optimized databases using multi-core cpus. Proc. VLDB Endowment **5**(1), 61–72 (2011)
10. Larson, P.-A.: Linear hashing with separators—a dynamic hashing scheme achieving one-access. ACM Trans. Database Syst. (TODS) **13**(3), 366–388 (1988)
11. Lehman, T.J., Carey, M.J.: A study of index structures for main memory database management systems. In: Conference on Very Large Data Bases, vol. 294 (1986)
12. Leis, V., Kemper, A., Neumann, T.: The adaptive radix tree: artful indexing for main-memory databases. In: 2013 IEEE 29th International Conference on Data Engineering (ICDE), pp. 38–49. IEEE (2013)
13. Litwin, W.: Linear hashing: a new tool for file and table addressing. In: VLDB, vol. 80, pp. 1–3 (1980)
14. Manegold, S., Kersten, M.L., Boncz, P.: Database architecture evolution: mammals flourished long before dinosaurs became extinct. Proc. VLDB Endowment **2**(2), 1648–1653 (2009)
15. Plattner, H.: The impact of columnar in-memory databases on enterprise systems. Proc. VLDB Endowment **7**(13), 1722–1729 (2014)
16. Ross, K.A.: Efficient hash probes on modern processors. In: IEEE 23rd International Conference on Data Engineering, ICDE, pp. 1297–1301. IEEE (2007)
17. Sidirourgos, L., Kersten, M.: Column imprints: a secondary index structure. In: Proceedings of the ACM SIGMOD International Conference on Management of Data, pp. 893–904. ACM (2013)

Benchmarking Replication in Cassandra and MongoDB NoSQL Datastores

Gerard Haughian[1], Rasha Osman[2(✉)], and William J. Knottenbelt[1]

[1] Department of Computing, Imperial College London, London SW7 2AZ, UK
{gh413,wjk}@imperial.ac.uk
[2] Faculty of Mathematical Sciences, University of Khartoum, Khartoum, Sudan
rosman@ieee.org

Abstract. The proliferation in Web 2.0 applications has increased the volume, velocity, and variety of data sources which have exceeded the limitations and expected use cases of traditional relational DBMSs. Cloud serving NoSQL data stores address these concerns and provide replication mechanisms to ensure fault tolerance, high availability, and improved scalability. In this paper, we empirically explore the impact of replication on the performance of Cassandra and MongoDB NoSQL datastores. We evaluate the impact of replication in comparison to non-replicated clusters of equal size hosted on a private cloud environment. Our benchmarking experiments are conducted for read and write heavy workloads subject to different access distributions and tunable consistency levels. Our results demonstrate that replication must be taken into consideration in empirical and modelling studies in order to achieve an accurate evaluation of the performance of these datastores.

1 Introduction

The volume, velocity and variety of data produced and consumed by organizations in recent years has outgrown the capabilities of traditional relational DBMSs, due to the explosion of the web generated content [10]. New data stores have been designed to accommodate this emerging landscape; some of which have even been designed to work exclusively in the cloud. A main feature of these cloud data stores is horizontal scalability and high availability. Horizontal scalability is achieved through linear expansion of the data store as the workload increases. High availability is achieved through replicating the data across different machines and data centers.

NoSQL data stores use eventual consistency protocols to ensure that replicated data in some time in the future will be consistent [1]. Each data store provides consistency guarantees to (1) control how the data is distributed between the nodes of the cluster, (2) define how read and write requests are handled, (3) determine when different copies of the data are updated, and (4) specify the accepted level of consistency of the data. The *replication factor* (RF) is the number of times a data item is duplicated across the cluster, which in most data store architectures reflects the number of physical nodes that hold a copy of the

S. Hartmann and H. Ma (Eds.): DEXA 2016, Part II, LNCS 9828, pp. 152–166, 2016.
DOI: 10.1007/978-3-319-44406-2_12

data item. The defined *consistency level* specifies how many of the replicas/nodes must respond to a request for the request to be considered valid.

Replication strategies and consistency levels impact the performance of the data store. Lower consistency levels provide lower latencies while stricter consistencies incur the overhead of inter-node communication and data passing. The performance comparison of replication and consistency guarantees is complicated by the different protocols implemented in NoSQL data stores. In this paper, we consider multi-master (Cassandra) and master-slave (MongoDB) replication and their corresponding consistency protocols.

There has been an increased interest in the benchmarking and performance of NoSQL data stores. However, the majority of the benchmarking studies in industry and academia do not consider the effect of replication in their studies. Further, different data access patterns are not investigated, as most depend on the uniform access of data and the disabling of consistency guarantees within their configurations. In contrast, this paper aims to fill a gap in the performance and benchmarking literature by presenting a benchmarking study in which we evaluate the impact of replication and consistency guarantees on the performance of Cassandra [2] and MongoDB [3]. This paper contributes the following.

- We illustrate the impact of replication on the performance of Cassandra and MongoDB NoSQL data stores using various cluster sizes in comparison to non-replicated clusters of equal sizes. Specifically, we analyze the impact of read and write heavy workloads under different levels of tunable consistency on the underlying optimizations and design decisions for each datastore.
- We provide insight into each data store's suitability to different industry applications by experimenting with three different data and access distributions, each simulating a different real-world use case.
- Our results demonstrate that replication and consistency levels have a direct impact on the performance of Cassandra and MongoDB. Therefore replication must be taken into consideration in empirical and modelling studies in order to achieve an accurate evaluation of the performance of these datastores.

This rest of this paper is organized as follows. Related work is presented in Sect. 2. Section 3 details the data stores benchmarked in this study. The experimental setup is described in Sect. 4. Benchmarking results are detailed in Sect. 5 and discussed in Sect. 6. Conclusions and future work are presented in Sect. 7.

2 Related Work

The development of the Yahoo! Cloud Serving Benchmark tool (YCSB) [4] has led to numerous benchmarking studies of NoSQL datastores. Cooper et al. [4] benchmarked HBase, Cassandra, PNUTS and *sharded* MySQL to illustrate the performance and scalability trade-offs of each system. Pirzadeh et al. [20] evaluated range query dominant workloads on Cassandra, HBase, and Voldemort. Rabl et al. [22] compared Redis, Cassandra and VoltDB in their ability to scale to support application performance management tools. The work in [21] compares

Voldemort and Cassandra for scalability, performance and focusing on failover characteristics under different throughputs.

Dede et al. [6] evaluated the use of Cassandra for Hadoop, discussing various features of Cassandra, such as replication and data partitioning which affect Hadoop's performance. The work evaluated different replication factors with a single consistency level on clusters of up to 8 nodes. The previous benchmarking studies evaluated NoSQL datastores with non-replicated or limited replication data configurations and thus evaluating the impact of replication and different consistency levels on performance was beyond their scope. In contrast to this work, most studies assumed uniform access and data distribution which does not accurately stress the datastore.

Industrial benchmarking studies [5,7,16,17,23], configured the data stores with constant replication factors with no comparisons to baseline configurations or assessment of different access and consistency levels. Some studies tackled a very narrow problem domain (i.e., [7,17]) by highly optimizing their studies for specific use cases or for specific data stores as in [23]. Similarly, performance modelling studies either considered configurations with no replication or replication with uniform distributions and access as in [8,18,19].

In this paper, we present a benchmarking study that examines the impact of replication, tuneable consistency levels and data and access distributions on the performance of two popular NoSQL datastores: Casandra and MongoDB. We investigate their performance using different replication factors selected based on the architecture of the data store using uniform, Zipfian and latest data and access distributions. We evaluate the impact of these configurations by comparing to non-replicated clusters of equal size with uniform data and access distributions.

To evaluate the effect of different consistency levels on performance we employ three different levels of consistency: (1) ONE: which indicates that only one node at most needs to reply to a request, (2) QUORUM: a specific number of nodes must reply before the request is considered valid, and (3) ALL: all nodes holding a copy of the data item must reply before a request is returned to the client. Each data store implements different replication strategies and thus these consistency levels may not be directly defined within the configuration parameters of the data store. For such cases, we have configured the data store to the closest possible configuration that produces the same level of consistency. In the following, we summarize the properties of Cassandra and MongoDB focusing on their replication strategies and consistency configurations.

3   Systems Under Investigation

3.1   Cassandra

Cassandra is a distributed extensible record data store, developed at Facebook [11] for storing large amounts of unstructured data on commodity servers. Cassandra's architecture is a peer-to-peer distribution model [10] with no single point of failure thus supporting high availability and horizontal scalability.

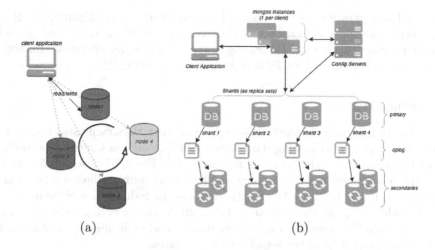

Fig. 1. The (a) Cassandra and (b) MongoDB architectures.

Data is distributed evenly across the cluster to guarantee load balancing. Cassandra offers tunable consistency settings for reads and writes, which provide the flexibility to make tradeoffs between latency and consistency [11]. For each read and write request, users choose one of the predefined consistency levels: ZERO, ONE, QUORUM, ALL or ANY. In this study, we investigated ONE, QUORUM and ALL.

Cassandra automatically replicates records throughout a cluster based on a user specified replication-factor which determines which nodes are responsible for which data ranges. Client applications can contact any node, which acts as a coordinator and forwards requests to the appropriate replica node(s) that store the data. This mechanism is illustrated in Fig. 1(a). A write request is sent to all replica nodes; however the consistency level determines the number of nodes required to respond for the transaction to be considered complete. For a read request, the coordinator contacts the number of replica nodes specified by the consistency level. Cassandra is optimized for large volumes of writes as each write request is treated like an in-memory operation, while all I/O is executed as a background process. In contrast, read requests require in-memory and I/O operations in addition to consistency checks between data returned from the replicas. Keeping the consistency level low makes read operations faster as fewer replicas are checked before returning the call.

For this study, Cassandra version 1.2.16 (the latest 1.X release available before commencing this study) was used based on the supported YCSB (see Sect. 4) Cassandra client driver with most of the default configurations. Hinted-handoff (a mechanism to ensure consistency of the cluster in the event of a network partition [10]) was disabled on all nodes within the cluster to avoid the hints building up rapidly within the cluster when a node fails. The tokens representing the data range for each node in each independent cluster configuration was pre-calculated and saved in separate copies of Cassandra configuration files. Finally, the RPC server type was changed to *hsha* to reduce the amount of memory used by each Cassandra node; this is ideal for scaling to large clusters. Justifications for these configurations and other Java JVM setting can be found in [9].

3.2   MongoDB

MongoDB is a document-oriented NoSQL data store that facilitates horizontal scalability by auto-partitioning data across multiple servers known as *sharding*. MongoDB's sharded architecture is represented in Fig. 1(b). Each shard exists as a *replica set* providing redundancy and high availability. Replica sets consist of multiple Mongo Daemon (*mongod*) instances, including an arbiter node[1], a master node acting as the primary, and multiple slaves acting as secondaries which maintain the same data set. If the master node crashes, the arbiter node elects a new master from the set of remaining slaves.

All write operations must be directed to a single primary instance. By default, clients send all read requests to the master; however, a *read preference* is configurable at the client level on a per-connection basis, which makes it possible to send read requests to slave nodes instead [15]. Varying read preferences offer different levels of consistency guarantees. Balancing is the automatic process used to distribute the data of a sharded collection evenly across a sharded cluster which takes place within the *mongos* App server (required in sharded clusters) [14].

In this study, we used MongoDB version 2.6.1 with all standard factory settings, with the exception that journaling (i.e., logging) was disabled since the overhead of maintaining logs to aid crash recovery was considered unnecessary in this work. We setup only one configuration server which resided on the same host as a single App server. Clients interacted with this App server exclusively. It has been shown that having only one configuration server is adequate for development environments [13]. In addition, we have observed that having both servers reside on the same host did not prove to be a bottleneck.

MongoDB replication operates by way of an *oplog*, to which the master node logs all changes to its data sets. Slave nodes then replicate the master's oplog, applying those operations to their data sets. This replication process is asynchronous, therefore slave nodes may not always reflect the most up to date data. Varying *write concerns* can be issued per write operation to determine the number of nodes that should process a write operation before returning to the client successfully. This allows for fine grained tunable consistency settings, including quorum and fully consistent writes [12].

MongoDB offers different write concerns for varying tunable consistency settings, of which NORMAL, QUORUM, and ALL write concerns where explored. MongoDB allows for concurrent reads on a collection, but enforces a single threaded locking mechanism on all write operations to ensure atomicity. In addition, all write operations need to be appended to the oplog on disk, which involves

---

[1] An arbiter node does not replicate data and only exists to break ties when electing a new primary if necessary.

greater overhead. In contrast, regardless of the requested *read concern* no additional consistency checks are performed between replicas on read operations.

4 Experimental Setup

YCSB Configuration. The Yahoo Cloud Serving Benchmark (YCSB) [4] was developed to support benchmarking cloud NoSQL data stores. We use the YCSB benchmark to execute our benchmarking experiments on Cassandra and MongoDB. However, for the purpose of this work, we have extended its functionality as described below.

Central to the YCSB tool is the YCSB Client, which when executed in load mode inserts a user specified number of randomly generated records of size 1 Kb into a specific data store with a specified distribution. In run mode, the chosen distribution determines the likelihood of certain records being read or updated. We use the following data and access distributions in the experiments, each simulating a different industry application use case [4]:

- *uniform*: items are chosen uniformly, this represents applications where the number of items associated with an event are variable, e.g., blog posts.
- *Zipfian*: items are chosen according to popularity irrespective of insertion time, this represents social media applications where popular users have many connections, regardless of the duration of their membership.
- *latest*: similar to the Zipfian distribution except items are chosen according to latest insertion time, this represents applications where recency matters, e.g., news is popular at its time of release.

In this study, one read-heavy and one write-heavy workload is used to stress the data stores. The read-heavy workload (referred to as G) is one of the default workloads provided within the YCSB Core Package; i.e., workload B comprising a 95/5 % breakdown of read/write operations. The write-heavy workload (referred to as H) was custom designed to consist of a 95/5 % breakdown of write/read operations. After preliminary tests, we configured the YCSB client to a fixed eight threads per CPU core, similar to [4]. For the Cassandra and MongoDB, which are not single threaded and can make use of all available CPU cores, a total of sixty-four threads were used. In order to accurately evaluate the effect of replication on data store performance, we did not increase the workload as the cluster size increased.

For MongoDB, the YCSB Client does not support write concerns or read preferences, therefore we extended the YCSB Client to facilitate this. A listing of these extensions are given in [9]. For all experiments the primary preferred read preference was used to favor queries hitting the master, however if the master was unavailable, requests would be routed to a replicated slave. For Cassandra, the configuration for the maximum number of concurrent reads and writes was increased to match the same number of threads used by the YCSB Client, i.e., sixty-four threads.

Further, we included an additional warm-up stage to the YCSB code base to improve results and comparative analysis by using the open-source[2] warm-up extension developed for the studies in [16,17]. Averages of the time for the data store to stabilize at or above the overall average throughput of a given experiment can be found in [9]. These warm-up times where subsequently passed as an additional configuration parameter to the YCSB Client for run phases only.

Table 1. Virtual machine specifications and settings.

Setting	Value
OS	Ubuntu 12.04
Word Length	64-bit
RAM	6 GB
Hard Disk	20 GB
CPU Speed	2.90 GHz
Cores	8
Ethernet	gigabit
Additional Kernel Settings	*atime* disabled

All experiments conducted in this study where carried out on a cluster of Virtual Machines (VM) hosted on a private cloud infrastructure within the same data center. Each VM had the same specifications and kernel settings as indicated in Table 1. A total of 14 VM nodes where provisioned for this study. One node was designated for the YCSB Client, and one additional node was reserved for MongoDB configuration and App servers which are required in *sharded* architectures to run on separate servers from the rest of the cluster. The remaining 12 nodes operated as standard cluster nodes which had both data stores installed but only one running at any given time. To ensure all nodes could interact effectively, each node was bound to a fixed IP address and each node was aware of the IP addresses of the other nodes.

Data Store Configuration. Each data store was configured and optimized for increased throughput, low latency, and where possible to avoid costly disk-bound operations. Each data store node hosted enough data to utilize a minimum of 80 % RAM. MongoDB was configured to have a constant replication factor of two replicas per *shard*, meeting the minimum recommended production settings. The number of shards were incremented from one shard with two replicas up to 4 shards each with two replicas, in order to directly explore the write-scalability of MongoDB. This corresponds to cluster sizes of three nodes up to 12 nodes. Cassandra, due to its multi-master architecture, was evaluated on 3 to 12 node clusters, in which the replication factor was increased with the increase in cluster

---

[2] Available at https://github.com/thumbtack-technology/ycsb.

size from two to 8. For both datastores experiments were also conducted on one node clusters with no replication.

To accurately evaluate the impact of replication on datastore performance, we conducted base line experiments for comparison. These base line experiments consisted of maintaining the same cluster sizes, with no replication, using the uniform distribution only. We limited ourselves to the uniform distribution as it has been used in previous benchmarking experiments and performance modelling papers to evaluate different scenarios. Each set of experiments was repeated a minimum of three times. For each experiment: there is a warm-up phase, and the main run phase for 10 minutes and a final cool down phase. To ensure all experiments and their iterations start with the same initial state, at the end of each iteration the entire cluster is torn down and a new cluster is reconfigured and loaded with data.

5   Experimental Results

In this Section, we report the results of our benchmarking experiments. For each data store we present results of replicated clusters for each workload under different consistency levels and compare with the corresponding non-replicated baseline clusters of equal size. Confidence intervals were calculated for all results and can be found in [9], however there were too tight to appear in the graphs. In addition, due to space limitations results for read and write latencies are available in [9].

5.1   Cassandra

Throughput. From Fig. 2, the effect of replication on the performance of Cassandra is very clear, as the trends of throughput for replicated clusters are directly opposite to those for non-replicated clusters. On a single non-replicated node, throughputs are 45.7 % higher for the write-dominated workload (H) than the read-dominated workload (G). This is expected due to Cassandra's write optimized architecture. Further, the throughput on non-replicated clusters for workload H consistently outperforms workload G by an average of 33.1 % per cluster size. In contrast, for replicated cluster sizes greater than one, we observe an average of 39.6 %, 37.9 %, and 30.3 % decrease in throughput for the write-heavy workload (H) compared to workload G, across all cluster sizes and consistency levels for uniform, Zipfian, and latest distributions respectively. This corresponds to a 19.5 %, 38.6 %, and 49.7 % decrease on average across all cluster sizes and distributions for ONE, QUORUM, and ALL consistency levels respectively.

Performance is most affected by the strictest consistency level ALL. This suggests that Cassandra is scalable at the cost of maintaining a lower level of consistency. However, stronger consistency levels tend to reduce scalability as the cluster size and replication factor increase. The QUORUM consistency level demonstrates a promising *sweet spot* in the consistency versus throughput

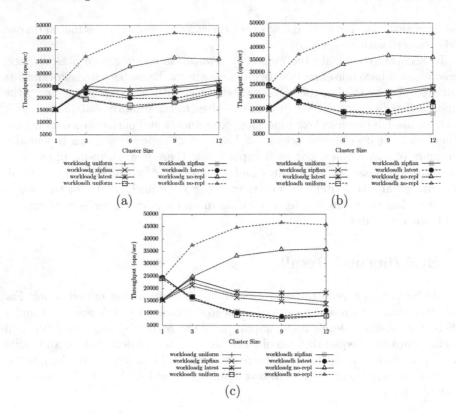

Fig. 2. Cassandra: overall throughputs per consistency level for all workloads and distributions: (a) ONE (b) QUORUM (c) ALL.

tradeoff battle. Moreover, stricter consistencies have a much greater impact on write-heavy workloads than on read-heavy workloads.

Access Distributions. For workload G, we observe that the uniform distribution on average outperforms the Zipfian and latest distributions by 4.2 % and 0.8 % respectively. Given that the YCSB client selects a node at random for forwarding requests, this is likely to impact relative performance between distributions, favoring the uniform distribution due to a stronger correlation in their random number generators. In addition, the uniform distribution will spread the requests more evenly throughout the cluster. However, for workload H the latest distribution on average outperforms the uniform and Zipfian distributions by 7.1 % and 9.7 % respectively. Zipfian's poorer performance could be related to high disk access due to one key being frequently updated.

Impact of Replication. From Table 2, when comparing replicated clusters to non-replicated clusters of equal size, we observe a consistent ordering of performance metrics for both workloads based on the consistency level. For workload

Table 2. Cassandra: the difference (%) In overall throughput between replicated and non-replicated clusters per workload.

		Workload G				Workload H			
cluster size		3	6	9	12	3	6	9	12
replication factor		2	4	6	8	2	4	6	8
Uniform	ONE	1.5	31.5	38.3	27.8	61.3	90.5	87.5	68.3
	QRM	8.6	46.4	49.6	36.8	70.5	105	114.2	129.6
	ALL	9.9	61.4	76	84.9	77.0	131.1	143.2	135.8
Zipfian	ONE	3.6	37.4	40.2	34	61.1	93.8	88.4	71.8
	QRM	7.1	49.2	51	42.2	68.8	115	120.6	110.7
	ALL	15.2	67.1	86.1	91.1	81.0	123.8	139	136.2
Latest	ONE	2.8	45.6	47.2	35.9	50.7	76.6	80	64.8
	QRM	6.3	53.2	56.7	43.6	68.3	106.6	107.3	87.1
	ALL	2.7	55.5	67.4	65.7	76.7	127	137.4	122.2

G, we see an average of 28.8 %, 55.1 % and 94.4 % decrease in throughput for consistency levels ONE, QUORUM and ALL, respectively for all distributions, cluster sizes and replication factors compared to non-replicated clusters of equal size. For workload H, there is an average decrease of 74.6 %, 104 %, and 120.7 % in throughput for consistency levels ONE, QUORUM and ALL respectively compared to non-replicated clusters of equal size. As the cluster size and replication factor increase more nodes are required to confirm each operation resulting in additional overhead and reduced performance. This trend is a reflection of Cassandra's architecture favoring availability and network partition tolerance over consistency. We note that the impact of replication on the write-heavy workload is more evident due to the overhead of updating data within the cluster.

5.2 MongoDB

Throughput. The effect of MongoDB's contrasting consistency checks for reads and writes is evident from Fig. 3 in which the throughput of the read-heavy workload (G) has on average an 89 % higher level of throughput than the write-heavy workload (H). This corresponds to 94.8,%, 84,%, and 87.2,% increases for uniform, Zipfian, and latest distributions respectively, on average across all consistency levels and cluster sizes. When broken down by consistency level, we can observe a 89.5, %, 87.1,%, and 89.5,% increase for ONE, QUORUM, and ALL consistency levels respectively. Figure 3 illustrates how this trend varies as the cluster size increases. For both workloads we observe a performance drop from cluster sizes 1 to 3. This is due to an additional replication factor of two being applied to the single shard in the 3 node cluster. The master node now needs to save data to an oplog on disk and manage two additional servers. As the cluster size increases above 3 nodes more shards distribute the load of reads

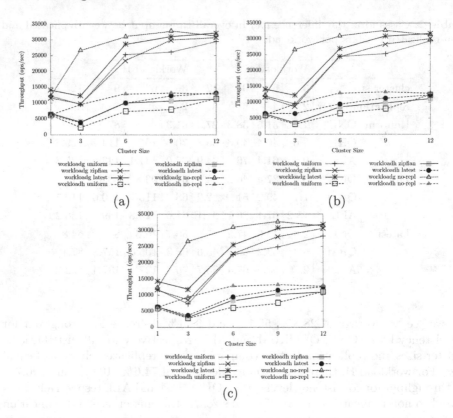

Fig. 3. MongoDB: overall throughputs per consistency level for all workloads and distributions: (a) ONE (b) QUORUM (c) ALL.

and writes and thus there is an increase in throughput following the trend of the baseline non-replicated clusters of equal size.

For all subsequent cluster sizes (6+), the average decrease in throughput is only 13.6 % and 40.3 % for workload G and H respectively in comparison to the non-replicated clusters. This suggests that replication has a lesser effect on performance for read-heavy workloads once the overhead of maintaining a small number of shards have been overcome. When comparing based on the consistency levels, we observe higher throughputs for a consistency level of ONE on average across all distributions and cluster sizes, with slight degradations for QUORUM and ALL consistency levels.

Access Distributions. The latest distribution outperforms the Zipfian and uniform distributions for both workloads. For workload G, the latest distribution has a 15 % and 17.9 % increase in throughput on average across all cluster sizes and consistency levels compared to the Zipfian and uniform distributions respectively. For workload H, the latest distribution has a 10.9 % and 27.9 % increase in throughput on average across all cluster sizes and consistency levels compared to Zipfian and uniform distributions respectively. This is expected as MongoDB stores all data on disk and reads data into RAM on a need to basis. The latest and Zipfian distributions would outperform the uniform distribution as accessed data would be in main memory after a short number of operations. Further, the warm-up stage added to the YCSB Client gives an added advantage to the latest and Zipfian distributions in this regard.

Impact of Replication. The impact of replication is more evident for write-heavy workloads due to the effect of consistency checks performed on writes. Table 3 shows the difference in percentages between replicated and non-replicated clusters of equal size for all experiments. From Table 3, the impact of replication on the performance of workload H in comparison to workload G, especially for cluster sizes 6+, is evident in the large differences between the throughput of workload H and that of the baseline non-replicated clusters of equal size. The effect of the access skew is clear when comparing to the baseline non-replicating clusters, as shown in Table 3. For the read-heavy workload, when comparing to the baseline non-replicated clusters of equal size, the Zipfian and latest distributions mitigate the overhead of replication due to the availability of data in main memory. This is not the case for the uniform distribution where the impact of replication is evident. When considering the write-heavy workloads, the increase in disk access on multiple replicas leads to the increased impact of replication, irrespective of access distribution, consistency level or cluster size.

Table 3. MongoDB: the difference (%) In overall throughput between replicated and non-replicated clusters per workload.

		Workload G				Workload H			
cluster size		3	6	9	12	3	6	9	12
replication factor		2	2	2	2	2	2	2	2
Uniform	ONE	94.1	20.4	22.5	6.2	120.4	54.1	49.6	12.9
	QRM	95.4	23.7	25.4	6.7	99	64.8	49	15.3
	ALL	113.6	31.5	27.1	10.5	107.2	71.4	54	15.5
Zipfian	ONE	93.6	28.2	9.3	1.7	78.4	24.9	21	14.6
	QRM	101.5	23.4	14.5	4.0	89	41	29	15.7
	ALL	102.5	30.1	15.5	2.5	97.8	43	24.3	16.2
Latest	ONE	73.8	7.8	5.6	2.5	83.6	25.1	7.1	1.7
	QRM	74	14	5.7	1	37.4	30.7	15.6	4.5
	ALL	77.6	19.4	6.0	1.6	76.7	127	137.4	122.2

6   Discussion

Throughput. For the read-heavy workload (G), MongoDB (averaging 21230 ops/sec) is only marginally better than Cassandra (which averages 20184 ops/sec) by 5.1 %. For workload H which is write dominated, the greatest difference is that Cassandra outperforms MongoDB by 72.5 %. This stark contrast is a clear indication of Cassandra's write optimized architecture.

For the read-heavy workload (G), MongoDB demonstrates better performance with the latest distribution, whereas Cassandra performs best with the uniform distribution. MongoDB outperforms Cassandra on all distributions, except for the uniform distribution in which Cassandra has better throughputs than MongoDB. Cassandra's better performance on read-heavy workloads with a uniform distribution is likely a result of a strong correlation between how the YCSB Client selects a node randomly for routing requests, spreading the requests more evenly across the cluster. Whereas the latest distribution would force the same set of nodes to constantly handle operations, causing a backlog of read-repairs to build up. When accessed with the latest distribution, MongoDB is only 1.1 times more performant than Cassandra.

For the write-heavy workload (H), the latest distribution once again outperforms all other distributions on average across all cluster sizes and consistency levels, followed by Zipfian, except for Cassandra which performs second best with the uniform distribution. When all data stores are accessed with the latest distribution, Cassandra is 2 times better than MongoDB. The reason we observe larger contrasts in relative performance compared to workload G, is because Cassandra is write optimized delaying consistency checks for read time. In contrast, MongoDB performs consistency checks at write time.

Replication. To assess the impact replication on data store performance, we compare two different replication strategies, i.e., the multi-master model used by Cassandra, and the replica set model used by MongoDB. We can observe that apart from the exception of consistency level ONE on workload G, for cluster sizes 6+, MongoDB's replica set replication model has less of an impact on throughput performance than Cassandra's multi-master replication model when compared to non-replicated clusters of equal size. Cassandra's replication model accounts for a 41.1 %, and 98 % throughput degradation for all consistency levels and distributions, averaged across all replicated clusters sizes for workload G and H respectively. In contrast, MongoDB's replication model only accounts for 33 % and 52 % degradation in throughput for workload G and H respectively. This suggests that MongoDB's master-slave replication architecture has less of an effect on cluster performance than Cassandra's multi-master architecture. This is a result of each master and slave being responsible for a single data partition leading to reduced access contention compared to the multi-master model used by Cassandra in which each node contains more than one unique partition on a single server.

Performance Summary. Write-heavy workloads on non-replicated Cassandra clusters are able to exploit Cassandras write-optimized architecture. In contrast, replication has a noticeable impact on the performance of write-heavy workloads in comparison to read-heavy workloads. Cassandra is scalable at the cost of maintaining a lower level of consistency, we observed 65 % and 75 % degradations in performance between consistency levels ONE and ALL for read-heavy and write-heavy workload respectively. Stricter consistency levels have a greater impact (9 %) on write-heavy workloads than on read-heavy workloads. Read-heavy workloads perform best when data access is random or close to random. For write-heavy workloads, memory resident datasets provide better performance (as represented by Zipfian and lastest distributions).

MongoDBs architecture is highly read-optimized, with read-heavy workloads outperforming write-heavy workloads on average by 90 % across all cluster sizes, distributions and consistency levels. An interesting observation is that replication has minimal impact on performance relative to non-replicated clusters of equal size once the overhead of maintaining a small number of shards have been overcome. In addition, stricter consistency levels have on average a 5 % impact on performance for both workloads. MongoDB performance is best when the entire or majority of the working data set can be kept in RAM as it would be for latest and Zipfian distributions.

7   Conclusions and Future Work

This study benchmarked replication in Cassandra and MongoDB NoSQL data stores, focusing on the effect of replication on performance compared to non-replicated clusters of equal size. To increase the applicability of this study to real-world use cases, a range of different data access distributions (uniform, Zipfian, and latest) were explored along with three tunable consistency levels: ONE, QUORUM, and ALL, and two different workloads: one read-heavy and one write-heavy. Our experiments have shown that master-slave type replication models, as exhibited by MongoDB tend to reduce the impact of replication compared to multi-master replication models exhibited by Cassandra. These results demonstrate that replication must be taken into consideration in empirical and modelling studies in order to achieve an accurate evaluation of the performance of these datastores. For future work, we plan to conduct a similar benchmarking study on the Amazon EC2 cloud, extending experiments to include larger data sets and cluster sizes while making use of solid-state disks to better reflect industry standard deployments.

References

1. Bailis, P., Venkataraman, S., Franklin, M.J., Hellerstein, J.M., Stoica, I.: Probabilistically bounded staleness for practical partial quorums. Proc. VLDB Endow. 5(8), 776–787 (2012)
2. Cassandra. http://cassandra.apache.org/

3. Chodorow, K.: MongoDB: The Definitive Guide. O'Reilly Media Inc, Sebastopol (2013)
4. Cooper, B.F., Silberstein, A., Tam, E., Ramakrishnan, R., Sears, R.: Benchmarking cloud serving systems with ycsb. In: Proceedings of the 1st ACM symposium on Cloud computing, pp. 143–154. ACM (2010)
5. Datastax Coperation. Benchmarking top NoSQL databases. A performance comparison for architects and IT managers (2013)
6. Dede, E., Sendir, B., Kuzlu, P., Hartog, J., Govindaraju, M.: An evaluation of cassandra for hadoop. In: IEEE Sixth International Conference on Cloud Computing (CLOUD), pp. 494–501. IEEE (2013)
7. Diomin and Grigorchuk. Benchmarking Couchbase server for interactive applications (2013). http://www.altoros.com/
8. Gandini, A., Gribaudo, M., Knottenbelt, W.J., Osman, R., Piazzolla, P.: Performance analysis of nosql databases. In: 11th European Performance Engineering Workshop (EPEW) (2014)
9. Haughian, G.: Benchmarking Replication in NoSQL Data Stores. Master's thesis, Imperial College London, UK (2014)
10. Hewitt, E.: Cassandra: The Definitive Guide. O'Reilly Media Inc., Sebastopol (2010)
11. Lakshman, A., Malik, P.: Cassandra: a decentralized structured storage system. ACM SIGOPS Oper. Syst. Rev. **44**(2), 35–40 (2010)
12. MongoDB Inc., MongoDB manual: Replication. http://docs.mongodb.org/manual/replication/
13. MongoDB Inc., MongoDB manual: Sharded cluster config servers. http://docs.mongodb.org/manual/core/sharded-cluster-config-servers/
14. MongoDB Inc., MongoDB manual: Sharded collection balancer. http://docs.mongodb.org/manual/core/sharding-balancing/
15. MongoDB Inc., MongoDB manual: Sharding. http://docs.mongodb.org/manual/sharding/
16. Nelubin and Engber. NoSQL failover characteristics: Aerospike, Cassandra, Couchbase, MongoDB (2013). http://www.thumbtack.net/
17. Nelubin and Engber. Ultra-high performance NoSQL benchmarking (2013). http://www.thumbtack.net/
18. Osman, R., Harrison, P.G.: Approximating closed fork-join queueing networks using product-form stochastic petri-nets. J. Syst. Softw. **110**, 264–278 (2015)
19. Osman, R., Piazzolla, P.: Modelling replication in NoSQL datastores. In: Norman, G., Sanders, W. (eds.) QEST 2014. LNCS, vol. 8657, pp. 194–209. Springer, Heidelberg (2014)
20. Pirzadeh, P., Tatemura, J., Hacigumus, H.: Performance evaluation of range queries in key value stores. In: IEEE International Symposium on Parallel and Distributed Processing Workshops and Phd Forum (IPDPSW), pp. 1092–1101. IEEE (2011)
21. Pokluda, A., Sun, W.: Benchmarking failover characteristics of large-scale data storage applications: Cassandra and Voldemort
22. Rabl, T., Gómez-Villamor, S., Sadoghi, M., Muntés-Mulero, V., Jacobsen, H.-A., Mankovskii, S.: Solving big data challenges for enterprise application performance management. Proc. VLDB Endowment **5**(12), 1724–1735 (2012)
23. Rogers, A.: VOLTDB in-memory database achieves best-in-class results, running in the cloud, on the YCSB benchmark, May 2014. http://tinyurl.com/VoltDB-YCSB. Last Accessed June 2016

τJSchema: A Framework for Managing Temporal JSON-Based NoSQL Databases

Safa Brahmia[1(✉)], Zouhaier Brahmia[1], Fabio Grandi[2], and Rafik Bouaziz[1]

[1] University of Sfax, Sfax, Tunisia
safa.brahmia@gmail.com,
{zouhaier.brahmia,raf.bouaziz}@fsegs.rnu.tn
[2] University of Bologna, Bologna, Italy
fabio.grandi@unibo.it

Abstract. Although NoSQL databases are claimed to be schemaless, several NoSQL database vendors have chosen JSON as agile data representation format and provide a JSON-based API or query facility to simplify the life of application developers. Whereas many applications require the management of temporal data, the JSON Schema language lacks explicit support for time-varying data. In this paper, for a systematic approach to the management of temporal data in NoSQL databases, we propose a framework called Temporal JSON Schema (τJSchema), inspired by the τXSchema framework defined for XML data. τJSchema allows defining a temporal JSON schema from a conventional JSON schema and a set of temporal logical and physical characteristics. Our framework guarantees logical and physical data independence for temporal schemas and provides a low-impact solution since it requires neither modifications of existing JSON documents, nor extensions to the JSON format, the JSON Schema language, and all related tools and languages.

Keywords: NoSQL databases · Document-oriented NoSQL databases · JSON · JSON schema · Temporal database · τXSchema · Conventional schema · Logical annotations · Physical annotations · Temporal schema · Temporal NoSQL databases

1 Introduction

NoSQL (Not Only SQL) databases [1–4] refer to nontraditional databases that usually do not require fixed schema, avoid join operations and typically scale horizontally. They emerged mainly to avoid some limitations of relational DBMSs related to scalability and storage performances when storing and analyzing large volumes of data or managing databases that are growing very fast. They are also considered as a very efficient support for managing big data [5–7] and running web applications in cloud computing environments [3, 8, 9]. Currently, more than 235 NoSQL database systems [4] are proposed as commercial or open source products. They could be classified in four main classes: (i) key-value NoSQL databases (e.g., DynamoDB, Riak, Redis) which store data as key-value pairs; (ii) column-oriented NoSQL databases (e.g., HBase, Cassandra, Hypertable) which store data tables as sections of columns rather than rows, like in relational

© Springer International Publishing Switzerland 2016
S. Hartmann and H. Ma (Eds.): DEXA 2016, Part II, LNCS 9828, pp. 167–181, 2016.
DOI: 10.1007/978-3-319-44406-2_13

databases; (iii) document-oriented NoSQL databases (e.g., MongoDB, Couchbase Sever, Elasticsearch) which store data as documents (e.g., in JSON or BSON format); (iv) graph-oriented NoSQL databases (e.g., Neo4j, InfoGrid, HyperGraphDB) which store data as directed graphs.

Furthermore, time has been always omnipresent in database applications [10, 11]. It allows timestamping data values when there is a need to track all changes on data and to have a complete history of the modeled reality. For that reason, dealing with temporal aspects of data has been since the 1980s one of the topics which interests several researchers of the database community. A lot of work has been and continues to be done on temporal databases [12–14]. Many temporal data models, query languages, and prototype systems have been proposed. Two times are used for managing temporal data: transaction time (i.e., the time when a datum is currently stored in the database) and valid time (i.e., the time when a datum was, is or will be valid in the real world). Data that are managed along both time dimensions are called bitemporal. Conventional data which are managed in a non-temporal manner (i.e., with destructive deletions and updates) are called snapshot.

Since modern computer science applications (e.g., social networks and collaborative web information systems) are changing very fast, NoSQL databases that are used by these applications (in addition to GUI, application source code and other components of such applications) must also evolve over time to reflect changes that rapidly occur in the real world. Therefore, also several NoSQL-based applications (e.g., e-commerce, e-government, and e-health applications) require keeping track of data evolution and versioning with respect to time and, thus, have to deal with time-varying NoSQL documents.

Unfortunately, although a continued interest in temporal and evolution aspects is exhibited by the research community [15–18], existing NoSQL data models and query languages but also state-of-the-art NoSQL DBMSs, APIs, and tools do not provide any built-in support for managing temporal data. In particular, the JSON format [19] and the JSON Schema language [20] lack explicit support for time-varying JSON documents, at both schema and instance levels, in document-oriented NoSQL databases. Although NoSQL databases are often claimed to be schemaless, several NoSQL database vendors have chosen JSON as agile data representation format and provide a JSON-based API or query facility to simplify the work of application developers. Thus, NoSQL Database Administrators (NSDBAs) relying on JSON must proceed in ad hoc manners when they need, for example, to specify a JSON schema for time-varying JSON data instances or to deal with temporal evolution of the JSON schema itself. In the rest of the paper, we define as NSDBA the person in charge of the maintenance of NoSQL databases.

According to what is presented in previous paragraphs, we think that if we would like to handle NoSQL database evolution over time in an efficient manner and to allow executing temporal queries on time-varying NoSQL instances, a built-in temporal support in NoSQL DBMSs is required. For that purpose, we propose in this paper a framework, called τJSchema (Temporal JSON Schema), for managing temporal JSON documents, through the use of a temporal JSON Schema extension. In fact, we want to introduce with τJSchema a principled and systematic approach to the temporal extension of JSON, similar to what Snodgrass and colleagues did to the XML language with τXSchema [21–23]. τXSchema is a framework (including a data model and a suite of tools) for managing temporal XML documents, well known in the database research community and, in particular, in the field of temporal XML [24]. Moreover, in our previous work [25–27], with the aim of completing the framework, we augmented τXSchema by defining necessary schema change operations acting on conventional schema, temporal schema, and logical and physical annotations (extensions which we plan to apply to τJSchema too).

Being defined as a τXSchema-like framework, τJSchema facilitates the creation of a temporal JSON schema from a conventional (i.e., non-temporal) JSON schema specification and a set of temporal logical and physical characteristics (or annotations). Temporal logical characteristics identify which components of a JSON document can vary over time; temporal physical characteristics specify how the time-varying aspects are represented in the document. By using temporal schema and characteristics to introduce temporal aspects in the conventional NoSQL setting, our framework (i) guarantees logical and physical data independence [28] for temporal schemas and (ii) provides a low-cost solution since it requires neither modifications of existing JSON documents already stored and used by applications, nor extensions to the JSON format, to the JSON Schema language, and to the JSON-based NoSQL systems (DBMSs and tools).

The rest of the paper is organized as follows. Section 2 motivates the need for an efficient management of time-varying JSON documents. Section 3 describes the τJSchema framework that we propose for extending NoSQL databases to capture temporal aspects: the architecture of τJSchema is presented and details on all its components and support tools are given. Section 4 provides a summary of the paper and some remarks about our future work.

2 Motivation

In this section, we present a motivating example that shows the limitation of the JSON Schema language [20] for explicitly supporting time-varying JSON data instances. Then, we provide the desiderata for a temporal JSON Schema extension which could accommodate time-varying instances in a systematic way.

2.1 Running Example

We assume to deal with a JSON-based NoSQL databases for managing data on Youtube channels. An example of a JSON document stored in such a database is presented in Fig. 1. It provides information of one Youtube channel having the name "Big Data videos" and created by the user "Safa" (owner). The number of subscriptions in such a channel is equal to 60,000. It contains two videos, one titled "Big Data Technologies" and the other is titled "Big Data Phenomenon"; the URL, and the number of likes, dislikes, and shares corresponding to each video are also maintained. Assume that information about this Youtube channel was added on November 10, 2015. Notice that we aim at providing a simple and intuitive example, even if it would not be the most significant to justify the choice of JSON.

```
{ "YouTubeChannels": [ {  "channelName": "Big Data videos",
                          "owner": "Safa",
                          "subscribedNumber": "60000",
                          "videos": [ {  "videoName": "Big Data Technologies",
                                         "URL": "https://www.youtube.com/watch?v=BDT",
                                         "likeNumber": "100",
                                         "dislikeNumber": "5",
                                         "shareNumber": "300" },
                                      {  "videoName": "Big Data Phenomenon",
                                         "URL": "https://www.youtube.com/watch?v=BDPP",
                                         "likeNumber": "50",
                                         "dislikeNumber": "2",
                                         "shareNumber": "100" } ] } ] }
```

Fig. 1. The "youtubeChannels.json" document on November 10, 2015

Suppose that on January 20, 2016, the owner modified the name of her Youtube channel from "Big Data videos" to "Big Data channel", the owner name from "Safa" to "S. Brahmia", and the name of the first video from "Big Data Technologies" to "Big Data Management: Current Approaches and Future Trends". Thus, the corresponding JSON document was revised to that shown in Fig. 2.

```
{ "YouTubeChannels": [ {  "channelName": "Big Data channel",
                          "owner": "S. Brahmia",
                          "subscribedNumber": "60000",
                          "videos": [ {  "videoName": "Big Data Management: Current Approaches and Future
                                                      Trends",
                                         "URL": "https://www.youtube.com/watch?v=BDT",
                                         "likeNumber": "100",
                                         "dislikeNumber": "5",
                                         "shareNumber": "300" },
                                      {  "videoName": "Big Data Phenomenon",
                                         "URL": "https://www.youtube.com/watch?v=BDPP",
                                         "likeNumber": "50",
                                         "dislikeNumber": "2",
                                         "shareNumber": "100"} ] } ] }
```

Fig. 2. The "youtubeChannels.json" document on January 20, 2016

In many JSON-based NoSQL database applications, bookkeeping of the whole history of JSON document changes is a fundamental requirement, since such a history allows recovering past document versions, tracking changes over time, and evaluating temporal queries. A τJSchema time-varying JSON document records the evolution of a conventional JSON document over time by storing all versions of the document in a way similar to that originally proposed for τXSchema [21].

Let us assume that the administrator of the Youtube database would like to keep track of the changes performed on our JSON document by storing both versions of Fig. 1 and of Fig. 2 in a single (temporal) JSON document. The result is the time-varying JSON document shown in Fig. 3, capturing the history of the specified information concerning Youtube channels.

```
{ "YouTubeChannels": [ { "versionedChannelName":  [ { "versionChannelName":{
                                                      "channelNameValidityStartTime":"2015-11-10",
                                                      "channelNameValidityEndTime":"2016-01-19",
                                                      "channelName":"Big Data videos" }},
                                          { "versionChannelName":{
                                                      "channelNameValidityStartTime":"2016-01-20",
                                                      "channelNameValidityEndTime":"now",
                                                      "channelName":"Big Data channel" }}],
                      "versionedOwner": [ { "versionOwner":{ "ownerValidityStartTime":"2015-11-10",
                                                      "ownerValidityEndTime":"2016-01-19",
                                                      "owner":"Safa" } },
                                          { "versionOwner": { "ownerValidityStartTime":"2016-01-20",
                                                      "ownerValidityEndTime":"now",
                                                      "owner":"S. Brahmia" } } ],
                      "subscribedNumber": "60000",
                      "videos": [ { "versionedVideoName": [ { "versionVideoName": {
                                                      "videoNameValidityStartTime":"2015-11-10",
                                                      "videoNameValidityEndTime":"2016-01-19",
                                                      "videoName":"Big Data Technologies" } },
                                              { "versionVideoName": {
                                                      "videoNameValidityStartTime":"2016-01-20",
                                                      "videoNameValidityEndTime":"now",
                                                      "videoName":"Big Data Management: Current
                                                            Approaches and Future Trends" } } ],
                          "URL": "https://www.youtube.com/watch?v=BDT",
                          "likeNumber": "100",
                          "dislikeNumber": "5",
                          "shareNumber": "300" },
                          { "versionedVideoName": [ { "versionVideoName": {
                                                      "videoNameValidityStartTime":"2015-11-10",
                                                      "videoNameValidityEndTime":"now",
                                                      "videoName":"Big Data Phenomenon" } } ],
                            "URL": "https://www.youtube.com/watch?v=BDPP",
                            "likeNumber": "50",
                            "dislikeNumber": "2",
                            "shareNumber": "100" } ] } } ] }
```

Fig. 3. The time-varying document of Youtube channel versions

In this example, we use valid-time to capture the history of such information. In order to timestamp the properties which can evolve over time, we use the following properties: channelNameValidityStartTime and channelNameValidityEndTime, for recording channel name evolution, ownerValidityStartTime and ownerValidityEndTime, for recording owner name evolution, and videoNameValidityStartTime and videoName-ValidityEndTime, for recording the video name history. The domain of channelName-ValidityEndTime, ownerValidityEndTime or videoNameValidityEndTime includes the value "now" [29]; the entity version that has "now" as the value of its validity end time property represents the current entity version until some change occurs.

Besides, the document presented in Fig. 4 represents the conventional (i.e., non-temporal) JSON schema for the JSON document presented in both Figs. 1 and 2. The conventional JSON schema is the schema for an individual version, which allows updating and querying individual JSON document versions.

```
{ "$schema": "http://json-schema.org/draft-04/schema#",
  "id": "http://jsonschema.net",
  "type": "object",
  "properties":
    { "YouTubeChannels": { "id": "http://jsonschema.net/YouTubeChannels",
                           "type": "array",
                           "items": { "type": "object",
                                      "properties": { "channelName": { "type": "string" },
                                                      "owner": { "type": "string" },
                                                      "subscribedNumber": { "type": "string" },
                                                      "videos": { "type": "array",
                                                                  "items": {"type": "object",
                                                                            "properties": {
                                                                                "videoName":{"type":"string"},
                                                                                "URL": {"type": "string" },
                                                                                "likeNumber":{"type":"string"}
                                                                                "dislikeNumber":{"type":"string"}
                                                                                "shareNumber":{"type":"string"}
                                                                                } } } },
                                      "required": [ "channelName", "owner", "subscribedNumber", "videos" ] },
              "required": [ "0" ] } },
  "required": [ "YouTubeChannels" ] }
```

Fig. 4. The "youtubeChannels.schema.json" JSON Schema document (the conventional schema)

The problem is that the time-varying JSON document (see Fig. 3) does not conform to the conventional JSON schema (see Fig. 4). Thus, to resolve this problem, we need a different JSON schema that can describe the structure of the time-varying JSON document. This new schema should specify, for example, timestamps associated to properties, time dimensions involved, and how the properties vary over time. This example will be continued in the Subsect. 3.2, in order to show how these problems can be solved in our proposed τJSchema framework.

2.2 Desiderata

There are several goals that can be fulfilled when augmenting the JSON Schema language to support time-varying data instances. Our approach aims at satisfying the following requirements:

- making easy the management of time for NSDBAs;
- supporting both transaction time and valid time in NoSQL databases;
- supporting temporal versioning of JSON Schema instances;
- keeping compatibility with existing JSON format and JSON Schema specifications, and editors, without requiring any changes to these models, languages, and tools;
- supporting existing applications that are already using JSON documents and JSON schema files;
- providing JSON data independence so that changes at the logical level are isolated from those performed at the physical level, and vice versa;
- proposing a variety of physical representations for time-varying JSON Schema instances.

Notice that similar goals have been fulfilled for XML with τXSchema and, thus, by defining τJSchema we aim at transferring their fulfillment to JSON.

3   The τJSchema Framework

In this section, we present our τJSchema framework for handling temporal JSON documents and provide an illustrative example of its use.

3.1   Architecture

In this subsection, we describe the overall architecture of τJSchema and the tools used for managing both τJSchema schema and τJSchema instances. Since τJSchema is a τXSchema-like framework, we were inspired by the τXSchema architecture and tools while defining the architecture and tools of τJSchema.

The τJSchema framework allows a NSDBA to create a temporal JSON schema for temporal JSON data instances from a conventional JSON schema, temporal logical characteristics, and temporal physical characteristics. Since it is a τXSchema-like framework, τJSchema use the following principles: (i) separation between the conventional schema and the temporal schema, and between the conventional instances and the temporal instances; (ii) use of logical and physical characteristics to specify temporal logical and temporal physical aspects, respectively, at schema level.

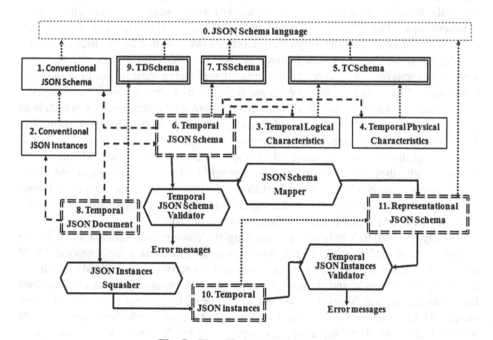

Fig. 5. The τJSchema architecture.

Figure 5 illustrates the architecture of τJSchema. Notice that only the components that are presented in the figure as rectangular boxes with one continuous line border (i.e., boxes 1, 2, 3, and 4) are specific to an individual time-varying JSON document and need to be supplied by a NSDBA. The framework is based on the JSON Schema language [20].

The NSDBA starts by creating the conventional JSON schema (box 1), which is a traditional JSON Schema document that models a given real world entity, without any temporal aspect. To each conventional JSON schema corresponds a set of conventional (i.e., non-temporal) JSON documents or JSON Schema instances (box 2). Any change to the conventional JSON schema is propagated to its corresponding instances.

After that, the NSDBA augments the conventional schema with temporal logical and temporal physical characteristics, which allow him/her to express, in an explicit way, all requirements dealing with the representation and the management of temporal aspects associated to the components of the conventional schema, as described below.

Temporal logical characteristics [23] allow the NSDBA to specify (i) whether a conventional schema component varies over valid time and/or transaction time, (ii) whether its lifetime is described as a continuous state or a single event, (iii) whether the component may appear at certain times (and not at others), and (iv) whether its content changes. If no logical characteristics are provided, the default logical characteristic is that anything can change. However, once the conventional schema is annotated, components that are not described as time-varying are static and, thus, they must have the same value across every conventional JSON document instance (box 2).

Temporal physical characteristics [23] allow the NSDBA to specify the timestamp representation options chosen, such as where the timestamps are placed and their kind (i.e., valid time or transaction time) and the kind of representation adopted. The location of timestamps is largely independent of which components vary over time. Timestamps can be located either on time-varying components (as specified by the logical characteristics) or somewhere above such components. Two JSON documents with the same logical characteristics will look very different if we change locations of their physical timestamps. Changing an aspect of even one timestamp can make a big difference in the representation. τJSchema supplies a default set of physical characteristics, which is to timestamp the root property with valid and transaction times. However, explicitly defining them can lead to more compact representations [23].

Although the two sets of temporal characteristics are orthogonal and can evolve independently, they are stored together in a single JSON document associated to the conventional schema which is a standard JSON document named the temporal characteristics document. The schema for the logical and physical characteristics is given by TCSchema (box 5) which is JSON Schema document [20].

Finally, the NSDBA finishes by annotating the conventional schema and asks the system to save his/her work. Consequently, the system creates the temporal JSON schema (box 6) providing the linking information between the conventional schema and its corresponding logical and physical characteristics. The temporal schema is a standard JSON document, which ties the conventional schema, the logical characteristics, and the physical characteristics together. In the τJSchema framework, the temporal JSON schema is the logical equivalent of the conventional JSON schema in a non-temporal environment. This document associates a series of conventional schema definitions with temporal characteristics, along with the time span during which the association was in effect. The schema for the temporal JSON schema document is given by TSSchema (box 7) which is JSON Schema document.

After creating the temporal schema, the system creates a temporal JSON document (box 8) in order to link each conventional JSON document (box 2), which is valid to a conventional JSON schema (box 1), to its corresponding temporal JSON schema (box 6), and more precisely to its corresponding logical and physical characteristics (which are referenced by the temporal JSON schema). A temporal document is a standard JSON document that maintains the evolution of a non-temporal JSON document over time, by keeping track of all the versions (or temporal slices) of the document with their corresponding timestamps and by specifying the temporal schema associated to these versions. This document associates a series of conventional JSON documents with logical and physical characteristics, along with the time span during which the association was in effect. Therefore, the temporal JSON document facilitates the support of temporal queries involving past JSON document versions or dealing with changes between JSON document versions. The schema for the temporal document is the JSON Schema document TDSchema (box 9).

Notice that, whereas TCSchema (box 5), TSSchema (box 7), and TDSchema (box 9) have been developed in this work, JSON Schema (box 0) corresponds to the existing language endorsed by the Internet Engineering Task Force (IETF) [20] for specifying the structure of JSON documents.

Similarly to what happens in the τXSchema framework, the temporal JSON schema (box 6) is processed by the **temporal JSON schema validator** tool in order to ensure that the logical and physical characteristics are (i) valid with respect to TCSchema, and (ii) consistent with the conventional schema. The temporal JSON schema validator tool reports whether the temporal JSON schema document is valid or invalid.

Once all the characteristics are found to be consistent, the **JSON schema mapper** tool generates the representational JSON schema (box 11) from the temporal JSON schema (i.e., from the conventional JSON schema plus the logical and physical characteristics); it is the result of transforming the conventional schema according to the requirements expressed through the different temporal characteristics. The representational JSON schema becomes the schema for temporal JSON data instances (box 10). These temporal instances could be obtained in four ways: (i) automatically from the temporal JSON document (box 8) (i.e., from non-temporal JSON instances (box 2) and the temporal JSON schema (box 6)), using the **JSON instances squasher** tool (such an operation is called "squash" in the original τXSchema approach); (ii) automatically from instances stored in a JSON-based NoSQL database, that is as the result of a "temporal query" or a "temporal view"; (iii) automatically from a third-party tool; (iv) manually: temporal JSON instances are directly added by the NSDBA to the τJSchema repository.

Moreover, temporal JSON instances are validated against the representational JSON schema through the **temporal JSON instances validator** tool, which reports whether the temporal JSON instances (box 10) are valid or invalid.

The four mentioned tools (i.e., Temporal JSON Schema Validator, Temporal JSON Instances Validator, JSON Schema Mapper, and JSON Instances Squasher) are currently under development. For example, the temporal JSON instances validator tool is being implemented as a temporal extension of an existing conventional JSON instances validator [30] based on the coding of the corresponding τXSchema tool.

3.2 Running Example Reprise

In order to show the functioning of the proposed approach, we continue in the following our motivating example of the Subsect. 2.1. In particular, we will show how management of temporal JSON document versions is dealt with in the τJSchema approach.

On November 10, 2015, the NSDBA creates a conventional JSON schema (box 1), named "youtubeChannels_V1.schema.json" (as in Fig. 4), and a conventional JSON document (box 2), named "youtubeChannels_V1.json" (as in Fig. 1), which is valid with respect to this schema. We assume that the NSDBA defines also a set of temporal logical and physical characteristics, associated to that conventional JSON schema; they are stored in a temporal characteristics document (boxes 3 and 4) titled "youtubeChannelsTemporalCharacteristics_V1.json" as shown in Fig. 6.

```
{ "temporalCharacteristicSet": { "logical": [ {  "target":"YouTubeChannels/channelName",
                                                 "validTime": {  "kind":"state",
                                                                 "content":"varying",
                                                                 "existence":"constant" } },
                                              {  "target":"YouTubeChannels/owner",
                                                 "validTime": {  "kind":"state",
                                                                 "content":"constant",
                                                                 "existence":"constant" } },
                                              {  "target":"YouTubeChannels/videos/videoName",
                                                 "validTime": {  "kind":"state",
                                                                 "content":"varying",
                                                                 "existence":"constant" } } ],
                                "physical": [ {  "target":"YouTubeChannels/0/videos/0/videoName",
                                                 "dataInclusion":"expandedVersion",
                                                 "stampKind": {  "timeDimension":"validTime",
                                                                 "stampBounds":"extent" } } ] } }
```

Fig. 6. The temporal characteristics document on November 10, 2015

After that, the system generates the temporal JSON schema (box 6) in Fig. 7, which ties "youtubeChannels_V1.schema.json" and "youtubeChannelsTemporalCharacteristics_V1.json" together; this temporal schema is saved in a JSON file titled "youtubeChannelsTemporalSchema.json". Consequently, the system uses the temporal JSON schema of Fig. 7 and the conventional JSON document in Fig. 1 to create a temporal JSON document (box 8) as in Fig. 8, which lists both versions (i.e., temporal "slices") of the conventional JSON documents with their associated timestamps. The squashed version (box 10) of this temporal document, which could be generated by the JSON Instances Squasher, is provided in Fig. 9.

{ "temporalSchema": { "convetionalSchema": { "scliceSequence": [{ "slice": {
                                          "location":"youtubeChannels_V1.schema.json",
                                          "begin":"2015-11-10" } }] },
                "temporalCharasteristicSet": { "scliceSequence": [{ "slice":{
                                          "location":"youtubeChannelsTemporalCharacteristics_V1.json",
                                          "begin":"2015-11-10" } }] } } }

Fig. 7. The temporal JSON schema on November 10, 2015

{ "temporalRoot": { "temporalSchemaLocation":"youtubeChannelsTemporalSchema.json" },
                "sliceSequence": [{ "slice": { "location":"youtubeChannels_V1.json",
                                          "begin":"2015-11-10" } }] }

Fig. 8. The temporal JSON document on November 10, 2015

{ "YouTubeChannels": [{ "channelName_RepItem": [.{ "channelName_Version": {
                                          "timestamp_ValidExtent": {
                                              "begin":"2015-11-10",
                                              "end":"now" },
                                          "channelName":"Big Data videos" } }],
                "owner_RepItem": [{ "owner_Version": {
                                          "timestamp_ValidExtent": {
                                              "begin":"2015-11-10",
                                              "end":"now" },
                                          "owner":"Safa" } }],
                "subscribedNumber": "60000",
                "videos": [{ "videoName_RepItem": [{ "videoName_Version": {
                                          "timestamp_ValidExtent": {
                                              "begin":"2015-11-10",
                                              "end":"now"},
                                          "videoName":"Big Data Technologies" } }],
                "URL": "https://www.youtube.com/watch?v=BDT",
                "likeNumber": "100",
                "dislikeNumber": "5",
                "shareNumber": "300" },
                { "videoName_RepItem": [{ "videoName_Version": {
                                          "timestamp_ValidExtent": {
                                              "begin":"2015-11-10",
                                              "end":"now", },
                                          "videoName":"Big Data Phenomenon" } }],
                "URL": "https://www.youtube.com/watch?v=BDPP",
                "likeNumber": "50",
                "dislikeNumber": "2",
                "shareNumber": "100" } }] } }

Fig. 9. The squashed document correponding to the temporal document on November 10, 2015.

On January 20, 2016, the NSDBA updates the conventional JSON document "youtubeChannels_V1.json" as presented in the subsection 2.1 to produce a new conventional JSON document named "youtubeChannels_V2.json" (as in Fig. 2). Since the conventional JSON schema (i.e., youtubeChannels_V1.schema.json) and the temporal characteristics document (i.e., youtubeChannelsTemporalCharacteristics_V1.json) are not changed, the temporal JSON schema (i.e., youtubeChannelsTemporalSchema.json) is consequently not updated. However, the system updates the temporal JSON document, in order to include the new slice of the new conventional JSON document, as shown in Fig. 10. The squashed version of the updated temporal JSON document is provided in Fig. 11.

```
{ "temporalRoot": { "temporalSchemaLocation":"youtubeChannelsTemporalSchema.json" },
  "sliceSequence": [ { "slice": { "location":"youtubeChannels_V1.json",
                                  "begin":"2015-11-10" } },
                     { "slice": { "location":"youtubeChannels_V2.json",
                                  "begin":"2016-01-20" } } ] }
```

Fig. 10. The temporal document on January 20, 2016

```
{ "YouTubeChannels": [ { "channelName_RepItem": [ { "channelName_Version": {
                                                      "timestamp_ValidExtent": {
                                                        "begin":"2015-11-10",
                                                        "end":"2016-01-19" },
                                                      "channelName":"Big Data videos" } },
                                                  { "channelName_Version": {
                                                      "timestamp_ValidExtent": {
                                                        "begin":"2016-01-20",
                                                        "end":"now" },
                                                      "channelName":"Big Data videos" } } ],
              "owner_RepItem": [ { "owner_Version": {
                                     "timestamp_ValidExtent": {
                                       "begin":"2015-11-10",
                                       "end":"2016-01-19"}
                                     "owner":"Safa"} },
                                 { "ownerVersion":" {
                                     "timestamp_ValidExtent": {
                                       "begin":"2016-01-20",
                                       "end":"now" },
                                     "owner":"S. Brahmia" } } ],
              "subscribedNumber": "60000",
              "videos": [ { "videoName_RepItem": [ { "videoName_Version": {
                                                      "timestamp_ValidExtent": {
                                                        "begin":"2015-11-10",
                                                        "end":"2016-01-19", },
                                                      "videoName":"Big Data Technologies" } },
                                                  { "videoName_Version": {
                                                      "timestamp_ValidExtent": {
                                                        "begin":"2016-01-20",
                                                        "end":"now" },
                                                      "videoName":"Big Data Management:
                                                                   Current Approaches and Future
                                                                   Trends"} } ],
                            "URL": "https://www.youtube.com/watch?v=BDT",
                            "likeNumber": "100",
                            "dislikeNumber": "5",
                            "shareNumber": "300" },
                          { "videoName_RepItem": [ { "videoName_Version": {
                                                      "timestamp_ValidExtent": {
                                                        "begin":"2015-11-10",
                                                        "end":"now"},
                                                      "videoName":"Big Data Phenomenon" } } ],
                            "URL": "https://www.youtube.com/watch?v=BDPP",
                            "likeNumber": "50",
                            "dislikeNumber": "2",
                            "shareNumber": "100" } } ] } ]
```

Fig. 11. The squashed document corresponding to the temporal document on January 20, 2016

Obviously, each one of the squashed documents (see Figs. 9 and 11) must conform to a particular schema, which is the representational JSON schema (box 11) that is generated by the JSON Schema Mapper from the temporal JSON schema shown in Fig. 7.

4   Conclusion and Future Work

In this paper, we proposed τJSchema, a τXSchema-like framework, which allows creating a temporal JSON schema from a conventional JSON schema and a set of temporal logical and temporal physical characteristics. It ensures logical and physical data independence, since it separates conventional schema, logical characteristics, and physical characteristics, allowing them to be changed independently and safely. Furthermore, the adoption of τJSchema provides a low-impact solution, since it requires neither modifications of existing JSON documents, nor extensions to JSON format, JSON Schema language, and available tools that are based on JSON/JSON Schema.

Currently, we are extending τJSchema to also support JSON schema versioning [27, 31], since (i) JSON schemata are also evolving over time to reflect changes in real-world applications, and (ii) keeping a full history of both JSON schema and instance changes is required by several NoSQL database applications [18]. Moreover, in the next future, we plan to develop a system prototype (as a temporal stratum on top of an existing JSON-based NoSQL DBMS, like MongoDB) showing the feasibility of our approach, and to study manipulation of temporal JSON data instances in the τJSchema framework, by proposing an extension of the JSONiq query language [32] to temporal and versioning aspects.

References

1. Cattell, R.: Scalable SQL and NoSQL data stores. ACM SIGMOD Rec. **39**(4), 12–27 (2010)
2. Tiwari, S.: Professional NoSQL. Wiley, Indianapolis (2011)
3. Pokorný, J.: NoSQL databases: a step to database scalability in web environment. Int. J. Web Inf. Syst. **9**(1), 69–82 (2013)
4. NoSQL Databases. http://www.nosql-database.org/
5. Barbierato, E., Gribaudo, M., Iacono, M.: Performance evaluation of NoSQL big-data applications using multi-formalism models. J. Future Gener. Comput. Syst. **37**, 345–353 (2014)
6. Gudivada, V.N., Rao, D., Raghavan, V.V.: NoSQL systems for big data management. In: Proceedings of the 2014 IEEE World Congress on Services (SERVICES 2014), Anchorage, AK, USA, 27 June–2 July, pp. 190–197 (2014)
7. Sharma, S., Tim, U.S., Gadia, S., Wong, J., Shandilya, R., Peddoju, S.K.: Classification and comparison of NoSQL big data models. Int. J. Big Data Intell. **2**(3), 201–221 (2015)
8. Grolinger, K., Higashino, W.A., Tiwari, A., Capretz, M.A.: Data management in cloud environments: NoSQL and NewSQL data stores. J. Cloud Comput. Adv. Syst. Appl. **2**(1), 22 (2013)
9. Ganesh Chandra, D.: A survey of cloud database systems. IT Prof. **16**(2), 50–57 (2014)
10. Jensen, C.S., Snodgrass, R.T.: Temporal data management. IEEE Trans. Knowl. Data Eng. **11**(1), 36–44 (1999)

11. Snodgrass, R.T.: Developing Time-Oriented Database Applications in SQL. Morgan Kaufmann Publishers Inc., San Francisco (1999)
12. Tansel, A., Clifford, J., Gadia, S., Jajodia, S., Segev, A., Snodgrass, R.T. (eds.): Temporal Databases: Theory, Design, and Implementation. Benjamin/Cummings Publishing Company, Redwood City (1993)
13. Etzion, O., Sripada, S., Jajodia, S. (eds.): Temporal Databases: Research and Practice. LNCS, vol. 1399. Springer, Heidelberg (1998)
14. Grandi, F.: Temporal databases. In: Koshrow-Pour, M. (ed.) Encyclopedia of Information Science and Technology, 3rd edn., pp. 1914–1922. IGI Global, Hershey (2015)
15. Monger, M.D., Mata-Toledo, R.A., Gupta, P.: Temporal data management in NoSQL databases. J. Inf. Syst. Oper. Manag. 6(2), 237–243 (2012)
16. Castelltort, A., Laurent, A.: Representing history in graph-oriented NoSQL databases: a versioning system. In: Proceedings of the 8th International Conference on Digital Information Management (ICDIM 2013), Islamabad, Pakistan, 10–12 September, pp. 228–234 (2013)
17. Hu, Y., Dessloch, S.: Defining temporal operators for column oriented NoSQL databases. In: Manolopoulos, Y., Trajcevski, G., Kon-Popovska, M. (eds.) ADBIS 2014. LNCS, vol. 8716, pp. 39–55. Springer, Heidelberg (2014)
18. Cuzzocrea, A.: Temporal aspects of big data management: state-of-the-art analysis and future research directions. In: Proceedings of the 22nd International Symposium on Temporal Representation and Reasoning (TIME 2015), Kassel, Germany, 23–25 September, pp. 180–185 (2015)
19. Internet Engineering Task Force (IETF): The JavaScript Object Notation (JSON) Data Interchange Format, Internet Standards Track, March 2014. https://tools.ietf.org/html/rfc7159
20. Internet Engineering Task Force (IETF): JSON Schema: core definitions and terminology, Internet-Draft, 31 January 2013. http://tools.ietf.org/html/draft-zyp-json-schema-04
21. Currim, F., Currim, S., Dyreson, C.E., Snodgrass, R.T.: A tale of two schemas: creating a temporal XML schema from a snapshot schema with tXSchema. In: Proceedings of the 9th International Conference on Extending Database Technology (EDBT 2004), Heraklion, Crete, Greece, 14–18 March, pp. 348–365 (2004)
22. Snodgrass, R.T., Dyreson, C.E., Currim, F., Currim, S., Joshi, S.: Validating quicksand: temporal schema versioning in τXSchema. Data Knowl. Eng. 65(2), 223–242 (2008)
23. Currim, F., Currim, S., Dyreson, C.E., Joshi, S., Snodgrass, R.T., Thomas, S.W., Roeder, E.: τXSchema: support for data- and schema-versioned XML documents. Technical report TR-91, TimeCenter, 279 p., September 2009. http://timecenter.cs.aau.dk/TimeCenter Publications/TR-91.pdf
24. Dyreson, C.E., Grandi, F.: Temporal XML. In: Liu, L., Özsu, M.T. (eds.) Encyclopedia of Database Systems, pp. 3032–3035. Springer, USA (2009)
25. Brahmia, Z., Bouaziz, R., Grandi, F., Oliboni, B.: Schema versioning in τXSchema-based multitemporal XML repositories. In: Proceedings of the 5th IEEE International Conference on Research Challenges in Information Science (RCIS 2011), Guadeloupe, French West Indies, France, 19–21 May, pp. 1–12 (2011)
26. Brahmia, Z., Grandi, F., Oliboni, B., Bouaziz, R.: Versioning of conventional schema in the τXSchema framework. In: Proceedings of the 8th International Conference on Signal Image Technology & Internet Systems (SITIS 2012), Sorrento, Naples, Italy, 25–29 November, pp. 510–518 (2012)
27. Brahmia, Z., Grandi, F., Oliboni, B., Bouaziz, R.: Schema change operations for full support of schema versioning in the τXSchema framework. Int. J. Inf. Technol. Web. Eng. 9(2), 20–46 (2014)

28. Burns, T., Fong, E., Jefferson, D., Knox, R., Mark, L., Reedy, C., Reich, L., Roussopoulos, N., Truszkowski, W.: Reference model for DBMS standardization, database architecture framework task group (DAFTG) of the ANSI/X3/SPARC database system study group. SIGMOD Rec. **15**(1), 19–58 (1986)

29. Clifford, J., Dyreson, C.E., Isakowitz, T., Jensen, C.S., Snodgrass, R.T.: On the semantics of "now" in databases. ACM Trans. Database Syst. **22**(2), 171–214 (1997)

30. Internet Engineering Task Force (IETF): JSON Schema: interactive and non interactive validation, Internet-Draft, 1 February 2013. http://tools.ietf.org/html/draft-fge-json-schema-validation-00

31. Roddick, J.F.: Schema Versioning. In: Liu, L., Özsu, M.T. (eds.) Encyclopedia of Database Systems, pp. 2499–2502. Springer, USA (2009)

32. Florescu, D., Fourny, G.: JSONiq: the history of a query language. IEEE Internet Comput. **17**(5), 86–90 (2013)

Multimedia Data

Enhancing Similarity Search Throughput by Dynamic Query Reordering

Filip Nalepa[✉], Michal Batko, and Pavel Zezula

Faculty of Informatics, Masaryk University, Brno, Czech Republic
f.nalepa@gmail.com

Abstract. A lot of multimedia data are being created nowadays, which can only be searched by content since no searching metadata are available for them. To make the content search efficient, similarity indexing structures based on the metric-space model can be used. In our work, we focus on a scenario where the similarity search is used in the context of stream processing. In particular, there is a potentially infinite sequence (stream) of query objects, and a query needs to be executed for each of them. The goal is to maximize the throughput of processed queries while maintaining an acceptable delay. We propose an approach based on dynamic reordering of the incoming queries combined with caching of recent results. We were able to achieve up to 3.7 times higher throughput compared to the base case when no reordering and caching is used.

Keywords: Stream processing · Similarity search

1 Introduction

Current digital media explosion results in huge amounts of unstructured data that lack any searchable metadata. In order to make such data findable, the content-based search must be applied that treats the data by similarity rather than exact match of their attributes. Such search then usually uses *k-nearest-neighbors queries* (*kNN*), which retrieve the k objects that are the most similar to a given query object.

To make things even more complex, some applications need to deal with a continuous stream of arriving data that are to be searched. For example, consider a text search-engine crawler that gathers images from the web and needs to annotate them by textual descriptions according to the image content, a spam filter that compares the incoming emails to some learned spam knowledge base so that spam messages can be detected, or a news notification system that needs to compare the newly published articles to the profiles of all the subscribed users to find out who should be notified.

All these applications require to process each and every data item that appeared in the stream by some form of content-based searching. Luckily, the data in such applications need not be processed immediately but some small delay is acceptable. The most important thing is the number of processed images

© Springer International Publishing Switzerland 2016
S. Hartmann and H. Ma (Eds.): DEXA 2016, Part II, LNCS 9828, pp. 185–200, 2016.
DOI: 10.1007/978-3-319-44406-2_14

in a given time, i.e. the *throughput* of such an application. The individual query search time can be improved by applying some similarity indexing technique, for which there are efficient algorithms based on the metric model of similarity [14]. As opposed to interactive applications focusing on the single query optimization, in our scenario, we can afford a slight delay of the single query processing if the overall throughput of the system is improved. Performance of such stream processing applications is studied in [7,8].

In our work, we exploit the fact that the order of the processed queries may have a significant impact on the processing time. We propose a novel approach based on dynamic reordering of the incoming queries combined with caching mechanism used to lower the I/O costs. The solution is based on the assumption that to evaluate two similar queries, similar data of the index need to be accessed. According to our experiments, the proposed technique results in significantly higher throughput.

The rest of the paper is organized as follows. First, we present some related work on caching and stream data throughput improvements. In Sect. 3 we formally define our problem. Components and the overall architecture of our technique is presented in Sect. 4. Section 5 presents the basic query ordering technique that is further enhanced in Sect. 6 so that the approach can deal with too high delays. Experimental evaluation of our approach can be found in Sect. 7. Finally, the paper is concluded in Sect. 8.

2   Related Work

The usage of a caching mechanism in similarity search has been proposed in several papers to reduce the amount of disk accesses. In [3], the authors deal with kNN queries to search for similar images in the metric space. They build their approach on the assumption that there exists a set of popular images which are queried by users significantly more often than the other images. They propose an approach where the result sets of individual kNN queries are stored in a cache, and they are reused to produce approximate results of subsequent queries. Unlike traditional caching, the proposed cache can manage not only exact hits, but also approximate ones that are solved by similarity with respect to the result sets of past queries present in the cache.

The concept of caching in similarity search is used also in [10] where it is applied to contextual advertising systems. For a kNN query q, if there is a cache miss, a larger set of objects than are actually needed is retrieved from the disk and stored in the cache. When a similar query to the cached query q comes to the system, the cached values of q are explored to obtain results for the new query. In this way, an approximate answer is returned.

These two approaches use the cache to speedup the processing of a query for the price of reduced accuracy. On the other hand, our proposed approach does not compromise the accuracy, and the result for each query is exactly the same as it would be without the cache.

Another way to improve the throughput of a stream of kNN queries, is to reorder the queries. In [12], the authors optimize nearest neighbor search for videos when each video is represented by a sequence of high-dimensional feature vectors. Given a query video containing n feature vectors, a search for each vector is performed, and the overall similarity is computed at last. The authors make use of the fact that nearby feature vectors in a video are similar, and they propose dynamic query ordering for advanced optimization of both I/O and CPU costs. They make an observation that the overlapped candidates of a previous query may help to further reduce the candidate sets of subsequent queries. The algorithm aims at progressively finding a query order such that the common candidates among queries are fully utilized to maximally reduce the total number of candidates.

The aforementioned techniques assume the existence of sequences of similar queries in the stream. In our approach, we use no such assumption, on the contrary, we reorder the queries so that we obtain the desired sequences.

A slightly different problem is dealt with in [13]. The stream of queries is given as a line segment in a space, and the task is to perform a continuous nearest neighbor search; that is to retrieve a nearest neighbor for every point on the line segment. The result contains a set of (*point, interval*) tuples such that the *point* is the nearest neighbor of all points in the corresponding *interval*. The proposed technique uses a single database traversal to identify all the split points which form the result set. A motivating use case is to find all nearest gas stations during a route between two places.

Another task in similarity search for streams is a classification problem when a set of classes are assigned to each data item of the stream. Usually, the class labels are predicted based on some training data which are correctly labeled. The challenge in the stream processing is to deal with a high rate of incoming data items and with the concept drift, i.e., the applications have to adapt to new trends (new classes may emerge; some classes may disappear; the definition of the classes may change in time). In [15], there is a proposal of a new indexing structure called Ensemble-tree (E-tree) which is a height balanced tree consisting of R-tree like structure storing the decision rules and a table structure storing classifier-level information. Other approaches to the stream classification can be found in [4,6,11]. We do not address the problem of classification itself in our work; we focus rather on improving the efficiency of a particular type of classification (kNN classification).

3   Problem Definition

Suppose there is a domain of complex objects D (e.g., images) and a large database containing such objects. Let $s = (q_1, q_2, \ldots)$ be a stream, i.e., a potentially infinite sequence of query objects of the same type, where $q_i \in D$ for each i. All the objects are indexed in the metric space which is a universal model of similarity [14]. There is defined a total distance function $d : D \times D \to R$. The distance between two objects corresponds to the level of their dissimilarity ($d(p, p) = 0$, $d(o, p) \geq 0$).

For each query object q_i in the stream s, a k-nearest neighbors query $NN(q_i, k)$ is executed which returns k nearest objects from the database to the query object. The goal is to achieve as high throughput as possible, i.e., to process a given set of queries as fast as possible.

It is allowed to locally change the order of the processed queries. More precisely, we define a buffer B containing at maximum N query objects. A query object of the stream is added to the buffer only if it is not full, i.e., it contains less then N queries. As a next query to be processed, a query object from the buffer is selected.

Using the local reordering described above, some queries may be a subject to starvation, i.e., they never get selected from the buffer for processing. Therefore, we introduce a constraint on the *maximum delay* (*MD*) and update the retrieval method as follows. As soon as an object is stored to the buffer, it is assigned a timestamp. Whenever an object is about to be selected from the buffer and there exists an object whose timestamp is older than the allowed maximum delay, the object with the oldest timestamp in the buffer has to be processed.

Formally, a buffer $B = \{(q_1, t_1), \ldots, (q_h, t_h)\}$ where $\{q_1, \ldots, q_h\} \subseteq D$; $|B| \leq N$; t_i is the entrance time of q_i to the buffer. We define two functions for the buffer.

$push(q, t)$ inserts the query q to the buffer and assigns it the timestamp t.

$pop(timeLimit) = (q, t)$ removes the pair $(q, t) \in B$ from the buffer and returns it on the output. If there exists a pair $(p, u) \in B$ such that $u \leq timeLimit$, then the following has to hold: $t = min(\{v | (r, v) \in B\})$.

The generic processing of the stream of queries is performed as follows. The buffer is filled with first N queries of the stream. Then there is a cycle in which a query is popped from the buffer, it is processed, and another query is pushed into the buffer. It means that there are always N queries in the buffer before a query is popped. The pseudocode can be seen in Algorithm 1.

Algorithm 1. Generic algorithm

---

   function processStream($(q_1, q_2, \ldots), N, MD$)
      $B \leftarrow$ new $Buffer()$
      $i \leftarrow 1$
      while $i \leq N$ **do**
         B.push$(q_i, now())$
         i++
      loop
         process(B.pop($now() - MD$))
         B.push$(q_i, now())$
         i++

To measure the throughput, we define the function $processingTime((q_1, q_2, \ldots), N, m) = T$ where T is the time since the algorithm was launched until m queries have been processed, i.e., until $i = N + m + 1$ in Algorithm 1. By repeatedly calling pop, a permutation $(q_{i_1}, q_{i_2}, \ldots)$ of the original stream is continuously generated. The ultimate goal is to specify the function pop so that T

is minimized. Note that *pop* can access only the queries which are in the buffer, and it cannot foresee any future ones.

4   Architecture

In this section, we describe the architecture of the whole system used for processing a stream of similarity search queries. There are two main parts: the metric index and the buffer of waiting queries. The schema is shown in Fig. 1.

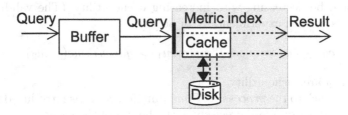

Fig. 1. Architecture

The buffer is used to temporarily store the incoming query objects which are waiting for processing. The metric index takes care of the query evaluation. It contains a disk where the database of objects is stored and a main memory cache used for storing recently loaded data from the disk.

Whenever there is a spare space in the buffer, the next query of the stream is loaded. When the metric index is ready for processing another query, a query is picked from the buffer according to a chosen strategy. During processing of the query, the metric index looks into the cache to possibly use any intermediate results obtained from evaluating recent queries. If the data are not in the cache, they are loaded from the disk.

We consider a generic metric index which uses data partitioning $P = \{p_1, \ldots, p_n\}$ where $p_i \subseteq D$. When evaluating a query, it needs to access a subset of the partitions $Q \subseteq P$. The partitions are typically stored on a disk [14].

The cache system is based on storing intermediate results of recent queries so that they can be reused by the metric index when processing similar queries. The cache is used to lower down the number of disk accesses when retrieving data partitions. The loaded partitions are kept in the cache so that they can be reused by later queries which access the same partitions. The cache is a set of partitions $cache = \{p_1, \ldots, p_m\} \subseteq P$. The size of the cache is limited by the number of objects within the cached partitions: $\sum_{p \in cache} |p| \leq cacheLimit$. To measure the utility of the cache, we define the function

$$cacheUtility(q, cache) = \frac{|I(q) \cap cache|}{|I(q)|} \tag{1}$$

where *cache* represents the content of the cache and $I(q) \subseteq P$ is the set of partitions accessed during the evaluation of q.

To keep track of the content of the cache, we define the function $updateCache(q, cache)$ returning the content of the cache after processing the query q where *cache* represents the content of the cache before executing q. In our implementation, we use the least recently used policy. In particular, the partitions with the oldest last access time are discarded and replaced with the new partitions of the last query while obeying *cacheLimit*.

The $queryTime(q, cacheUtility)$ represents the time to process the given query q using the given cache utility. The desired property of the function is that it should be decreasing with increasing cache utility. (The validity of the assumption is verified experimentally in Sect. 7.2.)

$$cu_1 \leq cu_2 \Leftrightarrow queryTime(q, cu_1) \geq queryTime(q, cu_2) \tag{2}$$

where cu_1, cu_2 are cache utility.

Let us get back to the *processingTime* function which is introduced in Sect. 3 to specify the function more precisely with the use of the cache.

$$processingTime((q_1, \ldots), N, m) = \sum_{j=1}^{m} (queryTime(q_{i_j}, cu_j)) + bt \text{ where} \tag{3}$$

$$cu_j = cacheUtility(q_{i_j}, cache_{j-1})$$
$$cache_0 = \{\}$$
$$cache_j = updateCache(q_{i_j}, cache_{j-1}) \text{ for } 1 \leq j \leq m$$

$(q_{i_1}, \ldots, q_{i_m})$ is the sequence of queries generated by the *pop* function. $cache_j$ for $1 \leq j \leq m$ represents the state of the cache right after q_{i_j} has been processed. bt is the overall time needed for the buffer management (*push* and *pop* calls).

Let us denote $sequenceTime((q_{i_1}, \ldots, q_{i_m})) = \sum_{j=1}^{m} (queryTime(q_{i_j}, cu_j))$. It is likely that $sequenceTime(s) \gg bt$ since the evaluation of a query is a costly operation. Therefore we will focus on minimization of *sequenceTime* in the next section.

5   Query Ordering

In this section, we discuss the optimal query ordering so that the minimal processing time is achieved, i.e., the throughput is maximized.

According to Formula 2 showing the dependency of the query time on the cache utility, we can suppose that the key to a good ordering of the queries is to maximize the cache utility. As for metric indexes, the following formula typically holds [14]:

$$d(q_1, q_2) \approx 0 \rightarrow I(q_1) \approx I(q_2) \tag{4}$$

That is, if the queries q_1, q_2 are close to each other, the sets of accessed partitions are approximately the same. In other words, if q_1 is processed, the $I(q_1)$ is stored in the cache, and subsequently q_2 is processed, then the cache utility is likely to be very high. The idea is to find such a sequence of queries where the neighboring ones are similar.

Let us start by defining a function returning the number of pairs of similar subsequent queries.

$$similarCount((q_1,\ldots,q_m)) = |\{i|d(q_{i-1},q_i) \approx 0 \land 2 \le i \le m\}| \qquad (5)$$

Our assumption is that by maximizing $similarCount$, we can achieve high cache utility and thus low processing time.

Let the function $overallCacheUtility$ return the fraction of all the partitions retrieved from the cache out of all the requested partitions for a given sequence of queries.

$$overallCacheUtility((q_1,\ldots,q_m), initialCache) = \frac{\sum\limits_{i=1}^{m} |I(q_i) \cap cache_{i-1}|}{\sum\limits_{i=1}^{m} I(q_i)} \qquad (6)$$

where $cache_0 = initialCache$ and $cache_i$ is the content of the cache after processing q_i for $1 \le i \le m - 1$.

Let r and s be sequences of queries of the same lengths; we assume the following to hold

$$similarCount(r) \le similarCount(s) \Rightarrow \qquad (7)$$
$$overallCacheUtility(r, \{\}) \le overallCacheUtility(s, \{\}) \Rightarrow$$
$$sequenceTime(r) \ge sequenceTime(s)$$

Let us define the function returning the indexes of the queries which are not similar to the previous query in the given sequence $s = (q_1,\ldots,q_m)$.

$$dissimilar(s) = (x_1,\ldots,x_h) \qquad (8)$$

so that x_i is in the tuple iff $x_i = 1 \lor d(q_{x_i}, q_{x_i-1}) \not\approx 0$. The function returns the elements of the tuple in the ascending order, i.e., $x_{i-1} < x_i$ for each i.

Now, the number of similar pairs can be computed by summing the number of queries between the dissimilar pairs:

$$similarCount(s) = \sum\limits_{i=1}^{h}(x_i - 1 - x_{i-1}) + (m - x_h) \qquad (9)$$

where $x_0 = 0$.

Let us denote $simSeq_i = x_i - 1 - x_{i-1}$. Then

$$similarCount(s) = \sum\limits_{i=1}^{h} simSeq_i + (m - x_h) \qquad (10)$$

Each $simSeq_i$ represents the number of subsequent pairs of similar queries. To achieve high $similarCount$, we propose a greedy algorithm which for each i tries to maximize $simSeq_i$.

First, the algorithm selects a highly populated region of query objects in the buffer. Subsequently, it processes all the queries in this region. By operating in dense regions, there is an increased chance of finding a long sequence of similar queries, hence achieving large $simSeq_i$.

A challenge is to efficiently search for dense regions. To achieve this, we cluster all the queries in the buffer based on their mutual distances in the metric space, i.e., any two queries in the same cluster should be close to each other. Therefore the task to find dense regions is reduced to searching for clusters containing large number of queries. The clustering has to be done on the fly as soon as a query enters the buffer and it is required not to impose a significant overhead so that it does not actually slow down the whole process. Formally, the cluster $C = \{(q_1, t_1) \ldots, (q_m, t_m)\} \subseteq B$ where $d(q_i, q_j) \approx 0$ for each i, j. It is reasonable to adapt the clustering technique to the used indexing mechanism. In our case, an adequate clustering technique is a pivot based approach [9]. Specifically, there is a set of reference objects (pivots). To assign a query to a cluster, the distances of the query object to all the pivots are computed. The pivots are then sorted by the distances and the obtained pivot permutation determines the cluster.

The pseudocode can be seen in Algorithm 2. The *pop* function is called repeatedly according to Algorithm 1. Its functionality depends on a state which keeps a reference to the content of the buffer and to the currently processed cluster. If there are no more query objects in $currentCluster$, it finds the most populated cluster of query objects in the buffer and sets it as $currentCluster$. Then a query object is taken from $currentCluster$ and returned to be processed.

Algorithm 2. Greedy algorithm

---

var $buffer, currentCluster$
function pop
   if empty($currentCluster$) **then**
      $currentCluster \leftarrow$ getMostPopulatedCluster($buffer$)
   $query \leftarrow currentCluster$.getNext()()
   removeFromBuffer($buffer, query$)
   return $query$

---

6   Delay Limit

In the definition of the problem, we have defined a limit for the maximum delay. Using the strategy for query ordering as stated above, there can be clusters which do not ever get to be processed. This is the case of clusters with low number of queries that are never selected as the cluster with the highest population. In this section, we show how to modify the query ordering so that the delay limit is obeyed.

To remind the delay limit, whenever a query is about to be picked from the buffer and there exists a query whose timestamp is older than the allowed delay, the query with the oldest timestamp in the buffer has to be processed.

Since a timed out query may be dissimilar to the previous queries, it is probable that the cache utility will be zero. The situation can be observed in Fig. 2b. The processing of the cluster A is interrupted by the timed out query in the cluster B. The question is how to proceed after processing the timed out query. Basically, there are two options. The first one is not to cache the results of the timed out query and continue with the processing of the cluster A. The second option is to cache the results of the timed out query and continue with the processing of the cluster B. According to the proposed greedy algorithm, we should select the option which maximizes the number of consecutive queries having small distances between them (maximizing $simSeq_i$, see Formula 10).

The option can be selected based on the number of query objects in the currently processed cluster and in the cluster containing the timed out query. Using the approach presented in the previous section, the cluster with the bigger number of queries should be chosen. However, selecting always the most populated cluster does not guarantee the sequence of nearby queries is maximized because processing of the queries may be interrupted by a timed out query from another cluster as we have seen in Fig. 2b. Therefore it is needed to consider also the timeouts which may occur during processing of a cluster.

The key question is how many queries in the selected cluster can be processed without any interruption, i.e., without encountering any dissimilar query object (x_i in Formula 8). The interruption happens in two situations: all the queries in the currently processed cluster are finished or there is a timeout outside the selected cluster. Therefore the number of subsequent similar queries is the minimum of the number of the queries in the cluster and the number of the queries which can be processed until the first timeout occurring in a different cluster.

To implement the greedy algorithm, we compute the time after which the processing of a given cluster would be interrupted. The algorithm selects the cluster with the largest time before an interruption. Using this approach, we can prioritize small clusters with early timeouts over large clusters with late timeouts.

Let B_t be the content of the buffer at time t and the cluster

$$C = \{(q_1, t_1), \ldots, (q_m, t_m)\} \subseteq B_t$$

For the sake of simplicity, let us suppose the query times are constant after processing the first query in the cluster since the cache utility is constant, i.e., $queryTime(q_i, cacheUtility) = c$ for $2 \leq i \leq m$, and $queryTime(q_1, 0) = d$. Then

$$clusterTime(C) = sequenceTime((q_1, \ldots, q_m)) = c \cdot (|C| - 1) + d \quad (11)$$

is the time needed to process the cluster C.

Now we derive the time after which there is a timeout outside the cluster C. Let MD be the maximum delay limit.

$$earliestTimeout(B_t, C) = max\{0, min\{MD - (t - e)|(q, e) \in B_t \wedge (q, e) \notin C\}\} \tag{12}$$

is the earliest time after which there is a timeout outside the cluster C. Note that all the computations are related to the state of the buffer at the time t since we cannot predict what queries will enter the buffer after t.

When put together, the time after which the processing of the cluster C is interrupted is $min\{clusterTime(C), earliestTimeout(B_t, C)\}$. The greedy algorithm selects the cluster which maximizes the interruption time.

The pseudocode can be found in Algorithm 3. If there exists a query object exceeding the delay limit, the oldest query in the buffer is chosen to be processed. The *currentCluster* and *timedOutCluster* are compared regarding the time until an interruption. If *currentCluster* has a larger time, the results of the timed out query are not cached since its cluster will not be processed in the next round and it would be useless. If there is no timeout and if *currentCluster* does not contain any more queries, a new cluster maximizing the time until interruption is selected from all the clusters in the buffer. Then a next query is taken from *currentCluster*; it is removed from the buffer, and processed. The *pop* function returns also the information whether the caching should be used during processing of the returned query. The individual situations are illustrated in Fig. 2.

(a) Another query in the cluster is processed.

(b) 2 is a timed out query; 3 can be either in the previous (A) or in the timed out cluster (B).

(c) A cluster is completely processed and a query in a new one is selected.

Fig. 2. Algorithm illustration; the numbers indicate the order of processed queries

Since Algorithm 3 can disable caching for a particular query, we have to slightly modify the definition of *similarCount* so that it skips the queries for which the caching was disabled.

$$similarCount((q_1, \ldots, q_m)) = |\{i|d(q_{prevCached(i)}, q_i) \approx 0 \wedge 2 \leq i \leq m\}| \tag{13}$$

where $prevCached(i) = max\{h|h < i \wedge (caching(q_h) \vee h = 1)\}$ and $caching(q)$ is true iff the caching was enabled when processing q.

Note that in case of a timed out query, we choose the next *currentCluster* from only two clusters, so it is in compliance with the greedy approach. Processing of the timed out query outside *currentCluster* generates an item

---

Algorithm 3. Full greedy algorithm considering the delay limit

---
  var $buffer, currentCluster$
  function pop($timeLimit$)
    $caching \leftarrow$ true
    if existsTimeout($buffer, timeLimit$) **then**
      $query \leftarrow$ getOldestQuery($buffer$)
      $timedoutCluster \leftarrow$ getClusterByQuery($query$)
      $currentCluster$ $\leftarrow$ findClusterWithLargestTimeUntilInterruption()
$[currentCluster, timedOutCluster]$
      if $timedOutCluster \neq currentCluster$ **then**
        $caching \leftarrow$ false
    else
      if isEmpty($currentCluster$) **then**
        $currentCluster$ $\leftarrow$ findClusterWithLargestTimeUntilInterruption()
$buffer$.getClusters()
      $query \leftarrow currentCluster$.getNext()
    removeFromBuffer($buffer, query$)
    return $query, caching$

---

in *dissimilar* (Formula 8) since $d(q_i, q_{i-1}) \not\approx 0$. We either cache the results of the processing and proceed with processing the cluster of the timed out query, hence using the cached values, or the results are not cached and we get back to the original *currentCluster* and use the already cached values. If a different cluster was selected, there would be $dissimilar(s) = (\ldots, i, i+1, \ldots)$ where i corresponds to the timed out query and $i+1$ corresponds to the first query of the new cluster. This would violate the greedy algorithm since there would be a zero-length sequence of similar items ($simSeq_k = 0$ in Formula 10 for the corresponding k).

Since the greedy algorithm always chooses just one cluster which maximizes the cache utility at that moment, it can be very efficient. On the other hand, the best global ordering of the queries may be missed that way, and algorithms which compute the best ordering of all the queries in the buffer could be considered. However, they are likely to be slower and since there are new queries continuously added to the buffer, the computed ordering is only relevant for the actual state of the buffer.

7 Experiments

In this section, we experimentally validate the hypotheses presented above.

7.1 Experiment Setup

We use the M-Index [9] technique to index the metric-space data. It employs practically all known principles of metric space partitioning, pruning, and filtering, thus reaching high search performance. The actual data are partitioned into buckets which are stored as separate files on a disk and read into the main memory during query evaluations. To partition the data, M-Index uses a set of pivots. To insert an object into the index, the pivots are sorted based on the distance to the object. In this way, a pivot permutation is obtained which identifies the bucket to insert the object. During a similarity search, mutual distances between the query object and the pivots are used to reduce the set of buckets which need to be accessed. The M-Index supports (among others) executing approximate kNN queries. One of the stop conditions of a query evaluation is given by the maximum number of accessed objects (the size of a candidate set). Such a stop condition is used in our experiments.

We use the Profimedia dataset of images [2] in the experiments. We created three different subsets of the images and extracted their visual-feature descriptors. The generated datasets are: 1 million Caffe descriptors [5] (4096 dimensional vectors), 1 million MPEG-7 descriptors [2] and 10 million MPEG-7 descriptors. Separately, we created streams of images represented by corresponding descriptors. During each experiment, the image descriptors of the stream are continuously sent to the buffer and processed by an approximate 10-NN query.

As the streamed queries enter the buffer, they are clustered using the pivots of the M-Index as described in Sect. 5. The clustering does not introduce a significant overhead since the same set of pivots is used also during the query evaluation and the computed distances can be reused.

The tested applications are implemented using Java programming language with the use of the MESSIF library [1] providing an implementation of the M-Index. The experiments were run on Intel Xeon 2.00 GHz with 8 GB RAM. The descriptors of the datasets are stored on a HDD.

7.2 Cache Utility

Figure 3a shows results of our experiments exploring the impact of the cache utility on the query time. We ran approximate 10-NN queries for each dataset and we were continuously changing the percentage of accessed buckets stored in the cache. The M-Index used candidate sets of size 10,000. The x-axis shows the percentage of the cached values; the y-axis represents the percentage of the time to process the query compared to the situation when the cache is not used. It can be observed that the processing time can be improved dramatically if the cache is filled with appropriate values, thus the assumption in Formula 2 is valid.

7.3 Buffer Size

In this group of experiments, we explore the impact of the size of the buffer. The experiments were conducted for all three datasets of descriptors. No maximal delay constraint was used. The maximum size of the cache was set to 90,000 descriptors for the 10 mil. MPEG-7 dataset (i.e., 0.9 % of the database); up to 40,000 descriptors were stored for the 1 mil. MPEG-7 and Caffe datasets (i.e., 4 % of the database). The size of the M-Index candidate set was 2,000 for the 10 mil. MPEG-7 dataset, and 1,000 for the others. With a growing size of the buffer, also

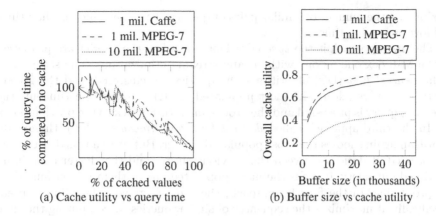

(a) Cache utility vs query time   (b) Buffer size vs cache utility

Fig. 3. Cache utility experiments

the cache utility and the throughput grow since the processed clusters are more populated and the cached values are reused more times, i.e., *similarCount* is higher, see Figs. 3b and 4a. The throughput speedup was computed as the ratio of the number of processed queries using a given buffer size and the number of processed queries without the usage of the proposed optimizations. The increase in the median delay of the processed queries (the time spent in the buffer) is shown in Fig. 4b. It was also measured there are just minor overhead costs connected with the buffer management (query clustering and reordering).

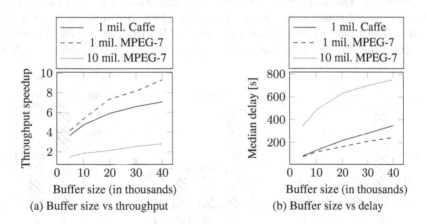

(a) Buffer size vs throughput   (b) Buffer size vs delay

Fig. 4. Buffer size analysis; every experiment ran 30 min

7.4   Greedy Algorithm

In this section we evaluate four different approaches to processing a stream of kNN queries, and we experimentally prove Formula 7, i.e., by greedily maximizing the number of similar pairs of queries, we achieve high cache utility and low processing time.

The first approach is the *base* (B) algorithm when the queries are processed by the M-Index one by one without any stream optimizations. The second one is the *densest first* (DF) algorithm. The queries are clustered, and the greedy algorithm always selects the most populated cluster. In case of a timeout, the timed out query is processed and the processing returns back to the previous cluster. In the third approach called *densest first with timeouts* (DFT), the greedy algorithm again chooses the most populated cluster. But after a timed out query is processed, it decides between the previous cluster and the cluster containing the timed out query. It picks the more populated one of them to continue with the processing. In the fourth approach, the *full* (F) greedy algorithm is used. Specifically, it maximizes the sequences of similar queries by computing the time until interruption for individual clusters (Algorithm 3).

In the first group of experiments, we used the dataset of 1 mil. Caffe descriptors. The approaches using the cache (DF, DFT, F) stored up to 40,000 descriptors in the cache (4 % of the DB) and used a buffer of size 8,000. The delay limit was set to 15 min, and 100,000 queries were processed for each approach. The size of the candidate set of the M-Index was set to 1,000 for each approximate kNN query.

See Fig. 5a for the results. The buffer-based approaches were able to process the queries much faster than the *base* algorithm. The individual optimizations pay off as expected, and the best results are obtained using the *full* greedy algorithm when the processing time was reduced 3.69 times compared to the *base* approach. The impact of individual optimizations can be observed also in the number of timed out queries: DF with 13408 timeouts vs F with 1491 timeouts. The cache utility of the *full* greedy algorithm was 0.62; its median delay (the time since a query enters the buffer until it is processed) was 128 s.

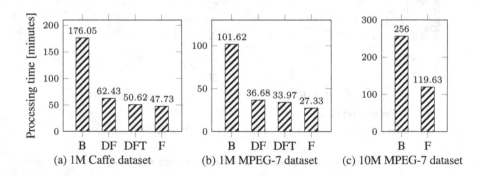

(a) 1M Caffe dataset   (b) 1M MPEG-7 dataset   (c) 10M MPEG-7 dataset

Fig. 5. Processing times of 100,000 queries

We have repeated the experiment for the dataset of 1 mil. MPEG-7 descriptors with the same settings. That is, the buffer 8,000, up to 40,000 descriptors in the cache, 15 min delay limit, and 100,000 queries processed with 1,000 candidate objects used for each query. The results are presented in Fig. 5b. We were able to reduce the processing time 3.72 times compared to the *base* approach. The cache utility of the *full* greedy algorithm was 0.66; its median delay was 79 s, and there were 169 timed out queries.

In Fig. 5c, we can see a comparison of processing times when the 10 mil. dataset of MPEG-7 descriptors is used. We used a buffer of size 10,000; up to 90,000 descriptors were stored in the cache (0.9 % of the DB); 100,000 queries were processed. The timeout limit was set to 30 min. Up to 2,000 candidate objects were used for each approximate query. The *full* greedy algorithm clearly outperforms the *base* approach as the processing was 2 times faster. The median delay of the *full* greedy algorithm was 548 s (9 min); the cache utility was 0.38; there were 2,838 timeouts.

Through the previous experiments, we verified the validity of our approach. That is, by finding long sequences of similar pairs of queries, we achieve high cache utility, and thus low processing time.

8 Conclusion

We have presented a novel approach to enhance the throughput of similarity search queries while obeying a given delay limit and processing every data item in the stream of queries. The technique is based on dynamic reordering of the incoming queries combined with appropriate caching strategy for partitions of the indexing structure. We have proposed three variants of the reordering by applying greedy heuristic approach to identify long sequences of similar queries. Two of the variants deal with starvation of the buffered queries by employing timeouts with different strategies of processing continuation. Our experimental evaluation proved our expectations and the most sophisticated technique was able to achieve up to 3.7 times higher throughput compared to the base case when no reordering is used.

Acknowledgements. This work was supported by the Czech national research project GA16-18889S.

References

1. Batko, M., Novak, D., Zezula, P.: MESSIF: metric similarity search implementation framework. In: Thanos, C., Borri, F., Candela, L. (eds.) Digital Libraries: Research and Development. LNCS, vol. 4877, pp. 1–10. Springer, Heidelberg (2007)
2. Budikova, P., Batko, M., Zezula, P.: Evaluation platform for content-based image retrieval systems. In: Gradmann, S., Borri, F., Meghini, C., Schuldt, H. (eds.) TPDL 2011. LNCS, vol. 6966, pp. 130–142. Springer, Heidelberg (2011)
3. Falchi, F., Lucchese, C., Orlando, S., Perego, R., Rabitti, F.: Similarity caching in large-scale image retrieval. Inf. Process. Manage. 48(5), 803–818 (2012)

4. Gaber, M.M., Zaslavsky, A., Krishnaswamy, S.: A survey of classification methods in data streams. In: Aggarwal, C.C. (ed.) Data Streams: Models and Algorithms. Advances in Database Systems, vol. 31, pp. 39–59. Springer, Heidelberg (2007)

5. Jia, Y., Shelhamer, E., Donahue, J., Karayev, S., Long, J., Girshick, R., Guadarrama, S., Darrell, T.: Caffe: convolutional architecture for fast feature embedding. In: Proceedings of the ACM International Conference on Multimedia, pp. 675–678. ACM (2014)

6. Law, Y.-N., Zaniolo, C.: An adaptive nearest neighbor classification algorithm for data streams. In: Jorge, A.M., Torgo, L., Brazdil, P.B., Camacho, R., Gama, J. (eds.) PKDD 2005. LNCS (LNAI), vol. 3721, pp. 108–120. Springer, Heidelberg (2005)

7. Mera, D., Batko, M., Zezula, P.: Towards fast multimedia feature extraction: hadoop or storm. In: 2014 IEEE International Symposium on Multimedia (ISM), pp. 106–109. IEEE (2014)

8. Nalepa, F., Batko, M., Zezula, P.: Performance analysis of distributed stream processing applications through colored petri nets. In: Kofron, J., Vojnar, T. (eds.) MEMICS 2015. LNCS, vol. 9548, pp. 93–106. Springer, Heidelberg (2016). doi:10.1007/978-3-319-29817-7_9

9. Novak, D., Batko, M., Zezula, P.: Metric index: an efficient and scalable solution for precise and approximate similarity search. Inf. Syst. **36**(4), 721–733 (2011)

10. Pandey, S., Broder, A., Chierichetti, F., Josifovski, V., Kumar, R., Vassilvitskii, S.: Nearest-neighbor caching for content-match applications. In: Proceedings of the 18th International Conference on World Wide Web, pp. 441–450. ACM (2009)

11. Pietruczuk, L., Duda, P., Jaworski, M.: A new fuzzy classifier for data streams. In: Rutkowski, L., Korytkowski, M., Scherer, R., Tadeusiewicz, R., Zadeh, L.A., Zurada, J.M. (eds.) ICAISC 2012, Part I. LNCS, vol. 7267, pp. 318–324. Springer, Heidelberg (2012)

12. Shao, J., Huang, Z., Shen, H.T., Zhou, X., Lim, E.P., Li, Y.: Batch nearest neighbor search for video retrieval. IEEE Trans. Multimed. **10**(3), 409–420 (2008)

13. Tao, Y., Papadias, D., Shen, Q.: Continuous nearest neighbor search. In: Proceedings of the 28th International Conference on Very Large Data Bases, VLDB Endowment, pp. 287–298 (2002)

14. Zezula, P., Amato, G., Dohnal, V., Batko, M.: Similarity Search: The Metric Space Approach. Advances in Database Systems, vol. 32. Springer, Heidelberg (2006)

15. Zhang, P., Zhou, C., Wang, P., Gao, B.J., Zhu, X., Guo, L.: E-tree: an efficient indexing structure for ensemble models on data streams. IEEE Trans. Knowl. Data Eng. **27**(2), 461–474 (2015)

Creating a Music Recommendation
and Streaming Application for Android

Elliot Jenkins and Yanyan Yang[✉]

School of Engineering, University of Portsmouth, Portsmouth, UK
elliot.jenkins@myport.ac.uk, linda.yang@port.ac.uk

Abstract. Music recommendation and streaming services have grown exponentially with the introduction of smartphones. Despite the large number of
systems, they all currently face a lot of issues. One issue is a cold start where a
user who is new to the system can't be made recommendations until the system
learns their tastes. They also lack context awareness to make truly personalised
recommendations to the user. This paper introduces a new recommendation and
streaming application, individual Personalised Music (iPMusic), for Android
which is specifically designed to address the issues. We examine the effectiveness of iPMusic based on real world users' feedback which shows positive
results.

Keywords: Music retrieval · Mobile computing · Recommendation system ·
Music streaming

1 Introduction

With the continual evolution of personal media players from Sony Walkman's, to
portable CD players, MP3 players and eventually the Apple iPod it is clear that individuals like to listen to music wherever they are. This is becoming an area that smartphones are largely moving into to help cope with the huge demand for music. The
amount of music being released has also been growing at ever larger rates with multiple
services such as Google Play Music and Deezer having in excess of 30 million songs [1].
With such huge libraries of music it is possible for users to become lost and struggle to
find new music that they like. As new songs are released users may not be informed so it
becomes problematic for them.

To help the users, recommendation systems have been created making use of
content-based and collaborative algorithms. There are however issues with these current recommendation algorithms, those used by Spotify [2] and Google Music [3]
require the user to first of all listen to music before any recommendation can be made to
them. The issue with this is if a new user joins then no recommendation can be made to
them. Similarly if a new song is added until users listen to it and rate it, collaborative
methods would rate the song lowly so it is unlikely to be recommended. A hybrid
approach was put forward by Wang et al. [4] that makes use of the user's context such
as location and listening history and combines this with content based methods to
overcome the issue of not providing recommendations to new users. By taking the
user's location into account it creates a more personalised recommendation compared

© Springer International Publishing Switzerland 2016
S. Hartmann and H. Ma (Eds.): DEXA 2016, Part II, LNCS 9828, pp. 201–215, 2016.
DOI: 10.1007/978-3-319-44406-2_15

to Spotify or Google Music but this can be further improved so that every recommendation is unique to that individual user. This provides an interesting challenge for developers to solve as the system needs to be able to cope with new users being added and to provide them with personalised recommendations from the beginning but also when a new song is added it is considered fairly when making recommendations.

The motivation for this work has been to provide users with an easy way to discover new music. We have developed an algorithm that will provide unique personalised recommendations based on the users' Twitter posts as well as their listening history. The system will pull song lyrics and other fields to generate a list of similar songs when creating the recommendation. Collaborative methods have been developed to find similar users in the system and to enhance the recommendation based on what similar users like. This hybrid approach combining information from a variety of sources will produce a more accurate and personalised recommendation to the user on the individual Personalised Music (iPMusic) App. It will then be possible for the user to stream the songs in the app or to play the music video through YouTube. At the time of writing, we are unaware of any music service that can produce recommendations without any prior knowledge of personal listening history.

The rest of the paper is organised into the following sections; in Sect. 2, we discuss the problem in more detail and related work that has already been carried out. Section 3 introduces the proposed system and real world usage scenarios. In Sect. 4, we describe the architecture and introduce the main components in more detail. In Sect. 5, we discuss the current implementation of the system and evaluate the findings from user feedback. We conclude in Sect. 6.

2 Problem Description and Related Work

The current issue with the music industry is there is a huge quantity of songs available to the user. The largest library of music is provided by Apple Music [1] which has in excess of 43 million songs. Work carried out by Nathan et al. [5] suggested that the average song length is 226 s which means to listen to 43 million songs would take 308 years. This is not possible and many users wouldn't like all of the music available to them so the music services have begun to provide the user with recommendations. Despite that when making recommendations they face an issue known as the cold start. The cold start issue is where a new user is added to the system so has no known tastes meaning a recommendation can't be made to them. So many current services require the user to start searching and playing music that they like before any recommendations can be made to them. A further weakness is that recommendations usually do not include any user context so are not truly personalised to the user. Another issue is when a new song is added the user may not be made aware of this. This is because collaborative methods take into account song ratings and how frequently they have been played, meaning a new song would have nothing for both most likely resulting in it not being recommended to the user.

Su et al. [6] propose a prototype music recommendation system that is designed to make use of the user's context and context information mining to offer recommendations that will suit the listener in their current situation. This would provide a personalised recommendation in a way no other system can by making inferences from patterns detected by the context miner. Despite providing better recommendations than existing systems there is still the major issue of a cold start that has not been addressed. Due to the nature of the algorithm a context log is required which describes multiple conditions about the user at different times of the day which would need to be collected prior to making a recommendation.

Narayanan et al. [7] presented a collaborative method making use of K-Nearest Neighbourhood (K-NN). This method allows predicting what one user will like based on another user who is similar to him. Although not directly solving any of the issues we are looking to overcome, the idea can be used to further enhance the accuracy of the recommendations as proven by Narayanan et al. work.

Adomavicius et al. [8] presented many approaches to recommendation systems detailing any advantages or limitations. The main issue highlighted is the cold start issue and their solution to solve it is using a hybrid approach of collaborative and content-based methods. However they discuss different hybrid approaches such as using the two methods separately and carrying forward the recommendation that is the most accurate. This approach can therefore remove any personalisation of the recommendation if just content-based methods are used.

Wheal et al. [9] developed CSRecommender which provides recommendations for different cloud based services that are currently available. Wheal's approach uses a hybrid recommender making use of collaborative methods taking into account the user and similar users and combining this with a content-based method that finds a similar service. By making use of both approaches it allows for an accurate recommendation to be made to the user. Despite recommending cloud services a similar approach can be taken to make song recommendations.

Twitter Music [10] is a system that pulls music from iTunes, Spotify, Rdio and Vine and then presents the best new music that is trending on Twitter. The recommendations being made by the service are not personalised and are instead based on the entirety of Twitter users. This is a reliable method to find what the most popular music is and is the only system to alleviate the cold start issue for a new user. However if a new song has been released and it doesn't trend on Twitter then it won't be recommended so it faces the same issue as the other systems.

This research work addresses the aforementioned issues by creating a hybrid approach that will take into account the majority of above methods in a unique recommendation algorithm. By using context information collected from Twitter combined with a K-NN approach the cold start issue can be addressed whilst offering a high level of personalisation.

3   System Overview

The system is based on a client server architecture which both communicate with one another as well as external sources which is shown in Fig. 1. It is possible for multiple clients using the iPMusic App to connect to the Server simultaneously and each will be handled by their own thread.

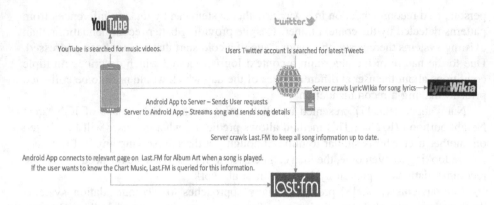

Fig. 1. Overview of the system

There are two phases that occur on the Server, initially a setup phase runs before users can connect and make requests. The setup phase crawls Lyric Wikia and Last.FM to create an up to date index for the initial library of music. Once the information has been collected the Server exits the setup phase and allows connections from the iPMusic App. The Server now remains in this stage so it is necessary for it to detect new music being added to the system and to obtain the information for new items and to correctly add them to the index.

From the users' point of view, they will start off by downloading iPMusic from the Google Play Store. Once downloaded and installed they can create an account which will then allow them full access to the Application. They will have the ability to see what music is in the current charts, to get a list of recommendations, search for a song and to display their favourite music to play back at any time. The following real world scenarios are an indication of how the system can be used:

- A new user may be wanting to discover new music so use the "Play me something" button. This would present the user with a list of uniquely generated recommendations and the ability to play any of these back. On playing a song and the user providing a rating it further improves the accuracy of future recommendations.
- The system has the capability to generate a unique and personalised recommendation created just-in-time so users may be wishing to take advantage of this feature that is not offered by other systems at such a personalised level.
- Although not the primary purpose, the search feature in the app not only searches song titles, artists and album names but also the lyrics. Therefore meaning if the user knows the lyrics of a song but is unsure what it is called then they can find out via the app.

4   System Architecture and Design

An overview of how all of the components on the server and client side will communicate is detailed below in Fig. 2.

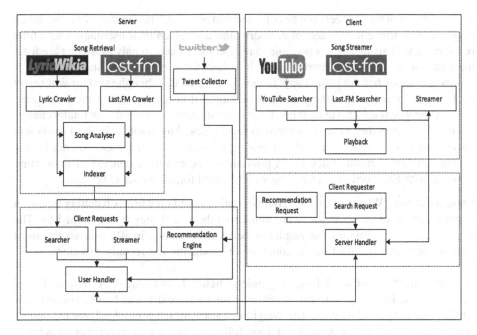

Fig. 2. Block diagram of server and client

4.1 Server Side

Last.FM Crawler. The purpose of the Last.FM crawler on the server side is to maintain an up to date list of all the music files in the system. When reading the tag fields from an .mp3 file it is possible some information is missing or it is wrong due to being entered incorrectly. Last.FM provides an API to access their different services and the crawler will make use of track.getInfo which returns the metadata for any given track. In doing this it will allow the recommendations being made to be more accurate since the tracks will contain more details such as release dates and album that may otherwise be missing.

Lyric Crawler. The main part of the recommendation algorithm is based on the song lyrics so it is necessary to find all of the song lyrics for the tracks in the system. LyricWikia provides lyrics for 1,798,797 different songs, so it is likely that any song searched for produces a result. So the purpose of the lyric crawler is to obtain the lyrics for every song in iPMusic. It is possible a song has no lyrics such as instrumental but it will remain in the system as the song title and album names could still hold relevance to a recommendation. When searching for song lyrics the sitemap will be queried for the current song and if there is a match the crawler will connect to the matching URL. Once connected to the web page the content will be parsed by Jsoup and the lyrics will get extracted and stored in the index. The process is highlighted in Algorithm 4.1.1.

Tweet Collector. Twitter has over 320 million monthly users so there is a high probability that the user will have a Twitter account. To get a more personalised recommendation their Tweets can be collected and indexed. The Twitter API provides an easy way to retrieve the Tweets of an individual user given their username and a time based context. This means it is possible to restrict the Tweets to only those that are from the past day as this would more accurately reflect the user's current state of mind. If however the user has not posted any Tweets in the past day then the system will revert back to any Tweets within the past week or month if necessary. When the Tweets are indexed the words are kept in their raw form since stemming the words can totally change the meaning when then being compared to song lyrics. Any numbers in the Tweets will also be indexed as they can be related to song titles or lyrics. If a user does not have a Twitter account then this stage is not possible so there is the extension of integrating iPMusic with Facebook at a later stage which would follow the same process.

Song Analyser. When making a recommendation based on a user's listening history it is necessary to know what songs are similar to those that they have listened to. The Song Analyser determines the similarity between all songs in iPMusic so the most similar songs can be taken into account when producing the recommendation.

A song is made up of the 5 following unique fields; Lyrics, Artist, Album, Title and Release Year. By using different weightings for each field it can be determined how similar one song is to another. The weightings used by the Song Analyser are shown below. These weightings were determined following the use of experimental weightings until the yielded results were liked by a set of users.

- 55 % - Lyric Similarity
- 20 % - Artist or Album matching
- 20 % - Title Similarity
- 5 % - Same release year

The process to generate a list of similar songs starts by using cosine similarity to compare the similarity of the song lyrics which is then multiplied by the weighting of 55 %. If the Artist or Album match then the score is increased by 20 %. The titles of the song being compared against is split into individual terms and the similarity of the two titles is calculated by seeing how many times the individual terms appear in the other title. For every match a counter is incremented and this can then be converted to a percentage for the overall similarity which is then multiplied by a weighting of 20 %. The remaining 5 % comes from the year that the songs were released as it may have some relevance to the recommendation. The formula to calculate similarity is shown in Eq. 2. The list of similar songs is then sorted into descending order and stored in the index.

$$\text{Cosine } Similarity(A, B) = \frac{A \cdot B}{|A||B|} \tag{1}$$

Song Similarity whereby LS is Lyric Similarity and TS Title Similarity

$$Son\ Similarity = (LS \times 0.55) + if\,(Album\ or\ Artist\ Match, 0.2, 0)$$
$$+ (TS \times 0.2) + if\,(Years\ Match, 0.05, 0) \tag{2}$$

Indexer. With all of the data gathered by the above methods it is necessary to index the data so it can be quickly queried and provides results to the user in the shortest amount of time. The index will be made up of the following fields:

- The Song and Artist ID's will allow for fast identification of songs without the need for searching for titles and artists and finding a match. Instead the system will be able to go the nth Artist and to that artists n^{th} Song therefore greatly increasing the speed and efficiency of the system.
- The Song Title, Artist and Album names the Term Frequency (*TF*) and Inverse Document Frequency (*IDF*) will be calculated using Eqs. 3 and 4 respectively. From this the *TF-IDF* can be calculated for each term in the fields and this will be stored as a posting in the Inverted File Index.

$$TF = \frac{Number\ of\ times\ t\ appears\ in\ a\ document}{Total\ number\ of\ terms\ in\ the\ document} \tag{3}$$

$$IDF = \log \frac{Total\ number\ of\ documents}{Number\ of\ documents\ with\ term\ t} \tag{4}$$

- The Release Date too but this can be in a variety of different formats such as an English date of DD/MM/YYYY compared to American format of MM/DD/YYYY. The date may need to be changed so that all of the formats are the same allowing for quicker comparisons when creating a recommendation.
- The Lyrics will be stored using TF-IDF but also the document vectors will be kept which will allow for getting the cosine similarity between two songs.
- The file location will indicate where the song is stored on the iPMusic Server so that the song can be quickly streamed to the smartphone. This allows for scalability as it is possible for songs to be stored in more than one location.
- The Similar Songs List will be generated by the Song Analyser and is stored in descending order to quickly find the most similar song.

Recommendation Engine. The recommendation engine goes through the following series of stages each of which are explained further below.

- K-Nearest Neighbour – 25 %
- Twitter – 25 %
- Similar Songs – 50 %

If the user's listening history is not empty the first stage is to use a collaborative approach using K-Nearest Neighbour. A user is considered a nearest neighbour based on the similarity between the users listening history and the neighbours taking into account their ratings as well as play count. Since nearest neighbours will have similar interests in music it can be assumed that if neighbour N likes song A then user U will also like song A so a recommendation can be made based upon this.

Algorithm 4.1.1 K-Nearest Neighbour

```
Get vector for User U
while there are Users to compare Uc do
  Get vector for Uc
  Calculate Cosine Similarity for U and Uc (Equation 1)
  Set counter and matches to 0
  while there are songs S in Uc Listening History do
    if S is in U listening history then
      Increment matches
      if ratings for songs U and Uc are above 3 then
        Increment counter
      end if
    end if
  end while
  Calculate rating similarity from counter / matches
  if rating similarity > cosine similarity then
    Add user to nearest neighbour list
  end if
end while
Sort the nearest neighbour list
for the top 5 nearest neighbours do
  While there are Songs S in listening history do
    if S rating is 5 then
      Add S to recommendation with score of 25
    else if S rating is 4 then
      Add S to recommendation with score of 20
    else if S rating is 3 then
      Add S to recommendation with score of 15
    end if
  end while
end for
```

The next step is to find songs similar to those that the user has already listened to. In the index the list of similar songs can be used to identify those that are similar to what the user has already listened to. This however creates an issue of how to handle the score if two or more songs have a similar song in common. If the scores were to be added then the results would be skewed and songs that aren't that similar could get higher scores than those that are. Or if the average was taken any outliers would bring the score down. So a method was designed to combine the scores without skewing the results as shown in Algorithm 4.1.2 by removing any scores outside the standard deviation of the average.

Algorithm 4.1.2 Similar Songs

```
while there are songs S in listening history do
  for the top 10 similar songs SS do
    Add SS to recommendation list
    Add variance to list of variances in recommendation list using equation 5
    Keep running total of variance squared and sum of variances for
    recommendation
  end for
end while
while there are recommendations R do
  Calculate standard deviation from equation 6
  Remove any variances below average minus standard deviation
  Set score to average variances
  Add weighting onto score from equation 7
end while
```

Following this the top terms from the users Tweets will be searched for amongst the song lyrics. Any songs that match the search query are then added to the list of song recommendations with a score determined by their TF-IDF score. The maximum score a song can get from this stage is 25, so the TF-IDF scores are normalised to range from 0–100 %. As previously mentioned each stage carries a different weighting towards the final score and for this stage it is 25 %. This was determined through testing of different weightings until the recommendation best reflected the test users taste. With the normalised score now as a percentage the final score can be calculated by multiplying it by the weighting of 25 %.

$$Variance = Similar\ Song\ Score - Average\ Song\ Score \qquad (5)$$

$$Standard\ Deviation = \sqrt{\frac{\sum Variances^2}{Number\ of\ Variances}} \qquad (6)$$

$$Weighting = \frac{100 - Maximum\ Score}{Maximum\ number\ of\ occurrences} \times Number\ of\ occurrences \qquad (7)$$

The final stage of the algorithm is optional and can restrict the songs to only those that have been released within the user's lifetime. The likelihood of the songs being restricted is calculated from the percentage of songs that have been released during the user's lifetime compared to those that have not. This means that the more music a user listens to that has been released during their lifetime the higher the likelihood of the restriction being put in place, if the majority of the music listened to is not in their lifetime there is a small likelihood of the restriction being in place.

Following all of these stages a recommendation can be made to the user and the top 10 are presented to the user for them to then pick which to listen to. If however the user has never listened to anything on the system then two of the stages are used to stop the cold start issue.

The K-NN approach is used again but instead focuses on the user's age. There is a high chance that if one user is the same age as another then they will like similar music. With the nearest 5 neighbours found all of their highest rated songs are added to the recommendation list.

The next stage using the same approach as before with Twitter. If the user does not have a Twitter account then this stage is skipped and the recommendations are given straight to the user. A future addition will be the integration of Facebook as well as other social networking platforms to further enhance the level of personalisation.

This hybrid approach combining K-NN with the content based methods using Twitter alleviates the cold start issue that other systems suffer from. As the system grows and the number of songs listened to increases the algorithm will be able to provide more accurate recommendations. If however there are no other users in the system and the user does not have Twitter then the algorithm would not work. So it will be necessary to populate the system with some default users that fit certain categories.

Searcher. The searcher will provide a fast way to search the index file and return back the most relevant results. The recommendation algorithm will make use of the searcher when searching lyrics and titles for Twitter keywords and will return back the most relevant results. Similarly if the user submits a search from the iPMusic App then the lyrics, artist, title and album will all be queried returning back the most relevant results. The relevancy of a result is calculated from multiplying the term frequency with inverse document frequency from the inverted index to get the TF-IDF weight w. The Vector Space Model is then used to rank the results whereby q is the term being queried in two documents. Equation 8 - Vector Space Model whereby N is the number of results, $w_{i,j}$ is the weight given to the *ith* word in document j and $w_{i,q}$ the ith word in document q

$$Result = \frac{\sum_{i=1}^{N} w_{i,j} w_{i,q}}{\sqrt{\sum_{i=1}^{N} w_{i,j}^2} \sqrt{\sum_{i=1}^{N} w_{i,q}^2}} \tag{8}$$

User Handler. The last component of the server is the user handler and its purpose is to service all of the requests coming from the iPMusic App. The requests will be sent over a TCP socket which then need to be handed to the correct part of the server. As there will be multiple users it will be necessary to first login and once logged in the user can request recommendations or search for songs. The server will then send back to the user the list of recommendations or search results.

4.2   iPMusic Client

YouTube Searcher. The purpose of the YouTube Searcher is to try and identify the correct music video for the song currently being played. This will then allow the user to watch the music video rather than just listening to the song. The searcher will use the YouTube API to create a query made up of the song name and the artist. This is likely to return the correct video but it is impossible to guarantee it.

Last.FM Searcher. When playing music, album art should be displayed which Last. FM hosts for the majority of songs. So the Searcher will be used to get find the album art and display it on the app when playing a song. The album art is then cached locally on the device so that it can be used again without downloading the image.

Streamer. The Streamer is responsible for downloading the songs from the server onto the client so that they can be played back. When a user requests a song to be played that song will be downloaded as well as the songs preceding and following it. This allows the user to either fast forward or rewind songs without having to wait for those songs to be downloaded. Prior to end of the current song being played the next song will be requested and downloaded ensuring it is ready to play with continuous playback.

Playback. The playback component is responsible for displaying the information provided by the Songs MP3 tags, the album art from Last.FM and playing the Songs MP3 file. It also needs to handle playing and pausing of songs as well as fast forwarding and rewinding tracks.

Recommendation Request. The recommendation request will ask the Server to produce a recommendation via the Server Handler, the response will contain a list of recommendations which will then be downloaded and stored on the client. These will then be displayed in panels on the iPMusic client so the user can select what to listen to first.

Search Request. The search request component will get the users search term from the input box and pass this onto the Server Handler. The rest of the component works in the same way as a recommendation request with Search Result in place of Recommendation.

Server Handler. The server handler will be responsible for communicating with the server. It will use TCP sockets to either send or receive messages. Like the user handler any messages will need to be given to the correct component such as Search Request so that it is handled correctly.

5 Implementation and Evaluation

At the time of writing, iPMusic has been released on the Google Play Store as a closed Alpha. Android was the platform of choice as following its introduction 8 years ago [11], it is estimated to have a 46.7 % [12] share of 6,931,000,000 [13] active mobile phones, the largest of any platform. The Server and Android Client have both been written in Java and make use of the following open source Java API's (Application Program Interface): Apache Lucene, Jaudiotagger, Jsoup, Twitter4 J and lastly YouTube. All of these API's provide services required by either the Server or the Android Application. A library of 1,200 songs has been imported into iPMusic and a small number of seed users have been created. The android app interface is shown below in Fig. 3.

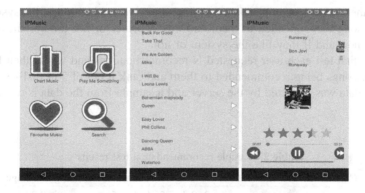

Fig. 3. Welcome screen, recommendations, playback, lyrics

Prior to being released as an Alpha on the Google Play Store internal testing was carried out by a selected group of users and by drawing comparisons between Last.FM [14] and iPMusic. This testing was very positive with the majority of recommendations being made being accurate and providing new recommendations to the user. When comparing the recommendations to Last.FM suggested songs there were some matches between the two. There are a few cases where the song and artist both matched from both systems but in more cases they both suggested songs from the same artists. A small sample is shown in Table 1. This highlights the fact that the recommendation engine works successfully. However due to the limited library compared to Last.FM a lot of songs suggested by Last.FM are not in the system so this is not a totally fair comparison.

Table 1. Bold indicates matches of song or artist between both systems, *Red* shows songs not in our system.

Song listened to	iPMusic recommendation	Last.FM recommendation
I Will Be – Leona Lewis	Better In Time – **Leona Lewis** The Best You Never Had – **Leona Lewis** Whatever It Takes – **Leona Lewis** Come In With The Rain Taylor Swift **Here I Am – Leona Lewis**	**Here I Am – Leona Lewis** *Yesterday – **Leona Lewis*** *I Still Believe – Mariah Carey* *Beautiful – Christina Aguilera*
Easy Lover – Phil Collins	If Leaving Me is Easy – **Phil Collins** **Two Hearts – Phil Collins** Wannabe – Spice Girls When You're Gone – Avril Lavigne	**Two Hearts – Phil Collins** *Something Happened On The Way to Heaven – **Phil Collins*** *Invisible Touch - Genesis* *Land Of Confusion - Genesis*

To further test the recommendation system users were asked to test the system and rank each song recommended to them and then whether or not the system provides something new and if they like the system or not.

For the first test each user requested N recommendations and would then listen to each of the songs being recommended to them and rank it out of 5 (1 dislike, 5 really like). This data was collected by the server and a sample from the data is displayed in Table 2.

Table 2. Sample recommendation test results

User ID	Number of recommendations	Average recommendation rating	Average rating
3	6	2, 2.4, 3.5, 4, 4, 3.5	3.23
4	5	0.8, 1.2, 1.3, 2.4, 3	1.74
8	5	3.6, 3.8, 4.3, 4.1, 4.5	4.06

The recommendation test results show that for most users the number of useful results started off low and as the system gained a better understanding of their tastes the recommendations being made to them improved. Despite that for user 4 the average

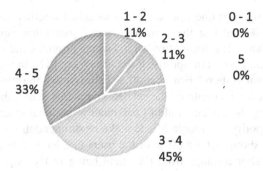

Fig. 4. Recommendation results

rating was 1.74 which is really low, this could be down to the fact the library of music was limited so there were few songs the user liked.

The average recommendation rating from 70 users is 3.4 which means the majority of the songs being recommended are liked by the users. 45 % of the ratings ranged between 3 and 4 which is the highest percentage for any score, and was closely followed by 33 % of the ratings being between 4 and 5. This shows positive results as the majority of the songs are highly rated by the users. Figure 4 shows a further breakdown of the scores and shows one area for improvement which is the fact that no user rated all 10 recommendations as 5 stars.

However further analysis of the results also show positive findings. iPMusic was designed to use the listening history of its' users when making recommendations and as a result the accuracy should increase over time. This is proven in Fig. 5 which shows that initially the songs being recommended are not as relevant as those suggested later during the users experiences. These results also show that the issues concerned with a cold start have been addressed since the average rating for the initial recommendations is 3.1. This is a satisfactory score showing that the users like the songs being recommended even when they are new to the system, if this score was lower, it would suggest the cold start issues had not been addressed.

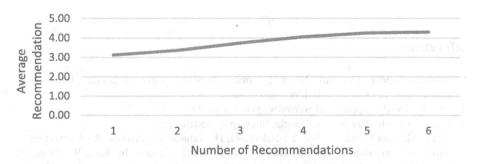

Fig. 5. Relationship between ratings and the number of recommendations

Following on to the next question, the user was asked whether the system provides something new. 60 % said that the system provided something new and one of the positive comments was "The integration with Twitter provides me more personalised recommendations than I get from Spotify, if the library could be increased substantially then it would be a great system." However 40 % believed that the system does not offer anything new. However a couple of the related comments stated these users did not have Twitter meaning the system couldn't personalise the recommendations so would be similar to how Spotify or Google Music make recommendations.

The final results show that the 80 % of the users that have tested the system liked using it despite the shortcomings from the client being in the Alpha stage of development. One user stated "The app is really simple to use and the real time recommendations come really quickly."

6 Conclusion

The music recommendation system is unique compared to the other music streaming services as it alleviates the cold start issue and provides much more personalised recommendations. With the ever growing amount of music and increasing number of individuals with smartphones there will be a greater need for advanced recommendation algorithms and this satisfies that demand.

The system has become very complex and many improvements can still be made to further improve the recommendations. The immediate goal is to integrate the system with Facebook in addition to Twitter which will provide more contextual information about the user. With a better profile built up about each user the K-Nearest Neighbour algorithm can find better matches further increasing the accuracy of the recommendations. Other immediate goals are to fix any remaining issues within the Android Client and to ensure maximum compatibility across devices.

With a unique approach to identifying similar songs and making recommendations, the project has the potential to become a marketable solution if music licencing laws are taken into account. If the system was migrated to iOS it would provide an easy way for any smartphone user to discover and to listen to new music.

References

1. Software Insider: Compare Music Streaming Services, Software Insider. http://music-streaming-services.softwareinsider.com/#main
2. Spotify: Spotify, Spotify. https://www.spotify.com/uk/
3. Google: Google Play Music, Google. https://play.google.com/
4. Wang, M., Kawamura, T., Sei, Y., Nakagawa, H., Tahara, Y., Ohsuga, A.: Context-aware music recommendation with serendipity using semantic relations. In: Kim, W., Ding, Y., Kim, H.-G. (eds.) JIST 2013. LNCS, vol. 8388, pp. 17–32. Springer, Heidelberg (2014)
5. Anisko, N., Anderson, E.: Average Length of Top 100 Songs on iTunes. StatCrunch (2012)
6. Su, J.-H., Yeh, H.-H., Yu, P.S., Tseng, V.S.: Music recommendation using content and context information mining. IEEE Intell. Syst. 25(1), 16–26 (2010)

7. Narayanan, S., Goswami, V.: K-Nearest Neighborhood based Music Recommendation System. http://www.cs.ucsb.edu/~sivabalan/cs240a/CS240A_Project_Report.pdf
8. Adomavicius, G., Tuzhilin, A.: Toward the next generation of recommender systems: a survery of the state-of-the-art and possible extension. IEEE Trans. Knowl. Data Eng. **17**, 734–749 (2005)
9. Wheal, J., Yang, Y.: CSRecommender: a cloud service searching and recommendation system. J. Comput. Commun. **3**, 65–73 (2015)
10. Twitter: #music, Twitter. https://music.twitter.com/
11. GSMArena: T-Mobile G1, GSMArena, October 2008. http://www.gsmarena.com/t_mobile_g1-2533.php
12. Canalys, Androdlib, Gartner: Android Phone Statistics. Statisticsbrain, 17 March 2015. http://www.statisticbrain.com/android-phone-statistics/
13. World Bank: Mobile Cellular Subscribers/Global & US. Statistic Brain, 17 March 2015. http://www.statisticbrain.com/mobile-cellular-subscribers-global-us/
14. Last.FM: Last.FM, Last.Fm (2015). http://www.last.fm/

A Score Fusion Method Using a Mixture Copula

Takuya Komatsuda[1(✉)], Atsushi Keyaki[2], and Jun Miyazaki[2]

[1] Hitachi, Ltd., Yokohama, Japan
takuya.komatsuda.ur@hitachi.com
[2] Department of Computer Science, School of Computing,
Tokyo Institute of Technology, Tokyo, Japan
keyaki@lsc.cs.titech.ac.jp, miyazaki@cs.titech.ac.jp

Abstract. In this paper, we propose a score fusion method using a mixture copula that can consider complex dependencies between multiple relevance scores in order to improve the effectiveness of information retrieval. The combination of multiple relevance scores has been shown to be effective in comparison with a single score. Widely used score fusion methods are linear combination and learning to rank. Linear combination cannot capture the non-linear dependency of multiple scores. Learning to rank yields output that makes it difficult to understand the models. These problems can be solved by using a copula, which is a statistical framework, because it can capture the non-linear dependency and also provide an interpretable reason for the model. Although some studies apply copulas to score fusion and demonstrate the effectiveness, their methods employ a unimodal copula, thus making it difficult to capture complex dependencies. Therefore, we introduce a new score fusion method that uses a mixture copula to handle the complicated dependencies of scores; then, we evaluate the accuracy of our proposed method. Experiments on *ClueWeb'09*, a large-scale document set, show that in some cases, our proposed method significantly outperforms linear combination and others existing methods that use a unimodal copula.

Keywords: Copulas · Information retrieval · Dependencies between relevance scores

1 Introduction

Given a user query, search systems calculate the relevance scores of documents with respect to the query and return a list of documents ranked by relevance. In order to improve search accuracy, many IR models that calculate relevance scores have been proposed [3, 22, 27, 30, 32–38]. Owing to the diverse and complex nature of the information needs of users, it is difficult to determine an appropriate IR model that always yields the most accurate search results. In order to address

---

T. Komatsuda—This work was conducted while the author was at Tokyo Institute of Technology.

S. Hartmann and H. Ma (Eds.): DEXA 2016, Part II, LNCS 9828, pp. 216–232, 2016.
DOI: 10.1007/978-3-319-44406-2_16

this challenge, many studies have combined multiple relevance scores obtained from multiple IR models [11,18,20].

Relevance scores can be combined using various approaches, such as function-based methods [2,9,10,41,43], learning to rank [6,7,21,28], and score fusion methods using a copula [15]. Linear combination, which is one of the representative function-based methods, cannot capture the non-linear dependency between relevance scores. In addition, the output of learning to rank is complex with respect to understanding the model. These problems can be solved with a copula, which is a statistical framework used for analyzing complex multi-dimensional dependencies [15]. A copula is a model that represents the relationship between a multidimensional distribution and the marginal distributions. By applying a copula to a score fusion method, we can build a model that captures the non-linear dependency and is easy to understand intuitively.

Existing score fusion methods using a copula [15] cannot capture complex dependencies easily because these methods employ a unimodal copula that is assumed to model a unimodal distribution. For example, Fig. 1 shows a distribution of two relevance scores. In the figure, each point denotes a document; the vertical axis represents the relevance scores x from an IR model X, and the horizontal axis represents the relevance scores y from a model Y. The set of documents exhibits some correlations locally around $(x, y) = (0.3, 0.5), (0.8, 0.1)$, and $(0.8, 0.8)$. From Fig. 1(a), the contour plot of the distribution estimated by using a unimodal copula cannot capture these correlations.

In this paper, we propose a score fusion method using a mixture copula that consists of multiple unimodal copulas. Mixture copulas can capture complex dependencies; therefore, they can estimate multimodal distribution accurately. Figure 1(b) shows that the distribution estimated by a mixture copula can capture complex dependencies between multiple relevance scores. Further, we evaluate the effectiveness of our proposed method by demonstrating that the consideration of complex dependencies can improve search accuracy.

Section 2 provides a basic introduction to copulas, and Sect. 3 reviews related work. Section 4 describes a score fusion method using a mixture copula, and Sect. 5 presents the evaluation method and results for the proposed method. In Sect. 6, the conclusion is stated, and plans for future work are described.

2   Copulas

Before applying copulas to an IR model, we provide a basic introduction of copulas that have been used in other research fields such as finance. For more detail, refer to the book by Nelsen [25].

2.1   Definitions and Properties

Copulas are models that describe the relationship between a multivariate distribution and the marginal distributions. Let X be a k-dimensional random

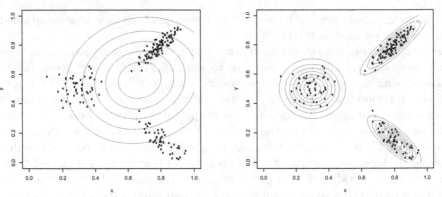

(a) Distribution using a Unimodal Copula   **(b)** Distribution using a Mixture Copula

Fig. 1. Examples of complex distribution

vector $X = (x_1, x_2, ..., x_k)$. Further, let a function $F_k(x)$ be a marginal cumulative distribution function for an element x_k of the random vector X, where $F_k(x) = P[X_k \geq x]$. Then, we can map X to a k-dimensional unit cube $[0,1]^k$ as $U = (u_1, u_2, ..., u_k) = (F_1(x_1), F_2(x_2), ..., F_k(x_k))$. A k-dimensional copula C is described as a joint cumulative distribution function of the normalized random vector U. Most importantly, it has been proved that there exists a copula C that satisfies $F(x_1, x_2, ..., x_k) = C(F_1(x_1), F_2(x_2), ..., F_k(x_k))$ in any k-dimensional joint cumulative distribution function $F(x_1, x_2, ..., x_k)$ [25]. This general fact indicates the high applicability of copulas. In addition, copulas facilitate our analysis of the structure of joint distribution because we separately estimate each marginal distribution $F_k(.)$ and the dependency structure between the marginal distributions.

2.2   Copulas and Dependency of Relevance Scores

Let us introduce the constraint for copulas, assuming that the dependency between relevance scores is for extreme conditions such as independent, completely positive correlation, and completely negative correlation.

When the dependency between relevance scores is independent, the copulas are described as independent copulas C_{indep}.

$$C_{indep}(U) = \exp(-\sum_{i=1}^{k} - \log u_i)$$

Thus, independent copulas are equivalent to the product of all elements of U. It must be noted that while independence is frequently assumed in IR theory, it is a naive assumption.

When the dependency between relevance scores is a completely positive correlation, the copulas can be represented by the formula below.

$$C_{coMono}(U) = min\{u_1, u_2, ..., u_k\}$$

When the dependency between relevance scores is a completely negative correlation, the copulas can be represented by the formula below.

$$C_{counterMono}(U) = max\{\sum_{i=1}^{k} u_i + 1 - k, 0\}$$

Copulas have parameters that provide an interpretable reason for the dependency structure of marginal distributions. Therefore, we can understand joint distribution clearly. The parameters can be estimated by using maximum likelihood estimation or the Monte Carlo method [8].

2.3 Typical Families of Copulas

Families of copulas are of various types, such as elliptical copulas, Archimedean copulas and empirical copulas.

- **Elliptical Copulas**
  An elliptical copula is a copula derived from standard distribution, such as Gaussian distribution and t distribution. Equation (1) shows the formula for a Gaussian copula.

$$C_{Gaussian}(U) = \Phi_{\Sigma}(\Phi^{-1}(u_1), ..., \Phi^{-1}(u_k)) \tag{1}$$

where Φ_{Σ} denotes a cumulative distribution function of standard normal distribution, and Φ^{-1} denotes its inverse function. A Gaussian copula requires a parameter $\Sigma \in R^{k \times k}$, which shows the observed covariance matrix.

- **Archimedean Copulas**
  Let ϕ be a continuous, strictly decreasing function from $\mathbf{I}$ to $[0, \infty]$ such that $\phi(1) = 0$. Then,

$$C_\phi(U) = \phi^{-1}(\phi(u_1) + \phi(u_2) + ... + \phi(u_k)), U \in (0, 1]^k$$

This formula represents a k-dimensional Archimedean copula, where ϕ is a generator of C_ϕ. For $\phi(t) = \frac{t^{-\theta} - 1}{\theta}, (-logt)^\theta, -log\frac{e^{\theta t} - 1}{e^\theta - 1}$, the copulas are called Clayton copulas, Gumbel copulas and Frank copulas, respectively. Further, Clayton, Gumbel and Frank copulas are defined by Eqs. (2), (3) and (4), respectively, and their corresponding contour plots are shown in Figs. 2(a), (b) and (c), respectively.

$$C_{Clayton}(U) = (1 + \theta(\sum_{i=1}^{k} \frac{1}{\theta}(u_i^{-\theta} - 1)))^{\frac{-1}{\theta}} \tag{2}$$

$$C_{Gumbel}(U) = \exp(-(\sum_{i=1}^{k} (-log(u_i))^\theta)^{\frac{1}{\theta}}) \tag{3}$$

$$C_{Frank}(U) = \frac{1}{\theta} log(1 + \frac{\prod_{i=1}^{k}(\exp(-\theta\ u_i) - 1)}{\exp((-\theta) - 1)^{k-1}}) \tag{4}$$

As seen in Fig. 2, different copulas have different features. For example, in Fig. 2(a), we assume that for a Clayton copula, the dependency of the lower region is strong whereas the dependency of the upper region is independent. The use of a Clayton copula is effective if the dependency of relevance scores is strong in cases where relevance scores is low.

– **Empirical Copulas**
An empirical copula refers to a copula that is derived from an empirical joint distribution whose marginal distributions are estimated by empirical distribution. Thus, an empirical copula is a nonparametric joint distribution that is based on observations, without assuming any specific distribution. A k-dimensional empirical copula $\hat{C}(U)$ is described as Eq. (5)

$$\hat{C}(U) = \frac{1}{N} \sum_{n=1}^{N} \prod_{i=1}^{k} \mathbf{1}\{t_i^n \leq u_i\} \tag{5}$$

where N denotes the number of observations to estimate the empirical copula, and t_i^n represents a score of the i-axis of the nth observation. The probability of a k-dimensional joint cumulative distribution derived from an empirical copula is calculated by dividing the number of training data, such that ($t_1^n \leq u_1, t_2^n \leq u_2, ..., t_k^n \leq u_k$) by the number of all training data N.

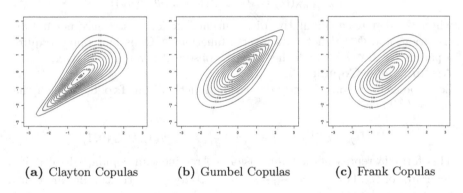

(a) Clayton Copulas   (b) Gumbel Copulas   (c) Frank Copulas

Fig. 2. Contour plots of 2-dimension joint distribution using different copulas

We need to select an appropriate model from the various types of copulas. A model of copulas can be selected based on certain criteria such as tail dependence coefficient and rank correlation coefficient [17]. The tail dependence coefficient is an indicator of the dependence structure at the end points of the probability, i.e., for probability around 0 or 1. If we use the tail dependence coefficient for the selection of a model, it implies that we focus on the dependency between high relevance (or low relevance). The rank correlation coefficient is an indicator of the dependence structure in the entire distribution. If we select a model based on the rank correlation coefficient, it implies that we focus on the average dependency in the overall distribution.

2.4   Mixture Copula

A mixture copula is a copula that is composed of several copulas. By using a mixture copula, we can build a multimodal joint distribution that enables us to capture a complex dependency.

A mixture copula is described as the weighted sum of k copulas as shown in Eq. (6).

$$C_{mix}(U) = \sum_{i=1}^{k} p_i C_i(U) \tag{6}$$

In order to construct a mixture copula, each copula C_i and its weight p_i should be estimated. These parameters can be estimated by using approaches based on clustering [12,40]. Thus, we can construct a mixture copula by performing the following steps: (1) the training data to estimate the mixture copula is split into k clusters, (2) the data in each cluster is fit to a unimodal copula. The number of clusters k is determined in advance. One of the methods to decide the value of k is the usage of an information criterion such as AIC (Akaike Information Criterion) [5].

3   Related Work

3.1   IR Models

Many IR models that calculate the relevance scores of a document have been proposed [3,22,36–38]. BM25, one of the classical probability models [30,32], has demonstrated high effectiveness [33]. Ponte and Croft proposed a probabilistic language model [27] which is developed in a mathematical framework, while vector space models [34,35] and classical probability models have been proposed as heuristic approaches.

3.2   Fusion of IR Models

Although many IR models exist, it is difficult to determine an appropriate model because the information needs of a user are diverse and complex. In order to address this challenge, various studies have focused on the fusion of multiple relevance scores calculated by several IR models. For example, some meta search engines have attempted to improve the accuracy by combining the results from multiple engines. These studies are known as score fusion of relevance scores [1,11,18,20,24].

The retrieval of structured documents such as XML posed a challenge in combining the structure information with the relevance scores of a document with respect to a query. Robertson et al. explain the difficulty in combination of document's structure information [31]. In the case of information retrieval for children, it is important to consider the credibility and readability of the document, as well as the relevance scores for a query [13].

3.3 Fusion of Relevance Scores

Advances of Score Fusion. Although score fusion is often achieved by obtaining the sums or products of results from individual systems [18], probabilistic approaches also exist [11,20]. Aslam and Montague proposed a probabilistic model based on ranking [1]. They improve their model by incorporating a majority method [24]. Further, they attempted to make the model robust against outliers by normalizing the scores [23].

Linear Combination. Vogt et al. introduced linear combination in information retrieval [41]. The suitability of linear combination has been demonstrated [2,9,10,43]. Gerani et al. applied nonlinear transformation to relevance scores before applying linear combination [19]. Gerani et al. showed that their method outperformed standard linear combination. This result demonstrates the need for a model that can capture complex and nonlinear dependency.

Learning to Rank. Learning to rank is a ranking model that uses machine learning and enables easy unification to obtain one score from a large number of document features [6,7,21,28]. This approach extracts the features of relevant documents from a set of documents that are labeled as either relevant or irrelevant. The disadvantage of this approach is the difficulty in understanding the resulting model.

Copulas. In general, copulas are widely used in quantitative finance and in portfolio management [4,17]. Some recent studies have applied copulas in other research fields [26,29,39]. Vrac et al. applied a mixture copula to a global climate dataset and showed that a mixture copula can group the climate of the world correctly in terms of meteorology [42].

Eickhoff et al. applied copulas to score fusion and their proposed method outperformed the baselines as a result of combining two relevant features in some cases [15]. They verified the effectiveness of the approach when the number of relevant features was increased from 2 to 136. The result showed that their proposed method is more effective than linear combination as the number of relevant criteria increases [14]. In addition, they applied copulas to language models in which independence is frequently assumed. Their proposed method showed that it has competitive performance when compared with naive language models and some learning to rank methods [16].

Although the approaches proposed by Eickhoff et al. have demonstrated the effectiveness of copulas, their methods cannot estimate a joint distribution correctly when the distribution is complex, as shown in Fig. 1(a), because the complexity of the dependency makes it difficult to estimate the joint distribution precisely. One of the solutions to the problem is the use of multiple copulas to estimate a multimodal distribution.

4   Proposed Approach

We propose a score fusion method that uses a mixture copula. We use mixture copulas to precisely estimate a joint distribution of relevant documents which has some strong correlations locally as shown in Fig. 1(b). In Fig. 1(b), for example, the distribution is accurately captured by three copulas, although its right two areas have strong correlations locally.

In our method, first, a mixture copula is estimated; then, models of score fusion are constructed by using the mixture copula. As we mentioned in Sect. 2.4, a mixture copula is composed of multiple unimodal copulas. During the process, clustering is useful in estimating a mixture copula [12,40]. We use a clustering approach in our method.

The process for constructing the proposed model is described below;

1. Apply a clustering algorithm to relevant documents.
2. In each cluster, estimate a joint distribution using a unimodal copula.
3. Combine the unimodal copulas estimated in the previous step and construct a mixture copula.
4. Create a score fusion method by using the estimated mixture distribution.

Figure 3(a) shows the process for constructing a model by using our method, and Fig. 3(b) shows the process that uses the method of Eickhoff et al. [15]. Our proposed method begins with the clustering of relevant documents and estimates joint distribution of relevant documents in individual clusters using unimodal copulas, whereas Eickhoff et al. estimate a joint distribution of relevant documents using one unimodal copula.

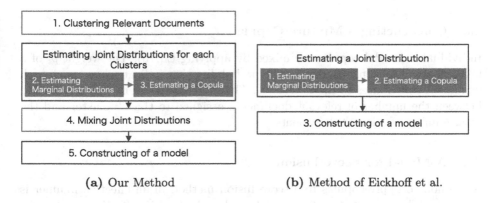

(a) Our Method   (b) Method of Eickhoff et al.

Fig. 3. Processes for constructing a model using copulas

4.1   Clustering Relevant Documents

A group-average agglomerative clustering method is used as a clustering algorithm. The group-average agglomerative clustering method merges two clusters whose distance is the closest until the number of clusters reaches k determined in advance. The distance between two clusters is defined as the average of distances between all pairs of documents except for pairs from the same cluster. Although we employ a group-average agglomerative clustering method, any clustering algorithms can be used in our method.

4.2   Estimating Joint Distribution in Individual Clusters

In individual clusters, joint distributions are obtained by performing the following two steps: **(1)** estimating the marginal distributions, and **(2)** estimating the dependency structure between the marginal distributions by using a copula.

We infer a marginal distribution by using a marginal distribution function. Equation (7) shows a Gaussian cumulative distribution function as an example of a marginal distribution function:

$$\hat{F}(x) = \frac{1}{\sqrt{2\pi}\sigma} \int_{-\infty}^{x} \exp(-\frac{(t-m)^2}{2\sigma^2})\, dt \tag{7}$$

where m denotes the mean of the distribution and σ represents the variance of the distribution. A cumulative distribution function is used because copulas require a random vector whose component is a cumulative score.

Next, we estimate the dependency between the marginal distributions by using copulas. The best copula is determined by comparing the performance of each copula. As mentioned in Sect. 2, although we must select an appropriate copula from various copulas, this challenge will be addressed in the future. In order to estimate the parameters of a copula, we use maximum likelihood estimation.

4.3   Constructing a Mixture Copula

Individual unimodal copulas are mixed by applying Eq. (6). The weight p_i of a mixture copula is considered to be the probability that indicates how much each document contributes to the ith cluster. Therefore, we set weight p_i as the ratio between the number of relevant documents assigned to the ith cluster and the total number of relevant documents.

4.4   A Model for Score Fusion

A method to apply copulas to a score fusion method in an effective manner is unknown. Thus, we propose two models and evaluate their effectiveness.

The first model is a cumulative joint distribution function estimated by a mixture copula, which is in Eq. (6). In comparison with a unimodal copula, a mixture copula can be precisely fitted to data when the dependency structure is complex. We evaluate its effect on accuracy improvement when estimating joint distribution more precisely.

The second model is the product of the likelihood of a normalized random vector U and a cumulative score derived from a mixture copula $C_{mix}(U)$, as shown in Eq. (8).

$$C_{mix-prod} = C_{mix}(U) \prod_{i=1}^{k} u_i \qquad (8)$$

The likelihood of U is calculated based on an assumption that individual components of U occur independently. This assumption is very naive. In order to consider the dependency of the components, we multiply the likelihood by the mixture copula $C_{mix}(U)$ for which the correlation among individual components is considered. Eickhoff et al. multiplied the likelihood by a single copula; however, we use a mixture copula instead of a single copula.

5   Evaluation

We evaluate the effectiveness of our models proposed in Sect. 4.4 when combining the relevance scores of two IR models.

5.1   Setup for Evaluation

- **Dataset**
  The dataset that we use is the Category B of *ClueWeb'09*[1], excluding Wikipedia documents. *ClueWeb'09* is a dataset used for the Web Track in TREC 2009-2012. Category B, a subset of *ClueWeb'09*, contains approximately 44 million English documents.
- **Queries**
  We used 45 out of 50 queries for ad-hoc tasks of the Web Track in TREC2011. We omitted five queries: four queries that do not have relevant documents in the dataset and one query that includes a numeric term.
- **Measures for Evaluation**
  The evaluation measures that we use are: Precision (P@k) and normalized Discounted Cumulative Gain (nDCG@k) in the top-k documents; Interpolated Precision (IP@i), where i is recall level; and Mean Average Interpolated Precision (MAIP). Further, we set $k = 5, 10, 15, 20$ and $i = 0.0, 0.1,..., 0.5$. P@k is the ratio between the number of relevant documents in the top-k documents and the total number of top-k documents; it is defined as:

$$P@k = \frac{|A \cap B|}{|A|} \qquad (9)$$

where A is a set of top-k documents, and B is a set of relevant documents.

---

[1] http://lemurproject.org/clueweb09/.

Equation (10) shows nDCG@k where iDCG@k is the maximum score of DCG@k, defined as Eq. (11). In Eq. (11), rel_i denotes a binary variable for the ith document, such that when the ith document is relevant, rel_i equal to 1; otherwise, rel_i is equal to 0. nDCG increases as relevant documents are ranked higher.

$$\text{nDCG@}k = \frac{\text{DCG@}k}{\text{iDCG@}k} \tag{10}$$

$$\text{DCG@}k = \sum_{i=1}^{k} \frac{2^{rel_i} - 1}{\log_2(i+1)} \tag{11}$$

IP@k is defined as shown in Eq. (12) where R@r is the value of Recall in the top r documents.

$$\text{IP@}i = \max_{r}\{P@r | R@r \geq i\} \tag{12}$$

MAIP is the average of 11 points of Interpolated Precision, as defined in Eq. (13).

$$\text{MAIP} = \frac{\sum_{i\in\{0,0.1,...,1\}} \text{IP@}i}{11} \tag{13}$$

- **Cross Validation**
  Some baselines and our proposed models have parameters that are estimated with training data. We trained the models with a part of the dataset and tested them with the other parts of the dataset. In our experiment, we divided the dataset into 5 parts of training data, then used each parts of training data for a test set. The accuracy of the models is calculated by the average of 5 test results.
- **Target Models for Combination**
  We combine two IR models: BM25 [33] and a query likelihood model [27]. Dirichlet smoothing is applied to the query likelihood model. BM25 parameters b, k, and smoothing parameter μ are set to 0.75, 1.2 and 110, respectively. During evaluation, we change the variations of marginal distributions, copulas, and the number of clusters for a mixture copula for as shown in Table 1. In our experiments, a cluster containing only one document is omitted as an outlier. The selection of an appropriate model is a task that will be considered in future work.

Table 1. Model parameters for evaluation

	Values used in the experiment
Marginal distributions	Gaussian and empirical distribution
Copulas	Clayton, Gumbel, Frank, Gaussian and empirical copula
The number of clusters	2–10

5.2   Baselines

We prepare five baselines for comparison with the performance of the two models shown in Eqs. (6) and (8). The component x_i of a random vector X denotes a normalized relevance score, and the component u_i of a random vector U denotes a score to which a cumulative distribution function $F_i(.)$ maps the x_i.

Linear combination:

$$LIN(X) = \sum_{i=1}^{k} \lambda_i x_i \qquad (14)$$

Harmonic mean:

$$HM(X) = \frac{k \cdot \prod_{j=1}^{k} x_j}{\sum_{i=1}^{k} \frac{\prod_{j=1}^{k} x_j}{x_i}} \qquad (15)$$

If random vector X contains at least one component whose score is low, the final score obtained by combination using harmonic mean tends to be low. For example, $HM((0.5, 0.5)) = 0.5$, whereas $HM((0.9, 0.1)) = 0.18$.

An independent copula is given by:

$$C_{indep}(U) = \prod_{i=1}^{k} u_i \qquad (16)$$

An independent copula denotes the product of U indicating a cumulative score of joint distribution based on the assumption that individual components U occur independently.

A joint distribution using a single copula is given by:

$$C_{mono}(U) = C(U) \qquad (17)$$

The product of a likelihood of U and a score of the cumulative distribution is given by:

$$C_{mono-prod} = C_{mono}(U) \prod_{i=1}^{k} u_i \qquad (18)$$

5.3   Statistical Testing

We test the statistical significances by using a Wilcoxon signed-rank test at two significance levels-0.01 and 0.05.

The three major observations are: **(1)** The models using a copula perform significantly better than the models using linear combination, which has shown high performance so far. **(2)** The models that consider dependency perform significantly better than the models that ignore dependency. **(3)** The models that use a mixture copula perform significantly better than the models that use a single copula.

In order to clearly demonstrate these three observations, we compare **(1)** LIN with C_{indep}, C_{mono}, $C_{mono-prod}$, C_{mix}, $C_{mix-prod}$, and $C_{mix-prod}$, and **(2)** C_{indep} with C_{mono}, $C_{mono-prod}$, C_{mix}, and $C_{mix-prod}$, and **(3)** C_{mono}, $C_{mono-prod}$ with C_{mix}, $C_{mix-prod}$.

5.4   Results

In order to determine the best combination of a marginal distribution, a copula, and the number of clusters, we compared the performance of the proposed models for each combination of parameters in Table 1.

We conducted preliminary experiments to determine the best marginal distribution models, copulas, and cluster sizes. Due to the page limitation, we only show the summary of the best combinations in Table 2.

Next, we compare our models with baselines. Table 3 shows the results of the performance. In Table 3, The symbols $*$, $\dagger$, $\ddagger$, and $\S$ indicate statistically significant improvements over LIN, C_{indep}, C_{mono}, and $C_{mono-prod}$, respectively. A single symbol indicates statistically significant improvements at the 0.01-level and a double symbol indicates statistically significant improvements at the 0.05-level. A cumulative function of a joint distribution estimated by a mixture copula C_{mix} gains of 10 % and 15 % over linear combination with respect to P@5 and nDCG@5, respectively. In particular, in terms of nDCG@5, C_{mix} shows a statistically significant improvement over LIN, C_{indep}, C_{mono}, and $C_{mono-prod}$. Among C_{indep}, C_{mono}, and C_{mix}, the performance of C_{mix} is the best, and the performance of C_{mono} exceeds that of C_{indep}. This result indicates that a multidimensional cumulative distribution function can retrieve more relevant documents in the top-5 results when considering the dependency between marginal distributions.

Table 2. Best combination for methods with a copula

Model	Marginal distribution	Copulas	Number of clusters
C_{mono}	Empirical distribution	Empirical copulas	-
$C_{mono-prod}$	Empirical distribution	Empirical copulas	-
C_{mix}	Gaussian distribution	Clayton copulas	3
$C_{mix-prod}$	Gaussiandistribution	Clayton copulas	6

However, in terms of P@$k(\geq 10)$, we do not observe a tendency that the performance of C_{mix} surpasses that of C_{mono}. C_{mix} is effective for the top-5 results, whereas $C_{mix-prod}$ is relatively effective when retrieving 20 % of relevant documents. In terms of IP@$i(= 0.1, 0.2)$, $C_{mix-prod}$ outperforms the other models and shows a 5 % improvement over LIN, C_{indep}, and $C_{mono-prod}$, C_{mix} is the worst model. From these discussions, we conclude that **(1)** the performance of C_{mix} tends to deteriorate when retrieving 10 or more documents, whereas it is effective in the top-5 results. **(2)** $C_{mix-prod}$ is effective when retrieving 20 % of relevant documents.

Table 3. Evaluation results

	LIN	HM	C_{indep}	C_{mono}	$C_{mono-prod}$	C_{mix}	$C_{mix-prod}$
IP@0.0	0.4326	0.4228	0.4392	0.4367	0.4388	0.4603	**0.4609**
IP@0.1	0.2779	0.198	0.2799	0.2756	0.2797	0.2725	**0.294**
IP@0.2	0.2032	0.1196	0.2049	0.198	0.2019	0.1708	**0.2207**
IP@0.3	**0.0846**	0.056	0.0852	0.0874	0.089	0.0471	0.0814
IP@0.4	0.0366	0.0216	0.0382	**0.0404**	0.0398	0.0186	0.0306
IP@0.5	0.0141	0.0068	0.0146	0.0149	**0.0156**	0.0077	0.0111
MAIP	0.0963	0.0751	0.0974	0.0966	0.0976	0.0892	**0.1003**
P@5	0.236	0.236	0.212	0.228	0.228	**0.26**†	0.24†
P@10	**0.232**	0.2	0.206	0.226†	0.226†	0.218	0.224†
P@15	0.2187	0.1893	0.2027	0.2147†	0.2133	0.204	**0.2227**††
P@20	0.219	0.18	0.205	**0.22**†	0.219	0.196	**0.22**
nDCG@5	0.1616	0.1595	0.1529	0.1613	0.1615	**0.1873***†‡‡§§	0.1685
nDCG@10	0.161	0.1472	0.1537	0.1613	0.1604	**0.166**	0.1644†
nDCG@15	0.1574	0.1442	0.1505	0.1576	0.1561	0.1595†	**0.1634**††
nDCG@20	0.1638	0.1426	0.1583	0.1661†	0.1663	0.1612	**0.1711**†

6   Conclusion

In this paper, we proposed a score fusion method that uses a mixture copula. Copulas, a family of robust statistical methods, can unify multidimensional relevance scores into a single score, capturing the non-linear dependency among relevance scores. In addition, copulas can provide an interpretable reason for the final result by decomposing a joint distribution into individual marginal distributions and their dependency structure. In the existing score fusion methods that use a copula, it is difficult to capture complex dependencies because these methods employ a unimodal copula, which is expected to be used for a unimodal joint distribution. In contrast, our proposed method can capture complex dependencies by using a mixture copula, which can accurately model a multimodal distribution.

We used more than 44 million documents in *ClueWeb'09*, to compare our method with linear combination and existing score fusion methods that use a copula. For nDCG with the top-5 documents, the proposed method showed a 15 % improvement in effectiveness when compared with linear combination.

In future work, the following challenges must be addressed: (1) In order to construct a mixture copula automatically, we must determine a method to find the appropriate number of copulas. For example, an information criterion such as AIC can be adopted; and (2) We must determine a method to choose the best family of copulas that precisely fits the documents by using certain criteria such as tail dependence correlation and rank correlation coefficient.

Acknowledgment. This work was supported by JSPS KAKENHI Grant Numbers 15H02701 and 15K20990.

References

1. Aslam, J.A., Montague, M.: Bayes optimal metasearch: a probabilistic model for combining the results of multiple retrieval systems. In: Proceedings of SIGIR, pp. 379–381 (2000)
2. Bordogna, G., Pasi, G.: A model for a SOft fusion of information accesses on the web. Fuzzy Sets Syst. **148**(1), 105–118 (2004)
3. Borlund, P.: The concept of relevance in IR. J. Am. Soc. Inform. Sci. Technol. **54**(10), 913–925 (2003)
4. Bouchaud, J.P., Potters, M.: Theory of Financial Risk and Derivative Pricing: From Statistical Physics to Risk Management. Cambridge University Press, Cambridge (2003)
5. Breymann, W., Dias, A., Embrechts, P.: Dependence structures for multivariate high-frequency data in finance (2003)
6. Burges, C., Shaked, T., Renshaw, E., Lazier, A., Deeds, M., Hamilton, N., Hullender, G.: Learning to rank using gradient descent. In: Proceedings of ICML, pp. 89–96 (2005)
7. Chen, K., Lu, R., Wong, C., Sun, G., Heck, L., Tseng, B.: Trada: tree based ranking function adaptation. In: Proceedings of CIKM, pp. 1143–1152 (2008)
8. Choroś, B., Ibragimov, R., Permiakova, E.: Copula estimation. In: Jaworski, P., Durante, F., Härdle, W.K., Rychlik, T. (eds.) Copula Theory and Its Applications. Lecture Notes in Statistics, vol. 198, pp. 77–91. Springer, Heidelberg (2010)
9. da Costa Pereira, C., Dragoni, M., Pasi, G.: Multidimensional relevance: a new aggregation criterion. In: Boughanem, M., Berrut, C., Mothe, J., Soule-Dupuy, C. (eds.) ECIR 2009. LNCS, vol. 5478, pp. 264–275. Springer, Heidelberg (2009)
10. Craswell, N., Robertson, S., Zaragoza, H., Taylor, M.: Relevance weighting for query independent evidence. In: Proceedings of SIGIR, pp. 416–423 (2005)
11. Cummins, R.: Measuring the ability of score distributions to model relevance. In: Salem, M.V.M., Shaalan, K., Oroumchian, F., Shakery, A., Khelalfa, H. (eds.) AIRS 2011. LNCS, vol. 7097, pp. 25–36. Springer, Heidelberg (2011)
12. Diday, E., Schroeder, A., Ok, Y.: The dynamic clusters method in pattern recognition. In: IFIP Congress, pp. 691–697 (1974)
13. Eickhoff, C., Serdyukov, P., De Vries, A.P.: A combined topical/non-topical approach to identifying web sites for children. In: Proceedings of WSDM, pp. 505–514 (2011)
14. Eickhoff, C., de Vries, A.P.: Modelling complex relevance spaces with copulas. In: Proceedings of CIKM, pp. 1831–1834 (2014)
15. Eickhoff, C., de Vries, A.P., Collins-Thompson, K.: Copulas for information retrieval. In: Proceedings of SIGIR, pp. 663–672 (2013)
16. Eickhoff, C., de Vries, A.P., Hofmann, T.: Modelling term dependence with copulas. In: Proceedings of SIGIR, pp. 783–786 (2015)
17. Embrechts, P., Lindskog, F., McNeil, A.: Modelling dependence with copulas and applications to risk management. In: Rachev, S. (ed.) Handbook of Heavy Tailed Distributions in Finance, pp. 329–384. Elsevier, Amsterdam (2003)
18. Fox, E.A., Shaw, J.A.: Combination of multiple searches. NIST SPECIAL PUBLICATION SP, pp. 243–243 (1994)

19. Gerani, S., Zhai, C.X., Crestani, F.: Score transformation in linear combination for multi-criteria relevance ranking. In: Baeza-Yates, R., de Vries, A.P., Zaragoza, H., Cambazoglu, B.B., Murdock, V., Lempel, R., Silvestri, F. (eds.) ECIR 2012. LNCS, vol. 7224, pp. 256–267. Springer, Heidelberg (2012)
20. Kanoulas, E., Dai, K., Pavlu, V., Aslam, J.A.: Score distribution models: assumptions, intuition, and robustness to score manipulation. In: Proceedings of SIGIR, pp. 242–249 (2010)
21. Liu, T.Y.: Learning to rank for information retrieval. Found. Trends Inf. Retr. **3**(3), 225–331 (2009)
22. Mizzaro, S.: Relevance: the whole history. J. Am. Soc. Inform. Sci. Technol. **48**(9), 810–832 (1997)
23. Montague, M., Aslam, J.A.: Relevance score normalization for metasearch. In: Proceedings of CIKM, pp. 427–433 (2001)
24. Montague, M., Aslam, J.A.: Condorcet fusion for improved retrieval. In: Proceedings of CIKM, pp. 538–548 (2002)
25. Nelsen, R.B.: An Introduction to Copulas. Springer Series in Statistics. Springer, New York (2006)
26. Onken, A., Grünewälder, S., Munk, M.H., Obermayer, K.: Analyzing short-term noise dependencies of spike-counts in macaque prefrontal cortex using copulas and the flashlight transformation. PLoS Comput. Biol. **5**(11), e1000577 (2009)
27. Ponte, J.M., Croft, W.B.: A language modeling approach to information retrieval. In: Proceedings of SIGIR, pp. 275–281 (1998)
28. Radlinski, F., Joachims, T.: Query chains: learning to rank from implicit feedback. In: Proceedings of SIGKDD, pp. 239–248 (2005)
29. Renard, B., Lang, M.: Use of a gaussian copula for multivariate extreme value analysis: some case studies in hydrology. Adv. Water Resour. **30**(4), 897–912 (2007)
30. Rijsbergen, C.J.V.: Information Retrieval, 2nd edn. Butterworth-Heinemann, Newton (1979)
31. Robertson, S., Zaragoza, H., Taylor, M.: Simple BM25 extension to multiple weighted fields. In: Proceedings of CIKM, pp. 42–49 (2004)
32. Robertson, S.E., Jones, K.S.: Relevance weighting of search terms. J. Am. Soc. Inform. Sci. Technol. **27**(3), 129–146 (1976)
33. Robertson, S.E., Walker, S., Jones, S., Hancock-Beaulieu, M.M., Gatford, M., et al.: Okapi at TREC-3, pp. 109–109. NIST SPECIAL PUBLICATION SP (1995)
34. Salton, G.: Automatic Text Processing: The Transformation, Analysis, and Retrieval of Information by Computer. Addison-Wesley, Reading (1989)
35. Salton, G., Wong, A., Yang, C.S.: A vector space model for automatic indexing. Commun. ACM **18**(11), 613–620 (1975)
36. Saracevic, T.: The concept of relevance in information science: a historical review. In: Saracevic, T. (ed.) Introduction to Information Science, pp. 111–151. R.R. Bowker, New York (1970)
37. Saracevic, T.: Relevance reconsidered. In: Proceedings of CoLIS, vol. 2, pp. 201–218 (1996)
38. Schamber, L., Eisenberg, M.B., Nilan, M.S.: A re-examination of relevance: toward a dynamic, situational definition. Inf. Process. Manage. **26**(6), 755–776 (1990)
39. Schoelzel, C., Friederichs, P., et al.: Multivariate non-normally distributed random variables in climate research-introduction to the copula approach. Nonlin. Process. Geophys. **15**(5), 761–772 (2008)
40. Scott, A.J., Symons, M.J.: Clustering methods based on likelihood ratio criteria. Biometrics **27**, 387–397 (1971)

41. Vogt, C.C., Cottrell, G.W.: Fusion via a linear combination of scores. Inf. Retr. **1**(3), 151–173 (1999)
42. Vrac, M., Billard, L., Diday, E., Chédin, A.: Copula analysis of mixture models. Comput. Stat. **27**(3), 427–457 (2012)
43. Wu, S., Crestani, F.: Data fusion with estimated weights. In: Proceedings of CIKM, pp. 648–651 (2002)

Personal Information Management

Axiomatic Term-Based Personalized Query Expansion Using Bookmarking System

Philippe Mulhem[1,2]($\boxtimes$), Nawal Ould Amer[1,2,3], and Mathias Géry[3]

[1] Univ. Grenoble Alpes, LIG, 38000 Grenoble, France
{philippe.mulhem,nawal.ould-amer}@imag.fr
[2] CNRS, LIG, 38000 Grenoble, France
[3] Univ. Lyon, UJM-Saint-Etienne, CNRS, Institut d Optique Graduate School,
Laboratoire Hubert Curien UMR 5516, 42023 Saint-Étienne, France
mathias.gery@univ-st-etienne.fr

Abstract. This paper tackles the problem of pinpointing relevant information in a social network for Personalized Information Retrieval (PIR). We start from the premise that user profiles must be filtered so that they outperform non profile based queries. The formal *Profile Query Expansion Constraint* is then defined. We fix a specific integration of profile and a probabilistic matching framework that fits into the constraint defined. Experiments are conducted on the Bibsonomy corpus. Our findings show that even simple profile adaptation using query is effective for Personalized Information Retrieval.

Keywords: Social network · Probabilistic retrieval · Profile selection · Axiomatic IR

1 Introduction

Personalized Information Retrieval (IR) systems aim at returning personalized results. Personalizing IR relies on modeling *User's profiles* (interests, behavior, history, etc.). Such profile may be used for *query expansion*, or for *re-ranking*. The query expansion-based integration keeps the benefit for all the experimental and theoretical results from the IR domain. A new field of IR has emerged with [7]: the axiomatic characterization of IR models. Such works define the expected behaviors of systems using "axioms".

This paper first defines an axiom (i.e. a heuristic constraint) that is supposed to be validated by a personalized IR system using a social bookmarking system, and second evaluates the impact of the constraint on the IR system. Section 2 presents related works. The Sect. 3 defines the proposed axiom, called PQEC (Personalized Query Expansion Constraint). Section 4 focuses on the personalized frameworks proposed, before presenting several query expansions in Sect. 5. The experiments on the Bibsonomy corpus are presented and discussed in Sect. 6, before concluding.

© Springer International Publishing Switzerland 2016
S. Hartmann and H. Ma (Eds.): DEXA 2016, Part II, LNCS 9828, pp. 235–243, 2016.
DOI: 10.1007/978-3-319-44406-2_17

2   State of the Art

Personalized IR may consider user's model (called *user's profile*) based on user's query logs [6], posts [11], tags and bookmarking [1]. Several works improve personalized document ranking by using both the user's information and other social information. Such search function, for bookmarking systems, is based on user's tag profiles which are derived from their bookmarks [3,12]. [1] selects terms related to the user query terms. Similarly, [4] defines a query expansion that exploits relationships between users, documents, and tags. [3,12] considers both the matching score between a query and the social annotations of the document, and the matching between the user's profile and the document. Other works personalize a user search using other users from the social network. For example, selecting users that have an explicit [9,11] or implicit [3,12] relationships with the query issuer. [11] proposes a collaborative personalized search model based on topic models to disambiguate the query. [3] integrates other users from the social network that have annotated the document.

These approaches use the whole user profile, decreasing the effectiveness of the search. Query expansions tackle this problem by selecting the terms to extend user query. Our proposal benefits of both query expansion-like approaches [1,4] and social retrieval [3], and we defend the idea that social networks are beneficial to personalized retrieval by: (i) adapting the user profile using social neighbors that are constrained by the query, and (ii) selecting a part of a user profile adapted to a query.

Our approach also focuses on defining axioms (heuristics), i.e. expected behaviors of personalized IR systems. Such axioms serve as a basis to (a) explain the role of the different elements that are used by an IR system, (b) compare approaches from the theoretical basis and (c) propose new approaches based on these axioms. For instance, Fang, Tao and Zhai defined in 2004 [7] the first steps of this field of IR, with constraints related to the roles of term frequency, inverse document frequency, and document length. Many works followed, like heuristics for semantic models for IR [8], or for Pseudo Relevance Feedback [5]. To the best of our knowledge, no axiomatization work did focus on personalization of IR.

3   Profile Query Expansion Constraint

We propose here: a) to show that, in social bookmarking networks, integrating a part of a user activity (i.e. his bookmarks) may help to personalize results, and b) to define a first axiomatic expression that respects the findings of (a).

3.1   Empirical Study

We studied a set of 200 users from the Bibsonomy corpus, according to the evaluation framework described in Sect. 6.1. We compute that, when a query is generated for a given user using a term from his profile, 100 % of the relevant documents are tagged by at least one other term of the user profile. This empirical result enforces the fact that at least a part of a user profile is relevant to be used when processing personalized IR.

We study then the topics of queries. We generated a Latent Dirichlet Allocation model [2] for the whole set of users of the corpus (see Sect. 6.1), with the number of topics chosen to be 100. Using a threshold of 0.1 when assigning a topic to a user, we find out that 77 % of the users have more than one center of interest. If we assume that a query deals with one topic, as in [11], it is then clear that we have to filter terms of the profile to expand the query. All these elements reinforce our initial idea that focusing on an *adequate* subset of the user profile may help to focus on relevant documents.

3.2   Notations

Here are the notation used in the remaining of the paper. G: The tagging social network; G is a graph: $G = << D, U, W >, R >$. D: the set of documents $d \in D$. U: the set of users of the network, with $u \in U$. W: the set of tags (words) assigned by users to documents. R: the tags assigned by the users to the documents ($R \subset D \times U \times W$). $c(w, d)$: the count of word w in document d. $RSV(d, q)$: the Retrieval Status Value of a document d for a query q. $Profile(u)$: the profile of a user u by all the tags he used. $Profile(u) = \{w | w \in W, d \in D, R_u(d, u, w)\}$. R_u: term-term relationship for user u. $(w, w') \in R_u$ means that w and w' are related for the user u. $Profile(u, q)$: the profile of a user u filtered for the query q. $R_{u-local}$: term-term relationship for user u based on u's tagging. $RSV(d, q, u)$: the RSV of a document d for a query q and for u. $u_{sn} \subset U$: the social neighborhood of u. $R_{u-social}$: term-term relationship for u considering u_{sn}.

3.3   Profile Query Expansion Constraint (PQEC)

This constraint assumes that the integration of "adequate" terms (related to the query, and satisfying the term-term relationship R_u) from a user profile is needed:

Profile Query Expansion Constraint (PQEC): Assume that a query $q = \{w\}$, a document d from a corpus C so that $c(w, d) > 0$, and a user u with a profile $Profile(u)$. If $\exists w' \in Profile(u)$ so that $R_u(w, w')$, then for any $d' \in D$ so that $c(w, d') \neq 0$ and $c(w', d') = 0$ then $RSV(d, q_u) \geq RSV(d', q_u)$, with $q_u = q \cup \{w'\}$.

This constraint heavily relies on the personalized term-term relationship R_u that obviously influences the overall results: if R_u does not link properly terms according to the user u, then ensuring the constraint will impact negatively the quality of the system. In the following, we will focus on social inputs to define the R_u relationship. Our concern differs from semantic term constraints of [8], as we consider that the data that we have about the user is of primary importance.

4 Personalized Information Retrieval

4.1 Classical Framework

Our proposal computes a Retrieval Status Value (RSV) of a document d for a query q submitted by a user u as: $RSV(d, q, u) \propto RSV(d, q_u)$ with q_u the expanded query using terms coming from u's profile: $q_u = q \cup \{w'|w' \in W, \exists w \in q; R_u(w, w')\}$. Each document d (tagged using a social tagging system) contains 2 facets: the actual content of the document, noted σd, and the user's tags that describe d, noted τd. We combine linearly these facets in the expression $RSV(d, q)$, as in [3], using probabilities $P(q|\sigma d)$ and $P(q|\tau d)$ that rely on the classical IR language models with Dirichlet priors:

$$RSV(d, q) = \lambda.P(q|\sigma d) + (1 - \lambda).P(q|\tau d) \tag{1}$$

4.2 Adapted Framework to Ensure PQEC

A simple way to modify the classical framework to ensure PQEC is to split the retrieval in four steps:

1. Evaluate $RSV(d, q)$, i.e. without personalization, leading to a results list L_{init} of couples $< doc, rsv >$. Assign the larger score for the documents of L_{init} to $Topscore_{init}$;
2. Evaluate $RSV(d, q_u \setminus q)$, i.e. the RSV of d for the expanded query q_u without the initial query, leading to a results list L_{exp} of couples $< doc, rsv >$;
3. Fuse L_{init} and L_{exp} respecting: (a) for any d in L_{init} and L_{exp}, the final RSV of d is the sum of its scores in both lists and of $Topscore_{init}$; (b) for any d in L_{init} and not in L_{exp}, its final RSV is its score in L_{init};
4. Rank the result according to the final scores of documents

The documents that match the profile expansion are ranked before the documents that match only the query in the result list, thus PQEC is ensured.

5 Personalized Query Expansion Terms

We propose several variations of u_{sn}, the social neighborhood of a user u, depending on adaptations of the social neighborhood of u and of the profiles of the users in u's neighborhood, according to the query q. We define several $R_u(w, w')$ (cf. Sect. 3.3) to assess the usefulness of the constraint. Our personalized query expansion may also use others users u' in the social network: we will study in Sect. 6 the impact of PQEC on several categories of neighbors u', and on the filtering of u's profiles added to the query. In the following, $Profile(u, q)$ is the personalized profile of u asking q. We have: $Profile(u, q) = \{w'|w' \in W, \exists w \in q; R_u(w, w')\}$. We define two relationships, namely $R_{u-local}$ and $R_{u-social}$, that depict two personalized profiles of u.

5.1   Local Tagging Expansion Using $R_{u-local}$

Assuming a query asked by a user u. The first simple element proposed is to add terms from u's profile to the query, relying only on u's tagging behavior. We select the tags from u's profile that were used jointly with a query term q by u to tag one document. The idea is then to be able to expand the query with terms that are related to one query term according to u. More formally, the relation R_u is then expressed as its variant $R_{u-local}$: $R_{u-local}(w, w') \Leftrightarrow \exists d \in D, R(d, u, w) \wedge R(d, u, w')$. Such approach cannot be used when a user u does not: (i) use a query term in his profile, and (ii) tag documents with multiple tags, and this is mandatory for this local expansion. We propose then a second way to support a query expansion.

5.2   Social Tagging Neighborhoods Using $R_{u-social}$

When considering other users than u for the social tagging network, we need to define which users are considered experts to support the query expansion. Such neighbors set is noted u_{sn}. These experts are chosen according to their familiarity with the query, and/or their similarity with the user u. In Sect. 6.2, we define several neighborhoods. We consider here a simple definition of the profiles of $u' \in u_{sn}$. These profiles are built the same way as the profile of u using $R_{u'-local}$, i.e. they are filtered to keep the terms of $Profile(u')$ that co-occur with at least one query term in one document tagged by u'. Finally the profile of u is computed using the following expression of $R_{u-social}$: $R_{u-social}(w, w') \Leftrightarrow \exists u' \in u_{sn}, R_{u'-local}(w, w')$

6   Experimental Evaluation

6.1   Bibsonomy Dataset and Evaluation Protocol

We consider here explicit annotations of documents provided by a user from a tagging social network, namely Bibsonomy[1], which is a social tagging network dedicated for users to share their documents (using text tags) with other users of the network. It contains tags assigned by identified users to scientific articles (DOI) and Web pages (URL). From the full original corpus, we considered only the Web pages that still exist in September 2015, leading to a set of 308'906 documents. 241'706 document d are tagged by 4'911 users u, with 1,5 million occurrences of 59'886 unique tags w. On average, each user used 263 tags and each document has 6 tags.

On this dataset, we use the evaluation protocol of [3], which selects randomly one user u, and one random tag t used by u, as a query. All the tagging made by u using t on documents are then removed from the dataset. Then, the documents d initially tagged by u are marked relevant. We created 200 single term queries using this protocol.

---

[1] http://www.bibsonomy.org.

Classical measures evaluate the quality of the retrieval: MAP, $P@5$, $P@10$. Two other measures detail the configurations studied: (a) $PQEC@10$ measures the level of validation of PQEC on the top-10 results the frameworks: it is the ratio of the top-10 documents that do not contain a user u tag and that are ranked before a document that is tagged by a tag used by u. A strict validation of PQEC (i.e. $PQEC@10 = 1.0$) is expected to lead to better results; and (b) the $Prof_{overlap}$ values that describes the amount of overlap between the extended query and the user u profile. Such value is in $[0, 1]$. All statistical significance tests are paired bilateral Student t-tests.

6.2   Tested Configurations

All the experiments are based on language models with Dirichlet priors using the default parameters of Terrier 4.0 [10] (english stoplist, Porter stemmer, $\mu = 2500$). Similarly to [3], we fix $\lambda = 0.5$ for the documents matching in Eq. (1). We tested four groups of configurations: baselines (without query expansion, or with the full user profile expansion), very dense, dense and sparse neighborhoods. They simulate different topologies of users networks.

Baselines - Our approach is compared to two baselines: (1) general profile retrieval, where the user profile is represented by all his tags in $Profile(u, q) = Profile(u)$, and (2) a non-personalized retrieval, where the initial query only is used.

The results are presented in Table 1 (runs a, b and c). Using the full user profile (runs a and b) clearly outperforms the run c without any profile. The MAP differences between a and b are not significant (p = 0.101), but they are significant between runs a and c (p = 7.95E-09), as well as between b and c (p = 1.32E-10). Moreover, the adaptation described in Sect. 4.2 outperforms the classical framework. We notice also that the run b has a relatively low value for $PQEC@10$: most of the time in the classical framework the constraint PQEC does not hold.

Very dense neighborhoods - Here, all the user set U is used as a neighborhood, so $u_{sn} = U$. We also study the fact that we filter, or not: (a) the users from u_{sn} according to the fact that they are related to the query (i.e. they tagged one document with one query term). When filtering these users, we obtain an average of 152 neighbors for u; (b) the profiles of the users u' from u_{sn}. When they are not filtered we use the $Profile(u', q) = Profile(u')$, when they are filtered the used profiles for u' are filtered according to Sect. 5.2.

These results are presented in Table 1 (runs d to i). The runs f and h (resp. g and i) have exactly the same values for MAP, $P@5$ and $P@10$, because the filtered u_{sn} already generates the full user profile (as $Prof_{overlap} = 1.0$). Here again the adapted frameworks outperform their respective classical ones. The filtered profiles from the neighbors outperform the unfiltered ones: choosing the "right" terms of the neighbors profiles has a positive impact. The best results are obtained with an average of 30 % terms of the user's profile, which fits wells to the fact that users have more than 2 topics on average (as seen in Subsect. 3.1).

Here again we do not conclude that there are statistically significant differences between MAP of adapted d or classical e runs (p = 0.099), we notice however that adapted filtered run d has significant MAP differences (with p < 0.001) with all the other runs in Table 1, where the classical filtered run e has no significant difference between runs f (and h) with a p value of 0.304.

Table 1. Retrieval performances for all the runs.

Run	Framework	u_{sn}	$Profile(u', q)$	PQEC@10	$Prof_{overlap}$	MAP	P@5	P@10
a	adapted	$\emptyset$	$Profile(u)$	1.0	1.0	**0.4950**	**0.1860**	**0.1260**
b	classical	$\emptyset$	$Profile(u)$	0.0521	"	0.4639	0.1570	0.0970
c	classical	$\emptyset$	$\emptyset$	/	0.0	0.2934	0.1010	0.0585
d	adapted	filtered	filtered	1.0	0.3086	**0.5528**	**0.2060**	**0.1285**
e	classical	filtered	filtered	0.0646	"	0.5205	0.1790	0.1095
f	adapted	filtered	unfiltered	1.0	1.0	0.4950	0.1860	0.1260
g	classical	filtered	unfiltered	0.0521	"	0.4639	0.1570	0.0970
h	adapted	unfiltered	unfiltered	1.0	1.0	0.4950	0.1860	0.1260
i	classical	unfiltered	unfiltered	0.0521	"	0.4639	0.1570	0.0970
k	adapted	filtered	filtered	1.0	0.2508	0.4015	0.1590	0.0925
l	classical	filtered	filtered	0.0608	"	0.3946	0.1380	0.0810
m	adapted	filtered	unfiltered	1.0	0.6770	**0.4779**	0.1770	**0.1195**
n	classical	filtered	unfiltered	0.0410	"	0.4497	0.1590	0.0925
o	adapted	unfiltered	unfiltered	1.0	0.8695	0.4413	**0.1820**	0.1065
p	classical	unfiltered	unfiltered	0.0224	"	0.4269	0.1560	0.0880
q	adapted	filtered	filtered	1.0	0.2286	0.3923	0.1500	0.0930
r	classical	filtered	filtered	0.1020	"	0.3799	0.1310	0.0760
s	adapted	filtered	unfiltered	1.0	0.6300	0.3559	0.1480	0.1030
t	classical	filtered	unfiltered	0.0757	"	0.3708	0.1330	0.0795
v	adapted	unfiltered	unfiltered	1.0	0.8150	**0.3960**	**0.1680**	**0.1015**
w	classical	unfiltered	unfiltered	0.0804	"	0.3755	0.1350	0.0790

Dense neighborhoods - We consider a relatively dense subset of U for u_{sn}. The social neighborhood of u is composed of users u' that share at least one tag with u' profile: $\{u'|Profile(u') \cap Profile(u) \neq \emptyset\}$. Here, each user has on average 872 neighbors. We filter these users according to the fact that they are related to the query or not; and we investigate the impact of filtering or not the profiles of these users. The filtering of users according to the query gives neighborhoods of 40 users in average.

The results are presented in Table 1 (runs k to p). Again, the adapted frameworks outperform the classical ones. We notice that the best results for MAP and $P@10$ are obtained when the profiles of the neighbors are not filtered (run m), with queries expansions containing 68 % of the user's profile, on average. For the runs o and p, increasing the overlap with the user's profile does not help, except for the $P@5$ value, slightly higher for the run o than the run m. For the best dense neighbors run of Table 1, m, the difference in MAP is not statistically significant with its classical counterpart n (p = 0.177), neither with the two unfiltered runs o (p = 0.115) and p (with p = 0.065). This is explained by some instability of the neighbors selected.

Sparse neighborhoods - The last set of configurations studied mimics sparse neighborhoods for a user u (inspired from [3]): the social network of u is composed of users u' that tagged at least one document that u tagged, whatever the tags are: $\{u'|\exists w \in W, \exists d \in D, R(u', d, w) \wedge R(u, d, W)\}$. Compared to other neighborhoods, the neighbors are here expected to be more similar to u, because they focused on the same document. There are 56 neighbors, on average. Moreover the filtering of users according to the query (as described before) leads to sets of 10 neighbors on average.

The results in Table 1 (runs q to w) show that, unlike the others neighborhoods, the best configuration is obtained by unfiltered neighbors and profiles, and the adapted framework. Another difference with previous results is that one adapted run, namely s, underperforms its respective classical run. This is due to an inadequate filtering of the neighbors, and then, applying subsequently the adapted framework degrades the quality of the results. The differences in MAP are small, this explains why we could not find statistically significant differences for the MAP values between these runs.

6.3   Discussion

The first point that we get from these experiments is that our frameworks that validate the PQEC constraint are consistently better than the classical framework (though without statistically significant differences taken one against one, but the repetitive outperformance is clear). Our adapted framework is very simple and should certainly be extended to tackles more clearly queries with multiple terms, but the current proposal already shows its interest on the quality of the results. The second point is that filtering the neighbors according to the query, using a very large set of potential users (i.e., very dense neighborhoods) seems to lead to better results than filtering *a priori* users (i.e. dense or sparse neighborhoods). Processing very dense neighbors necessitates, for each query, to process the whole set of users. However, if users' profiles are represented as documents in a classical IR system, retrieving users that match a query is fast.

7   Conclusion

This paper proposes a probabilistic framework that exploits the profile of a user u asking a query q, in order to improve the search results. The profile is filtered regarding the query. We investigated two parameters that help in selecting the relevant parts of u's profile: one that exploits the query to select a useful subset of social neighbors of u, and one that uses sub-profiles of neighbors of u according to q. The main conclusion drawn from our experiments on the Bibsonomy corpus is that adapting the set of all users to the query and filtering u's profile according to the query improves the results. Short term extensions of this work will study the use of real friendship relations as social neighbors. Other future works will focus on users u with empty profiles that do not benefit from the proposed profile adaptations. We will then explore how social neighbors may be used to consider terms that do not belong to the initial profile of u.

Acknowledgements. This work is partly supported by the Guimuteic project funded by Fonds Européen de Développement Régional (FEDER) of région Auvergne Rhône-Alpes projects and by the ReSPIr project of the région Auvergne Rhône-Alpes.

References

1. Biancalana, C., Gasparetti, F., Micarelli, A., Sansonetti, G.: Social semantic query expansion. ACM Trans. Intell. Syst. Technol. **4**(4), 60:1–60:43 (2013)
2. Blei, D.M., Ng, A.Y., Jordan, M.I.: Latent dirichlet allocation. JMLR **3**, 993–1022 (2003)
3. Bouadjenek, M.R., Hacid, H., Bouzeghoub, M.: Sopra: a new social personalized ranking function for improving web search. In: SIGIR Conference, pp. 861–864 (2013)
4. Bouadjenek, M.R., Hacid, H., Bouzeghoub, M., Daigremont, J.: Personalized social query expansion using social bookmarking systems. In: SIGIR Conference, pp. 1113–1114 (2011)
5. Clinchant, S., Gaussier, E.: Information-based models for ad hoc ir. In: SIGIR Conference, pp. 234–241 (2010)
6. Dou, Z., Song, R., Wen, J.R.: A large-scale evaluation and analysis of personalized search strategies. In: Conference on World Wide Web, pp. 581–590 (2007)
7. Fang, H., Tao, T., Zhai, C.: A formal study of information retrieval heuristics. In: SIGIR Conference, pp. 49–56 (2004)
8. Fang, H., Zhai, C.: Semantic term matching in axiomatic approaches to information retrieval. In: SIGIR Conference, pp. 115–122 (2006)
9. Khodaei, A., Sohangir, S., Shahabi, C.: Personalization of web search using social signals. In: Ulusoy, Ö., Tansel, A.U., Arkun, E. (eds.) Recommendation and Search in Social Networks, pp. 139–163. Springer, Switzerland (2015)
10. Ounis, I., Amati, G., Plachouras, V., He, B., Macdonald, C., Lioma, C.: Terrier: a high performance and scalable information retrieval platform. In: SIGIR Workshop on Open Source Information Retrieval (2006)
11. Vosecky, J., Leung, K.W.T., Ng, W.: Collaborative personalized twitter search with topic-language models. In: SIGIR Conference, pp. 53–62 (2014)
12. Xu, S., Bao, S., Fei, B., Su, Z., Yu, Y.: Exploring folksonomy for personalized search. In: SIGIR Conference, pp. 155–162 (2008)

A Relevance-Focused Search Application for Personalised Ranking Model

Al Sharji Safiya[✉], Martin Beer, and Elizabeth Uruchurtu

Communication and Computing Research Institute, Sheffield Hallam University,
153 Arundel Street, Sheffield S1 2NU, UK
Safiya.M.AlSharji@student.shu.ac.uk

Abstract. The assumption that users' profiles can be exploited by employing their implicit feedback for query expansion through a conceptual search to index documents has been proven in previous research. Several successful approaches leading to an improvement in the accuracy of personalised search results have been proposed. This paper extends existing approaches and combines the keyword-based and semantic-based features in order to provide further evidence of relevance-focused search application for Personalised Ranking Model (PRM). A description of the hybridisation of these approaches is provided and various issues arising in the context of computing the similarity between users' profiles are discussed. As compared to any traditional search system, the superiority of our approach lies in pushing significantly relevant documents to the top of the ranked lists. The results were empirically confirmed through human subjects who conducted several real-life Web searches.

Keywords: User profile · Keyword-based features · Semantic-based features

1 Introduction

The use of Implicit Feedback (IF) is proven to improve the performance of retrieval systems [4], allowing relevant documents matching both the user's inputted keywords (i.e. queries) and particular needs to be retrieved. We build upon these ideas to construct users' interest profiles which are used to infer relevant documents. Ranking functions are then crafted based on both the relevance and interest scores of these documents leading to the generation of a relevance-focused personalised search. Query expansion technique is employed through WordNet[1] ontology to integrate terms which are not directly expressed in the users' queries.

The requirements for personalised search models include a learning process to extract users' information (i.e. interaction activities) meeting their individual information needs. We employ users' clicked documents to build and maintain their interest profiles. The rank algorithm takes into account the learned patterns together with the active users' profiles [10] to develop a PRM based on which search results are ranked to represent the users' interests [10]. The main argument is that IR can be employed by the PRM to provide ranked lists of the documents based on individual user's interests.

---

[1] https://wordnet.princeton.edu.

© Springer International Publishing Switzerland 2016
S. Hartmann and H. Ma (Eds.): DEXA 2016, Part II, LNCS 9828, pp. 244–253, 2016.
DOI: 10.1007/978-3-319-44406-2_18

It is thus investigated whether users' interests can be identified through implicit interactions in digital web documents. The main challenge addressed is how query keywords and their related concepts can be used to identify users' individual interests (i.e. relevant documents); and how acquired feedback is preserved over time in order to include representation of both the users' interests and modelling.

2 Related Work

Personalised searches differ in the type of data and approaches used to build the user profile [10] both of which play a major role in personalised search approaches. A recent study [13] uses spreading mechanism through ontology to provide inherent relationships between terms/concepts appearing in their respective bag-of-word representation in order to extend the semantic similarity concept between two entities. However, it is still an open research question whether a mechanism could be devised to control and correct the integration of ontology terms in the query expansion. This would match the users' information needs thereby guaranteeing that recall is improved during the phase without degrading precision as a result of this process. A technical report by William [14] presented the idea of indexing material at the sentence and phrase level to support improved information access so that the content of an individual sentence or phrase could be located in response to a specific description of need. To identify appropriate concepts within annotated audio text, Khan [5] has also presented an automatic disambiguation algorithm which could prune as many irrelevant concepts as possible while at the same time retaining the largest possible number of relevant concepts. While these studies provide the techniques adopted to improve the performance of Information Retrieval (IR) systems in terms of precision or recall or both, they do not however detail the effects of such integration with regards to different levels of keyword mixtures of the terms in both queries and ontology during the matching process. Following on from [1], this paper presents such effects.

3 Relevance-Focused Search

This section outlines our two models representing users' interests and preferences in a formal way, such that both approaches can be checked for validity to form customised views of a relevance-focused search application for personalised search.

3.1 Keyword-Based Features

Users' profiles are often defined by storing the content of documents clicked after being collected over time [10]. Given a set of users' Web search logs, any search documents clicked are archived for each user whose representations are determined based on these documents. For our purpose, a feature can be considered as an attribute of text content (i.e. document or query content) which is used to make decisions related to it. Thus, to determine a relevant document means to extract its important features that can determine factors which are important to a user searching for such a document. These features are then used to craft the ranking predictors which are often combined together to improve the retrieval process.

Assuming there is a set m of users represented by $U = \{u_1, u_2, \ldots u_m\}$ and a set n of documents represented by $D = \{d_1, d_2, \ldots d_n\}$, a profile for user $u \in U$ can be represented as an ordered pair of n-dimensional vectors using Eq. 1 [10].

$$u^{(n)} = \langle (d_1, s_u(d_1)), (d_2, s_u(d_2)), \ldots (d_n, s_u(d_n)) \rangle \tag{1}$$

where each $d_j \in D$ and s_u is the function for user u which assigns interest scores (i.e. interest score) to documents.

Since each document $d_j \in D$ can represent an HTML document in the context where the focus is to capture the implicit feedback related to the document clicked, Eq. 1 might be used to represent the user's profile. Each document d_j can then be represented as an attribute vector of k-dimensional features where k is the total number of features extracted [10]; and the feature weight associated with the document is represented by its corresponding dimension in a feature vector which is given by: $d_j = \langle fw_j(f_1), fw_j(f_2), \ldots, fw_j(f_k) \rangle$, where $fw_j(f_p)$ is the weight of the pth feature in $d_j \in D$, for $1 \leq p \leq k$. Since the features extracted are the textual content of pages represented in Bag-of-Words (BOW), i.e. a set of pairs, denoted as $\{t_i, w_i\}$, where t_i is a term describing the content of the page (i.e. document) such that $t_i \in d_j$, and w_i is its weight found by using the normalised $tf \bullet idf$ term values [9], each document can thus be represented by sets of term-score pairs (e.g., sport (cricket; 0:54); (baseball; 0:39); (soccer; 0:45)^2) leading to the user profile represented as a feature vector using the terms of documents as features.

Given a user profile UP containing v interest vectors for a user u_k, an overall interest vector is often determined by combining all interest vectors for that user [9]. Assuming T_i is the set terms in the $i^{th}(i \in [1, v])$ interest vector, the set of terms of the overall interest vector T can be found as $T = \cup_{i=1}^{v} T_i$. For every term $t \in T$, its overall interest vector can be calculated as $s_u(t) = \sum_{i=1}^{v} s_i(t) \bullet w_i$, where $s_i(t)$ is the score (relevance score) of t in the ith interest vector ($s_i(t) = 0$, if $t \notin T_i$) and w_i is the actual weight of the ith interest vector.

3.2 Semantic-Based Features

The spreading approach can be adopted [13] in order to perform the automatic query expansion [13] by appending terms that are conceptually related to the original set of terms in documents. We build on this earlier work and provide a conclusive empirical analysis when related terms are considered and the degree of their contribution to improve the performance of IR systems. Although there are many overlaps between the current research and the latter approach aimed at providing semantic similarities through ontologies, in terms of classification technique employed to create users'

---

^{2} Figures based on a different experiment and given here solely for illustrative purposes.

profiles to describe the contents of Web documents clicked, this project applies both term weight (i.e. term frequency factor) and dwell weight[3] directly as a dimensional feature to enrich the users' models [1, 2]. For instance, not only was it shown in these surveys that the performance of the PRM improved, but it was also demonstrated that it could be used as a complementary feature for the system to rely on when the keyword feature proves unsuccessful in identifying the relevance of documents.

Given ontology O and term t_i, spreading process might employ the ontology O to spread document d_j, to determine the terms that are related to t_i, so that any terms related to the original terms of the document can be included. Denoting these terms as $RelO(t_i)$, the results of spreading the document d_j, is an expanded document $\hat{d}_j$ such that the set of terms $\hat{d}_j = \{t_1, \ldots, t_n, t_{11}, \ldots, t_{mn}\}$ and $d_j \subseteq \hat{d}_j$ where $\forall t_{ij}/t_{ij} \in RelO(t_i)$ and a path exists from t_i to t_j.

This spreading process is an iterative process; and the terms from the previous iterations that are related to the original terms are joined to the document at the end of the iteration. The spreading process terminates when there are no related terms to spread the document with, or simply when $\forall t_i \in d_j/RelO(t_i) = 0$.

3.3 Cosine Similarity Measure

For the purpose of this work, in order to compute the vector similarities determining the user's interest in a particular document, the cosine similarity measure is adopted [9] as the technique to represent the user model.

Given a user profile $UP = s_{u_k}(d_j)$ and a document $d_j = \{t_1, \ldots, t_n, t_{11}, \ldots, t_{mn}\}$ for a given search (document containing a set of texts where each t_i is a k-dimensional vector in the space of content features), the binary cosine similarity [9] denoted as $Sim(UP, d_j)$ can be determined using Eq. 2. Such similarity between the two sets of texts clearly indicates the relevance of the document in the keyword-based approach which can be applied to the respective vectors.

$$Sim(UP, d_j) = \frac{|UP \cap d_j|}{|UP| \times |d_j|} \tag{2}$$

where $|UP \cap d_j|$ represents the number of keywords in both UP and d_j, and $|UP|$ and $|d_j|$ are respectively the number of keywords in UP and d_j.

3.4 Semantic Similarity Measure

Similarity can be determined to be equal to the inverse of distance in its simplest form or some other mathematical function based on ontological distance. Semantic similarity can thus be inversely proportional to the distance between concepts whereby the closer two concepts are in the ontological representation; the higher the similarity score

---

[3] This dual technique was thoroughly explained previously and authors do not claim this contribution in the current paper.

between them is [6]. Any similarity between two concepts can then be determined by taking the cosine angle between the two corresponding vectors [8]. Mathematically, semantic similarity is determined here by employing a fuzzy ontology value [7], whereby increasing distance between two consecutive terms is inversely proportional to an increase in semantic similarity. Here it is important to recall that words which have been integrated are not directly related to the keyword queries, thus, it is not feasible to apply the cosine similarity measure directly. The application of fuzzy ontology values as shown in Eq. 4 [2] addresses this problem. Thus, based upon this similarity measure (i.e. fuzzy ontology values) a set of relevant documents are obtained. However, expanded documents are still those documents matching the users' queries at first place as demonstrated elsewhere; therefore, after constructing the semantic document vectors in this way, the normal binary cosine similarity measures are applied to refine the ranking function.

Given a user profile with a set of texts $UP = s_{u_k}(d_j)$ and a document $\hat{d}_j = \{t_1, \ldots, t_n, t_{11}, \ldots, t_{mn}\}$ for search (expanded document containing a set of texts where each t_{ij} is a k-dimensional vector in the space of content features); cosine similarity denoted as $Sim(UP, \hat{d}_j)$ which is determined following Eq. 2 can be applied to represent the user's interests.

$$F_{jk} = \frac{c_{jk}}{X^2} \tag{3}$$

where c_{jk} is the distance between keywords t_{ij} and t_{ik} or the frequency of the keywords/concepts appearing consecutively in the keyword list, and X is the total number of t_{mn} terms (i.e. keywords) in that document.

3.5 Query Processing and Ranking

Users' queries expressed in keywords to represent their information needs can be considered as short documents. Thus, for each user u_k, a BOW representation for each query issued by the user in a particular session must also be created and compared with its set of corresponding documents. This comparison is based on the similarity between both the query and the targeted documents. Thus, Eq. 4 is applied to calculate the cosine similarity measure between the query vector, the vectors of the matching documents and the vectors of the matching user profile respectively.

$$Sim(q_i, d_j) = \frac{|q \cap UP \cap d_j|}{|q| \times |UP| \times |d_j|} \tag{4}$$

where $|q \cap UP \cap d_j|$ represent the number of keywords in q, UP and d_j, and $|q|$, $|UP|$ and $|d_j|$ are respectively the number of keywords in q, UP and d_j.

The highest similarity values are used to establish our relevance-focused search application. They are provided by Eq. 4 when considering the keyword-based features as well as the semantic-based features when the document is integrated with ontology terms. Thus they represent the most similar documents between the query, the user profile and the available documents.

3.6   Search Result Personalisation

The personalisation of search results to a large degree lies in merging the models that provide them. A description of the linear combination adopted in the current research can be found in [2]. Here, as outlined in the following section, the aim is to test the system' models on a deeper level and to investigate their real world problems as closely as possible. A set of the experiments performed by using human subjects (i.e. 729 query keywords[4]) while conducting real-life Web searches is thus presented to validate each model individually. Such evaluation enabled us not only to obtain the system's performance based on each model, but also to evaluate real collections based on different terms integration with different terms of query keywords.

4   Evaluation

The experimental results are presented in this subsection. For simplicity, the proposed personalised search approach is referred to as experimental system while the search approach which is not personalised is referred to as Baseline. There are two main sets of experiments: (1) Implicit Feedback vs. No-Feedback. Its relative experimental results are presented in Table 1 and visualised in Fig. 1. (2) Keyword-Based vs. Semantic-Based. Its relative experimental results are shown in Fig. 2.

4.1   Experimental Set up

Assuming a given user $u_k \in U$ clicked the document d_j after issuing a query containing the word t, then the document d_j is considered useful and relevant to t for user u_k, and documents that are not retrieved, are judged as non-relevant by the user [12]. To evaluate the search accuracy of the two models, sets of documents $d_j \in D$ containing the word t selected by u_k were checked whether they are highly ranked in the ranked list generated by the personalised search solution.

Implicit Feedback vs. No-Feedback. In this experiment, it was investigated how a sample of real data collected during interaction between users and the system can affect the performance of the personalised search. This includes investigating how useful the acquired feedback is when preserved over time in the form of user profiles [11] to include the representation of their interests. If the experimental system generates accurate ranked lists in terms of higher precision in the lower ranks, then it can be considered to perform better.

---

[4] A detail description of this data set can be found in [2].

Table 1. Average of Precision at Rank 5 and Rank 10

Precision	Baseline	Experimental system			
		Keyword-based	P(paired t-test)	Semantic-based	P(paired t-test)
System @ Rank 5	0.79	0.83	0.006 %	0.94	0.005
System @ Rank 10	0.56	0.75	0.50 %	0.85	0.78 %

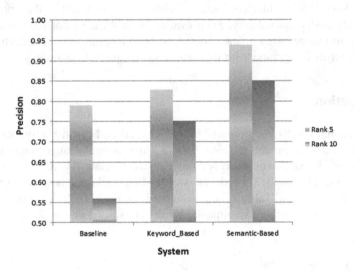

Fig. 1. System performance

A system's performance is often assessed in terms of search results and by its ability to push relevant documents to the lower ranks. Thus, to compare the performances of two systems - here, experimental and baseline systems - ranked lists of search results obtained by the user need to be considered for both systems. The one that is better able to *push* relevant documents to the top of the ranked lists of search results is the more efficient. Table 1 gives the overall precision obtained at rank 5 and 10 of both systems. It is important to recall that precision is obtained by dividing the number of relevant documents - for each user - among the top 5/10 documents by 5 or 10 accordingly. Here, results to the first page (i.e. 10 documents) are considered.

From Table 1, the overall averages of the precision at rank 5 and at rank 10 for the experimental system when employing the semantic-based approach, clearly indicate that out of 5 documents, the system can rank more than 4 documents based on their relevancy to the query (0.94*5 = 4.70 and 0.85*5 = 4.25). While the performance of the system is more or less constant at rank 5 by employing the keyword-based approach, it is poorer at rank 10, since out of 5 documents, it can only rank 3.75 (0.75*5 = 3.75) documents. The worst performance can be observed from the baseline, as its overall averages of the precision at rank 5 and at rank 10 indicate that having 5 documents, the system is able to rank, based on their relevancy to the query, less than 4 documents (0.79*5 = 3.95) and less than 3 documents (0.56*5 = 2.80) respectively.

Overall, the experiments showed that the personalised system outperforms the baseline with a statistically significant (paired t-test) difference between them.

Keyword-Based vs. Semantic-Based. The goal of this experiment was to use the same idea with the same data set to study whether the semantic-based approach is superior to relying on the keyword-based approach with regards to a personalised search. Here, it should be recalled that in the semantic-based approach, the spreading mechanism was used to incorporate the concept terms into the documents, however, the same statistics were used in both models. Therefore, the semantic-based approach is the expansion of the keyword-based approach with the integration of content semantics expressed in ontology terms so that an enriched user model (i.e. user profile) is generated. This experiment will test the effects of combinations of keywords from the ontology terms with the keywords from the query to enrich the user model, so that the effect of mixing different keywords to generate ranked lists can be investigated.

Each of the participant collections was thus indexed individually into document vector files. Figure 2 shows a representation of the distribution of document indices (here, the values of interest vector - denoted as $s_u(t)$) according to different combinations of the query keywords[5] with its related concepts[6] mixtures. Here, $kxny$ means x keyword(s) and y concept(s) or ontology terms are employed in the user model. For example, $k2n2$ and $k2n3$ are respectively the keywords employed in the iterations in which two and three ontology terms are integrated into the user model for the second keyword of the query. The threshold interest vector values are the values represented by $kxn\theta$, meaning that only keyword-based is employed and no ontology terms have yet been added to the documents.

As can be seen from Fig. 2, the semantic-based layout showed the best results when a document is integrated with 3 and 4 keywords (at $kxn3$ and $kxn4$) regardless of the original number of terms (i.e. keywords) contained in the query. The presentation given here is related to only one query, but statistical evidence (ANOVA p value = 6.80 %) indicated that out of 729 keyword queries, this observation is consistent across more than 650 keyword queries.

However, expanding the document with 1 or 2, 5 and 7 keyword(s) showed some slight improvements for most documents. On the other hand, integrating the document with 6 and 8 keywords showed worse performance (represented by $kxn6$ and $kxn8$ in Fig. 2), which might be due to the inclusion of keywords not related to the original term meaning.

---

[5] According to [3], on the average, a query contains 2.21 terms.

[6] It was demonstrated in [13] that the computation process of terms' weight during document expansion turns monotonic after the third iteration. In current work, this computation turns monotonic after the document is expanded with the eighth term concept.

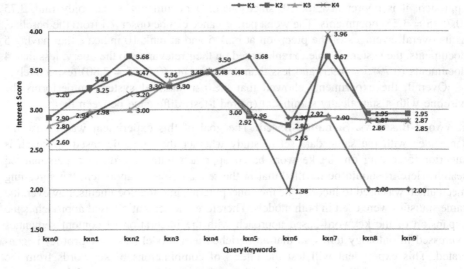

Fig. 2. Comparisons of mixtures of query keywords with ontology terms

Overall, employing keyword-based features alone showed poorer performance than employing semantic-based features if the spreading or query expansion integrates 3 or to 4 keywords into the document.

5   Conclusions

Derived from several existing techniques, this paper has presented an effective personalised search model that exploits users' profiles by employing their implicit feedback for query expansion through a conceptual search to index documents. Empirical validation confirmed the reliability of our system. A combination of the keyword-based and semantic-based features to provide further evidence of relevance-focused search application for each individual user was validated by using human subjects conducting real-life Web searches. The findings of the experiments demonstrated that, compared to any traditional search system, our approach can push significantly higher number of relevant documents to the top of the ranked lists.

A series of two different web search experiments was performed using different keywords from real users. For each search session, a list of personalised webpage re-ranking over the search results returned by Google was generated. Both the evaluation metric parameters of precision and recall were adopted to measure the ranking quality of the personalised search engine in order to determine the relevance of the results according to their order of relevance.

Acknowledgement. This research was supported by the Ministry of Manpower in Oman which has granted the funding for the survey of this research.

References

1. Safiya, A.-S., Martin, B., Elizabeth, U.: A dwell time-based technique for personalised ranking model. In: Chen, Q., Hameurlain, A., Toumani, F., Wagner, R., Decker, H. (eds.) DEXA 2015. LNCS, vol. 9262, pp. 205–214. Springer, Heidelberg (2015)
2. Safiya, A.-S., Martin, B., Elizabeth, U.: Enhancing the degree of personalisation through vector space model and profile ontology. In: IEEE RIVF International Conference Computing and Communication Technologies, Research, Innovation and Vision for Future (RIVF), pp. 248–252. IEEE (2013)
3. Jansen Bernard, J., Amanda, S., Tefko, S.: Real life, real users, and real needs: a study and analysis of user queries on the web. Inf. Process. Manage. 36(2), 207–227 (2000)
4. Diane, K., Jaime, T.: Implicit feedback for inferring user preference: a bibliography. In: ACM SIGIR Forum, vol. 37(2), pp. 18–28. ACM (2003)
5. Latifur, K., McLeod, D.: Disambiguation of annotated text of audio using onologies. In: Proceeding of ACM SIGKDD Workshop on Text Mining (2000)
6. Claudia, L., Chodorow, M.: Combining local context and WordNet similarity for word sense identification. In: Fellbaum, C. (ed.) WordNet: An Electronic Lexical Database, vol. 49(2), pp. 265–283. MIT Press (1998)
7. Dario, L., Morara, R.: First: fuzzy information retrieval system. J. Inf. Sci. 17(2), 81–91 (1991)
8. Thusitha, M., Lee, M.C., Cohen-Solal, E.V.: An ontology-based similarity measure for biomedical data-application to radiology reports. J. Biomed. Inform. 46(5), 857–868 (2013)
9. Manning Christopher, D., Prabhakar, R., Hinrich, S.: Introduction to Information Retrieval. Cambridge University Press, Cambridge (2008)
10. Mobasher, B.: Data mining for web personalization. In: Brusilovsky, P., Kobsa, A., Nejdl, W. (eds.) Adaptive Web 2007. LNCS, vol. 4321, pp. 90–135. Springer, Heidelberg (2007)
11. Gauch, S., Speretta, M., Chandramouli, A., Micarelli, A.: User profiles for personalized information access. In: Brusilovsky, P., Kobsa, A., Nejdl, W. (eds.) Adaptive Web 2007. LNCS, vol. 4321, pp. 54–89. Springer, Heidelberg (2007)
12. Jaime, T., Dumais, S.T., Horvitz, E.: Characterizing the value of personalizing search. In: Proceedings of the 30th Annual International ACM SIGIR Conference on Research and Development in Information Retrieval, pp. 757–758. ACM (2007)
13. Rajesh, T., Manjunath, G., Stumptner, M.: Computing Semantic Similarity Using Ontologies. Hewlett-Packard (HP) Development Company, L.P, Labs Technical Report HPL-2008-87 (2008)
14. Woods William, A.: Conceptual Indexing: A Better Way to Organize Knowledge. A technical report of Sun Microsystems, Inc. (1997)

Aggregated Search over Personal Process Description Graph

Jing Ouyang Hsu[1]([✉]), Hye-young Paik[1], Liming Zhan[1], and Anne H.H. Ngu[2]

[1] University of New South Wales, Sydney, NSW, Australia
{jxux494,hpaik,zhanl}@cse.unsw.edu.au
[2] Texas State University, Austin, TX, USA
angu@txstate.edu

Abstract. People share various processes in daily lives on-line in natural language form (e.g., cooking recipes, "how-to guides" in eHow). We refer to them as *personal process descriptions*. Previously, we proposed Personal Process Description Graph (PPDG) to concretely represent the personal process descriptions as graphs, along with query processing techniques that conduct exact as well as similarity search over PPDGs. However, both techniques fail if no single personal process description satisfies all constraints of a query. In this paper, we propose a new approach based on our previous query techniques to query personal process descriptions *by aggregation* - composing fragments from different PPDGs to produce an answer. We formally define the *PPDG Aggregated Search*. A general framework is presented to perform aggregated searches over PPDGs. Comprehensive experiments demonstrate the efficiency and scalability of our techniques.

1 Introduction

People are engaged in all kinds of *processes* all the time, such as cooking a dish, or filing a tax return. Although the area of business process management (BPM) has produced solutions for modelling, automating and managing many of the business organizational workflows, still significant portion of the processes that people experience daily exist outside the realm of these technologies.

These experiences are often shared on the Web, in the form of how-to guides or step-by-step instructions. Although these are primarily describing a workflow, without the formal modelling expertise, they are written in natural language. To distinguish these texts from the conventional organizational workflow models, we refer to them as *personal process descriptions*. Many examples of personal process descriptions are found in cooking recipes, how-to guides or Q&A forums.

The natural language texts are not precise enough to be useful in utilizing the process information presented in them. For example, the state-of-the-art for search technologies over the existing personal process descriptions are still keyword/phrase-based and users would have to manually investigate the results.

In our previous work [11], we proposed a simple query language designed to perform exact-match search over the personal process descriptions. The language is supported by a graph-based, light-weight process model called PPDG

© Springer International Publishing Switzerland 2016
S. Hartmann and H. Ma (Eds.): DEXA 2016, Part II, LNCS 9828, pp. 254–262, 2016.
DOI: 10.1007/978-3-319-44406-2_19

(Personal Process Description Graph) which concretely represents the personal process description texts. We further extended our query technique to return *similar* process descriptions to a query input in [4]. Using these techniques, we can perform a *process-aware* search over PPDGs such as showing dependencies between data and actions.

However, when there is no single PPDG in the repository that satisfies all constraints in a query, these techniques cannot produce an answer. To overcome this limitation, we present a new approach to querying PPDGs, which can still produce an answer when a single PPDG cannot satisfy all query constraints. This technique, *Query By Aggregation*, involves decomposing a query into subqueries, matching multiple fragments over different PPDGs. The answer to a query is then generated by composing these fragments according to ranking criteria. This approach allows the user not only to better utilise existing process information in the PPDG repository, but also to discover and reuse process fragments to compose his/her own processes. We summarise our contributions below:

- We formally define the *PPDG Aggregated Search*.
- We present a general framework to perform an aggregated search over PPDGs, including: (i) a query decomposition technique to break down the query into two categories of subqueries - constant query and anonymous query, (ii) a tri-level index scheme based on our previous search techniques [4,11] to reduce the search cost, and (iii) a ranking method to aggregate the matched fragments to obtain the closest query answers.
- We perform comprehensive experiments to demonstrate the efficiency and scalability of our techniques.

Due to the space limits, we do not include all the technical details and experimental results in the paper. See our technical report [5] for the further details.

2 Querying PPDG by Aggregated Search

Aggregated search is the task of searching and assembling information from a variety of sources, placing it into a single interface [6]. Our approach to PPDG aggregated search is based on the notion of graph aggregation problem [7], and aggregated search problem in BPMN[1] models [9]. In both of them, the answer of a query graph can be represented as aggregation of fragments from different processes which are stored in the process repositories. In our work, we define aggregated search based on PPDG as follows.

Definition 1 (PPDG Aggregated Search). *Given a PPDG query q and a set of PPDGs $\mathcal{P} = \{P_1, P_2, \ldots, P_n\}$, the problem of PPDG aggregated search is to find a subset $\mathcal{S} = \{P'_1, P'_2, \ldots, P'_m\}$ of $\mathcal{P}(m \leq n)$ and join their fragments $f_{P'_1}, f_{P'_2}, \ldots, f_{P'_m}$ to obtain a set of ranked PPDGs $\mathcal{R} = \{R_1, R_2, \ldots, R_k\}$, where each R matches the query q. That is, for each $R \in \mathcal{R}$, $R = f_{P'_1} \bowtie f_{P'_2} \bowtie \ldots \bowtie f_{P'_l} \mid l \leq m$ where f is a fragment (subgraph) of $P \in \mathcal{S}$.*

---

[1] Business Process Model and Notation, www.bpmn.org.

We take the PPDG query shown in Fig. 1 as an example. It describes a user query where the user wants to know what to do *before* booking the academic dress online and what needs to be done *between* getting the dress and attending ceremony. The prefix symbol "@" in the node label indicates an anonymous node[2] (i.e. "@D" for an anonymous data node, "@V" for an anonymous action node).

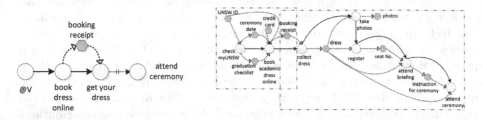

Fig. 1. Example of PPDG query   **Fig. 2.** Answer of PPDG aggregated query

Assume that the PPDG repository consists of two personal processes shown in Fig. 3. Both of these two processes fail to match the said query if we matched it with each PPDG separately. There is no information about getting dress or attending ceremony in the process on the left of Fig. 3. Similarly, the process on the right of Fig. 3 does not mention booking dress online.

In the aggregated search approach, we decompose the original query into subqueries, match them individually against the PPDGs in the repository, and aggregate the results to form the answers. The answer of the said query over the two sample PPDGs is shown in Fig. 2.

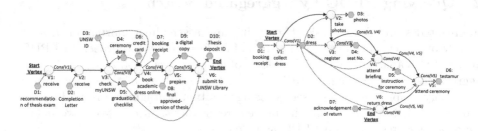

Fig. 3. Two PPDGs to query

---

[2] An anonymous node denotes an unnamed node in the query.

3   PPDG Query Decomposition

We can decompose a PPDG query into a set of subqueries which are classified into two categories as follows.

- A series of constant queries $\mathcal{Q}_c$: where each query has two constant (i.e., explicitly named) nodes connected by a direct flow edge.
- A series of anonymous queries $\mathcal{Q}_a$: where each query has (i) an anonymous node (i.e., unnamed node), or (ii) two constant nodes connected by a *path* edge.

Unlike traditional graph, PPDG is a directed graph with two types of nodes and edges, so it is not straightforward to deploy the general decomposition methods to solve our problem. We fully decompose PPDG into atomic fragments, because the size of most personal processes are not very big. The detail of the descriptions and algorithm can be found in our technical report [5].

4   PPDG Query Processing

After query decomposition, we use the two series of queries to find matched PPDG fragments and aggregate them to obtain the results. The framework of the aggregated search processing is presented as follows:

- **Constant Query:** We use each constant query in $\mathcal{Q}_c$ to perform similarity search on PPDGs to get the constant fragments with corresponding PPDGs,
- **Anonymous Query:** We use anonymous queries in $\mathcal{Q}_a$ based on the results in *Constant Query* phase to find the matched fragments,
- **Aggregating:** We rank and aggregate fragments to obtain the answers ordered by ranking score.

The experiments of our own proposed search techniques to process graph query over PPDGs showed that the main cost of the query processing is on matching the nodes between the query and PPDGs. Therefore, we first designed an indexing technique to speed up the node matching process as part of the framework. Then we use the index to obtain the matched fragments of constant queries and anonymous queries, and assemble the fragments to obtain the query answer.

4.1   Indexing PPDGs

The PPDG index based on the definition of label similarity in [4] has three levels "word(L1)-word set(L2)-PPDG(L3)". All the PPDGs entries are stored in L3. The word sets from each PPDG are extracted and stored in L2. Each word set entry points to the PPDG that it originates from. We cluster those word sets which have common words by choosing one of them as the center. The similarity between each word set and the center is more than a given correlation radius η. All the word sets in one cluster can be merged to construct a new word set entry in L2, which points to multiple PPDGs. L1 is an inverted index. When a word is given, we can use the inverted index to find the corresponding word set entry in L2.

4.2   Processing Constant Queries

To process constant queries, we launch our similarity query search proposed in [4] for each constant query to obtain the *similar* fragments. There is a small cost to do one similarity search. Since there are many constant queries generated for aggregated search, the total cost could be quite high. Therefore, we use the index built in Sect. 4.1 to reduce the query processing time.

Algorithm 1. *Constant Queries Search* $(\mathcal{Q}_c, \mathcal{P})$

Input : a series of Constant Queries $\mathcal{Q}_c$, a set $\mathcal{P}$ of PPDGs
Output: $\mathcal{Q}_c$ filled with matched fragments

```
 1  for each q ∈ 𝒬c do
 2  │   n = q.n_action;
 3  │   Get word set of n and obtain all matched PPDG graphs 𝒫m by index;
 4  │   if q.type is "C" then
 5  │   │   n' = q.n'_action;
 6  │   │   Get word set of n', and obtain all matched PPDG graphs 𝒫'm by index;
 7  │   │   𝒫''m = 𝒫m ∩ 𝒫'm;
 8  │   │   for each P ∈ 𝒫''m do
 9  │   │   │   f ← matched fragment;
10  │   │   │   score = similarity score between q and f;
11  │   │   └   q.results ← {f, score, P};
12  │   else
13  │   │   n' = q.n_data;
14  │   │   for each connected data node n_d in each P ∈ 𝒫m do
15  │   │   │   if n_d matches n' then
16  │   │   │   │   f ← matched fragment;
17  │   │   │   │   score = similarity score between q and f;
18  │   │   │   └   q.results ← {f, score, P};
19  return 𝒬c;
```

Algorithm 1 illustrates the details of how to use the PPDG index to process the similarity search of the constant queries. We match each query q in the constant query set $\mathcal{Q}_c$ to process the similarity search over PPDG. For each query result, we record the fragment f, the similarity score between f and q, and the corresponding PPDG P, which comprise a result tuple $\{f, score, P\}$. For each constant query q, we put its key node n_{action} into n, and get the word set of n to obtain the matched PPDGs $\mathcal{P}_m$ by index in Lines 2-3. Then we check the type of q in Line 4. If the second node n' is an action node, we use the similar method to get its matched PPDGs $\mathcal{P}'_m$ by index in Lines 5-6. The two matched PPDGs $\mathcal{P}_m$ and $\mathcal{P}'_m$ are joined to obtain a new PPDG set $\mathcal{P}''_m$ containing both action nodes n and n'. We get matched fragment from each $P \in \mathcal{P}''_m$ and put the result tuple into $q.results$ from Line 8 to 11. If the second node n' is data node, we check each connected data node in $P \in \mathcal{P}_m$ to obtain the matched fragment from Line 13 to 18. Note that the edge direction must be matched when matching data node in Line 15. The fragment results are also put into $q.results$ in Line 18. After traversing all constant queries, the result is returned in Line 19.

4.3 Processing Anonymous Queries

Each anonymous query has at least one known action node, so we can use the same similarity matching technique in Sect. 4.2 to obtain the PPDGs set containing the node, and then find the matched anonymous node or path. According to the decomposition technique in Sect. 3, an anonymous query contains at least one constant node, which exists in a constant query, and we have already processed the constant node in Sect. 4.2. These results can be stored in a map M. Then we search M first when the constant node in an anonymous query is given. If we cannot find the matched node in M, the normal index lookup is invoked to find the related PPDGs. The algorithm of constant queries can be adjusted to process anonymous queries. For details, please refer to [5].

4.4 Aggregating Fragments

After obtaining all matched fragments, we need to assemble them to obtain the required PPDGs. There are many ways to combine the fragments, therefore, we need to rank the possible aggregated results efficiently and recommend the user a list of aggregated PPDGs in an descending order by "score". In this section, we explain how to calculate such a score.

When we decompose the aggregated query into subqueries, the position of each query is kept, so we can assemble the aggregated result from each query tuple according to the position of the query. When we process the constant query or the anonymous query, we store the similarity *score* between the query and the fragment result. Intuitively, we could choose fragments with the highest score from each query tuple, and aggregate them to obtain the query result. Then the similarity score (SS) of a result R is calculated as $SS(R) = \prod q.score(f)$, where $q.score(f)$ represents the similarity score of a selected fragment f in R.

There may be some fragments coming from the same PPDG. The fragments from the same PPDG are preferred because intuitively they would form more coherent PPDG when put together, so we give the case a higher rank. For each possible aggregated result, we count the number n of fragments originated from each related PPDG. Then the similarity score of a result R can be adjusted as follows, which is called adjusted similarity score $ASS(R) = \prod(\prod_{i=1}^{n} q.score(f_i) \times C^{n-1})$, where C $(C \geq 1)$ represents a weighting factor for a scoring fragment and n is the number of fragments which come from the same PPDG containing f. Note that if the factor C is set to 1, ASS degenerates into SS.

5 Experiments

Now we present the results of a performance study to evaluate the efficiency and scalability of our proposed techniques. Two algorithms are evaluated. We use techniques proposed in [4] as Baseline, and the index based algorithms in this paper as INDEX.

Datasets. We have evaluated our aggregated search techniques on both synthetic and real datasets. The synthetic datasets are generated by randomization techniques as same as [4]. We select 100 process graphs and get their subgraphs to make 100 queries in our experiment. Some nodes in each query are randomly set as anonymous nodes. The average processing time of the 100 queries on each dataset represents the performance of our query processing mechanism. The factor C is set to 2 in all experiments.

The real dataset consists of 42 PPDGs about PhD programs collected from the Web and manually created by the authors. The dataset includes personal process descriptions on processes such as research degree admission, scholarship applications, and attending graduation ceremony. In this dataset, the queries are chosen manually.

Performance Tuning. The performance of our techniques is effected by the index. Especially, The correlation radius η between the data and the center of clusters in the index impacts the processing time of our algorithm. As expected, Fig. 4 shows the processing time drops when η increases, because the wordsets with similarity η are clustered in one index entry. On the other hand, if the η drops, the index degenerates. When η is equal to 0.2, there is no significant improvement between the algorithm INDEX and the algorithm Baseline. We notice that the performance of INDEX does not change much when η increases from 0.6 to 0.8, therefore, we use 0.6 as the default setting of η in the experiment.

Real vs Synthetic. We evaluate the performance of our techniques over the real and synthetic data. Due to the limited quantity of real process graphs, we magnify the result on the real data by 200 times in Fig. 5. It is shown that our techniques give the similar performance on both datasets, and the index is very effective and reduces the processing time.

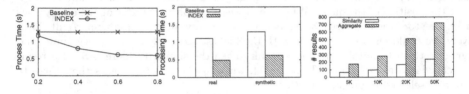

Fig. 4. Index tuning   **Fig. 5.** Real vs synthetic   **Fig. 6.** Varying p

Aggregate Output. Finding a way to systematically measure the quality of possible aggregates for a given query is still an open research issue [6] and is one of our immediate future work plans. In this paper, we evaluate the average output size of our techniques compared with similarity search techniques in [4]. It is clear that the output size of the two approaches increases when the number of process graphs (p) rises as demonstrated in Fig. 6. The figure also shows regardless of the parameter changes, the aggregate approach outputs about three times more results than similarity approach.

We also study the scalability of our algorithms with regards to the different experiment setting [5], which show that the indexing technique is effective and reduces the processing time.

6   Related Work

In the domain of Business Process Management (BPM), queries are processed over BPMN (Business Process Modelling Notation) or equivalent notations. The main purpose of query processing is to extract actions (i.e., control flows). For example, the BPMN-Q is a visual language to query repositories of process models [1]. It processes the queries by matching a process model graph, converted from BPMN, to a query graph [8]. In [3], the authors describe the problem of retrieving process models in the repository that most closely resemble a given process model but they have focused on developing the similarity metrics rather than efficient implementation of algorithms. In the graph querying domain, subgraph similarity search is to retrieve the data graphs that approximately contain the query. For example, Grafil [12] defines graph similarity as the number of missing edges in a maximum common subgraph. One of the closely related work to ours is Cooking Graphs [10]. A cooking graph describes a cooking process with cooking actions and relevant ingredients information. However, cooking graphs are tailored to recipes. None of the above business process querying or graph querying approaches addresses aggregated search on graphs. A similar work presented in [2], however, they do not address querying of both control and data flow. Due to the structure of PPDG and flexible attribute of personal processes, two approaches of aggregated graph search proposed in [7,9] are not suitable for applying directly to query PPDG repositories.

7 Conclusion

In this paper, we have investigated aggregated search over Personal Process Description Graph (PPDG). We formally define the PPDG aggregated search and propose a novel approach based on our previous query techniques to query personal process descriptions by aggregation. A comprehensive experimental study over both real and synthetic datasets demonstrates the efficiency and scalability of our techniques.

References

1. Awad, A.: BPMN-Q: a language to query business processes. In: The 2nd International Workshop on EMISA, pp. 115–128 (2007)
2. Awad, A., Sakr, S., Kunze, M., Weske, M.: Design by selection: a reuse-based approach for business process modeling. In: Jeusfeld, M., Delcambre, L., Ling, T.-W. (eds.) ER 2011. LNCS, vol. 6998, pp. 332–345. Springer, Heidelberg (2011)
3. Dijkman, R., Dumas, M., van Dongen, B.F., Käärik, R., Mendling, J.: Similarity of business process models: metrics and evaluation. Inf. Syst. 36(2), 498–516 (2011)
4. Hsu, J.O., Paik, H., Zhan, L.: Similarity search over personal process description graph. In: Wang, J., et al. (eds.) WISE 2015, Part I. LNCS, vol. 9418, pp. 522–538. Springer, Heidelberg (2015). doi:10.1007/978-3-319-26190-4_35
5. Hsu, J.O., Paik, H., Zhan, L., Ngu, A.H.H.: Aggregated search over personal process description graph. UNSW Computer Science and Engineering Technical report no. UNSW-CSE-TR-201608 (2016)
6. Kopliku, A., Pinel-Sauvagnat, K., Boughanem, M.: Aggregated search: a new information retrieval paradigm. ACM Comput. Surv. 46(3), 41:1–41:31 (2014)
7. Le, T.-H., Elghazel, H., Hacid, M.-S.: A relational-based approach for aggregated search in graph databases. In: Lee, S., Peng, Z., Zhou, X., Moon, Y.-S., Unland, R., Yoo, J. (eds.) DASFAA 2012, Part I. LNCS, vol. 7238, pp. 33–47. Springer, Heidelberg (2012)
8. Sakr, S., Awad, A.: A framework for querying graph-based business process models. In: WWW, pp. 1297–1300 (2010)

9. Sakr, S., Awad, A., Kunze, M.: Querying process models repositories by aggregated graph search. In: La Rosa, M., Soffer, P. (eds.) BPM Workshops 2012. LNBIP, vol. 132, pp. 573–585. Springer, Heidelberg (2013)
10. Wang, L.: CookRecipe: towards a versatile and fully-fledged recipe analysis and learning system. Ph.D. thesis, City University of Hong Kong (2008)
11. Xu, J., Paik, H., Ngu, A.H.H., Zhan, L.: Personal process description graph for describing and querying personal processes. In: Sharaf, M.A., Cheema, M.A., Qi, J. (eds.) ADC 2015. LNCS, vol. 9093, pp. 91–103. Springer, Heidelberg (2015)
12. Yan, X., Yu, P.S., Han, J.: Substructure similarity search in graph databases. In: SIGMOD, pp. 766–777 (2005)

Inferring Lurkers' Gender by Their Interest Tags

Peisong Zhu[1], Tieyun Qian[1(✉)], Zhenni You[1], and Xuhui Li[2]

[1] State Key Laboratory of Software Engineering, Wuhan University, Wuhan, China
{zhups24,qty,znyou}@whu.edu.cn
[2] School of Information Management, Wuhan University, Wuhan, China
lixuhui@whu.edu.cn

Abstract. Gender prediction has evoked great research interests due to its potential applications like personalized search, targeted advertisement and recommendation. Most of the existing studies rely on the content texts to build feature vector. However, there is a large number of lurkers in social media who do not post any message. It is unable to extract stylistic or syntactic features for these users as they do not have content information. In this paper, we present a novel framework to infer lurkers' gender by their interest tags. This task is extremely challenging due to the fact that each user only has a few (usually less than 10) and diverse tags. In order to solve this problem, we first select a few tags and classify them into conceptual classes according to social and psycholinguistic characteristics. Then we enlarge the conceptual class using an association mining based method. Finally, we use the conceptual class to condense the users feature space. We conduct experiments on a real data set extracted from Sina Weibo. Experimental results demonstrate that our proposed approach is quite effective in predicting lurkers' genders.

Keywords: Lurker's gender prediction · Interest tags · The conceptual class

1 Introduction

Gender classification has received considerable attention in recent years [3,5,8, 16,18,24]. Almost all existing methods use the content texts for classification. However, there is a large number of users in social media who register only for browsing, i.e., they do not have contents. For instance, a sample of 1.0 million users from Sina Weibo in China shows that about 7.4 % users do not post any message. We call this group of users the lurkers. In this work, we study the problem of inferring lurkers' gender in Sina Weibo. This problem is important because advertisement targeting or media audience analysis need to understand the full user population. Clearly, the group of lurkers cannot be ignored if its large population is taken into consideration.

Existing approaches on gender classification rely heavily on the lexical [3,4,19,20], syntactic [16], or stylistic features [5,7–9]. Without the content information, we are unable to extract the above mentioned features. Fortunately, the users in Sina Weibo have other information besides their posts.

S. Hartmann and H. Ma (Eds.): DEXA 2016, Part II, LNCS 9828, pp. 263–271, 2016.
DOI: 10.1007/978-3-319-44406-2_20

For example, about 56 % of users give their interest tags in their profile. Hence we aim to predict the gender of lurkers by their interest tags. This task is quite challenging since there are only a few tags for each user. Sina Weibo allows users to use at most ten tags and 63.37 % of the users have less than six tags indeed. Furthermore, the tags are often diverse. For instance, one user may use "Ferrari" and "Michael Schumacher"while the other chooses "Mclaren" and "Lewis Hamilton" as tags although both of them are male autofans. All these make the user's tag space very sparse and the classification task difficult. To deal with this problem, we propose a novel method to condense the tag feature space by the use of conceptual class. Specifically, we first select a few tags and classify them into conceptual classes according to social and psycholinguistic character-istics. Then we enlarge the conceptual class using an association mining based method. Finally, we add the conceptual class to each user's feature space.

The most related work to ours is the work of Bergsma and Durme [3], which also applied the conceptual classes to gender classification. However, their two conceptual classes are gender based and built on the top of syntactic analysis on the texts. At least 40 tweets are required for each user in their experiment. Since we do not have contents for lurkers, many of the words in their conceptual classes, for instance, wife and ex-boyfriend, are unlikely to appear as tags of a user. Finally, we expand the conceptual class using association mining rather than the mutual information method used in [3]. The computation of mutual information needs the class label information while association mining is independent of any class. This means that we can explore the tremendous unlabeled data in social media. We conduct experiments on a real data set from Sina Weibo. The results demonstrate the effectiveness of our proposed approach, both on the expansion of conceptual class and on the improvement of the classification.

The contributions of this paper are as follows:

1. We introduce an important problem to the field, i.e., inferring the lurkers' gender who do not have any content texts.
2. We propose a novel framework for gender classification which only uses the interest tags in Sina Weibo.
3. We design a conceptual class based method to effectively condense the sparse tag space, and then show on the concept level that users' interests are cate-gorized by their gender.

2   Related Work

We review the literature in gender detection in this section, organized by the data source and the feature set.

Data Source. Gender classification has been investigated in the contexts of various media. Early work focused on blogs [9,16,21], emails [5], telephone con-versations [8] and chat texts [19], and online reviews [18]. Recently, more interests are paid to the interactive social media like Facebook [11], YouTube [7,24], and twitter [2–4,14,17,20,23]. Different media have their own characteristics. In this work, we perform our study on Sina Weibo, which has distinct tag features.

Feature set. Almost all existing studies for gender classification use the content information. Among which, the word or character n-grams are the most widely used features [3–5,7,19]. There are also a number of stylistic features extracted from the content, including ratio of punctuation, capital letters, unique words [7], slang words [9], word or sentence length [7,9], conceptual class [3], and the POS sequence [16]. While previous studies show that the performance of gender classification can be enhanced by using the above features, a main hinder is that they mainly exist in the content texts. This means that they are not applicable to the special group of users in our study. Several studies extract features from the users profile such as first name [14], full name [4], description [4], and the screen name or user name [3,4]. All these methods are combined with the word based n-grams. The only exception is the one used in [2], which uses the first name, the screen name and profile color to construct feature space without borrowing any information from the content. Their evaluations show that the accuracy results obtained with first names are higher than those with colors and user names. There is no first name in Sina Weibo. Instead, we use the interest tags to infer the users' gender.

3   A Conceptual Class Based Method

3.1   Data Source and Tag Feature Vector

Our data is collected from Sina Weibo, one of the largest micro-blogging services in China. Each user in Sina Weibo has a profile, which has several fields, such as userid, screen name, gender, tags, description, the number ofbreak followers, followees, and messages. We start from a public domain dataset (see http:// www.nlpir.org/?action-viewnews-itemid-232) containing the profile information of 1 million users. From which we randomly collect 1000 certificated celebrities (500 female and 500 male) who have at least 1 tag for our experiments. The data are randomly split into two parts: 70 % for training and the rest 30 % for test. We extract the interest tags from each user's profile. The total number of tags in the training set is 81407.

While each tag is treated as a word in content texts when building the tag feature vector, the interest tags as a whole are significantly different from the microblogs. For example, the expressions in microblogs are often casual. There are a lot of emoticons, internet slangs, abbreviations in microblogs. Hence many stylistic features can be extracted from the texts. In contrast, people tend to use normal expressions for their interest tags and there are very few stylistic features. Futhermore, a long feature vector can be built by the aggregation of the multiple microblogs (if she has). This contradicts to the short tag feature vector.

Overall, the tag vector for each user is very short and the tag space is sparse. As a result, a number of users in test data may not appear in the tag space. Tag expansion is a possible way to deal with the problem. In the literature, a number of methods have been proposed for tag expansion such as association rule mining [10], neighbor voting [12], and bipartite graphs [22]. However, some of the methods need extra information like documents [13,22]. This makes them not applicable to our task. Although the association mining approach is effective for tag expansion in our work, it can not be directly used either. The reason is that the interest tags in Sina Weibo follow a power lower distribution. Our experiment shows that several tags have very high frequency but more than 99.4 % tags have a support value smaller than 0.001. The setting of minimum support becomes very difficult. If it is set high, only a few items can be found; and if low, too many items will bring combination explosion. Hence we develop a novel expansion method based on the conceptual class.

3.2   Building the Initial Conceptual Class

We first select about 1000 the most frequent tags from the data set. Then we classify them into conceptual classes based on the social and psycho-linguistic theory or observations.

1. Females pay more attentions to the family, and males pay more attentions to social and political affairs [15].
2. Females often define their jobs with gender information like "office lady" or "radio hostess", and males prefer to merely tell their jobs like "founder" or "CEO".
3. Females are the main consumers of the fashion, beauty and cosmetic products.
4. Males are more likely to become fans of cars, high technology, basket ball and football, and games than females.
5. There are some gender-reference tags for both male and female such as "material girl" and "diamond geezers".

We also add a few classes which are traditionally regarded as gender related such as sentiment and police law. Finally we define 27 conceptual classes (*CCs*) in total. The detailed categories will be given in the experimental part.

3.3   Expanding Conceptual Class

Since there are tremendous unlabeled data in social media, we can explore the unlabeled data to effectively expand our conceptual classes. In our case, we use all the tags in the raw data set as the unlabeled data U. Note that the tags in each user is treated as a transaction, and then an Apriori algorithm [1] is used to mine the frequent itemsets. The algorithm is shown in Algorithm 1.

Algorithm 1. Expanding Conceptual Class (ECC)

Input: The unlabeled data U, the initial conceptual class CC built in Sect. 4.1
Parameters: the maximum and minimum support threshold α, δ, the number of top ranked tags with high confidence K
Output: the expanded conceptual class CC
Steps:

1. Extract from U the tag vectors U' containing at least one tag in CC
2. Mine frequent itemsets FI from U', where $FI = \{x | x.supp > \delta, x$ is a tag$\}$
3. For each tag x, $x \notin CC$
4. $x.conf = 0$
5.   For each tag y, y $\in$ CC
6.   if $(x, y) \in FI$ and $x.supp < \alpha$
7.   then $x.conf \mathrel{+}= Pr(x|y)$
8.   if $x.conf > \beta$
9.   then $CC = CC \bigcup \{x\}$

We first use the tags in a conceptual class to filter the unrelated tag vectors, and then mine the frequent itemset meeting the minimum support threshold. For those tags not in the conceptual class, we sum their conditional probability with all tags already in the conceptual class. And the tags with a high confidence are merged into the conceptual class.

The maximum (δ) and minimum (α) support is set to 0.50 and 0.05 in our experiments. The rationale for the threshold of δ is similar to the widely used term frequency - inverse document frequency (tf_idf) statistic in information retrieval: if a word appears in a lot of documents, then it will be a common word and contributes little to the classification. In our case, if a tag is used by more than half of the total users, we then discard it as it is too general. The setting of α is determined after several trials. We try 0.01, 0.03, 0.05, 0.08 and 0.10. We find that 0.05 is the most appropriate value in our experiments. As for the confidence threshold β, we need manually check to decide a proper value for β. This is because it is hard to set one universal value for different conceptual classes. The other reason is that this process is actually very trivial for an annotator to filter. Hence we let this procedure to be manually done. This approach has also been adopted in the previous work [3].

Finally, we use the expanded conceptual class to condense the tag space for both the training and test data. We do this by treating each conceptual class as a pseudo word. For each tag in the class, if it appears in a vector, we then add one count for the pseudo word of the class. The rationale is to use the conceptual class to represent every tag it contains.

4 Experimental Evaluation

All our experiments use the libsvm classifier [6] with default parameter settings. We report the classification accuracy as the evaluation metric.

4.1 Effects of Minimum Support

We compare the number of tags in several expanded conceptual classes under different minimum supports. The results are listed in Table 1.

Table 1. The number of tags in expanded conceptual classes under different minimum supports

	$\delta = 1\%$	$\delta = 3\%$	$\delta = 5\%$	$\delta = 8\%$	$\delta = 10\%$
Basketball	1546	131	106	65	61
Beauty	2141	103	71	28	6
Car	2536	354	116	53	34

On one hand, when δ is set to a small value like 1 %, a great number of tags are added to the conceptual class. For example, there are 2536 tags in the *Car* conceptual class. This not only adds many noises into the conceptual class but also results extra efforts for checking. On the other hand, when δ is set to a big value, only a few tags can be founded. Hence we decide to use a value of 5 % in the following experiments.

4.2 Effects of Class Expansion

We compare the number of tags between the original and the expanded conceptual classes. The results are listed in Table 2.

Table 2. The comparison of the number of tags in conceptual classes

Class	Ori	Exp	Class	Ori	Exp	Class	Ori	Exp
family	29	85	beauty	48	83	fasion	25	73
cosmetic	23	111	sentiment	42	45	luxury	23	50
gender(F)	77	77	job(F)	17	47	hobby(F)	63	69
gender(M)	43	43	job(M)	53	53	hobby(M)	41	46
sports(F)	12	15	sports(M)	58	78	politics	13	35
game	26	52	basketball	8	20	football	7	18
social	32	171	science	8	8	culture	17	54
car	79	152	comic	14	22	e-business	19	121
technology	37	83	police-law	6	69	pets	11	17

Most of the conceptual classes are enlarged a lot. For example, the number of tags in *Social* class is augmented from 32 to 171. There are several classes unexpanded, including *Science*, *job(F)*, *gender(F)*, and *gender(M)*. The initial tags in these three classes are already very detailed and thus it is difficult to find counterparts for them. We further give sample conceptual classes in Fig. 1. We can see that tags in the expanded classes are highly correlated with those in the initial class. This shows that with the help of the conceptual class, the tag expansion is restricted in a narrow field and thus capable of finding very similar or related tags.

Concept Class	Initial Tags	Expanded Tags
Beauty	美白 (whitening) 化妆 (makeup) 臭美 (funky) 面膜 (facial mask) 保湿 (moisturizing)	瘦身 (slimming) 健康 (health) 祛斑 (freckle-removing) 化妆 (makeup) 丰胸 (breast augmentation) 护肤品 (skin-care product) 整形 (plastic surgery) 整形美容 ((plastic surgery and facial) 瘦脸减肥 (face-slimming) 美发 (hairdressing)
Family	双胞胎 (twins) 亲子 (parenthood) 育儿 (child rearing) 宝宝 (baby) 两性关系 (sexual relations)	妈妈 (mother) 早教 (early education) 绘本 (picture book) 母婴用品 (maternal and child products) 怀孕 (pregnancy) 孕妇 (pregnant woman) 生活 (life) 婴儿 (infant) 胎教 (fetal education) 育儿知识 (parenting knowledge)
Cosmetic	雅诗兰黛 (Estee Lauder) 兰蔻 (Lancome) 欧莱雅 (Loreal) 曼秀雷敦 (Mentholatum) 美即 (MG)	面膜 (facial mask) 护肤品 (skin-care product) 相宜本草 (INOHERB) 美白 (whitening) 欧树 (NUXE) 药妆 (cosmeceutical) 迪奥 (Dior) 资生堂 (Shiseido) 古素 (GUSUBUY) 欧舒丹 (L'OCCITANE)
Car	朗逸 (Lavida) 速腾 (Sagitar) 轩逸 (Sylphy) 捷达 (Jetta) 凯越 (Excelle)	cc polo 一汽大众 (FAW Volkswagen) 上海大众 (Shanghai Volkswagen) 锐志 (Reiz) 皇冠 (CROWN) 花冠 (COROLLA) 东风日产 (Dongfeng-Nissan) 丰田 (Toyota) 雅力士 (Yaris)
E-business	软件外包 (software outsourcing) SEO 专利 (patent) SAP 外贸 (Foreign trade) B2C	移动互联网 (mobile internet) lbs 物流 (logistics) b2c 微博营销 (microblog marketing) 电商 (E-commerce) 互联网创业 (internet entrepreneurs) 互动营销 (interactive marketing) it 培训 (IT training) 网赚项目 (Money-making project on Internet)
Technology	计算机 (computer) 网络广告 (online advertising) HTC 电子产品 (electronic product) Java	移动互联网 (mobile internet) seo java 云计算 (cloud computing) mysql mac ubuntu 软件 (software) 科技 (science and technology) 程序员 (programmer)

Fig. 1. Sample of expanded conceptual classes

4.3 Effects of Conceptual Class on Gender Classification

We show the effects conceptual class on gender classification in Table 3. For comparison, we use the character n-gram (n=1..3) in users' screenname and their tags as the baselines.

We can see that 1-gram is the best among the n-grams, both for the screenname and tags. We also find that, while the tag vector performs worse than tag 1-gram, its performance is greatly improved and beats tag 1-gram after using conceptual class. This clearly demonstrate that the proposed approach is very effective to enhance the performance of gender classification.

Table 3. Effects of conceptual class on gender classification

Screenname		Tag		Tag vector using SVM	
char 1-gram	63.67	char 1-gram	68.33	no conceptual class	65.33
char 2-gram	30.33	char 2-gram	65.33	original conceptual class	70.67
char 3-gram	30.33	char 3-gram	63.67	expanded conceptual class	71.33

5   Conclusion

We study a new problem of inferring the gender of lurkers by their interest tags and present a novel framework to solve this problem. We first initialize a set of conceptual classes. We then expand these classes by applying association rule mining on the unlabeled data. Finally, we use the conceptual class to condense the tag vector in this class. The results demonstrate that our expansion approach is very effective in finding similar tags. More importantly, the conceptual class can significantly improve the classification accuracy. We hope this study will inspire the research interests in user profiling of lurkers.

Acknowledgment. The work described in this paper has been supported in part by the NSFC projects (61272275, 61272110, 61572376), and the 111 project(B07037).

References

1. Agrawal, R., Imielinski, T., Swami, A.: Mining association rules between sets of items in large databases. In: Proceedings of SIGMOD, pp. 207–216 (1993)
2. Alowibdi, J.S., Buy, U.A., Yu, P.: Empirical evaluation of profile characteristics for gender classification on twitter. In: Proceedings of the 12th ICMLA, pp. 365–369 (2013)
3. Bergsma, S., Durme, B.V.: Using conceptual class attributes to characterize social media users. In: Proceedings of ACL, pp. 710–720 (2013)
4. Burger, J.D., Henderson, J., Kim, G., Zarrella, G.: Discriminating gender on twitter. In: Proceedings of EMNLP, pp. 1301–1309 (2011)
5. Cheng, N., Chen, X., Chandramouli, R., Subbalakshmi, K.P.: Gender identification from e-mails. In: Proceedings of CIDM, pp. 154–158 (2009)
6. Fan, R.E., Chang, K.W., Hsieh, C.J., Wang, X.R., Lin, C.J.: Liblinear: A library for large linear classification. J. Mach. Learn. Res. **9**, 1871–1874 (2008)
7. Filippova, K.: User demographics and language in an implicit social network. In: Proceedings of EMNLP-CoNLL, pp. 1478–1488 (2012)
8. Garera, N., Yarowsky, D.: Modeling latent biographic attributes in conversational genres. In: Proceedings of IJCNLP, pp. 710–718 (2009)
9. Goswami, S., Sarkar, S., Rustagi, M.: Stylometric analysis of bloggers age and gender. In: Proceedings of ICWSM, pp. 214–217 (2009)
10. Heymann, P., Ramage, D., Garcia-Molina, H.: Social tag prediction. In: Proceedings of SIGIR, pp. 531–538 (2008)
11. Kosinski, M., Stillwell, D., Graepel, T.: Private traits and attributes are predictable from digital records of human behavior. PNAS **110**, 5802–5805 (2013)

12. Li, X., Snoek, C.G., Worring, M.: Automatic tag recommendation algorithms for social recommender systems. IEEE Trans. Multimedia **11**, 1310–1322 (2009)
13. Liu, D., Hua, X.S., Yang, L., Wang, M., Zhang, H.J.: Tag ranking. In: Proceedings of WWW, pp. 351–360 (2009)
14. Liu, W., Ruths, D.: Whats in a name? using first names as features for gender inference in twitter. In: Proceedings of AAAI Spring Symposium on Analyzing Microtext (2013)
15. Macaulay, R.K.: Talk that counts: Age, Gender, and Social Class Differences in Discourse. Oxford University Press, New York (2005)
16. Mukherjee, A., Liu, B.: Improving gender classification of blog authors. In: Proceedings of EMNLP, pp. 207–217 (2010)
17. Nguyen, D., Trieschnigg, D., Dogruoz, A.S., Grave, R., Theune, M., Meder, T., de Jong, F.: Why gender and age prediction from tweets is hard: Lessons from a crowdsourcing experiment. In: Proceedings of COLING, pp. 1950–1961 (2014)
18. Otterbacher, J.: Inferring gender of movie reviewers: Exploiting writing style, content and metadata. In: Proceedings of CIKM, pp. 369–378 (2010)
19. Peersman, C., Daelemans, W., Vaerenbergh, L.V.: Predicting age and gender in online social networks. In: Proceedings of SMUC, pp. 37–44 (2011)
20. Rao, D., Yarowsky, D., Shreevats, A., Gupta, M.: Classifying latent user attributes in twitter. In: Proceedings of SMUC, pp. 37–44 (2010)
21. Schler, J., Koppel, M., Argamon, S., Pennebaker, J.W.: Effects of age and gender on blogging. In: Proc. of AAAI Spring Symposium on Computational Approaches for Analyzing Weblogs, pp. 199–205 (2005)
22. Song, Y., Zhang, L., Giles, C.L.: Automatic tag recommendation algorithms for social recommender systems. ACM TWEB **5**, 4 (2011)
23. Wei, G., Lim, E.P., Zhu, F.: Characterizing silent users in social media communities. In: Proceedings of ICWSM, pp. 140–149 (2015)
24. Xiao, C., Zhou, F., Wu, Y.: Predicting audience gender in online content-sharing social networks. J. Am. Soc. Inf. Sci. Technol. (JASIST) **64**, 1284–1297 (2013)

Semantic Web and Ontologies

Data Access Based on Faceted Queries over Ontologies

Tadeusz Pankowski[✉] and Grażyna Brzykcy

Institute of Control and Information Engineering,
Poznań University of Technology, Poznań, Poland
{tadeusz.pankowski,grazyna.brzykcy}@put.poznan.pl

Abstract. We propose a method for generating and evaluating faceted queries over ontology-enhanced distributed graph databases. A user, who only vaguely knows the domain ontology, starts with a set of keywords. Then, an initial faceted query is automatically generated and the user is guided in interactive modification and refinement of successively created faceted queries. We provide the theoretical foundation for this way of faceted query construction and translation into first order monadic positive existential queries.

1 Introduction

In recent years, there is an increasing interest in developing database systems enriched with ontologies. The terminological component of the ontology can be used as a global schema providing an integrated global view over a set of local databases. A crucial issue is then a query language and a query paradigm. A standard way for querying graph databases (including RDF repositories) is SPARQL [13]. However, it is not a suitable language for end-users. Moreover, in order to formulate structural queries (e.g., in SPARQL or SQL), users have to know both the structure of the underlying ontology and the query language. In order to gain knowledge about the ontology, there is a need to query metadata. Only then, the metadata can be used to formulate queries concerning data. It can be expected that in order to progressively improve queries, the process of querying data and metadata can be iterative, can make a lot of trouble and be time consuming.

To avoid the aforementioned inconveniences, another query paradigms have been proposed, such as *keyword search* [10,17], and *faceted search* [14]. Keyword search is the most popular in information retrieval systems, but lately we observe also a widespread application of keyword search paradigm to structured and semistructured data sources [3]. Faceted search has emerged as a foundation for interactive information browsing and retrieval and has become increasingly prevalent in online information access systems, particularly for e-commerce and site search [14]. Especially significant in the faceted search is implementation of the browsing paradigm, allowing for exploring and expressing information needs in interactive and iterative ways [7,16]. Most importantly, the browsing and exploring concerns both the data and metadata.

S. Hartmann and H. Ma (Eds.): DEXA 2016, Part II, LNCS 9828, pp. 275–286, 2016.
DOI: 10.1007/978-3-319-44406-2_21

In this paper, we follow both the keyword and the faceted search paradigms proposing a method of creating faceted queries starting from a keyword query. As a keyword query we assume a partially ordered set consisting of keywords being ontology concepts (unary predicates) and constants. The partial ordering is induced by the order of keywords in the query. The response to a keyword query is a subgraph of the ontology graph covering the given set of keywords and preserving partial ordering of keywords in the keyword query. The subgraph is used to generate an initial faceted query, which is presented to the user in a form of a faceted interface. A user can interactively modify and refine the faceted query browsing and exploring the ontology by means of the faceted interface. The final faceted query is translated into a first order (FO) query which is evaluated in local databases storing the extensional component of the ontology (in a form of graph databases).

Besides providing a global schema, an ontology is used for: (a) guiding the creation of faceted queries, (b) supporting translation of faceted queries into FO queries, (c) query rewriting, (d) dealing with labeled nulls, (e) deciding about query propagation, and to (f) control consistency [11]. It can be shown, that a faceted query is equivalent to a FO monadic positive existential query in a tree-shaped form. This allows for very efficient execution with polynomial combined complexity (considering the size of ontology rules, sizes of local graph databases and size of queries) [1,11].

Related work: The faceted search has been surveyed in [5,14]. This paradigm was used for querying documents, databases and semantic data, e.g., [4,7,12,18]. Our work mostly relates to the results reported in [16] and [1]. In [16], the authors focused on browsing-oriented semantic faceted search supporting users in addressing their imprecise (fuzzy) needs. To this order, an extended facet tree has been proposed, which compactly captures both facets and facet values. In this case, faceted queries are equivalent to a subclass of FO conjunctive queries. Our formalization of faceted queries is rooted in [1], where faceted queries are equivalent to a subclass of FO positive existential queries. In [11], we proposed a way of answering faceted queries in a multiagent system. We discussed, how local agents consult with each other while evaluating queries, and we have shown that it is enough to propagate only boolean queries during this cooperation. The efficiency of query execution can be increased by asynchronous and parallel processing.

Contribution: The main novelties of the paper are: (1) we propose a method of generating faceted queries starting from a keyword query, and (2) we define semantics of faceted queries by translating them into FO faceted queries (FOFQ).

Paper outline: The paper is organized as follows. In Sect. 2, we review preliminaries and define the class of ontology under interest. A motivating running example and architecture of the system are presented in Sect. 3. In Sect. 4, we propose the way of defining faceted queries. Formal syntax and semantics of faceted queries are studied and illustrated in Sect. 5. In Sect. 6, we summarize the paper.

2 Preliminaries

Let UP, BP and Const be countably infinite sets of, respectively, *unary predicates* (denoted by A, B, C), *binary predicates* (denoted by R, S, T) and *constants* (denoted by a, b, c). In BP we distinguish type (to denote the relation *"type of"*) and = (to denote equality relation between constants). In Const we distinguish a subset LabNull of *labeled nulls*. For constants, which are not in LabNull, the *Unique Name Assumption* (UNA) holds, i.e., different constants in Const\LabNull represent different values (nodes). For labeled nulls the UNA is not required, i.e., different labeled nulls may represent the same value (node) [6].

A *graph database* is a finite edge-labeled and directed graph $\mathcal{G} = (\mathcal{N}, \mathcal{E})$, where $\mathcal{N} \subseteq$ Const $\cup$ UP is a finite set of *nodes*, and $\mathcal{E} \subseteq \mathcal{N} \times$ BP $\times \mathcal{N}$ is a finite set of labeled edges (or *facts*), such that: if $(n_1, R, n_2) \in \mathcal{E}$ and $R \in$ BP $\setminus \{$type$\}$, then $n_1, n_2 \in$ Const, if $(n_1, $type$, n_2) \in \mathcal{E}$ then $n_1 \in$ Const, and $n_2 \in$ UP. In first order (FO) logic, we use the following notation: $A(n)$ for $(n, $type$, A) \in \mathcal{E}$; $n_1 = n_2$ for $(n_1, =, n_2) \in \mathcal{E}$, and $R(n_1, n_2)$, for $(n_1, R, n_2) \in \mathcal{E}$.

Let $\Sigma = \Sigma_E \cup \Sigma_I$ be a finite subset of UP $\cup$ BP. An *ontology* with *signature* Σ is a triple $\mathcal{O} = (\Sigma, \mathcal{R}, \mathcal{G})$, where $\mathcal{R}$ and $\mathcal{G}$ are, respectively, a finite set of rules and a finite database graph, over Σ and Const. The pair $(\Sigma, \mathcal{R})$ is called the *terminological* component (or a TBox) of the ontology, while $\mathcal{G}$ is called the *assertional* component (or the ABox) of the ontology [2]. Predicates occurring in $\mathcal{G}$ are referred to as *extensional predicates*, and are denoted by Σ_E. The set $\Sigma_I = \Sigma \setminus \Sigma_E$ of predicates which are not in $\mathcal{G}$ are called *intentional predicates*. In practice, an ontology conforms to one of OWL 2 profiles [9]. In this paper, we restrict ourselves to rules of categories (1)–(11) listed in Table 1, last category, (12), is the category of *integrity constraints* [8].

A FO formula is a *monadic positive existential query* (MPEQ), if it has exactly one free variable and is constructed only out of: (a) atoms of the form $A(v)$, $R(v_1, v_2)$ and $v = a$; (b) conjunction ($\wedge$), disjunction ($\vee$), and existential quantification ($\exists$). A query $Q(\mathbf{x})$, where $\mathbf{x}$ is a tuple of free variables (empty for boolean queries), is *satisfiable* in $\mathcal{O} = (\Sigma, \mathcal{R}, \mathcal{G})$, denoted $\mathcal{O} \models Q(\mathbf{x})$ if there is a tuple $\mathbf{a}$ (empty for boolean queries) from Const, such that $\mathcal{G} \cup \mathcal{R} \models Q(\mathbf{a})$, where $\mathcal{G} \cup \mathcal{R}$ denotes all facts deduced from $\mathcal{G}$ using rules from $\mathcal{R}$. Then $\mathbf{a}$ is an *answer* to $Q(\mathbf{x})$ with respect to $\mathcal{O}$ (the empty tuple $\mathbf{a}$ denotes TRUE).

3 Running Example and Architecture

Let $\mathcal{O} = (\Sigma, \mathcal{R}, \mathcal{G})$, where: (1) $\mathcal{G} = \mathcal{G}_1 \cup \mathcal{G}_2 \cup \mathcal{G}_3$ (Fig. 1); (2) rules in $\mathcal{R}$ are of categories listed in Table 1, some of them are given in Table 2; (3) $\Sigma = \Sigma_E \cup \Sigma_I$ is clear from the context.

Table 1. Categories of ontology rules

	Rule	Name	Representation
1	$B(x) \rightarrow A(x)$	subtype (subsumption)	$sub(B, A)$
2	$R(x, y) \rightarrow A(x)$	domain	$dom(R, A)$
3	$R(x, y) \rightarrow B(y)$	range	$rng(R, B)$
4	$R(x, a) \rightarrow A(x)$	specialization (by a constant)	$spec1(R, a, A)$
5	$R(x, y) \wedge B(y) \rightarrow A(x)$	specialization (by a type)	$spec2(R, B, A)$
6	$S(x, z) \wedge T(z, y) \rightarrow R(x, y)$	chain	$chain(S, T, R)$
7	$B(x) \wedge C(x) \rightarrow A(x)$	conjunction	$conj(B, C, A)$
8	$B(x) \wedge R(x, y) \rightarrow A(y)$	range (conditional)	$rngc(B, R, A)$
9	$S(y, x) \rightarrow R(x, y)$	inversion	$inv(S, R)$
10	$A(x) \wedge B(y_1) \wedge B(y_2) \wedge$ $R(x, y_1) \wedge R(x, y_2) \rightarrow$ $y_1 = y_2$	functionality	$func(A, R, B)$
11	$A(x_1) \wedge A(x_2) \wedge B(y) \wedge$ $R(x_1, y) \wedge R(x_2, y) \rightarrow$ $x_1 = x_2$	key (functionality of inversion)	$key(A, R, B)$
12	$A(x) \rightarrow \exists y \ R(x, y)$	existence	$exists(A, R)$

A system providing *data access based on faceted queries over ontologies* (DAFO) (Fig. 2) belongs to a class of Ontology-Based Data Access (OBDA) systems [15], and follows so called *single ontology approach*. Data in different local DAFO databases complement each other, can overlap but do not contradict one another. The union of all local databases is a consistent database.

A user interacts with the system using a faceted query interface (FQ Interface) (step 1) and is guided by an ontology $\mathcal{O} = (\Sigma, \mathcal{R}, \mathcal{G})$, stored in part in the global schema, $Sch = (\Sigma, \mathcal{R})$, and partly in local databases, DB_i, $1 \le i \le k$. As a result of the interaction, a faceted query is created. The query is translated into FOFQ query Q, and rewritten into Q' (step 2).

Table 2. Sample rules in $\mathcal{O}$ conforming to categories from Table 1

$org(x, ACM) \rightarrow ACMConf(x)$
$authorOf(x, y) \rightarrow Author(x) \wedge Paper(y)$
$atConf(x, y) \wedge ACMConf(y) \rightarrow ACMPaper(x)$
$authorOf(x, y) \wedge ACMPaper(y) \rightarrow ACMAuthor(x)$
$atConf(x, y) \wedge cyear(y, z) \rightarrow pyear(x, z)$
$authorOf(x, y) \rightarrow writtenBy(y, x)$
$Paper(x_1) \wedge Paper(x_2) \wedge String(y) \wedge title(x_1, y) \wedge title(x_2, y) \rightarrow x_1 = x_2$
$Author(x) \rightarrow \exists y \ authorOf(x, y)$

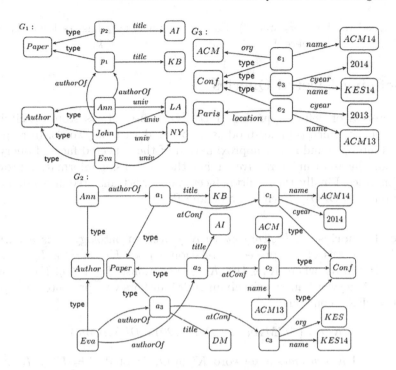

Fig. 1. A sample graph database consisting of three graphs

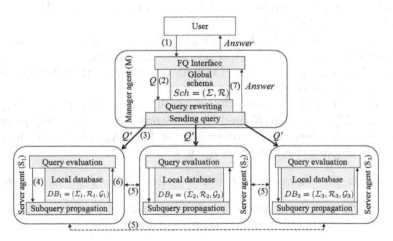

Fig. 2. Architecture of DAFO system

Then, Q' is sent to all server agents (step 3). An agent do some local database specific rewritings and evaluations (step 4), propagates (if necessary) some boolean requests to partner agents (step 5), and gathers local answers (step 6). Finally, answers obtained from server agents are collected by the manager agent and returned to the user (step 7). Each server agent S_i has its local database $DB_i = (\Sigma_i, \mathcal{R}_i, \mathcal{G}_i)$, where $\Sigma_i \subseteq \Sigma$, and $\mathcal{R}_i \subseteq \mathcal{R}$, $1 \leq i \leq k$.

4   Defining Faceted Queries

In the process of defining queries in DAFO, a user starts from specifying a keyword query, which is understood as an ordered set of keywords. In response, a faceted interface and a first approximation of the expected faceted query, are generated. The user can interactively refine the query using information provided by the interface. Finally, the resulting faceted query is translated into a first order faceted query (FOFQ), which is a monadic PEQ.

Keyword Queries. A *keyword query* KQ over an ontology $\mathcal{O}$ is a partially ordered (by means of the preceding relation $\prec$) set $KQ = (K_0, K_1, \ldots, K_q)$, $q \geq 0$, of keywords, where $K_0 \in \mathsf{UP}$, $K_i \in \mathsf{UP} \cup \mathsf{Const}$, $1 \leq i \leq q$. For example, the following keyword query asks about ACM authors who presented a paper in 2014 at a DEXA conference.

$$KQ = (ACMAuthor, Paper, 2014, DEXAConf). \tag{1}$$

A keyword K *subsumes* a keyword K' in $\mathcal{O}$, denoted $\mathcal{O} \models K' \sqsubseteq K$, iff: (1) if K and K' are constants, then $K = K'$; (2) if $K', K \in \mathsf{UP}$, then: (a) $K = K'$, or (b) $\mathcal{O} \models sub(K', K)$ (rule (1) Table 1), or (c) there is $A \in \mathsf{UP}$ such that $\mathcal{O} \models K' \sqsubseteq A$ and $\mathcal{O} \models A \sqsubseteq K$.

A sequence $s = (K_1, R_1, K_2, \ldots, R_{m-1}, K_m)$ is a path in $\mathcal{O}$ from K_1 to K_m, denoted $s \in path_\mathcal{O}(K_1, K_m)$, if: (1) $m = 1$; (2) if $K_i \in \mathsf{UP}$ then K_i is a domain of R_i, and a range of R_{i-1}; (3) if $K_i = a_i \in \mathsf{Const}$, then $\exists x R_i(a_i, x)$ and $\exists x R_{i-1}(x, a_i)$ are satisfied in $\mathcal{O}$.

We assume, that if $s \in path_\mathcal{O}(K_1, K_m)$, then also $s \in path_\mathcal{O}(K_1', K_m')$, for each K_1' and K_m', such that K_1 and K_m subsume K_1' and K_m', respectively, in $\mathcal{O}$. A sequence s preserves the ordering $K_i \prec K_j$ if K_i precedes K_j in s, and violates this ordering if K_j precedes K_i in s.

Definition 1. *The answer to KQ in $\mathcal{O}$ is a set $PSet$ of paths in $\mathcal{O}$ such that:*

- *any path starts with K_0 and ends with some $K \in KQ$,*
- *any path preserves ordering of keywords induced by KQ.*

Example 1. For the keyword query (1), $PSet$ can have five paths:
$s_1 = (ACMAuthor)$, $s_2 = (ACMAuthor, authorOf, Paper)$,
$s_3 = (ACMAuthor, authorOf, Paper, atConf, DEXAConf)$,
$s_4 = (ACMAuthor, authorOf, Paper, pyear, 2014)$,
$s_5 = (ACMAuthor, authorOf, Paper, atConf, Conf, cyear, 2014)$.
s_5 violates the preceding $2014 \prec DEXAConf$, and is removed from $PSet$.

From a given *PSet*, a set *TSet* representing *PSet* is created. Each path from *PSet*, longer than 1, is represented by a set of triples, and a path with length 1, with itself:

$$TSet = \cup\{tset(s) \mid s \in PSet'\},$$

$$tset(s) = \{A \mid s = (A)\} \cup \{(A, R, B) \mid (A, R, B) \in s\} \cup \{(A, R, a) \mid (A, R, a) \in s\}.$$

For example, $TSet = \{ACMAuthor, (ACMAuthor, authorOf, Paper),$ $(Paper, atConf, DEXAConf), (Paper, pyear, 2014), \dots\}$.

Creating Faceted Interface and Faceted Queries. Now, we discuss the way of creating faceted interfaces (FIs) and faceted queries (FQs) from a set *TSet* of triples representing the answer to a keyword query *KQ*. A FQ arises from a FI by selecting among alternatives offered by the FI.

Algorithm 1 specifies creation of FI (the upper part (FI)) and a selection procedure constituting a FQ (the bottom part (FQ)). In result, both FI and FQ are represented by the labeled tree T defined as the output of the algorithm.

A labeled tree in Fig. 3 represents a FI, and its underlined (selected) elements represent a FQ. The selection is done either by default (e.g., $\vee$) or is determined by the content of the underlying keyword query.

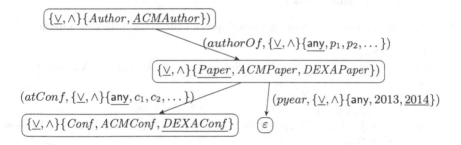

Fig. 3. FI and initial FQ generated by Algorithm 1 for the keyword query KQ (1)

In Fig. 4, there is the interface implemented in DAFO system. First, a keyword query is defined. Next, a FI and a FQ are generated as the answer to the keyword query. The query can be interactively modified by a user. For example, the labeled tree viewed in Fig. 4, represents the faceted query presented in Fig. 3 after some modifications (refinement).

The textual form of FQ in Fig. 4 is:

$$\Gamma = T_1[B_1 \wedge B_2/T_2[B_3/T_3 \wedge B_4]], \tag{2}$$

where: $T_1 = \{ACMAuthor\}$, $B_1 = (univ, \wedge\{NY, LA\})$, $B_2 = (authorOf, \{any\})$, $T_2 = \{Paper\}$, $B_3 = (atConf, \{any\})$, $T_3 = \vee\{ACMConf, DEXAConf\}$, $B_4 = (pyear, \{2014\})$.

Algorithm 1. Creating a faceted interface and a faceted query

---

Input: $\mathcal{O}$ – an ontology, $TSet$ – a set of triples being an answer to a keyword query $KQ = (K_0, K_1, \ldots, K_q)$

Output: A labeled tree $T = (r, V, E, \lambda_V, \lambda_E)$ representing a faceted interface (FI part) and a faceted query (FQ part), corresponding to $TSet$ and $\mathcal{O}$, where:

- V – a set of nodes, $r \in V$ – a distinguished root node,
- $E \subseteq V \times V$ – a set of ordered edges,
- λ_V – node labeling function,
- λ_E – edge labeling function.

(FI) Labeling functions for creating faceted interface:

1. $\lambda_V(r) = \{\vee, \wedge\}\{A \mid \mathcal{O} \models K_0 \sqsubseteq A\}$ – the set of all supertypes of K_0;
2. Let $e = (v_1, v_2) \in E$, $(A, R, B_i) \in TSet, 1 \leq i \leq k, \lambda_V(v_1) = \{\vee, \wedge\}X$, and $A \in X$, then
   - $\lambda_V(v_2) = \{\vee, \wedge\}\{B \mid \mathcal{O} \models rng(R, B)\}$ – the set of all ranges of R;
   - $\lambda_E(e) = (R, \{\vee, \wedge\}\{\text{any}\} \cup X)$, where $X = \{a \mid \mathcal{O} \models \exists x (A(x) \wedge R(x, a))\}$ – the set of all possible values of R.
3. Let $e = (v_1, v_2) \in E$, $(A, R, a_i) \in TSet, 1 \leq i \leq k, \lambda_V(v_1) = \{\vee, \wedge\}X$, and $A \in X$, then
   - $\lambda_V(v_2) = \varepsilon$;
   - $\lambda_E(e) = (R, \{\vee, \wedge\}\{\text{any}\} \cup X)$, where $X = \{a \mid \mathcal{O} \models \exists x (A(x) \wedge R(x, a))\}$ – the set of all possible values of R.

(FQ) Labeling functions for creating faceted query (selections in faceted interface):

1. $\lambda_V(r) = \vee\{K_0\}$;
2. Let $e = (v_1, v_2) \in E$, $(A, R, B_i) \in TSet, 1 \leq i \leq k, \lambda_V(v_1) = \vee X$, and $A \in X$, then
   - $\lambda_V(v_2) = \vee\{B_1, \ldots, B_k\}$;
   - $\lambda_E(e) = (R, \vee\{\text{any}\})$.
3. Let $e = (v_1, v_2) \in E$, $(A, R, a_i) \in TSet, 1 \leq i \leq k, \lambda_V(v_1) = \vee X$, and $A \in X$, then
   - $\lambda_V(v_2) = \varepsilon$;
   - $\lambda_E(e) = (R, \vee\{a_1, \ldots, a_k\})$.

Intuitively, $\vee\{ACMConf, DAXAConf\}$ denotes conferences, classified either as ACM or $DEXA$ conferences; $(atConf, \{\text{any}\})$ denotes papers which have been presented at any conference; $(univ, \wedge\{NY, LA\})$ denotes authors representing both universities, i.e., NY and LA university.

5   Formal Syntax and Semantics of Faceted Queries

Complex faceted queries, like (2), are built of *simple faceted queries* defined by Definition 3, which in turn refer to *simple faceted interfaces*.

Definition 2. *Simple faceted interfaces over an ontology $\mathcal{O}$ are:*

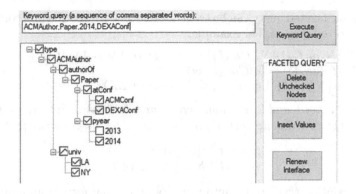

Fig. 4. A sample graphical form of a faceted query in DAFO

1. $F = (\text{type}, \{\vee, \wedge\}X)$, where $X \subseteq \mathsf{UP}$ – a simple type-based faceted interface,
2. $F = (R, (\{\vee, \wedge\}\{\text{any}\} \cup X))$, where $R \in \mathsf{BP}$, and $X \subseteq \{a \mid \mathcal{O} \models \exists x R(x, a)\}$ – a simple BP-based faceted interface.

Definition 3. *Simple faceted queries over simple faceted interfaces are:*

1. $\circ L$, where $L \subseteq X$ – over $F = (\text{type}, \{\vee, \wedge\}X)$, $\circ \in \{\vee, \wedge\}$ *denotes disjunctive* $(\vee)$ *and conjunctive* $(\wedge)$ *query;*
2. $(R, \{\text{any}\})$ *and* $(R, \circ L)$, where $L \subseteq X$ – over $F = (R, (\{\vee, \wedge\}\{\text{any}\} \cup X))$.

Definition 4. *Let T and B be simple* type- *and BP-based FQs, respectively. A (complex) FQ is an expression conforming to the syntax:*

$$\Gamma :: = T \mid T[\Delta]$$
$$\Delta :: = B \mid B/\Gamma \mid \Delta \wedge \Delta$$

Note, that the FQ Γ in (2) conforms to the above definition. A FQ in a tree form generated by means of Algorithm 1, can be translated into a FQ in the textual form defined by the grammar given in Definition 4. The translation is specified in Definition 5.

Definition 5. *Let $T = r((v_1, T_1), \ldots, (v_k, T_k))$ be a tree form of FQ, where T_i is a subtree with a root v_i. Translation $\tau(T)$ is defined recursively as follows:*

$$\tau(T) = \lambda_V(r)[\kappa(e_1, T_1) \wedge \cdots \wedge \kappa(e_k, T_k)], \quad \text{where } e_i = (r, v_i),$$

$$\kappa(e, T) = \begin{cases} \lambda_E(e) & \text{if } \lambda_V(T) = \varepsilon, \\ \lambda_E(e)/\tau(T) & \text{otherwise.} \end{cases}$$

In order to define semantics, the faceted queries will be represented by means of *atomic faceted queries*.

Definition 6. *An atomic faceted query is an unary predicate $A \in \mathsf{UP}$, a pair (R, any), and a pair (R, a), where $R \in \mathsf{BP}$.*

Any simple faceted query can be translated into a disjunction or a conjunction of atomic faceted queries. For example:

- $tr(\vee\{ACMConf, DAXAConf\}) = ACMConf \vee DAXAConf$,
- $tr((atConf, \{\text{any}\})) = (atConf, \text{any})$,
- $tr((univ, \wedge\{NY, LA\})) = (univ, NY) \wedge (univ, LA)$.

Definition 7. *Let t and b be atomic* type- *and* BP-*based FQs, respectively. A (complex) FQ in the atomic normal form is defined by the grammar ($\circ \in \{\vee, \wedge\}$):*

$$\alpha :: = t \mid t[\beta] \mid \alpha \circ \alpha \mid (\alpha)$$
$$\beta :: = b \mid b/\alpha \mid \beta \circ \beta \mid (\beta)$$

The translation $tr(\Gamma)$ of (2) into the atomic normal form results in:

$$\sigma = t_1[b_1 \wedge b_2 \wedge b_3/t_2[b_4/(t_3 \vee t_4) \wedge b_5]], \tag{3}$$

where: $t_1 = ACMAuthor$, $b_1 = (univ, NY)$, $b_2 = (univ, LA)$, $b_3 = (authorOf, \text{any})$, $t_2 = Paper$, $b_4 = (atConf, \text{any})$, $t_3 = ACMConf$, $t_4 = DEXAConf$, $b_5 = (pyear, 2014)$,.

Semantics for FQs is defined by means of the semantic function $[\![\alpha]\!]_x$ that assigns to a FQ in the atomic normal form a first order monadic positive existential query, referred to as FOFQ. x is then the only free variable in FOFQ.

Definition 8. *The semantic function $[\![\alpha]\!]_x$ for FQs conforming to the grammar given in Definition 7, is as follows ($\circ \in \{\vee, \wedge\}$):*

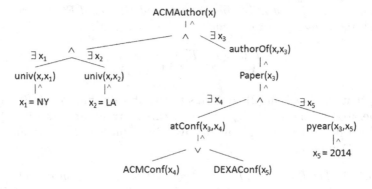

Fig. 5. Syntactic tree of FOFQ $[\![\sigma]\!]_x$, where σ is defined in (3)

$$
\begin{aligned}
[\![t]\!]_x &= t(x) & [\![(R, \text{any})]\!]_{x,y} &= R(x, y) \\
[\![t[b]]\!]_x &= [\![t]\!]_x \wedge \exists y([\![b]\!]_{x,y}) & [\![(R, a)]\!]_{x,y} &= R(x, y) \wedge y = a \\
[\![t[b/\alpha]]\!]_x &= [\![t]\!]_x \wedge \exists y([\![b/\alpha]\!]_{x,y}) & [\![b/\alpha]\!]_{x,y} &= [\![b]\!]_{x,y} \wedge [\![\alpha]\!]_y \\
[\![t[\beta_1 \circ \beta_2]]\!]_x &= [\![t[\beta_1]]\!]_x \circ t[\beta_2]]\!]_x & [\![\beta_1 \circ \beta_2]\!]_{x,y} &= \circ([\![\beta_1]\!]_{x,y}, [\![\beta_2]\!]_{x,y}) \\
[\![t[(\beta)]]\!]_x &= ([\![t[\beta]]\!]_x) & [\![(\beta)]\!]_{x,y} &= ([\![\beta]\!]_{x,y}) \\
[\![\alpha_1 \circ \alpha_2]\!]_x &= \circ([\![\alpha_1]\!]_x, [\![\alpha_2]\!]_x) \\
[\![(\alpha)]\!]_x &= ([\![\alpha]\!]_x)
\end{aligned}
$$

In general, a FQ σ in atomic normal form, can be expressed as a FOFQ $[\![\sigma]\!]_x$ of the form $A(x) \wedge \varphi(x)$, where $\varphi(x)$ is referred to as the *qualifier of the query*. For σ in (3), $[\![\sigma]\!]_x = ACM Author(x) \wedge \varphi(x)$, with the syntactic tree presented in Fig. 5.

In [11], we proposed a method for evaluating FOFQs in a multiagent system. Then a set of server agents (see Fig. 2) cooperate in answering the query.

6   Summary and Conclusions

We proposed a method of creating and evaluating faceted queries in an ontology-enhanced database. The ontology under consideration belongs to the class determined by OWL 2 RL profile, and serves many purposes (mainly, as the global schema, to query rewriting and to decide about query propagation). A user formulates a request starting from a keyword query which is used to generate an initial faceted query. The faceted query can be next modified and refined in interactive and iterative way. Finally, the query is translated into a first ordered query and answered by cooperating local agents. The main issue for future work concerns the way of presenting and browsing the information content in the process of faceted query creation. In particular, there is a need for: (1) creating hierarchies of value clusters, (2) inventing a way of presenting objects represented by null values, (3) adopting a method of compact representation of complex structures or complex contents. We are also planning to verify our approach in real-world applications. This research has been supported by Polish Ministry of Science and Higher Education under grant 04/45/DSPB/0149.

References

1. Arenas, M., Grau, B.C., Kharlamov, E., Marciuska, S., Zheleznyakov, D.: Faceted search over ontology-enhanced RDF data. In: ACM CIKM 2014, pp. 939–948. ACM (2014)
2. Baader, F., Calvanese, D., McGuinness, D., Nardi, D., Petel-Schneider, P. (eds.): The Description Logic Handbook: Theory. Implementation and Applications. Cambridge University Press, New York (2003)
3. Chen, Y., Wang, W., Liu, Z., Lin, X.: Keyword search on structured and semi-structured data. ACM SIGMOD **2009**, 1005–1010 (2009)
4. Dörk, M., Riche, N.H., Ramos, G., Dumais, S.T.: Pivotpaths: Strolling through faceted information spaces. IEEE Trans. Vis. Comput. Graph. **18**(12), 2709–2718 (2012)
5. Dumais, S.T.: Faceted Search. Encyclopedia of Database Systems. Springer, Heidelberg (2009)
6. Gottlob, G., Orsi, G., Pieris, A.: Ontological queries: Rewriting and optimization (extended version), pp. 1–25. CoRR abs/1112.0343 (2011)
7. Heim, P., Ertl, T., Ziegler, J.: Facet Graphs: Complex semantic querying made easy. In: Aroyo, L., Antoniou, G., Hyvönen, E., ten Teije, A., Stuckenschmidt, H., Cabral, L., Tudorache, T. (eds.) ESWC 2010, Part I. LNCS, vol. 6088, pp. 288–302. Springer, Heidelberg (2010)

8. Motik, B., Horrocks, I., Sattler, U.: Bridging the gap between OWL and relational databases. J. Web Semant. **7**(2), 74–89 (2009)
9. OWL 2 Web Ontology Language Profiles: www.w3.org/TR/owl2-profiles
10. Pankowski, T.: Keyword search in P2P relational databases. In: Agent and Multi-Agent Systems: Technologies and Applications (KES-AMSTA 2015). Smart Innovation Systems and Technologies, vol. 38, pp. 325–335. Springer, Heidelberg (2015)
11. Pankowski, T., Brzykcy, G.: Faceted query answering in a multiagent system of ontology-enhanced databases. In: Jezic, G., Jessica Chen-Burger, Y.-H., Howlett, R.J., Jain, L.C. (eds.) Agent and Multi-Agent Systems: Technology and Applications. SIST, vol. 58, pp. 3–13. Springer, Heidelberg (2016)
12. Papadakos, P., Tzitzikas, Y.: Hippalus: Preference-enriched faceted exploration. In: Workshops of the EDBT/ICDT. CEUR Workshop Proceedings, vol. 1133, pp. 167–172. CEUR-WS.org (2014)
13. SPARQL Query Language for RDF: (2008). http://www.w3.org/TR/rdf-sparql-query
14. Tunkelang, D.: Faceted Search. Morgan & Claypool Publishers, San Rafael (2009)
15. Wache, H., Vögele, T., Visser, U., Stuckenschmidt, H., Schuster, G., Neumann, H., Hübner, S.: Ontology-based integration of information - a survey of existing approaches. IJCAI **2001**, 108–117 (2001)
16. Wagner, A., Ladwig, G., Tran, T.: Browsing-oriented semantic faceted search. In: Hameurlain, A., Liddle, S.W., Schewe, K.-D., Zhou, X. (eds.) DEXA 2011, Part I. LNCS, vol. 6860, pp. 303–319. Springer, Heidelberg (2011)
17. Wang, H., Aggarwal, C.C.: A survey of algorithms for keyword search on graph data. In: Aggarwal, C.C., Wang, H. (eds.) Managing and Mining Graph Data. ADS, vol. 40. Springer, Heidelberg (2010)
18. Zhuge, H., Wilks, Y.: Faceted search, social networking and interactive semantics. World Wide Web **17**(4), 589–593 (2014)

Incremental and Directed Rule-Based Inference on RDFS

Jules Chevalier[✉], Julien Subercaze, Christophe Gravier,
and Frédérique Laforest

Laboratoire Hubert Curien UMR 5516, Univ Lyon, UJM-Saint-Etienne, CNRS,
42023 Saint Etienne, France
{jules.chevalier,julien.subercaze,christophe.gravier,
frederique.laforest}@univ-st-etienne.fr

Abstract. The Semantic Web contributes to the elicitation of knowledge from data, and leverages implicit knowledge through reasoning algorithms. The dynamic aspect of the Web pushes actual batch reasoning solutions, providing the best scalability so far, to upgrade towards incremental reasoning. This paradigm enables reasoners to handle new data as they arrive. In this paper we introduce Slider-p, an efficient incremental reasoner. It is designed to handle streaming expanding data with a growing background knowledge base. Directed reasoning implemented in Slider-p allows to influence the order of inferred triples. This feature, novel in the state of the art at the best of our knowledge, enables the adaptation of Slider-p's behavior to answer at best queries as the reasoning process is not over. It natively supports ρdf and RDFS, and its architecture allows to extend it to more complex fragments with a minimal effort. Our experimentations show that it is able to influence the order of the inferred triples, prioritizing the inference of selected kinds of triples.

Keywords: Incremental reasoning · Rule-based reasoning · Directed inference

1 Introduction

Reasoning is inherently a complex process, and while a large body of work in the area of reasoning algorithms and systems work and scale well in confined environment [5,6,11], the distributed and dynamic nature of Web creates new challenges for reasoning. This calls for new techniques to replace batch processing – where the arrival of new data re-initiates the reasoning process form scratch – to incremental reasoning [1]. This allows to handle new data as soon as they arrive, without re-inferring the previously inferred knowledge. In this paper, we consider reasoning in a forward chaining mode, with materialisation.

Several solutions have been proposed to optimise the incremental materialisation of ontologies. [4] proposes a technique to maintain the classification of ontologies as they evolve, and provides encouraging results. However, it is not a

© Springer International Publishing Switzerland 2016
S. Hartmann and H. Ma (Eds.): DEXA 2016, Part II, LNCS 9828, pp. 287–294, 2016.
DOI: 10.1007/978-3-319-44406-2_22

viable solution in case of static hierarchy of ontologies – i.e., if the hierarchy is not affected by modification. Moreover, it is not adapted for ontologies with a high number of nominals. [8] handles both addition and deletion in the setting of incremental classification and is generic to the fragment used for the inference. It is however limited to the classification on the TBox, and dedicated to a specific ruleset.

The major drawbacks that state-of-the art approaches suffer from is the inability to deal with complex ontologies and the fact they are not tailored to deal with large amount of dynamic RDF data and particularly large A-Boxes.

To overcome these drawbacks, we introduced Slider [3], an efficient reasoner to perform incremental reasoning. It is a parallel solution, generic face to the inference rules used. It limits the impact of duplicates generation and its data structures are optimised for both performance and space consumption. Finally, Slider allows to prioritize some triples based on their kind during the inference. Its core features that stand it apart from the previous approaches are as follows.

Parallel and Scalable Execution: We implement a parallel and scalable method to perform incremental reasoning. Each inference rule is mapped to an independent module. These modules receive triples according to defined rules and distribute infered triples to other modules for further processing.

Fragment's Customization: Slider natively supports both RDFS and ρdf [9] fragments, and its architecture allows it to be extended to any other fragment.

Duplicates Limitation: Reasoning could result in a massive amount of duplicate data that causes decrease in performance. To avoid such situation we use a vertical partitioning approach along-with multiple indexing (on predicates, subjects and objects) technique.

Dedicated Data Structures: To enhance scalability, the triplestore has been designed to minimize the space needed for the triples storage and to allow fast triples retrieval thanks to the vertical partitioning. The triples elements (subject, predicate and object) are stored as integers and a dictionary allows to retrieve original values.

In this paper, we present Slider-p, an extension of Slider [3] that allows rules prioritization. Incremental reasoning aims to be executed in a continuous process, updating the materialisation of the knowledge base as new data are sent to the system. In this case, it is impossible to guarantee that any implicit knowledge contained in the knowledge base has been inferred when a punctual query occurs. Slider-p proposes an inference mode, to prioritize the kinds of knowledge inferred first, in order to maximize the probability to answer correctly the query.

The rest of the paper is organized as follows: Sect. 2 describes the system architecture. Section 3 presents the directed inference principle that allows to prioritize rules. Section 4 presents the experimental results of a large evaluation campaign including batch, incremental and directed inference over standard datasets. Finally Sect. 5 concludes.

2   System Architecture

Figure 1 shows the high-level architecture of the system with three inference rules R_1, R_2 and R_3. Incoming triples are send to both the triplestore and the buffer of certain rule modules – where each module accepts the triples according to rules' predicates. Once a buffer exceeds its predifined size or timeout, it initiates a new instance of a rule module that applies its rule on buffered triples using also relevant triples stored in the triplestore. Newly created instances are managed by the thread pool for load distribution and scalability of the system. The distributor collects inferred data to be used as an input, and identifies the modules that require the resulted data for their inference process based on a rules dependency graph – thus ensuring completeness. For instance, the result of R_1 is used by R_2, R_3 and R_1 itself. Distributors and buffers play an important role in the architecture. Buffers orchestrate the load distribution of triples and the creation of rule modules instances when required.

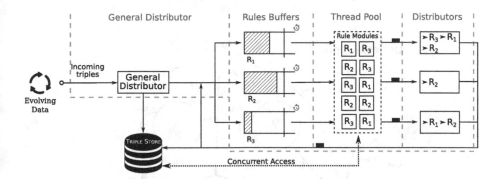

Fig. 1. Global architecture of Slider

A more detailed description of the architecture can be found in [3].

3   Directed Inference

The classic method for reasoning aims to explicit all the implicit triples from the knowledge base as fast as possible. This method is not sufficient in a continuous reasoning system. Reasoning is a complex process and, as new data constantly arrive, it is not conceivable to wait for the end of the process to answer a punctual request.

To address this issue, we propose directed inference. It has been implemented in Slider-p. Its goal is to infer first the triples used by a future punctual query, to maximize the probability to answer correctly this query. The execution of the rules inferring target triples are prioritized. To do so, a smaller buffer size and a smaller timeout are assigned to these rules, leading to a more frequent execution of them.

More practically, a hierarchy is defined between rules of the fragment, based on a level assigned to each rule. The following steps are used to associate a level to all the fragment's rules, using the rules dependency graph:

1. The rules that directly infer the targeted triples have level **1**;
2. the rules parent of the rules with the highest level l have level $l+1$;
3. repeat step 2 until there is no rule without a level parent of a rule with a level;
4. the remaining rules with no level have the highest level assigned plus one.

Rules with level **1** are the most important ones, and rules with the biggest level are the less important ones. Figure 2 shows an example of hierarchy to prioritize the triples with **type** as predicate on ρdf. The rules **prp-dom**, **prp-rng** and **cax-sco** can infer triples with predicate **type**. Their level is set to **1**. The rules parents of the rules of level **1** are **scm-sco** and **prp-spo1**. Their level is set to **2**. There are no rules without level parent of a rule with the level **2**. The remaining rules, **scm-spo**, **scm-dom2** and **scm-rng2**, get level **3**.

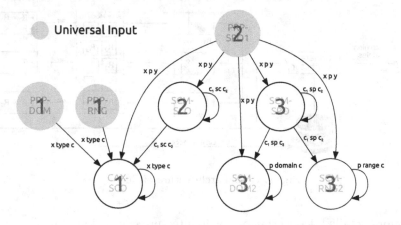

Fig. 2. Rules dependency graph with level to prioritize **type** triples on ρdf

Once a level is assigned to each rule, the Eq. 1 is used to compute the size of the buffer for a rule with a level $i > 1$. $BufferSize_1$ determines the size of the buffer for a rule with the level 1, i is the level of the rule, and α is a coefficient. The timeout is calculated similarly with Eq. 2.

$$BufferSize_i = \lfloor \alpha \; BufferSize_1 \; log(i) + BufferSize_1 \rfloor \qquad (1)$$

$$Timeout_i = \lfloor \alpha \; Timeout_1 \; log(i) + Timeout_1 \rfloor \qquad (2)$$

The coefficient α allows to tune the importance of the prioritization. The bigger it is, the higher the difference of buffer size and timeout between two levels is important.

4 Experimentations

Our experimental settings use four ontologies, listed in Table 1. The first one is generated using the Berlin SPARQL Benchmark (BSBM) [2] to generate a 5 million triples ontology. This ontology shows Slider-p ability to handle high rates of data by producing smaller set of inferred triples during reasoning process. The second ontology is made of a chain of a thousand `subClassOf` relations. This ontologies is easy to generate but provides the utmost practical interest due to its complexity. The chain of n rules produce $O(n^2)$ unique triples, however commonly used iterative rules schemes produce $O(n^3)$ triples [11]. The last category of ontologies contains real-world ontologies: one based on Wikipedia, and the other based on WordNet [10].

These ontologies are representative of synthetic data (issued from BSBM benchmark generator tool), extensive closure computation [7] (chained subsomptions), and ontologies of practical interest (Wordnet and Wikipedia).

We ran our experimentations on a standalone machine under Linux Ubuntu 12.04, with an AMD processor with 4 1.4 GHz cores, and 16 GB RAM.

Table 1. Ontologies used for the experimentations, with number of triples before inference, and after inference on ρdf and RDFS

Ontology	Input	Inferred with ρdf	Inferred with RDFS
BSBM5M	5000000	43212	1449107
wikipedia	458369	191574	555653
wordnet	473589	0	634692
subClassOf1000	2000	498501	499505

4.1 Incremental Performances

In this section we compete the incremental reasoning and the batch reasoning, using Slider-p. The inference is done on each ontology (i) incrementally by 10 % step of the total ontology by using the previously inferred knowledge and (ii) by restarting each time the inference process from the beginning, as in batch reasoning. We consider the ontologies BSBM5M, SubClassOf1000, Wikipedia and Wordnet in these experimentations. For these experimentations, all the rules have level **1**.

Figures 3 and 4 show the results of this experimentation. The shown inference times are the average for the four used ontologies. For each step, the sum of the inference times is also shown for the incremental reasoning.

The use of incremental reasoning is always interesting, and its performances are linear through the addition of new triples. The sum of the inference times for incremental reasoning is higher than for batch reasoning, but follows the same evolution as batch reasoning.

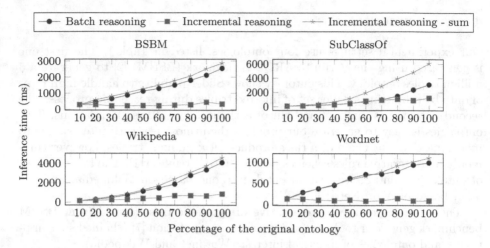

Fig. 3. Inference time comparison between batch and incremental reasoning, on ρdf

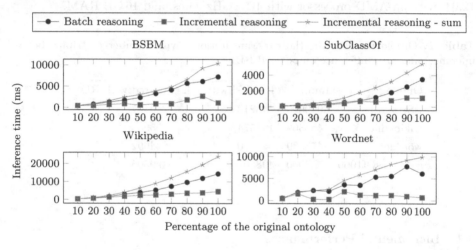

Fig. 4. Inference time comparison between batch and incremental reasoning, on RDFS

These results shows that incremental reasoning allows to update inferred knowledge faster than with batch reasoning –which has to restart the process from the beginning– while having a total cost similar to the cost of the batch processing.

4.2 Directed Inference

This section describes the experimentations conducted to validate the prioritization of the knowledge inferred during the reasoning process. We compare the number of triples inferred per time unit without prioritization, then prioritizing the triples with predicate type and finally prioritizing the triples with predicate `subClassOf`. We used the Wikipedia ontology for this experimentation: the inference on this ontology only generates triples with `type` or `subClassOf`. The experimentations have been executed for both ρdf and RDFS, with α set to 50.

The results of this experimentation are shown in Fig. 5. Without prioritization, the two kinds of triples are generated uniformly during the reasoning. With prioritization, prioritized triples are inferred first in all our experimentations. Certain cases enhance the performance of the reasoner, e.g., the prioritization of `subClassOf` triples on RDFS improves the inference time by about 600 %.

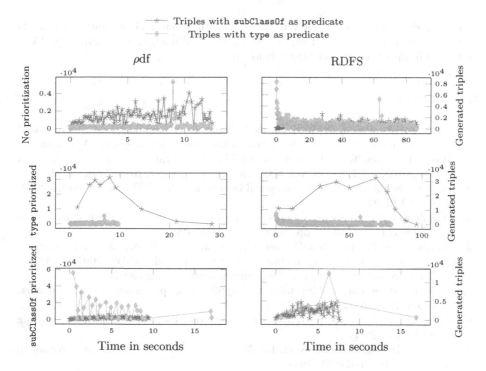

Fig. 5. Amount of triples inferred through time on ρdf and RDFS depending on the prioritized triple kind

5   Conclusion and future work

In the frame of reasoning on evolving triples, few proposals enable to continuously infer new knowledge as new explicit triple are sent to the reasoner. Most of the solutions limit the amount of data in the knowledge base by eliminating former triples. Instead of firing a full inference at a regular interval of time, we propose Slider-p, a reasoner that handles triples flows at the very core of its architecture. Its source code is available here: http://juleschevalier.github. io/slider. We evaluate Slider-p against 4 ontologies: an ontology generated from

BSBM, a specific one to the worst-case reasoning on subsumption relationship, and finally real field ontologies. Our experimentations on the incremental performances show that this paradigm is faster than batch reasoning, and has the same cost. We also proved experimentally that the directed inference allows to select a kind of knowledge to be inferred in priority. For our future endeavours, we will focus on two main aspects of Slider-p. First, we will implement more complex inference rules, to implement reasoning over more complex fragments. Second, we will implement a just-in-time optimisation of the rules execution's scheduling.

Acknowledgments. This work was funded by the French Fonds national pour la Société Numérique (FSN), and supported by Pôles Minalogic, Systematic and SCS, under the frame of the project OpenCloudware.

References

1. Barbieri, D.F., Braga, D., Ceri, S., Della Valle, E., Grossniklaus, M.: Incremental reasoning on streams and rich background knowledge. In: Aroyo, L., Antoniou, G., Hyvönen, E., ten Teije, A., Stuckenschmidt, H., Cabral, L., Tudorache, T. (eds.) ESWC 2010, Part I. LNCS, vol. 6088, pp. 1–15. Springer, Heidelberg (2010)
2. Bizer, C., Schultz, A.: The berlin sparql benchmark. Int. J. Semant. Web Inf. Syst. 5(3), 1–22 (2009)
3. Chevalier, J., Subercaze, J., Gravier, C., Laforest, F.: Slider: an efficient incremental reasoner. In: SIGMOD, pp. 1081–1086. ACM (2015)
4. Cuenca Grau, B., Halaschek-Wiener, C., Kazakov, Y.: History Matters: incremental ontology reasoning using modules. In: Aberer, K. (ed.) ASWC 2007 and ISWC 2007. LNCS, vol. 4825, pp. 183–196. Springer, Heidelberg (2007)
5. Dean, J., Ghemawat, S.: MapReduce: simplified data processing on large clusters. Commun. ACM 51(1), 107–113 (2008)
6. Hoeksema, J., Kotoulas, S.: High-performance distributed stream reasoning using s4. In: ISWC (2011)
7. Jagadish, H., Agrawal, R., Ness, L.: A study of transitive closure as a recursion mechanism. In: SIGMOD (1987)
8. Kazakov, Y., Klinov, P.: Incremental reasoning in OWL EL without bookkeeping. In: Alani, H. (ed.) ISWC 2013, Part I. LNCS, vol. 8218, pp. 232–247. Springer, Heidelberg (2013)
9. Muñoz, S., Pérez, J., Gutierrez, C.: Minimal deductive systems for RDF. In: Franconi, E., Kifer, M., May, W. (eds.) ESWC 2007. LNCS, vol. 4519, pp. 53–67. Springer, Heidelberg (2007)
10. Snasel, V., Moravec, P., Pokorny, J.: WordNet ontology based model for web retrieval. In: Web Information Retrieval and Integration (2005)
11. Urbani, J., Kotoulas, S., Maassen, J., van Harmelen, F., Bal, H.: OWL reasoning with WebPIE: calculating the closure of 100 billion triples. In: Aroyo, L., Antoniou, G., Hyvönen, E., ten Teije, A., Stuckenschmidt, H., Cabral, L., Tudorache, T. (eds.) ESWC 2010, Part I. LNCS, vol. 6088, pp. 213–227. Springer, Heidelberg (2010)

Top-k Matching Queries for Filter-Based Profile Matching in Knowledge Bases

Alejandra Lorena Paoletti$^{(\boxtimes)}$, Jorge Martinez-Gil, and Klaus-Dieter Schewe

Software Competence Center Hagenberg, Hagenberg, Austria
{Lorena.Paoletti,Jorge.Martinez-Gil,kd.schewe}@scch.at

Abstract. Finding the best matching job offers for a candidate profile or, the best candidates profiles for a particular job offer, respectively constitutes the most common and most relevant type of queries in the Human Resources (HR) sector. This technically requires to investigate top-k queries on top of knowledge bases and relational databases. We propose in this paper a top-k query algorithm on relational databases able to produce effective and efficient results. The approach is to consider the partial order of matching relations between jobs and candidates profiles together with an efficient design of the data involved. In particular, the focus on a single relation, the matching relation, is crucial to achieve the expectations.

1 Introduction

A profile describes a set of skills either, a person posses detailed in a curricula vitae (CV) or, a job advertisement details via the job description. In this line, profile matching concerns to measure how well a *given* profile matches a *requested* profile. Although, profile matching is not only concerned to the HR sector but a wide range of other application areas: real state domain, matching system configurations to requirements specifications, image similarity, etc.

With respect to querying knowledge bases in the HR domain, the commonly investigated approach is to find the best k (with $k \geq 1$) matches for a given profile, either a CV or a job offer [1]. This constitutes what is commonly known as top-k queries. Top-k queries have been thoroughly investigated in the field of databases, usually in the context of the relational data model [2,7]. The study of such queries in the context of knowledge bases has also been researched [6].

Top-k queries in relational databases are in general addressed by associating weights or aggregates acting as ranking the part of data relevant to the user's

The research reported in this paper was supported by the Austrian Forschungsförderungsgesellschaft (FFG) for the Bridge project "Accurate and Efficient Profile Matching in Knowledge Bases" (ACEPROM) under contract [**FFG: 841284**]. The research reported in this paper has been supported by the Austrian Ministry for Transport, Innovation and Technology, the Federal Ministry of Science, Research and Economy, and the Province of Upper Austria in the frame of the COMET center SCCH [**FFG: 844597**].

S. Hartmann and H. Ma (Eds.): DEXA 2016, Part II, LNCS 9828, pp. 295–302, 2016.
DOI: 10.1007/978-3-319-44406-2_23

needs, a potential join of the relevant relations involved and, a ranking (or sorting) of the tuples that constitutes the expected result set. Computing all these steps in a query is a high resource consuming process, depending on the design and the nature of data. The contribution of this paper takes benefits of the data structure supporting subsumption hierarchy as in rings and spiders known from network databases and reviewed in object-oriented databases. We make use of these structures, known for excellent performance in supporting queries that exploit hierarchical data structuring, in order to minimize the selection of tuples and the join of relations as well as eliminating the weighting and sorting of tuples on the query, by means of pre-weighting on the partial order of concepts of knowledge bases and pre-ordering of tuples via the matching measures.

The research in this paper is in line with a previous work [3] where we addressed techniques on improving profile matching and the novel idea of blow-up operators in knowledge bases (KB) in the HR sector. The starting point of this research is based on [4] where Popov and Jebelean exploited the representation of taxonomies by defining an asymmetric matching measure based on filters in KB. This has been further investigated in [5] by extending the ontology hierarchy with cross relations in the form of weighted directed edges.

The paper is organized as follows: In Sect. 2 we cover some aspects of our theory on profile matching introduced in [3] relevant to this work. Our contribution on data organization for top-k queries is introduced in Sect. 3. In Sect. 3.1 we introduce a relational database schema to implement top-k queries and in Sect. 3.2 we show an algorithm implementing our approach of top-k queries.

2   Preliminaries

We present in this section basic concepts from [3] that are fundamental to the representation of profile matching in the HR domain.

Concepts C_i in a TBox of a KB define a *lattice* $(\mathcal{L}, \leq)$ and we refer to concepts C_i in $\mathcal{L}$ to denote concepts C_i in a given KB. Thus, the terms TBox and lattice are used as synonyms from now on.

A *filter* in a lattice $(\mathcal{L}, \leq)$ is a non-empty subset $\mathcal{F} \subseteq \mathcal{L}$ such that for all C, C' with $C \leq C'$ whenever $C \in \mathcal{F}$ holds, then also $C' \in \mathcal{F}$ holds.

If $P \subseteq \mathcal{I}$ is a *profile*, P defines in a natural way a filter $\mathcal{F}$ of the lattice $\mathcal{L}$ of concepts. Therefore, for determining matching relations we can concentrate on filters $\mathcal{F}$ in a lattice.

If $(\mathcal{L}, \leq)$ is a lattice, and $\mathbb{F} \subseteq \mathcal{P}(\mathcal{L})$ denote the set of filters in this lattice. A *matching measure* is a function $\mu : \mathbb{F} \times \mathbb{F} \to [0,1]$ such that $\mu(\mathcal{F}_1, \mathcal{F}_2) = m(\mathcal{F}_1 \cap \mathcal{F}_2)/m(\mathcal{F}_2)$ with $\mathcal{F}_1, \mathcal{F}_2 \in \mathbb{F}$. If w is a *weight* associated to every concept $C \in \mathcal{L}$ then, a matching measure μ is defined by weights $w(C) = m(\{C\}) \in [0,1]$ such that

$$\mu(\mathcal{F}_1, \mathcal{F}_2) = \sum_{C \in \mathcal{F}_1 \cap \mathcal{F}_2} w(C) \cdot \left(\sum_{C \in \mathcal{F}_2} w(C) \right)^{-1} \tag{1}$$

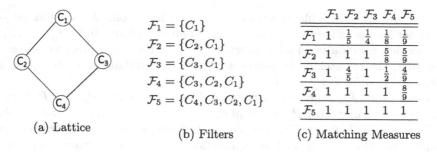

$$\mathcal{F}_1 = \{C_1\}$$
$$\mathcal{F}_2 = \{C_2, C_1\}$$
$$\mathcal{F}_3 = \{C_3, C_1\}$$
$$\mathcal{F}_4 = \{C_3, C_2, C_1\}$$
$$\mathcal{F}_5 = \{C_4, C_3, C_2, C_1\}$$

	$\mathcal{F}_1$	$\mathcal{F}_2$	$\mathcal{F}_3$	$\mathcal{F}_4$	$\mathcal{F}_5$
$\mathcal{F}_1$	1	$\frac{1}{5}$	$\frac{1}{4}$	$\frac{1}{8}$	$\frac{1}{9}$
$\mathcal{F}_2$	1	1	1	$\frac{5}{8}$	$\frac{5}{9}$
$\mathcal{F}_3$	1	$\frac{4}{5}$	1	$\frac{1}{2}$	$\frac{4}{9}$
$\mathcal{F}_4$	1	1	1	1	$\frac{8}{9}$
$\mathcal{F}_5$	1	1	1	1	1

(a) Lattice   (b) Filters   (c) Matching Measures

Fig. 1. A lattice, its filters and the matching measures

Example 1. A lattice with four elements: $\mathcal{L} = \{C_1, C_2, C_3, C_4\}$ defines up to five filters $\mathbb{F} = \{\mathcal{F}_1, \mathcal{F}_2, \mathcal{F}_3, \mathcal{F}_4, \mathcal{F}_5\}$, as shown in Fig. 1(a) and (b) respectively.

If we give some weights to the elements of $\mathcal{L}$, for instance $w(C_1) = \frac{1}{10}, w(C_2) = \frac{2}{5}, w(C_3) = \frac{3}{10}$, and $w(C_4) = \frac{1}{2}$ and calculate the matching measure $\mu(\mathcal{F}_i, \mathcal{F}_j)$ and $\mu(\mathcal{F}_j, \mathcal{F}_i)$ (for $1 \leq i, j \leq 5$) with the Formula in (1), we obtain the result shown in Fig. 1(c).

Note that in general, the matching measures are not symmetric. If $\mu(\mathcal{F}_g, \mathcal{F}_r)$ expresses how well a given filter $\mathcal{F}_g$ matches a required filter $\mathcal{F}_r$, then $\mu(\mathcal{F}_r, \mathcal{F}_g)$ measures the excess of skills in the given filter $\mathcal{F}_g$ that are not required in $\mathcal{F}_r$. And clearly, $\mu(\mathcal{F}_i, \mathcal{F}_j) = \mu(\mathcal{F}_j, \mathcal{F}_i) = 1$, when $i = j$ for $1 \leq i, j \leq 5$.

Example 2. Take for instance, $\mathcal{F}_r = \mathcal{F}_3$ as a required profile and two given filters $\mathcal{F}_{g_1} = \mathcal{F}_3$ and $\mathcal{F}_{g_2} = \mathcal{F}_4$. They are both equally and highly qualified for the requirements in $\mathcal{F}_r$ given their matching measures: $\mu(\mathcal{F}_{g_1}, \mathcal{F}_r) = \mu(\mathcal{F}_{g_2}, \mathcal{F}_r) = 1$. Although, if we consider the measures $\mu(\mathcal{F}_r, \mathcal{F}_{g_1}) = 1$ and $\mu(\mathcal{F}_r, \mathcal{F}_{g_2}) = \frac{1}{2}$, $\mathcal{F}_3$ matches better than $\mathcal{F}_4$ as C_2 is not part of the required skill set.

3   Internal Structure of Profile Matching

In a modeled selection process where there is a set of profiles $\mathbb{P}$, i.e., job and applicants profiles, defined by filters in a lattice $\mathcal{L}$, we denote by φ the conditions to be met by profiles in order to be selected, then $\mathbb{P}_\varphi$ denotes the set of profiles in $\mathbb{P}$ satisfying φ and $P_r \in \mathbb{P}$ is a *required* profile driving the selection by holding the conditions φ.

Note that, when referring to matching measures we refer to Formula (1) that includes weighting on the elements of the lattice $\mathcal{L}$, as shown in Example 1.

Definition 1. *For all $P \in \mathbb{P}_\varphi$ and $P' \in (\mathbb{P} - \mathbb{P}_\varphi)$, select λ profiles, where $\lambda \geq k$, out of the set of profiles $\mathbb{P}_\varphi$ such that P is selected and P' is not selected if $\mu(P, P_r) > \mu(P', P_r)$ and no subset of $\mathbb{P}_\varphi$ satisfy this property.*

In order to obtain the best-k matching profiles (either job or applicant profiles) we first need to query for filters representing those profiles.

Consider $\mathcal{F}_r$ being a filter representing the required profile P_r, a requested job profile for instance. Then, consider l being a number of filters in $\mathbb{F}$ $(\mathcal{F}_{g_1}, \ldots, \mathcal{F}_{g_l})$ representing candidates profiles matching P_r in a certain degree, satisfying φ such that, their matching measures are above a threshold $t_i \in [0,1]$ this is, $\mu(\mathcal{F}_{g_x}, \mathcal{F}_r) \geq t_i$ for $x = 1, \ldots, l$.

Then, every $\mathcal{F}_{g_x}$ represents a finite number j of profiles $(P_{g_1}, \ldots, P_{g_j})$, candidates profiles matching P_r, where $\mu(P_{g_y}, P_r) \geq t_i$ for $y = 1, \ldots, j$ and $j \leq k$.

The relation between filters in $\mathcal{L}$ and the number of *related* profiles represented by filters is defined by a function $\nu : \mathbb{N} \to \mathbb{N}$ where $\nu(x) = j$ and $\sum_{x=1}^{l} \nu(x) = \lambda$. Then any $\mathcal{F}_{g_{l+1}}$ *is not selected* as the matching value $\mu(\mathcal{F}_{g_{l+1}}, \mathcal{F}_r) < t_i$.

As for the second part of the definition, each filter $\mathcal{F} \in \mathbb{F}$ is uniquely determined by its minimal elements such that, we can write $\mathcal{F} = \{C_1, \ldots C_r\}$. Then, every profile represented by a filter is also uniquely determined by the elements in $\mathcal{F}$. Therefore, for any profile P'' in a subset of $\mathbb{P}_{\varphi}$ the matching value is $\mu(P'', P) < t_i$ then P'' does not satisfy the property.

Note that we used λ instead of k in function $\nu(x)$ as we consider the profiles $\lambda - k$ in $\mathbb{P}_{\varphi}$ need to be consider as well, as they satisfy $\mu(\mathcal{F}_{g_x}, \mathcal{F}_r) \geq t_i$. In other words, we choose to select all profiles with the same matching measure right above the threshold rather than cutting off exactly on k.

Example 3. Assume to have a job offer A and four candidates profiles $\{B, C, D, E\}$ that meet the requirements in A, where the five profiles are represented by the filters in Example 1 such that:

Filters representing Profiles	Matching Measures
$\mathcal{F}_4$ represents $\{A, B\}$	$\mu(B, A) = 1$
$\mathcal{F}_2$ represents $\{C\}$	$\mu(C, A) = 0.63$
$\mathcal{F}_3$ represents $\{D, E\}$	$\mu(D, A) = \mu(E, A) = 0.5$

If we choose $k = 3$ with $t_i = 0.5$ the final output is $\{B, C, D, E\}$ with $\lambda = 4$, providing all profiles represented by $\mathcal{F}_3$.

In order to obtain the best l filters satisfying φ, we first need to know the minimum matching value representing l filters. Thus, we start by selecting any t_i. If less than l solutions are found, we decrease t_i (t_{i-1}). If more than l solutions are found, we increase t_i (t_{i+1}). The search stops when the l filters satisfying $\mu(\mathcal{F}_{g_x}, \mathcal{F}_r) \geq t_i$ for $x = 1, \ldots, l$ are found.

With the optimum t_i, we query for the related k profiles where $\mu(P_g, P_r) \geq t_i$. This assumes to be given the matching measures between all filters in $\mathcal{L}$ and ultimately, between all profiles represented by filters.

As exposed in Example 1, matching measures between filters define a matrix as depicted in Fig. 1(c). In theory, also matching measures between profiles should do it. Although, we focus on filters in order to achieve a faster and more efficient search of profiles as the number of filters is assumed to be smaller than the number of profiles ($l \leq k$).

Definition 2. *A Matrix $\mathcal{M}$ is a structure of matching measures μ_ϕ between filters in $\mathbb{F}$ where columns represent the required filters $\mathcal{F}_r$ and rows represent the given filters $\mathcal{F}_g$.*

Obtaining the k solutions in $\mathcal{M}$ involves referring either to one column or to one row. The process is analogous although, the perspective is different. While reading the measures from the columns provides the so called *fitness* between profiles $\mu(P_g, P_r)$, the measures read from rows are the inverted measure $\mu(P_r, P_g)$ denoted as *overqualification*. This should be considered as emphasized in Example 2 where the requirements maybe subject to a second ranking with respect to the inverted measure.

If we focus on columns, when querying for a particular $\mathcal{F}_r$, there would be $\mathcal{F}_{g_x}$ ($x = 1, \ldots, l$) where $\mu(\mathcal{F}_{g_x}, \mathcal{F}_r) \geq t_i$. We assume all elements are in total order according to the $\leq$ relation of $\mu(\mathcal{F}_{g_x}, \mathcal{F}_r)$. The advantage is that when searching for any given l and t_i we only need to point to the right element in the column and search for the next consecutive $l - 1$ elements in descending order of μ. The process is analogous if searching on rows.

We explain next how we organize profiles. We first assume an identification label ρ_i representing the number i of rows and, σ_i representing the number i of columns in $\mathcal{M}$ where $i > 0$.

Definition 3. *Given a required filter $\mathcal{F}_r$, for every element μ_i in column σ_i in $\mathcal{M}$ representing $\mu(\mathcal{F}_{g_x}, \mathcal{F}_r)$ there is a profile record*

$$(\mu_i, n_i^>, n_i^=, n_i^<, \text{next}, \text{prev}, p)$$

describing the matching profiles P_g where:

$n_i^>$ denotes the number of profiles P_g where $\mu(P_g, P_r) > \mu_i$,
$n_i^=$ denotes the number of profiles P_g where $\mu(P_g, P_r) = \mu_i$,
$n_i^<$ denotes the number of profiles P_g where $\mu(P_g, P_r) < \mu_i$,
next is a reference to the next matching value in σ_i where $\mu(\mathcal{F}_{g_{(x+1)}}, \mathcal{F}_r) \geq \mu_i$,
prev is a reference to the next matching value in σ_i where $\mu(\mathcal{F}_{g_{(x-1)}}, \mathcal{F}_r) \leq \mu_i$,
p is a reference to a linked-list of filters matching $\mathcal{F}_r$.

The numbers $n_i^>, n_i^=, n_i^<$ are significantly important when determining the number of profiles represented by a filter without actually querying for them, i.e., if $(n_i^> + n_i^=) \geq k$ for a given μ_i we get all profiles needed.

References *Next* and *Prev* make possible to track the following greater or smaller matching value by following the references. Every profile record contains additionally a reference p to the related profiles in a σ_i column where they are organized in a transitive closure structure, ordered by the $\leq$ elements of μ.

Example 4. Figure 2 shows a representation of profile records corresponding to $\mathcal{F}_4$ (filter representing profile A) from Example 3. Note that only the relation to filters and profiles of $\mu = 0.5$ are shown in here in order to simplify the graph.

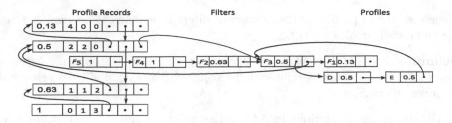

Fig. 2. Linked list of matching measures

Organizing data in this rings and spiders structure known from network databases leads to a better performance on the search for the top-k elements. Starting by fetching a column σ_i and together with $n_i^>$, $n_i^=$, $n_i^<$ calculate the profile records needed to get the k matching profiles. Then, following the ordered linked-list structure of filters, and profiles afterward, until the $k(\lambda)$ elements are found. The definition of profile records on rows ρ_i of $\mathcal{M}$ is analogous to Definition 3.

The following section shows an implementation of $\mathcal{M}$, profile records and linked-list of profiles in a relational database schema. In Sect. 3.2 we show an algorithm that implements our definition of top-k queries.

3.1 Implementation of Top-K Profile Matching

Our implementation approach of top-k queries as described in Sect. 3 is designed on a relational database schema for the storage and maintenance of filters, profiles and matching measures of an instance of $\mathcal{L}$. The relational schema is composed of 9 relations although, we present only two relations: *ProfileRecords* and *MatchingProfiles* that describe profile records as in Definition 3 and the linked-lists of matching profiles, respectively. Figure 3 shows an example of the relations representing the elements involved in Example 3. For every *RequiredFilter* in *ProfileRecords* there is a number of matching measure, that represent the number of elements per column σ_i in $\mathcal{M}$. The attribute *NextID* in *ProfileRecords* is a reference to another tuple (ID) in the relation defining the $\leq$ relation of elements of *Fitness*. Attributes *GreaterFitness*, *EqualFitness* and *LesserFitness* represent, respectively, the elements $n_i^>$, $n_i^=$, $n_i^<$ from profile records as in Definition 3.

ID	Required Filter	Fitness	Greater Fitness	Equal Fitness	Lesser Fitness	Next ID
1	$\mathcal{F}_4$	1	0	1	3	2
2	$\mathcal{F}_4$	0.63	1	1	2	3
3	$\mathcal{F}_4$	0.50	2	2	0	4
4	$\mathcal{F}_4$	0.13	4	0	0	null

(a) Relation *ProfileRecords*

ID	Required Filter	Required Profile	Given Filter	Given Profile	Fitness	Next ID
1	$\mathcal{F}_4$	A	$\mathcal{F}_4$	B	1	null
2	$\mathcal{F}_4$	A	$\mathcal{F}_2$	C	0.63	null
3	$\mathcal{F}_4$	A	$\mathcal{F}_3$	D	0.5	4
4	$\mathcal{F}_4$	A	$\mathcal{F}_3$	E	0.5	null

(b) Relation *MatchingProfiles*

Fig. 3. Example of relations *ProfileRecords* and *MatchingProfiles*

In turns, for every *RequiredFilter*($\mathcal{F}_r$) in *ProfileRecords* there is a finite number of *GivenFilters* in *MatchingProfiles* that match the requirements in $\mathcal{F}_r$. The attribute *NextID* is a reference to another tuple (ID) in the relation defining the $\leq$ relation of elements of *Fitness*. Note that a value of *null* represents the end of the list for the *GivenFilter*.

3.2   Querying the Top-K Candidate Profiles

Filters in a lattice $\mathcal{L}$ represent the properties of profiles via the hierarchical dependency of concepts in $\mathcal{L}$. Thus, for every *required* profile P_r in $\mathbb{P}$ there is a *required* filter $\mathcal{F}_r \in \mathcal{L}$ representing the profile. Then retrieving the top-k candidate profiles for a required filter from our relational schema is mainly performed by querying on relations *ProfileRecords* and *MatchingProfiles*.

Algorithm. Top-k
Input:
      required filter: $\mathcal{F}_r$, number of matching profiles: k, matching threshold: μ
Output:
      matching profiles: $P_{g_1}, \ldots, P_{g_k}$, measures: $\mu_1, \ldots, \mu_k$
Begin
1   CREATE relation *Results* = (GivenProfile, Fitness, NextID)
2   (fitness, count, nextid) := $\pi_{(3,5,7)} \left(\sigma_{\substack{(2=\mathcal{F}_r, \\ max(Fitness))}} (\textit{ProfileRecords}) \right)$
3   WHILE (count < k) OR (fitness $\geq \mu$) DO
4   *Results* $\leftarrow \pi_{(5,6,7)} \left(\sigma_{\substack{(6=\text{fitness}, \\ 2=\mathcal{F}_r)}} (\textit{MatchingProfiles}) \right)$
5   next := *Results.NextID*
6   WHILE (next IS NOT NULL) DO
7   *Results* $\leftarrow \pi_{5,6,7} \left(\sigma_{(2=\mathcal{F}_r)} (\textit{MatchingProfiles}) \underset{1=3}{\bowtie} \textit{Results} \right)$
8   END WHILE
9   (fitness, total, nextid) := $\pi_{(3,5,7)} \left(\sigma_{(1=\text{nextid})} (\textit{ProfileRecords}) \right)$
10   count := count + total
11   END WHILE
12   RETURN ($\pi_{1,2}(\textit{Results})$)
End

The algorithm Top-k returns an ordered list of top-k profiles matching a given filter. We use relational algebra notation thus, σ, π and $\bowtie$ are the *selection*, *projection* and *natural join* operators, respectively. Numeric subscripts are used to denote relation attributes. For instance, $\pi_1(\textit{MatchingProfiles})$ is the projection of attribute *ID* of relation *MatchingProfiles*.

The algorithm accepts as inputs: the required filter $\mathcal{F}_r$, the number k of matching profiles and the minimum matching value μ to search for. The outputs are: the k matching profiles $P_{g_1}, \ldots, P_{g_k}$ and their matching measures $\mu_1, \ldots, \mu_k$.

With $\mathcal{F}_r$, the algorithm fetches the tuple with the greatest value of *Fitness* in *ProfileRecords* (line 2) and follows the references on *NextID* (line 9) until the k tuples are reached or $\mu(\mathcal{F}_g, \mathcal{F}_r) < t_i$ (line 3). Then, for every $\mathcal{F}_g$ in *MatchingProfiles*, the algorithm queries on the linked-list of profiles (lines 6–8) and appends them in the temporary relation *Results* (line 7). Note that by using "fitness $\geq \mu$" in line 3, we include the $\lambda - k$ elements as in Definition 1. The algorithm finishes by returning the elements of *GivenProfile* and *Fitness* on the tuples of *Results*.

An implementation of B-Tree indexes on elements of *Fitness* (*ProfileRecords* and *MatchingProfiles*) in order to access the sorted elements, as well as indexes on *RequiredFilter* (*ProfileRecords* and *MatchingProfiles*) for random access is expected to improve performance. The implementation of a parallel processing on the search of matching profiles given the required profile records by calculating $(n_i^> + n_i^=)$ is another point of improvement.

Note that over-qualification has not been considered in Sects. 3.1 and 3.2. Although, it has been thought as an analogous process shown with *Fitness*. Additionally, a couple of straightforward algorithms to update profile changes and in consequence, matching values, linked-list of profiles and in particular the elements $n_i^>$, $n_i^=$, $n_i^<$ in profile records have been considered.

4   Conclusion

We presented in this paper an algorithm to address top-k queries where we use a transitive closure structure on top of relational databases for implementation. The identification of missing requirements on profiles, essential on selecting the best candidates, has still to be consider. This implies an investigation of gap queries on grounds of matching measures that is the focus of our future research.

References

1. Chakrabarti, K., Ortega-Binderberger, M., Mehrotra, S., Porkaew, K.: Evaluating refined queries in top-k retrieval systems. IEEE Trans. Knowl. Data Eng. **16**(2), 256–270 (2004)
2. Ilyas, I.F., Aref, W.G., Elmagarmid, A.K.: Supporting top-k join queries in relational databases. VLDB J. **13**(3), 207–221 (2004)
3. Paoletti, A.L., Martinez-Gil, J., Schewe, K.-D.: Extending knowledge-based profile matching in the human resources domain. In: Chen, Q., Hameurlain, A., Toumani, F., Wagner, R., Decker, H. (eds.) DEXA 2015. LNCS, vol. 9262, pp. 21–35. Springer, Heidelberg (2015)
4. Popov, N., Jebelean, T.: Semantic matching for job search engines: a logical approach. Technical report 13-02, RISC Report Series, University of Linz, Austria (2013)
5. Rácz, G., Sali, A., Schewe, K.-D.: Semantic matching strategies for job recruitment: a comparison of new and known approaches. In: Gyssens, M., Simari, G. (eds.) FoIKS 2016. LNCS, vol. 9616, pp. 149–168. Springer, Heidelberg (2016). doi:10.1007/978-3-319-30024-5_9
6. Straccia, U., Madrid, N.: A top-k query answering procedure for fuzzy logic programming. Fuzzy Sets Syst. **205**, 1–29 (2012)
7. Theobald, M., Weikum, G., Schenkel, R.: Top-k query evaluation with probabilistic guarantees. In: (e)Proceedings of the Thirtieth International Conference on Very Large Data Bases, Toronto, Canada, pp. 648–659, 31 August–3 September 2004

FETA: Federated QuEry TrAcking
for Linked Data

Georges Nassopoulos[✉], Patricia Serrano-Alvarado, Pascal Molli,
and Emmanuel Desmontils

LINA Laboratory, Université de Nantes, Nantes, France
{georges.nassopoulos,patricia.serrano-alvarado,
pascal.molli,emmanuel.desmontils}@univ-nantes.fr

Abstract. Following the principles of Linked Data (LD), data providers
are producing thousands of interlinked datasets in multiple domains
including life science, government, social networking, media and publications. Federated query engines allow data consumers to query several datasets through a federation of SPARQL endpoints. However, data
providers just receive subqueries resulting from the decomposition of the
original federated query. Consequently, they do not know how their data
are crossed with other datasets of the federation. In this paper, we propose FETA, a Federated quEry TrAcking system for LD. We consider that
data providers collaborate by sharing their query logs. Then, from a federated log, FETA infers Basic Graph Patterns (BGPs) containing joined
triple patterns, executed among endpoints. We experimented FETA with
logs produced by FedBench queries executed with Anapsid and FedX
federated query engines. Experiments show that FETA is able to infer
BGPs of joined triple patterns with a good precision and recall.

Keywords: Linked data · Federated query processing · Log analysis ·
Usage control

1 Introduction

Linked Data (LD) interlinks massive amounts of data across the Web in multiple
domains like life science, government, social networking, media and publications.
Federated query engines [1–3,5,9,11] allow data consumers to execute SPARQL
queries over a decentralized federation of SPARQL endpoints maintained by LD
providers. But, data providers are not aware of users' federated queries; they
just observe subqueries they receive. Thus, they do not know when and which
datasets are joined together in a single query. Consequently, the federation does
not hold enough meta-information to ensure services, such as, efficient materialization to improve joins, activation of query optimization techniques, discovering data providers partnership, etc. Knowing how provided datasets are queried
together is essential for tuning endpoints, justify return of investment or better
organize collaboration among providers.

© Springer International Publishing Switzerland 2016
S. Hartmann and H. Ma (Eds.): DEXA 2016, Part II, LNCS 9828, pp. 303–312, 2016.
DOI: 10.1007/978-3-319-44406-2_24

A simple solution for this problem is to consider that data consumers publish their federated queries. However, public federated queries cannot be considered as representative of real data usage because they may represent a small portion of really executed queries. Only logs give evidences about real execution of queries.

Thus, in this paper, we address the following problem: if data providers share their logs, can they infer the Basic Graph Patterns (BGP) of federated queries executed over their federation? Many works have focused on web log mining [6,7], but none has addressed reversing BGPs from a federated query log.

The main challenge is the concurrent execution of federated queries. If we find a function f, to reverse BGPs from isolated traces of one federated query, is f able to reverse the same BGP from traces of concurrent federated queries?

We propose FETA to implement f, a Federated quEry TrAcking system that computes BGPs from a federated log. Based on subqueries contained in the log, FETA deduces triple patterns and joins among triple patterns with a good precision and recall. Our main contributions are:

1. the definition of the problem of reversing BGPs from a federated log,
2. the FETA algorithm to reverse BGPs from federated logs,
3. an experimental study using federated queries of the benchmark FedBench[1]. From execution traces of these queries, FETA deduces BGPs under two scenarios, queries executed in isolation and in concurrence.

The paper is organized as follows. Section 2 introduces a motivating example and our problem statement. Section 3 presents FETA and its heuristics. Section 4 reports our experimental study. Section 5 presents related work. Finally, conclusions and future work are outlined in Sect. 6.

2   Motivating Example and Problem Statement

In Fig. 1, two data consumers, C_1 and C_2, execute concurrently federated queries $CD3$ and $CD4$ of FedBench. They use Anapsid or FedX federated query engines to query a federation of SPARQL endpoints composed of *LMDB, DBpedia InstanceTypes (IT), DBpedia InfoBox (IB)* and *NYTimes (NYT)*. Data providers hosting these endpoints receive only subqueries corresponding to the execution of physical plans of $CD3$ and $CD4$. For example, $CD3$ can be decomposed into $\{tp_1^{@IT}.(tp_2.tp_3)^{@IB}.(tp_4.tp_5)^{@NYT}\}$, and NYT just observes tp_4 and tp_5: it does not know these triple patterns are joined with tp_1 from IT and (tp_2, tp_3) from IB. So, NYT does not know the real usage of data it provides.

More formally, we consider that an execution of a federated query FQ_i produces a partially ordered sequence of subqueries SQ_i represented by $E(FQ_i) = [SQ_1, ..., SQ_n]$. Subqueries are processed by endpoints of the federation at given times represented by timestamps. We suppose that endpoints' clocks are synchronized, i.e., timestamps of logs can be compared safely. Timestamps of subqueries in a federated log are partially ordered because two endpoints can receive

---

[1] http://fedbench.fluidops.net/.

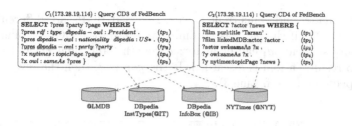

Fig. 1. Concurrent execution of FedBench queries CD3, CD4 over a federation of end-points.

queries at same time. Query execution with a particular federated query engine, qe_i, is represented by $E_{qe_i}(FQ_i)$. In addition, we represent a concurrent execution of n federated queries by $E(FQ_1 \parallel ... \parallel FQ_n)) = [SQ_1, ..., SQ_n]$. This work addresses the following research question: *if data providers share their logs, can they rebuild the BGPs annotated with the sources that evaluated each triple pattern?* From the previous example, we aim to extract two BGPs: one corresponding to $CD3$ $\{tp_1^{@IT}.(tp_2.tp_3)^{@IB}.(tp_4.tp_5)^{@NYT}\}$ and another to $CD4$ $\{(tp_1.tp_2.tp_3)^{@LMDB}.(tp_4.tp_5)^{@NYT}\}$. Next definitions formalize this problem.

Definition 1 (BGPs' reversing). *Given a federated log corresponding to the execution of one federated query $E(FQ_i)$, find a function $f(E(FQ_i))$ producing a set of BGPs $\{BGP_1, ..., BGP_n\}$, where each triple pattern is annotated with endpoints that evaluated it, such that $f(E(FQ_i))$ approximates ($\approx$) the BGPs existing in the original federated query. Thus, if we consider that $BGP(FQ_i)$ returns the set of BGPs of FQ_i then $f(E(FQ_i)) \approx BGP(FQ_i)$.*

In our motivating example, if C_1 and C_2 have different IP addresses, then it is straightforward to apply the reversing function separately on each execution trace. However, in the worst case, if they share the same IP address, we expect that $f(E(CD3 \parallel CD4)) \approx f(E(CD3) \cup f(E(CD4)$ as defined next.

Definition 2 (Resistance to Concurrency). *The reversing function f should guarantee that BGPs obtained from execution traces of isolated federated queries, approximate ($\approx$) results from execution traces of concurrent federated queries: $f(E(FQ_1)) \cup ... \cup (f(E(FQ_n)) \approx f(E(FQ_1 \parallel ... \parallel FQ_n))$.*

3   FETA, a Reversing Function

Finding a reversing function f requires to join IRIs, literals or variables from different SPARQL subqueries. We propose FETA as a system of heuristics to implement the reversing function f. Figures 2a, b present two endpoints, each providing some triples. Figure 2d has the federated log corresponding to the execution of queries $Q_1 = SELECT\ ?z\ ?y\ WHERE\ \{?z\ p1\ o1\ .\ ?z\ p2\ ?y\}$ and $Q_2 = SELECT\ ?x\ ?y\ WHERE\ \{?x\ p1\ ?y\}$. Figure 2c shows reversing results according to different *gap* values described below. Depending on the *gap*, on verifications made, and concurrent traces, reversed BGPs are different. In the example of Fig. 2c, if the *gap* has no limit, we obtain the BGP of Line 1, if the *gap* is 1, only sq_3 and sq_4 can be analyzed together (cf. Line 2). As the join on $?y$ gives no results, a join is discarded. If the *gap* is 2, then the reversed BGPs are the expected ones (cf. Line 3).

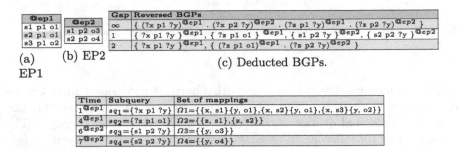

(a) EP1   (b) EP2   (c) Deducted BGPs.

(d) Federated log.

Fig. 2. Motivating example.

We assume pairwise disjoint infinite sets $\mathcal{B}$, $\mathcal{L}$, $\mathcal{I}$ (blank nodes, literals, and IRIs respectively). We also assume an infinite set $\mathcal{S}$ of variables. A mapping μ is a partial, non surjective and non injective, function that maps variables to RDF terms $\mu : \mathcal{S} \rightarrow \mathcal{BLI}$. A set of mappings is represented by Ω. See [8] for more explanations. Next, we present the input and output of FETA.

Given: a federation of endpoints Φ, a federated input log $\mathcal{Q} = \{\langle q, t, ep, ip \rangle\}$, a federated output log $\mathcal{A} = \{\langle \{\mu\}, t, ep, ip \rangle\}$, and a user-defined *gap*,

Find $\mathcal{G} = \{\langle g, ip \rangle\}$, the set of connected graphs corresponding to the BGPs of the federated queries processed by Φ, such that:

- $g = \langle V, E \rangle$ is an undirected connected graph where $V = \{tp\}$ is an unordered set of distinct triple patterns, (annotated with the endpoints that processed tp and $\mathcal{T}$ the set of timestamps given by the endpoints), and E is an unordered set of edges representing the joins between triple patterns.
- ip is the IP address of the client query engine that sent g.

3.1 FETA Algorithms

FETA has 4 main phases. From input logs and a predefined *gap*, a graph of subqueries $\mathbb{G}$ is constructed in the first phase. Then, this graph is reduced and transformed into a graph of triple patterns $\mathcal{G}$, where, from a big set of subqueries, frequently only one triple pattern is obtained. In a third phase, joins between triple patterns executed through nested-loops are identified. Finally, symmetric hash joins, possibly made at the federated query engine, are identified. Next sections present these algorithms at high level of abstraction.

Graph Construction. This phase executes two main functions, (a) *LogPreparation* and (b) *CommonJoinCondition*. This module builds $\mathbb{G} = \{g\}$, a set of graphs, where $g = \langle V', E', ip \rangle$ is an undirected connected graph, with different semantics than $\mathcal{G}$. In $\mathbb{G}$, nodes are queries and arcs are labeled with the number of common variables between each pair of queries. *LogPreparation*, prepares and cleans the input log. ASK queries are suppressed and identical subqueries or differing only in their offset values are aggregated in one single query. Timestamp of such aggregated query becomes an interval. *CommonJoinCondition*, incrementally constructs $\mathbb{G}$, by joining queries depending on the given *gap* and having common projected variables or triple patterns with common IRI or literal on their subjects or objects. In general, subqueries are joined on their common projected variables. However, we consider also IRIs and literals, even if they can produce some false positives. In our example, with an infinite *gap*, two graphs are constructed as shown in Fig. 3: $\mathbb{G} = \{g_1, g_2\}$, where $g_1 = \langle \{sq_1, sq_3, sq_4\}, \{(sq_1, sq_3), (sq_1, sq_4), (sq_3, sq_4)\} \rangle$ and $g_2 = \langle \{sq_2\} \rangle$. To simplify, all annotations to sq_i are omitted.

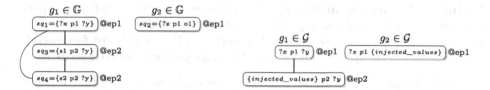

Fig. 3. Deduced graphs in $\mathbb{G}$ after *GraphConstruction*, for an infinite *gap*.

Fig. 4. Deduced graphs in $\mathcal{G}$ after *GraphReduction*, for an infinite *gap*.

Graph Reduction. The graph of queries is transformed into a graph of patterns. This heuristic aggregates triple patterns, produced from mappings of the outer dataset towards the inner dataset, into one big aggregated pattern (that we call inner pattern). This pattern, for instance, has the form of $\langle injected\_values, predicate, object \rangle$, if mappings of the outer dataset are injected into the subject. Graph reduction significantly reduces the size of each $g \in \mathbb{G}$, because nested-loops can be executed with hundreds of subqueries. Figure 4, illustrates $\mathcal{G}$ after the graph reduction phase, for our motivating example.

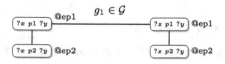

Fig. 5. Deduced graphs in $\mathcal{G}$ after *NestedLoopDetection*, for an infinite *gap*.

Algorithm 1. NestedLoopDetection($\mathcal{G}, \mathcal{A}, gap$)

```
input  : G, A, gap
output: G
1  foreach g ∈ G do
2  |   foreach tp_j ∈ g do
3  |   |   foreach (tp_j ∈ g) ∨ (tp_j ∈ g' : g' ∈ G, g' ≠ g) do
4  |   |   |   if (t_{tp_j}^{max} − t_{tp_i}^{min}) ≤ gap ∧ (μ^{-1}(tp_j, A) ∈ var(tp_i)) then
5  |   |   |   |   dp ←Association(tp_i, tp_j)
6  |   |   |   |   G ←Update(G, tp_j, dp,' nestedLoop')
```

Nested-Loop Detection. This heuristic analyzes existing graphs in $\mathcal{G}$ to identify nested-loops. From n subqueries, it obtains two joined triple patterns by nested-loop. To do this, Algorithm 1, Lines 3–6, associates the pattern that pushes the outer dataset (that we call outer pattern) towards the inner pattern. This association is made by searching for a matching, between the injected values of the inner pattern and the variable mappings of the outer, with the function of *inverse mapping* that we propose below.

Definition 3 (Inverse mapping). *We define the inverse mapping as a partial, non surjective and non injective, function* $\mu^{-1} : \mathcal{BLI} \to \mathcal{S}$ *where* $\mu^{-1} = \{(val, s) \mid val \in \mathcal{BLI}, s \in \mathcal{S}\}$ *such that* $(s, val) \in \mu$. $\mathcal{B}$ *is considered for generalization reasons even if blank nodes cannot be used for joins between datasets.*

NestedLoopDetection is the most challenging heuristic of FETA because μ^{-1} may return more than one variable, when the same value was returned for more than one variable. This depends on the similarity of concurrent federated queries and the considered *gap*. Thus, some times, identifying the variable that appears in the original query is uncertain. Figure 5, illustrates $\mathcal{G}$ after *NestedLoopDetection* for our motivating example with an infinite *gap*. We observe that graphs g_1 and g_2 are merged.

Symmetric Hash Detection. This heuristic verifies that *(i)* edges of $g \in \mathcal{G}$ that were not produced by an exclusive group or a nested-loop, are on same ontologically concepts for their common projected variables, and *(ii)* their join has a not null result set. From this, symmetric hash joins are identified, otherwise joins are removed. Symmetric hash detection produces false positives as it infers all possible joins that may be made at the query engine. If a star-shape set of triple patterns exists, all possible combinations of joins will be deduced instead of the subset of joins chosen by the query engine. Consequently, FETA privileges recall to the detriment of precision. For our example, this phase has no impact.

3.2   Time Complexity of FETA

The computational complexity of the global algorithm of FETA is, in the worst case, $O(N^2 + N * M + M^2)$, where N is the number of queries of $\mathbb{G}$, and M is the number of triple patterns of $\mathcal{G}$. The overload produced by FETA is high but we underline that the size of the log corresponds to a *sliding window of time* and that the log analysis can be made as a batch processing.

4   Experiments

To the best of our knowledge, a public set of real federated queries executed over the LD does not exists, thus we evaluated FETA using the queries and the setup of FedBench [10]. We used the collections of Cross Domain (CD) and Life Science (LS), each one has 7 federated queries. We setup 19 SPARQL endpoints using Virtuoso OpenLink[2] 6.1.7. We executed federated queries with Anapsid 2.7 and FedX 3.0. We configured Anapsid to use Star Shape Grouping Multi-Endpoints (SSGM) heuristic and we disabled the cache for FedX. We captured http requests and answers from endpoints with justniffer 0.5.12[3]. FETA is implemented in Java 1.7 and is available at https://github.com/coumbaya/feta.

The goals of the experiments are : (i) to evaluate the precision and recall of FETA with federated queries executed *in isolation* and (ii) to evaluate the precision and recall of FETA with federated queries executed *concurrently* under a worst case scenario, i.e., when BGPs of different federated queries cannot be distinguished as they share the same IP address. All results are available at: https://github.com/coumbaya/feta/blob/master/experiments_with_fedbench.md.

To analyze traces of federated queries in isolation, we executed CD and LS collections. We captured 28 sequences of subqueries used as input for FETA one by one. In average, we obtained 94,64 % of precision and 94,64 % of recall of *triple patterns deduction*. We obtained 79,40 % of precision and 87,80 % of recall for *joins deduction*. Deducing *sets of joined triple patterns*, i.e., BGPs, is more challenging. From Anapsid traces, BGPs deduced correspond to CD and LS queries, except for Union queries, i.e., CD1, LS1 and LS2. These queries have two BGPs but a join is possible between them locally at the query engine, and FETA deduces a symmetric hash join. All other problems of deduction come from *NestedLoopDetection*. False triple patterns are deduced from FedX traces that decreases precision. This is because μ^{-1} may return more than one variable and more than one triple pattern may be deduced. But as right triple patterns are in general well deduced, recall is good. FETA succeeds in deducing 11 out of 14 exact BGPs from Anapsid traces, and 7 out of 14 from FedX traces. It finds 18/28 exact BGPs, i.e., 64 %. If we include Union queries where all triple patterns are deduced, FETA finds (18+3)/28 BGPs, i.e., 75 % BGPs of FedBench.

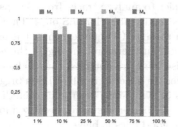

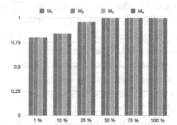

Fig. 6. Recall of joins from ANAPSID MX traces, by *gap*.

Fig. 7. Recall of joins from FedX MX traces, by *gap*.

---

[2] http://virtuoso.openlinksw.com/.

[3] http://justniffer.sourceforge.net/.

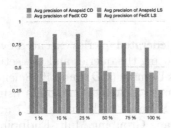

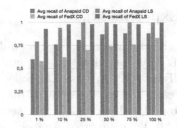

Fig. 8. Average of precision of joins, for four mixes by *gap*.

Fig. 9. Average of recall of joins, for four mixes by *gap*.

To analyze traces of concurrent federated queries, we implemented a tool that shuffles logs of queries executed in isolation to produce different sequences of $E(FQ_1 \parallel ... \parallel FQ_n)$. These traces vary in (i) the order of queries, (ii) the number of subqueries, of the same federated query, appearing continuously (blocks of 1 to 16 subqueries), and (iii) the delay between each subquery (1 to 16 units of time). In our experiments, *gap* varies from 1 % to 100 % of the total time of each mix. We measured precision and recall of deductions made by FETA, from traces of federated queries in isolation against our mixes of traces of concurrent queries.

If FETA can distinguish triple patterns of concurrent federated queries, precision and recall by join are perfect when the *gap* is big enough. We analyzed a set of chosen queries having distinguishable triple patterns that we named MX: CD3, CD4, CD5, CD6, LS2 and LS3. We produced 4 different mixes of traces of these queries ($M_1, ..., M_4$) that were analyzed by FETA under 6 different *gaps* (1 %, 10 %, etc.) producing 6 groups of deductions. We obtained 100 % of precision of joins from traces of Anapsid and FedX since the smallest *gap*. Figures 6 and 7 show recall of joins from Anapsid and FedX traces respectively. We get 100 % of recall with a *gap* of 50 % from traces of both query engines.

If triple patterns of concurrent queries are the same or syntactically similar, it is hard for FETA to obtain good precision and recall of joins. We produced four different and concurrent mixes by queries' collection (4 for CD and 4 for LS). We analyzed them by query engine and by *gap*. Figure 8 shows the average of precision of joins, each bar concerns 4 mixes. We can see that for FETA it is easier to analyze traces from Anapsid than from FedX. Moreover, CD queries are more distinguishable than LS ones. That is because triple patterns of LS queries vary less than those of CD queries, thus it is less evident to separate LS queries from their mixed traces. Furthermore, the bigger the *gap* the smaller the precision. That is because more false joins are detected thus reducing precision. Figure 9 shows the average of recall of joins. In general, recall of LS is bigger than recall of CD because LS queries generate lots of symmetric hash joins including the good ones. Unlike precision, the bigger the *gap*, the bigger the recall because more joins are detected thus the possibility of finding the good ones is bigger.

5   Related Work

Extracting information from logs is traditionally a data mining process [4]. As a log of subqueries is in fact a log of accessed resources, data log mining algorithms could be used to solve our problem, where each item is an accessed predicate or triple pattern.

Sequential pattern mining [7] focuses on discovering frequent subsequences (totally or partially ordered) from an ordered sequence of events. An event is a collection of unordered items, an item is a literal, and a set of items composes an alphabet. In our context, we focus on sequential pattern mining algorithms able to operate on non-transactional logs such as WINEPI or MINEPI [6]. WINEPI decomposes a temporal sequence in overlapping windows of a user-defined size n and counts the frequency of episodes in all windows. Episodes can be of size 1 to n. MINEPI instead, looks for minimal occurrences of episodes. It identifies in a sequence, the set of time intervals of minimal occurrences of episodes according to the maximum user-defined window size. The number of minimal occurrences of an episode is called support. The minimum frequency (for WINEPI), the minimal support (for MINEPI) and the maximum window size (for both), are thresholds defined by the user. The difference of these approaches, is that WINEPI can be interpreted as the probability of encountering an episode from randomly chosen windows, while MINEPI counts exact occurrences of episodes.

We think that searching for BGPs in a federated log is not like searching for frequent episodes in a temporal log. First, the alphabet of events in a federated log can be proportional to the cardinality of data in the federation. A nested-loop operator can generate thousands of different subqueries as we observed with FedX. Managing huge alphabets is challenging for sequential pattern algorithms. FETA uses heuristics to reduce the alphabet by deducing hidden variables. Second, frequency of events in a federated log is related to the selectivity of operations and can confuse sequential pattern algorithms. Suppose, two queries Q_1 : $\{?x\ p1\ o1\ .\ ?x\ p2\ ?y\}$ and Q_2 : $\{?x\ p1\ ?y\ .\ ?y\ p3\ ?z\}$. The federated query engine executes the joins with a nested-loop. So, $?x\ p1\ o1$ and $?x\ p1\ ?y$ will appear once in the log, while patterns with $IRIs\ p2\ ?y$ and $IRIs\ p3\ ?z$ will appear many times according to the selectivity of the triple patterns on $p1$. Searching for frequent episodes will raise up episodes with $p2$ and $p3$ but joins were between $p1, p2$ and $p1, p3$.

Limitations of sequential pattern mining algorithms have been pointed out in process mining [12]. Process mining algorithms recompute workflow models from logs. However, queries are not workflows and federated logs are not process logs. In a process log, events corresponds to identified tasks which is not the case in our context. The number of different subqueries can be proportional to the cardinality of the federated datasets. Moreover, in a federated log, a subquery cannot be the cause of another; in general, join ordering is decided according to the selectivity of subgoals in the original query.

6 Conclusions and Future Work

Federated query tracking allows data providers to know how their datasets are used. In this paper we proposed **FETA**, a federated query tracking approach that reverses federated Basic Graph Patterns (BGPs) from a shared log maintained by data providers. **FETA** links and unlinks variables from subqueries of the federated log by applying a set of heuristics to decrypt behavior of physical join operators.

Even in a worst case scenario, **FETA** extracts BGPs that contain original BGPs of federated queries executed with Anapsid and FedX. Extracted BGPs, annotated with endpoints, give valuable information to data providers about which triples are joined, when and by whom.

We think **FETA** opens several interesting perspectives. First, heuristics can be improved in many ways by better using semantics of predicates and answers. Second, we can improve **FETA** to make it agnostic to the federated query engine.

Third, FETA can be used to generate a transactional log of BGPs from a temporal log of subqueries. Analyzing frequency of BPGs in a transactional log allows to discriminate false positive deductions of FETA.

Acknowledgments. This work was partially funded by the French ANR project SocioPlug (ANR-13-INFR-0003), and by the DeSceNt project granted by the Labex CominLabs excellence laboratory (ANR-10-LABX-07-01).

References

1. Acosta, M., Vidal, M.-E., Lampo, T., Castillo, J., Ruckhaus, E.: ANAPSID: an adaptive query processing engine for SPARQL endpoints. In: Aroyo, L., Welty, C., Alani, H., Taylor, J., Bernstein, A., Kagal, L., Noy, N., Blomqvist, E. (eds.) ISWC 2011, Part I. LNCS, vol. 7031, pp. 18–34. Springer, Heidelberg (2011)
2. Basca, C., Bernstein, A.: Avalanche: putting the spirit of the web back into semantic web querying. In: International Semantic Web Conference (ISWC) (2010)
3. Görlitz, O., Staab, S.: SPLENDID: SPARQL endpoint federation exploiting VOID descriptions. In: International Workshop on Consuming Linked Data (COLD) (2011)
4. Han, J., Kamber, M., Pei, J.: Data Mining: Concepts and Techniques. Elsevier, London (2011)
5. Hartig, O., Bizer, C., Freytag, J.-C.: Executing SPARQL queries over the web of linked data. In: Bernstein, A., Karger, D.R., Heath, T., Feigenbaum, L., Maynard, D., Motta, E., Thirunarayan, K. (eds.) ISWC 2009. LNCS, vol. 5823, pp. 293–309. Springer, Heidelberg (2009)
6. Mannila, H., Toivonen, H., Verkamo, A.I.: Discovery of frequent episodes in event sequences. Data Min. Knowl. Discovery 1(3), 259–289 (1997)
7. Mooney, C.H., Roddick, J.F.: Sequential pattern mining-approaches and algorithms. ACM Comput. Surv. (CSUR) 45(2), 19 (2013)
8. Pérez, J., Arenas, M., Gutierrez, C.: Semantics and complexity of SPARQL. ACM Trans. Database Syst. (TODS) 34(3), 16:1–16:45 (2009)
9. Quilitz, B., Leser, U.: querying distributed RDF data sources with SPARQL. In: Bechhofer, S., Hauswirth, M., Hoffmann, J., Koubarakis, M. (eds.) ESWC 2008. LNCS, vol. 5021, pp. 524–538. Springer, Heidelberg (2008)
10. Schmidt, M., Görlitz, O., Haase, P., Ladwig, G., Schwarte, A., Tran, T.: FedBench: a benchmark suite for federated semantic data query processing. In: Aroyo, L., Welty, C., Alani, H., Taylor, J., Bernstein, A., Kagal, L., Noy, N., Blomqvist, E. (eds.) ISWC 2011, Part I. LNCS, vol. 7031, pp. 585–600. Springer, Heidelberg (2011)
11. Schwarte, A., Haase, P., Hose, K., Schenkel, R., Schmidt, M.: FedX: optimization techniques for federated query processing on linked data. In: Aroyo, L., Welty, C., Alani, H., Taylor, J., Bernstein, A., Kagal, L., Noy, N., Blomqvist, E. (eds.) ISWC 2011, Part I. LNCS, vol. 7031, pp. 601–616. Springer, Heidelberg (2011)
12. Van Der Aalst, W.: Process Mining: Discovery Conformance and Enhancement of Business Processes. Springer, Heidelberg (2011)

Database and Information System
Architectures

Dynamic Power-Aware Disk Storage Management in Database Servers

Peyman Behzadnia[1], Wei Yuan[2], Bo Zeng[3], Yi-Cheng Tu[1(✉)], and Xiaorui Wang[4]

[1] Department of Computer Science and Engineering, University of South Florida,
Tampa, FL, USA
{peyman,ytu}@cse.usf.edu

[2] Department of Industrial and Management Systems Engineering, University of South Florida,
Tampa, FL, USA
weiyuan@mail.usf.edu

[3] Department of Industrial Engineering, University of Pittsburgh, Pittsburgh, PA, USA
bzeng@pitt.edu

[4] Department of Electrical and Computer Engineering, The Ohio State University, Columbus,
OH, USA
wang.3596@osu.edu

Abstract. Energy consumption has become a first-class optimization goal in design and implementation of data-intensive computing systems. This is particularly true in the design of database management system (DBMS), which was found to be the major consumer of energy in the software stack of modern data centers. Among all database components, the storage system is the most power-hungry element. In this paper, we present our research on designing a power-aware data storage system. To tackle the limitations of the previous work, we introduce a DPM optimization model to minimize power consumption of the disk-based storage system while satisfying given performance requirements. It dynamically determines the state of disks and plans for inter-disk fragment migration to achieve desirable balance between power consumption and query response time. We evaluate our proposed idea by running simulations using several synthetic workloads based on popular TPC benchmarks.

1 Introduction

Data centers, criticized as the SUVs of the IT world, consume massive and growing amount of energy. A recent report shows that, in 2013, data centers in the Unites States consumed an estimated 91 billion kilowatt-hours (kWh) of electricity (which costed roughly 7.5 billion US dollars) and are on-track to reach 140 billion kWhs by 2020 [1]. In a typical data center, Database Management System (DBMS) is the largest power consumer among all software modules deployed. And, among all components of a database server, storage system is the most energy hunger constituent. Disk storage system is estimated to consume 25–35 % of total energy consumption in a data center [2]. Another report [3] shows that power consumed by storage in large online transaction processing (OLTP) systems is more than 70 % of the total power of all IT equipment. Power consumption rate of storage systems will grow even larger in the next years - an

© Springer International Publishing Switzerland 2016
S. Hartmann and H. Ma (Eds.): DEXA 2016, Part II, LNCS 9828, pp. 315–325, 2016.
DOI: 10.1007/978-3-319-44406-2_25

annual growth of 60 % has been reported in [4]. Given this strong demand for energy reduction in storage systems, we tackle the problem of designing a power-aware database disk storage system in this paper. Note that the use of SSD drives simplifies the problem since they are highly energy efficient compared to HDDs, but, as of today, SSDs are still not in a position to replace all magnetic disks in large-scale database systems, especially those handling today's big data applications. In previous work, Dynamic Power Management (DPM) algorithms are normally used to save energy in disk storage systems. Such algorithms make real-time decisions on when to transition magnetic disks to lower-power modes with the price of longer response time to data access requests. Many modern hard disks have two power states: active and stand-by. Disks in stand-by mode stop rotation completely thus consume significantly less energy than in active state. However, it incurs a remarkable energy and time cost to spin up to active mode in order to serve a request. Figure 1 shows the detailed specifications related to the power and transition time among different states of a typical multi-mode disk (model Ultra-star 7k6000 from IBM) [5]. In order to amortize the aforementioned penalty cost of disk state change, effective DPM techniques extended the idle period of disks by either controlling the I/O intervals [6–10] or migrating data among disks [11–16]. The first set of works usually considers single-disk systems and utilizes energy-efficient caching or pre-fetching techniques to prolong the idle periods in the I/O trace. The second set of works basically consolidates the most frequently accessed data (called "hot" data in literature) on subset of disks to allow "cold" disks sleep longer. Therefore, they perform corresponding inter-disk data migration in order to achieve the hot data consolidation goal.

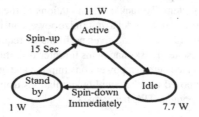

Fig. 1. Power modes and their power consumption of the IBM Ultra-Star 7k6000

As the major limitation, work of this type cannot efficiently handle the dynamic I/O traces where arrival rate of data requests changes significantly with respect to time. Furthermore, they do not provide efficient disk state configuration or inter-disk data migration. In this paper, we tackle the limitations of the previous work. The best known algorithm that tries to handle dynamic environment is named Block Exchange (BLEX) presented in [15]. However, we believe BLEX, again, does not efficiently adapt to dynamicity in the workload. The reason is that it maintains some data in stand-by disks and therefore, it incurs significant penalties related to spinning stand-by disks up and down in order to adapt to dynamic changes in data request arrival rates. We address this issue by introducing an optimization model that integrates the Model Predictive Control (MPC) strategy to accommodate dynamic scenarios by enabling optimization actions in an online fashion. Our experimental results clearly show that our proposed model outperforms the BLEX algorithm significantly in terms of both energy savings and response time.

Our contributions and roadmap are summarized as follows: (1) We introduce an integrated DPM optimization model extended with MPC strategy that dynamically determines state (power mode) adjustment and efficient fragment migration to achieve the optimal tradeoff between power consumption and the query response time; (2) We conduct experimental simulations using extensive set of synthetic workloads based on popular TPC benchmarks to evaluate our solution in terms of power saving and response time compared with those of the BLEX algorithm. Our proposed DPM optimization model outperforms the BLEX algorithm significantly in terms of both energy savings and response time in data access. The remainder of this paper is organized as follows: Sect. 2 provides a survey on the related work in the literature; Sect. 3 illustrates the proposed DPM optimization model in detail; Sect. 4 discusses our experimental evaluation; and Sect. 5 concludes the paper.

2 Related Work

DPM algorithms are the most popular techniques to achieve energy savings in disk storage systems. Intuitively, the core idea of an effective DPM algorithm is to prolong the idling period of disks in order to allow them sleep longer in the lower-power mode and thus, boost power saving opportunity. We classify algorithmic techniques extending disks idleness period into three different categories: (1) the *first* approach taken in DPM algorithms is *data packing* that consolidates the frequently accessed data (hot fragments) into fewer disks (hot disks) in order to help other disks stay in idle mode longer. An efficient algorithm named Block Exchange (BLEX) is introduced in [15] that dynamically achieves load consolidation and performs block exchange between disks. To the best of our knowledge, BLEX is the most effective algorithm in literature that tries to handle the dynamic I/O traces. Therefore, we will frequently make comparisons to BLEX in describing our solutions in the remainder of this paper. Our experiments will also use BLEX as the baseline. Other similar proposals that exploit data packing are found in [12–14, 16]. They assume RAID layouts which is not the focus of our work; (2) the *second* approach to extend disk inactivity period is to manage I/O intervals via power-aware caching and prefetching algorithms. The main idea is to deploy energy-aware policy in cache data management algorithm (or in prefetching techniques) to redirect some I/O requests to cache in order to change I/O intervals towards longer idle times. Work presented in [6, 7, 10] tackles this method to achieve energy conservation; (3) the *third* class of research works extending disk idleness period tackle energy proportionality in data parallel computing clusters whose files systems maintain a set of replicas for each data block. Papers in [21–23] are classified under this category for energy savings in data parallel clusters. Some other miscellaneous research proposals along with more details on the aforementioned related work are provided in our more thorough survey over the literature in [20].

3 Proposed DPM Optimization Model

In this section, we show the design of a DPM optimization model towards balance between energy consumption and performance impact. It is well-known that the arrival rate of data requests changes significantly in respect to time in I/O traces of database servers. This is particularly true in scientific database servers and OLTP servers. The SSDS SkyServer is a famous scientific database server that clearly shows significant changes in the server traffic rate [24]. Also, [17] shows workload changes in an OLTP trace that demonstrates notable arrival rate changes in respect to time. The major problem of previous contributions is that they cannot efficiently adapt to dynamic I/O workloads. We solve this issue by integrating Model Predictive Control (MPC) strategy in an optimization model to enable optimization actions in an online fashion. Section 3.4 describes in detail how our optimization model integrates the MPC technique in order to capture the dynamic changes in data access frequency. Given such significant arrival rate changes in dynamic I/O workloads, we partition the planning horizon into multiple periods where the arrival rate in each period can be modeled by a constant. We formulate a model as a (nonlinear) mixed integer program (shown in Sect. 3.1) where the objective function is the overall cost from all energy consumption elements in the storage system during one epoch. At the beginning of each epoch, based on the observed I/O and the predicted workload for the epoch, the model configures the optimal disk state setting and corresponding inter-disk fragment migration that minimize the energy consumption (aforementioned objective function) during the epoch while maintaining query response time quality. In order to avoid the disk overloading problem, the model performs load balancing between the overloaded disk (s) and other active disks at the beginning of each epoch. In addition to the MPC strategy implemented in our model, another advantage of the DPM model is that we explicitly include fixed charge penalty on disk status change to avoid excessive spin up and down operations (with expensive response time and energy costs), while it is rather considered subjectively in BLEX.

The length of the epoch should be short enough to capture changing arrival rates and also long enough to accommodate disks transition cost and data migration periods, and to impose tolerable number of on/off actions on disks in order to not damage their lifetime services. Considering arrival rate change patterns existing in database I/O traces, we

Table 1. DPM model parameters

Name	Description	Name	Description
i	Index of disks, $i = 1,...,I$	ed_i	Energy to spin down disk i
j	Type of fragmentation, $j = 1,...,J$	ep_i	Energy to spin up disk i
$\lambda_{j,t}$	Hotness level/popularity of fragment type j in period t	$p_{i,k}^t$	power consumption of disk i at k spinning state in period t
k	State of disk	Γ	Response time penalty parameter
Sc_i	Storage capacity of disk i	$maxfrag$	Disk maximum no. of fragments
c_j	Migration cost of fragment type j	λ^{max}	Maximum fragment popularity
b_j	Block size of fragment type j	M	Maximum no. of blocks in a disk

verified different epoch length values to determine an efficient value that fulfills the above requirements. Based on our sensitivity analysis in [20], the energy saving ratio is insensitive to the epoch lengths larger than 30 min. Therefore, we determined the epoch length to be 30-min long since it captures arrival rate changes effectively while exploiting energy savings. Table 1 introduces the main parameters and indices used in the model development. Table 2 introduces the list of decisions variables used in our DPM optimization model including binary, integer, and continuous variables.

Table 2. Decision variables

Name	Type and description
$x_{i,j}^t$	Integer- Quantity of j type fragment on disk i in period t
y_{j,i_1,i_2}^t	Integer- Quantity of j type fragments migrated from $i1$ to $i2$ at the end of period t
$s_{i,k}^t$	Binary- Equals to 1 if disk i is in state k in period t
u_i^t	Binary-Equals to 1 if disk i should be spun up in period t
d_i^t	Binary- Equals to 1 if disk i should be spun down in period t
t_i^k	Continuous- Response time of disk i
T_i^k	Continuous- Response time penalty of disk i

3.1 Formulation of DPM Optimization for Multi-state Disks

Our objective is to minimize the energy consumption within each epoch period. The total energy consumption during an epoch consists of four elements. The first part is the basement energy that relates to disk state (rotation speed) and number of disks spinning in each state. It is independent of the migration operations. The second part is the energy consumed during the migration time which strictly depends on the total fragment size of migration. And, the rest of energy consumption includes energy costs for disk spin-up and spin-down operations. The objective function is shown in the following equation:

$$
\begin{aligned}
min \sum_{t=1}^{T}\sum_{i=1}^{I}\sum_{k} p_{i,k}^t s_{i,k}^t + \sum_{t=1}^{T}\sum_{j=1}^{J}\sum_{i_1=1}^{I}\sum_{(i_2 \in I, i_2 \neq i_1)} c_j y_{j,i_1,i_2}^t + \sum_{t=1}^{T}\sum_{i=1}^{I} ep_i u_i^t \\
+ \sum_{t=1}^{T}\sum_{i=1}^{I} ed_i d_i^t + \sum_{t=1}^{T}\sum_{i=1}^{I}\sum_{k} \Gamma \cdot s_{i,k}^t T_i^k
\end{aligned}
\tag{1}
$$

The physical and logical constraints are as follows: (1) Fragments stored in a disk can never exceed the disk capacity; (2) Disks during an epoch period must stay in a certain state; (3) During any epoch t, there must be at least one active disk serving the data requests; (4) Any fragment can only migrate once in a certain epoch t; (5) A disk in stand-by mode is not considered as source or destination for data migration; (6) There is a limit for data migration (H) that represents the data transfer limit for any disk within an epoch. The migration limit by default is set to half of the epoch. The following equations represent the aforementioned constraints respectively:

$$\sum_{j=1}^{J} b_j x_{i,j}^t \leq Sc_i \quad \forall i, t \tag{2}$$

$$\sum_{k=1}^{K} s_{i,k}^t = 1 \quad \forall i, t \tag{3}$$

$$\sum_{i=1}^{I} s_{i,k=1}^t \leq I - 1 \quad \forall t \tag{4}$$

$$\sum_{i_2} y_{j,i,i_2}^t \leq x_{i,j}^t \quad \forall i, t, j(i_2 \neq i) \tag{5}$$

$$y_{j,i_1,i_2}^t \leq M \cdot \sum_{k=2}^{K} s_{i_1,k}^t \quad \forall i, t, j(i_2 \neq i_1)$$

$$\sum_{j} (\sum_{i_2} y_{j,i,i_2}^t + \sum_{i_2} y_{j,i_2,i}^t) \leq H \quad \forall i, t, j(i_2 \neq i) \tag{6}$$

Also, the migration equation that links $x_{i,j}^t$ and y_{j,i_1,i_2}^t is:

$$x_{i,j}^t + \sum_{i_1 \in I, i_1 \neq i} y_{j,i_1,i}^t = x_{i,j}^{t+1} + \sum_{i_2 \in I, i_2 \neq i} y_{j,i,i_2}^t \quad \forall i, t \geq 1, j \tag{7}$$

And, in order to determine the binary indicating variables related to spin up and down of disks, the following equations are used in the model:

$$\sum_{k} k s_{i,k}^t - \sum_{k} k s_{i,k}^{t+1} \leq u_i^t \tag{8}$$

$$\sum_{k} k s_{i,k}^{t+1} - \sum_{k} k s_{i,k}^t \leq d_i^t \tag{9}$$

3.2 Two-State Optimization Model

It is easy to obtain the model formulation for two-state disk (active and stand-by) storage by setting two values for parameter k (1 or 2) in the general formulation provided in the previous section for multi-mode disk. The equations related to two-mode optimization model are provided in detail in [20]. The general DPM optimization model assumes 10 levels of popularity (hotness) for data fragments based on the observed data request arrival rate. We believe that having 10 levels is sufficient to accurately classify data blocks based on the hotness level (if more resolution would be needed, the model can certainly have more levels that indeed reduce the MPC computational time). An important feature of two-state model is that the least and the second least popular data stay in original disks. This will help to minimize the migration cost.

3.3 Response Time Modeling

The expected response time of a disk is a function of its spinning state and the total arrival rate. Thus, if we consider the state of disk constant, the response time of the disk is a convex function with respect to its hotness level with increasing first derivative order. We modeled this function by using Piecewise linear (PWL) functions for our optimization model since they are widely used to approximate any arbitrary function (specially convex functions) with high accuracy. The input of PWL function is relative hotness of a disk. The relative hotness of a disk is calculated by following equation:

$$\lambda_{t,i} = \frac{\sum_j \lambda_j x_{i,j}}{\lambda^{max} maxfrag} \tag{10}$$

where $0 \le \lambda_{t,i} \le 1, 1 \le \lambda_j \le 10$ is the popularity (arrival rate) of fragment type j, $maxfrag$ is maximum number of fragments in a disk and λ^{max} is upper bound for popularity. We define L as the number of linear functions to approximate the response time. It is well known that PWL functions can represent arbitrary functions to any accuracy by simply increasing the number of segments (L) to the point of desired accuracy. Therefore, we verified different L values for approximation of the response time convex function. We decided to use 9-piece-linear function shown in Fig. 2 for two-state disk storage system since it approximates the convex function with high accuracy.

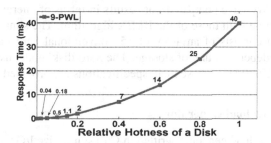

Fig. 2. 9 PWL function of response time model

3.4 Model Predictive Control (MPC)

The presented optimization model is rather static while our real system works in a dynamic online enviroment. Thus, we extend the model to accommodate dynamic scenarios by using Model Predictive Control (MPC) techniques to solve this issue. MPC, also known as receding horizon control (RHC) or rolling horizon control, is a form of control strategy to integrate optimization. Specifically, the current control action is obtained in an online fashion where, at each sampling instant, a finite horizon optimization problem (which is (1)–(10)) is solved and its optimal solution in the first stage is applied as the current control decision while remaining solutions will be disregarded. Such procedure repeats along the whole control process. Therefore, all controllable variables (such as disk status and response time) for the first period are implemented in MPC.

It has been observed that MPC is a very effective control strategy with reasonable computational overhead [18]. The prediction information on workload arrival rate is provided to the MPC optimization model. This plays a key role in developing an accurate underlying mixed integer program model since any mis-prediction on data request arrival rates could cause the model to produce a solution with a less desired quality. However, as observed in many other applications of MPC, since only the first stage solution will be implemented and remaining parts will be ignored, MPC is robust to poor predictions and has a strong adjustment capability [26].

3.5 Solving Strategy

Our initial attempt to find solutions to the two-state model is to implement and solve the model in the well-known Cplex solver. The solver is installed on a server which is connected to the server running the widely used disk simulator, *Disksim* [19], which is utilized as an accurate and reliable simulation platform by many related works. In other words, the model solution is integrated in the storage system simulated in Disksim. Technical details regarding the experimental simulations are provided in Sect. 4.

4   Empirical Evaluation

We conducted simulations under extensive set of dynamic I/O workloads to validate our proposed method. We have compared our results in terms of energy saving ratio and average response time with those of the BLEX algorithm. The simulated disk storage system in Disksim consists of an array of 15 conventional hard disks; each disk is configured as in independent unit of storage. The hard disk model used in simulations is IBM Ultrastar 7K6000 [5] whose main specifications are provided in [20].

4.1   Synthetic Workload Generator

We developed a workload generator written in C to synthesize I/O workloads for disks based on popular database TPC benchmarks. We follow the well-known b/c model in generating a workload of a series of random data read operations (b % of all read operations is against c % of the data) [25]. It is well known that database tuple access pattern is highly skewed and can be described as an 80/20 or even a 90/10 model [20]. Zipf probability distribution is used in the generator to produce b/c model. The default b/c model used in simulations is set to 80/20. We have used Gamma distributions in our workload generator to reflect the dynamic behavior of database I/O disk trace. Given the data correlations among database tuples in queries, the access frequency change pattern of each data fragment type is represented by a Gamma distribution.

4.2   Experimental Platform

Our model is integrated in the disk simulator as well as BLEX, as the comparison target. We enhanced Disksim with a multi-speed disk power model where the power consumption rate is proportional to disk rotation speed. Also, it is augmented with extra features such as dynamic disk spin up (and down), disk state adjustment and inter-disk data migration during the simulation. The predicted access frequency (hotness level) for each fragment type for the next k epochs is provided to the model along with the observed fragment type frequencies in the previous epoch. The prediction is performed by the prediction and autoregressive modeling methods in MATLAB. In particular, based on the observed data access frequency, autoregressive modeling tool develops an identified model. Then, the prediction method forecasts fragments access frequency for k epochs ahead based on the identified model and the observed fragments frequency.

4.3   Simulation Results and Comparisons

In this section, we describe our experimental results in terms of energy saving and average response time under extensive set of dynamic traces.

Energy Saving Results. Figure 3(a) shows energy saving for various I/O traces with different mean arrival rates. Figure 3(a) clearly shows that the DPM optimization model significantly outperforms the BLEX algorithm by saving energy up to 60 %. The proposed model outperforms BLEX with the difference of minimum 16 % and up to 23 % in energy savings. Based on the results, it saves 19 % more energy on average than BLEX. Figure 3(b) shows the total power consumption of the disk storage system for each power saving method compared to that of no power saving (NPS) method applied, where all disks constantly run in active mode. Such results are shown for several I/O traces. We can conclude that DPM optimization model is dominant in power saving.

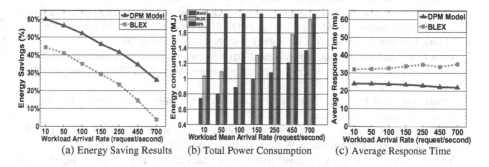

(a) Energy Saving Results   (b) Total Power Consumption   (c) Average Response Time

Fig. 3 Experimental results under dynamic I/O traces with different mean arrival rates

Average Response Time Results. It is important to measure the response time effected by power saving schemes to ensure high quality of service for queries. Figure 3(c) shows the average response time for DPM model and BLEX algorithm under several workloads with various mean arrival rates. Note that the computational time to obtain the solution for both power saving schemes is up to a second, which is apparently ignorable comparing to the epoch length (30 min), and thus it is excluded from the response time computations above. The results show that optimization model provides significantly better response time than BLEX. The reason, in addition to response time consideration in its optimal power-performance tradeoff, is that it takes into account the predicted information on data access frequency for the next epoch in its solutions.

5 Conclusion

Power consumption has increased greatly in data centers, and DBMS is the major energy consumer. Disk storage systems are the most power-hungry components among all in DBMS. Thus, we present our proposals in this paper on designing a power-aware disk storage system that improves on the limitations of previous contributions. We introduced a DPM optimization model extended with the MPC strategy that can be adapted to any multi-speed disk storage system. We developed the two-state DPM optimization model for two-mode disk storage systems since most of the modern disks in the market run in two modes. We evaluated our proposed method by experimental simulations using extensive set of synthetic I/O traces based on popular TPC benchmarks.

Acknowledgments. The project described was supported by grants IIS-1117699 and IIS-1156435 from the National Science Foundation (NSF) of USA. Equipments used in the experiments are partially supported by a grant (CNS-1513126) from NSF.

References

1. http://www.nrdc.org/energy/data-center-efficiency-assessment.asp
2. Gurumurthi, S., Sivasubramaniam, A., Kandemir, M., Franke, H.: Reducing disk power consumption in servers with DRPM. J. Comput. **36**(12), 59–66 (2003)
3. Poess, M., Nambiar, R.O.: Energy cost, the key challenge of today's data centers: a power consumption analysis of tpc-c results. In: VLDB 2008 Proceedings. ACM Press (2008)
4. Moore, F.: more power needed. In: Energy User News, November 2002
5. www.hgst.com/hard-drives/enterprise-hard-drives/enterprise-sas-drives/ultrastar-7k6000
6. Zhu, Q., Zhou, Y.: Power-aware storage cache management. IEEE Trans. Comput. **54**(5), 587–602 (2005)
7. Papathanasiou, A.E., Scott, M.L.: Energy efficient prefetching and caching. In: USENIX Annual Technical Conference, Boston (2004)
8. Li, D., Wang, J.: Eeraid: power efficient redundant and inexpensive disk arrays. In: 11th Workshop on SIGOPS European Workshop, Belgium (2004)
9. Yao, X., Wang, J.: RIMAC: a novel redundancy-based hierarchical cache architecture for energy efficient, high performance storage systems. In: ACM SIGOPS OS Review (2006)
10. Zhu, Q., David, F.M., Devaraj, C.F., Li, Z., Zhou, Y., Cao, P.: Reducing energy consumption of disk storage using power-aware cache management. In: IEEE Proceedings of Software (2004)
11. Pinheiro, E., Bianchini, R.: Energy conservation techniques for disk array-based servers. In: Proceedings of ICS 2004 (2004)
12. Colarelli, D., Grunwald, D.: Massive arrays of idle disks for storage archives. In: ACM/IEEE Conference on Supercomputing, pp. 1–11 (2002)
13. Weddle, C., Oldham, M., Qian, J., Wang, A., Reiher, P., Kuenning, G.: PARAID: A gear-shifting power-aware RAID. ACM Trans. Storage (TOS) 3(3) (2007). 13

14. Verma, A., Koller, R., Useche, L., Rangaswami, R.: SRCMap: energy proportional storage using dynamic consolidation. In: Proceedings of FAST 10, vol. 10
15. Otoo, E., Rotem, D., Tsao, S.-C.: Dynamic data reorganization for energy savings in disk storage systems. In: Gertz, M., Ludäscher, B. (eds.) SSDBM 2010. LNCS, vol. 6187, pp. 322–341. Springer, Heidelberg (2010)
16. Guerra, J., Pucha, H., Glider, J., Belluomini, W., Rangaswami, R.: Cost effective storage using extent based dynamic tiering. In: Proceedings of FAST, pp. 273–286 (2011)
17. Zhu, Q., Chen, Z., Tan, L., Zhou, Y., Keeton, K., Wilkes, J.: Hibernator: helping disk arrays sleep through the winter. In: ACM SIGOPS Operating Systems Review (2005)
18. Garcia, C.E., Prett, D.M., Morari, M.: Model predictive control: theory and practice- a survey. In: Automatica 1989
19. http://www.pdl.cmu.edu/DiskSim/
20. Behzadnia, P., Tu, Y.-C., Zeng, B., Yuan, W., Wang, X.: Dynamic Power-Aware Disk Storage Management in Database Server. Technical report CSE/15-123., Department of Computer Science and Engineering, University of South Florida (2015). http://msdb.csee.usf.edu/E2DBMS/tech-report-123.pdf
21. Kim, J., Chou, J., Rotem, D.: iPACS: power-aware covering sets for energy proportionality and performance in data parallel computing clusters. J. Parallel Distrib. Comput. Elsevier 74(1), 1762–1774 (2014)
22. Kim, J., Chou, J., Rotem, D.: Energy proportionality and performance in data parallel computing clusters. In: 23rd SSDBM Conference, July 2011
23. Chou, J., Kim, J., Rotem, D.: Energy-aware scheduling in disk storage systems. In: Proceedings of ICDCS 2011 (2011)
24. The SDSS DR1 SkyServer. http://skyserver.sdss.org/dr1/en/skyserver/paper/
25. Nicola, M., Jarke, M.: Performance modeling of distributed and replicated databases. IEEE Trans. Knowl. Data Eng. (TKDE) 12(4), 645–672 (2000)
26. http://cepac.cheme.cmu.edu/pasilectures/lee/LecturenoteonMPC-JHL.pdf

FR-Index: A Multi-dimensional Indexing Framework for Switch-Centric Data Centers

Yatao Zhang, Jialiang Cao, Xiaofeng Gao$^{(\boxtimes)}$, and Guihai Chen

Shanghai Key Laboratory of Data Science, Department of Computer Science and Engineering, Shanghai Jiao Tong University, Shanghai 200240, China
confidentao@gmail.com, mintyck@gmail.com,
{gao-xf,gchen}@cs.sjtu.edu.cn

Abstract. Data center occupies a decisive position in business of data management and data analysis. To improve the efficiency of data retrieval in a data center, we propose a distributed multi-dimensional indexing framework for switch-centric data centers with tree-like topologies. Taking Fat-Tree as a representative, which is a typical switch-centric data center topology, we design FR-Index, a two-layer indexing schema fully taking advantage of the Fat-Tree topology and R-tree indexing technology. In the lower layer, each server indexes the local data with R-tree, while in the upper layer the distributed global index depicting an overview of the whole data set. To improve the efficiency of query processing, we also provide special techniques to reduce the dimensionality of the index. Experiments on Amazon's EC2 show that our proposed indexing schema is scalable, efficient and lightweight, which can significantly promote the efficiency of query processing.

1 Introduction

An attractive challenge for large scale distributed storage systems is how to retrieve specified data from massive data set efficiently. Empirically, designing appropriate and effective index is a typical solution for this challenge. [8–10] designed a distributed indexing scheme. Each of them deploys a P2P network on the distributed server cluster as an overlay for data mapping and routing queries. However, running P2P networks must consume some extra overhead.

In recent years, as a kind of infrastructure, *Data Centers* are playing increasingly vital role in cloud services. A new challenge is to construct efficient indexing systems for storage systems deployed on data centers. Servers and switches in a data center is connected by a well-designed data center network (DCN), which may be helpful for indexing. Gao et al. [3–5,7] have done some network-aware

This work has been supported in part by the China 973 project (2014CB340303), the Opening Project of Key Lab of Information Network Security of Ministry of Public Security (The Third Research Institute of Ministry of Public Security) Grant number C15602, the Opening Project of Baidu (Grant number 181515P005267), the Open Project Program of Shanghai Key Laboratory of Data Science (No. 201609060001) and National Natural Science Foundation of China (No. 61472252, 61133006).

S. Hartmann and H. Ma (Eds.): DEXA 2016, Part II, LNCS 9828, pp. 326–334, 2016.
DOI: 10.1007/978-3-319-44406-2_26

indexing techniques. Those works presented excellent proposals, yet they have not discussed the multi-dimensional indexing in switch-centric DCN's.

This paper will present our well-designed proposal for constructing distributed multi-dimensional indexing framework on switch-centric DCN's. Taking Fat-Tree [1] as a representative, we design **FR-Index**, a two-layer indexing system fully taking advantage of the Fat-Tree topology and R-tree indexing technology. In the lower layer, each server indexes the local data with R-tree. In the upper layer, the distributed global index depicts an overview of the whole data set.

To improve the efficiency of query processing, we also provide special techniques to reduce the dimensionality of the FR-Index. Efficient query processing method is also put forward based on this indexing system. Finally, we conduct experiments on Amazon's EC2 to evaluate the performance of FR-Index.

2   Related Works

As a subclass of switch-centric DCN's, tree-like DCN's connect devices by links similar to a multi-rooted tree. Most tree-like DCN's tend to divide lower layer switches and servers into some substructures, like *pod* in Fat-Tree. A k-pod Fat-Tree consists of three layer of k-port switches. In each pod, the $k/2$ aggregation layer switches and the $k/2$ edge layer switches interconnect as a complete bipartite graph. Every switch in aggregation layer connects to $k/2$ switches in core layer. Every switch in the edge layer connects to $k/2$ servers. Thus, a k-pod Fat-Tree can support connecting $k^3/4$ servers. [1] designs different IP addressing rules for switches and servers. For pod switches, the form *10.pod.swi.1* acts as their IP addresses, where $pod \in [0, k-1]$ denotes the pod number, and *swi* denotes the position of the switch in the pod (in $[0, k-1]$, starting from left to right, bottom to top). The address of a server is *10.pod.swi.ID* where *pod* and *swi* follows the address of the edge switch which the server connects, and *ID* (in $[2, k/2 + 1]$, starting from left to right) denotes the server's position in that subnet. Figure 1 illustrates a Fat-Tree topology with 4 pods and examples of the addressing scheme.

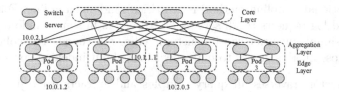

Fig. 1. A Fat-Tree topology with 4 pods

3   FR-Index

All servers in the data center participate in constructing index and maintaining the consequent FR-Index. Besides servers in the data center, we set an individual server as a historical data collector (called **collector** for short) of FR-Index. The collector will collect some historical data as the basis of some decisions we made.

An FR-Index system is composed of a set of *index instance*s denoted by $\mathcal{I}=\{I_1, I_2, \cdots, I_w\}$. An index instance I_i indexes an *"indexing space"* denoted by $I_i.space$, which is composed of several selected dimensions of the data set. The FR-Index collector generates $I_i.space$ and informs all servers of $I_i.space$. Each server builds a local R-tree to index its local data on the dimensions contained by $I_i.space$. Then each server publishes a portion of local R-tree index nodes to different servers based on our proposed mapping schema to compose the global index. So, we get a distributed global index and each server maintains a portion of the global index. An index instance I_i can be regarded as a combination of all local indexes and the distributed global index which are built on $I_i.space$.

Suppose that a data set is composed of d attributes. Every element in the data set can be regarded as an object in a d-dimensional space which can be denoted by $D=\{D_0, \cdots, D_{d-1}\}$. Thus, each $I_i.space$ is a subset of D. Additionally, a multi-dimensional query is denoted as $Q(Ctr)$, where $Ctr=\{ctr_1, \cdots, ctr_u\}$ is a set of query criteria on u dimensions. We take a set $Qd=\{qd_1, \cdots, qd_u\}$ to represent the u dimensions. Obviously, Qd is a subset of D.

3.1   Selecting Indexing Dimensions

In most cases, a query would not be related to too many dimensions of the data, which means that for a query $Q(Ctr)$, the cardinality of the corresponding Qd wouldn't be too large. Thus, it is necessary to reduce the dimensions of the proposed multi-dimensional index. However, a single index which is built on a few dimensions might not facilitate processing all queries. Hence, we build a set of index instances $\mathcal{I}$, as mentioned above. To better manage the indexing system, we set each $|I_i.space|$ as a fixed value in an FR-Index. For example, we will show how to determine a set of indexing spaces with 3 dimensions for an FR-Index.

The FR-Index collector collects query samples by requesting servers for their logs. An optional sampling method is stratified random sampling. For example, we can send log requests to all servers or to servers in some randomly selected pods. The sample quantity can be customized or self-tuned. The collector then analyses those query samples to make decisions about indexing spaces:

1. The collector traverses all query samples and extracts each query's Qd. A histogram is maintained to record the occurrence frequency P_i of every different Qd_i. Then, all different Qd's are sorted by the occurrence frequency in descending order, denoted as a collection $\mathcal{D}=\{Qd_1, Qd_2, \cdots, Qd_m\}$.
2. The collector selects the first x sets in $\mathcal{D}$ by calculating an integer x which satisfies $\sum_{j=1}^{x} P_j \geqslant P_{thr}$, where $P_{thr} \in [0,1]$ is a threshold to control the performance of FR-Index. Usually, a higher P_{thr} might incur more index

instances accompanied by more maintaining costs and higher query processing efficiency, while a lower P_{thr} may lead to opposite results. Now $\mathcal{D}$ is
pruned into $\mathcal{D}_p=\{Qd_1, Qd_2, \cdots, Qd_x\}$. We regard that $\mathcal{D}_p$ depicts the feature of $P_{thr} \times 100$ percent of all historical queries as well as all subsequent
queries.

3. Based on $\mathcal{D}_p$, the collector finds a collection $\mathcal{D}_{ans}=\{Dc_1, Dc_2, \cdots, Dc_y\}$ which
   has the following three properties:
   (a) $\forall i \in \{1, 2, \cdots, y\}$, $Dc_i \subseteq \boldsymbol{D}$ and $|Dc_i| = 3$.
   (b) $\forall j \in \{1, 2, \cdots, x\}$, if $|Qd_j| < 3$, $\exists i \in \{1, 2, \cdots, y\}$, such that $Qd_j \subseteq Dc_i$.
   (c) $\forall j \in \{1, 2, \cdots, x\}$, if $|Qd_j| \geqslant 3$, $\exists i \in \{1, 2, \cdots, y\}$, such that $Dc_i \subseteq Qd_j$.

Each set in $\mathcal{D}_{ans}$ will become an 3-dimensional indexing space on which an
index instance will be built. All of these index instances constitute an FR-Index
system. Since $\mathcal{D}_p$ depicts the feature of $100P_{thr}$ percent of all historical queries,
properties (b)(c) of $\mathcal{D}_{ans}$ guarantee that our FR-Index can efficiently facilitate
processing $100P_{thr}$ percent of all subsequent queries.

3.2   Partitioning Indexing Space

The information of a selected indexing space will be sent to all servers by the
FR-Index collector. Once a server received the information, it will build a local R-
tree index on the dimensions contained by the indexing space. To better illustrate
our proposal, we will take one index instance as an example to show that how
our system works, since we build different index instances independently.

As we mentioned above, a server needs to maintain a portion of the global
index. A new challenge is to determine the range of the global index that a server
should be responsible for. In the following discussion, we denote this range as
Potential Indexing Range (PIR). As a tree-like data center network, the hierarchical structure of Fat-Tree offers us a convenient and efficient way to partition
the indexing space such that we can generate PIR for every server.

In the whole data center, all multi-dimensional data forms a data boundary
denoted as $\boldsymbol{B}=(B_0, B_1, B_2, \cdots, B_{d-1})$, which is a d-dimensional rectangle as the
bounding box of the spatial data objects. Here each B_i is a closed bounded interval $[l_i, u_i]$ describing the range which is covered by the data objects along dimension D_i. Suppose that we had chosen an indexing space $\boldsymbol{I}_j.space=(D_0, D_1, D_2)$.
Since $\boldsymbol{B}'=(B_0, B_1, B_2)$ is the "meaningful" subspace of $\boldsymbol{I}_j.space$ for our work,
we will consider that $\boldsymbol{I}_j.space=\boldsymbol{B}'=(B_0, B_1, B_2)$ in the following discussion.

In a k-pod Fat-Tree, we code a server S_t by $t = (k/2)^2 pod+(k/2)swi+(ID-2)$,
where *pod*, *swi* and *ID* are parameters in the IP address of this server.

Intuitively, we partition the indexing space into k (the number of pods) parts
along the first dimension, then $k/2$ (the number of edge switches in each pod)
parts along the second dimension, then $k/2$ (the number of servers connecting with each edge server)parts along the third dimension. Now, we gain $k^3/4$
equal-sized partitions of the indexing space and map each partition to a server.
Therefore, for server S_t, its PIR denoted by pir_t can be expressed by Eq. (1).

$$pir_t = \{[l_0^t, u_0^t], [l_1^t, u_1^t], [l_2^t, u_2^t]\} = \left\{ \begin{bmatrix} \frac{kl_0 + pod(u_0 - l_0)}{k}, \frac{kl_0 + (pod+1)(u_0 - l_0)}{k} \end{bmatrix}, \\ \begin{bmatrix} \frac{kl_1 + 2swi(u_1 - l_1)}{k}, \frac{kl_1 + 2(swi+1)(u_1 - l_1)}{k} \end{bmatrix}, \\ \begin{bmatrix} \frac{kl_2 + 2(ID-2)(u_2 - l_2)}{k}, \frac{kl_2 + 2(ID-1)(u_2 - l_2)}{k} \end{bmatrix} \right\} \quad (1)$$

3.3   Publishing in FR-Index

To build the global index, a server, say, S_t, adaptively selects a set of R-tree nodes, $\mathbf{IN_t} = \{In_t^1, \cdots, In_t^n\}$, from its local index and publishes them into the global index. The index nodes in $\mathbf{IN_t}$ should cover all data stored in S_t. Each In_t^i in $\mathbf{IN_t}$ will be published as a format (ip_t, mbr_i), where ip_t is the IP address of S_t and mbr_i is the bounding box of In_t^i, which represents a hypercube.

Wang et al. [8] proposed a novel mapping schema to regulate the publishing process and the consequent query processing. We will transplant this schema to our system to determine the servers which will store a published R-tree node as a portion of the global index.

For an R-tree node In_t^i to be published, we take its center $In_t^i.c$ and radius $In_t^i.r$ of its bounding box as the criteria for mapping. We first map the node to the server whose PIR contains $In_t^i.c$, say S_x. Then S_x compares $In_t^i.r$ with a predefined threshold, say r_{max}. If $In_t^i.r > r_{max}$, then the node will be sent to each server whose PIR intersects with mbr_i. Otherwise, the node will be stored by S_x only. The impact of r_{max}'s value will be discussed in Sect. 4.

4   Query Processing

Suppose that we have deployed an FR-Index which is composed of several 3-dimensional index instances on the data center. Given a multi-dimensional query $Q(Ctr)$ where $Ctr = \{ctr_1, \cdots, ctr_u\}$, generally, ctr_j is a key or range criterion along qd_j $(1 \leqslant j \leqslant u)$. Considering that $u \geqslant 3$, we first prune Ctr into an indexing space $I_i.space$, where $I_i.space \subseteq Qd$ such that $Q(Ctr)$ is converted to $Q(Ctr')$. (If $u < 3$, we expand Ctr to match an indexing space $I_i.space$ where $Qd \subseteq I_i.space$.) Then we take advantage of the index instance I_i to process $Q(Ctr')$ and retrieve a result set. At last, we prune the result set according to criteria in the set Ctr/Ctr'. Furthermore, we call $Q(Ctr')$ a point query if the elements in Ctr' are all key criteria. Otherwise, we call $Q(Ctr')$ a range query.

A point query $Q(Ctr')$ is denoted as $Q(key)$. We first forward it to the server S_{init} whose potential indexing range pir_{init} contains key. S_{init} generates a hypersphere at the point key with radius r_{max} (the threshold we defined in Sect. 3.3). This hypersphere is defined as the search space of $Q(key)$, denoted by $key.searchspace$. S_{init} forwards $Q(key)$ to servers whose potential indexing ranges overlap with $key.searchspace$. Then, S_{init} and those servers search the global index buffered in their memory to find those published R-tree nodes whose bounding box contains key. The servers which publish those R-tree nodes will get $Q(key)$ and search on the local indexes to get the query results. At last, the query results will be pruned according to criteria in the set Ctr/key.

A range query is denoted as $Q(range)$, where $range=\{[a_1, b_1], \cdots, [a_{u'}, b_{u'}]\}$. We define $range.c=(\frac{a_1+b_1}{2}, \cdots, \frac{a_{u'}+b_{u'}}{2})$ and $range.r=\frac{1}{2}\sqrt{\sum_{i=1}^{u'}(b_i - a_i)^2}$. The search space for $Q(range)$ is a hypersphere with the center at $range.c$ and its radius is $range.r + r_{max}$. The following processing is similar with point query processing. [8] has proved that for $Q(key)$ or $Q(range)$, if we search the search space we generated, we can guarantee the completeness of the results.

Up to now, we can discuss the impact of r_{max} on the FR-Index system. According to Sect. 3.3, a smaller r_{max} incurs more index node replicas, which increases the maintenance cost. In the other hand, according to the query processing strategy, a larger r_{max} means that we must search more servers to retrieve complete results for a query, which reduces the efficiency of query processing.

5   Index Updating

In our proposal, each server chooses some index nodes from local R-tree index and publishes them into global index. A high-level R-tree node may incur less update costs, but it generates more false positives. Besides, its bounding box may overlap with more servers' potential indexing ranges, which increases storage cost and query processing complexity. Therefore, it is crucial to choose "proper" index nodes to publish. At the first time we construct FR-Index, we have no knowledge about the query pattern and the data updating pattern. Therefore, in an h-level local R-tree, we publish the index nodes in the $h - 1$ level (the h level nodes is the leaf nodes). After running the system for a period of time, by analysing the query pattern and node updating pattern, we adopt a cost model which is similar with the model in our former work [3] to choose some local index nodes with lower cost to update some published index nodes with higher cost.

Another updating requirement for our system is to update the index instances we have constructed. We propose a simple and efficient strategy to deal with this requirement. Each server stores a histogram to maintain the accessing status for every index instance. The FR-Index collector will request the histograms from the servers and adopt Least Recently Used algorithm to delete obsolete index instances and add new index instances.

6   Performance Evaluation

We evaluate our proposed indexing framework on Amazon's EC2 platform. We organize EC2 computing units into a simulative data center with the Fat-Tree topology. Experimental data set is YearPredictionMSD [2], achieved from UCI Machine Learning Repository [6]. We conduct several experiments to evaluate our proposal. Table 1 lists some common experiment settings.

Table 1. Common experiment settings

Configuration Items	Setting
Size of Data Center	4-pod(16 servers) or 6-pod(54 servers)
Number of Stored Data Items	20 k, 40 k, 60 k, 80 k or 100 k on each server
Indexed Dimensions	2nd, 3rd and 4th attributes of the dataset
Query Criteria	2nd, 3rd, 4th, 5th and 6th attributes of the dataset
Number of Queries	1 k, 2 k, 3 k, 4 k or 5 k
r_{max}	larger than the radius of 70 % of published index nodes

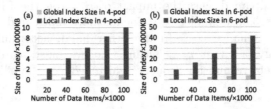

Fig. 2. Size of global index and local index

Evaluation for Index Construction: We randomly placed some data items on each server. Then, we build a 3-dimensional FR-Index instance on the data center. Figure 2 shows that the size of the global index is almost 10x smaller than that of local indexes under the same setting, which indicates that FR-Index is lightweight. Moreover, this advantage can retain when the data center or data capacity becomes larger, which verifies that FR-Index is scalable.

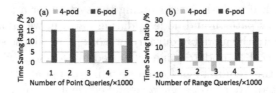

Fig. 3. *TSR* in query processing

Evaluation for Query Processing: We place 20,000 data items on each server randomly. Then a 3-dimensional FR-Index instance is built to facilitate query processing. We set two query processing strategies: (1) Strategy designed in Sect. 4 with the assistance of the FR-Index instance. (2) Broadcast queries to all servers and each server searches locally. Suppose that it costs T_1 time with the first strategy and T_2 time with the second one to process the same set of queries. We define $(T_2 - T_1)/T_2$ as time saving ratio of our proposal.

Figure 3(a) and (b) show time saving ratio (*TSR*, for short) in point/range query processing respectively. In the 4-pod data center, due to the extra cost

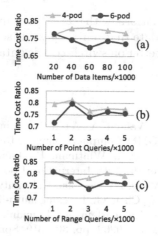

Fig. 4. FR-Index VS RT-CAN

for query forwarding and storage accessing, FR-Index behaves not very well. However, in the 6-pod data center, our proposal can reduce the time cost for query processing by nearly 20 %, which verifies the time-saving advantage of FR-Index will enhance with the increasing of the scale of the data center.

Comparison with RT-CAN: RT-CAN is the most relevant work to our proposal we have learned about. We follow [8] to implement the RT-CAN and make two comparisons between RT-CAN and FR-Index: (1) Suppose that it costs T_1 time to construct an FR-Index instance and T_2 time to construct an RT-CAN index. Figure 4(a) shows the value of $T_1:T_2$ under different configurations. (2) We place 20,000 data items on each server randomly and build an FR-Index instance and a RT-CAN index on the same indexing space. Then we adopt each of them to process queries. Suppose that FR-Index costs T_3 time and RT-CAN costs T_4 time to process the same set of queries. Figure 4(b)(c) respectively show the value of $T_3:T_4$ under different configurations. Figure 4 indicates that FR-Index saves 18 %–30 % of time cost for index constructing and 18 %–28 % of time cost for query processing than RT-CAN. Additionally, the performance difference between RT-CAN and FR-Index becomes larger with the increasing of the scale of the data center. It verifies that the FR-Index behaves better on the data center with tree-like switch-centric topology than RT-CAN.

7 Conclusion

This paper presents a distributed multi-dimensional indexing framework for data center with tree-like switch-centric data center network. We design a two-layer multi-dimensional indexing system FR-Index. Query processing and index updating strategies are also proposed based on FR-Index. Additionally, we design some tuning techniques for performance improvement. We evaluate the performance of

FR-Index on Amazon EC2 platform with real data set and compare FR-Index with RT-CAN. Experiments validate that our proposal is scalable, efficient and light-weight, which can behaves better on switch-centric data center.

References

1. Al-Fares, M., Loukissas, A., Vahdat, A.: A scalable, commodity data center network architecture. ACM SIGCOMM Comput. Commun. Rev. **38**(4), 63–74 (2008)
2. Bertin-Mahieux, T., Ellis, D.P., Whitman, B., Lamere, P.: The million song dataset. In: Proceedings of the 12th International Society for Music Information Retrieval Conference (ISMIR), pp. 591–596, University of Miami (2011)
3. Gao, L., Zhang, Y., Gao, X., Chen, G.: Indexing multi-dimensional data in modular data centers. In: Chen, Q., Hameurlain, A., Toumani, F., Wagner, R., Decker, H. (eds.) DEXA 2015. LNCS, vol. 9262, pp. 304–319. Springer, Heidelberg (2015)
4. Gao, X., Li, B., Chen, Z., Yin, M., Chen, G., Jin, Y.: FT-Index: a distributed indexing scheme for switch-centric cloud storage system. In: IEEE International Conference on Communications (ICC), pp. 301–306 (2015)
5. Hong, Y., Tang, Q., Gao, X., Yao, B., Chen, G., Tang, S.: Efficient R-tree based indexing scheme for server-centric cloud storage system. IEEE Trans. Knowl. Data Eng. **28**(6), 1503–1517 (2016)
6. Lichman, M.: UCI machine learning repository (2013). http://archive.ics.uci.edu/ml
7. Liu, Y., Gao, X., Chen, G.: A universal distributed indexing scheme for data centers with tree-like topologies. In: Chen, Q., Hameurlain, A., Toumani, F., Wagner, R., Decker, H. (eds.) DEXA 2015. LNCS, vol. 9261, pp. 481–496. Springer, Heidelberg (2015)
8. Wang, J., Wu, S., Gao, H., Li, J., Ooi, B.C.: Indexing multi-dimensional data in a cloud system. In: ACM International Conference on Management of data (SIGMOD), pp. 591–602 (2010)
9. Wu, S., Jiang, D., Ooi, B.C., Wu, K.L.: Efficient B-tree based indexing for cloud data processing. Int. Conf. Very Large Data Base (VLDB) **3**(1–2), 1207–1218 (2010)
10. Wu, S., Wu, K.L.: An indexing framework for efficient retrieval on the cloud. IEEE Data Eng. Bull. **32**(1), 75–82 (2009)

Unsupervised Learning for Detecting Refactoring Opportunities in Service-Oriented Applications

Guillermo Rodríguez[(✉)], Álvaro Soria, Alfredo Teyseyre, Luis Berdun,
and Marcelo Campo

ISISTAN Research Institute (CONICET-UNICEN), Campus Universitario,
Paraje Arroyo Seco, B7001BBO Tandil, Bs. As., Argentina
{guillermo.rodriguez,alvaro.soria,alfredo.teyseyre,
luis.berdun,marcelo.campo}@isistan.unicen.edu.ar

Abstract. Service-Oriented Computing (SOC) has been widely used for building distributed and enterprise-wide software applications. One major problem in this kind of applications is their growth; as size and complexity of applications increase, the probability of duplicity of code increases, among other refactoring issues. This paper proposes an unsupervised learning approach to assist software developers in detecting refactoring opportunities in service-oriented applications. The approach gathers non-refactored Web Service Description Language (WSDL) documents and applies clustering and visualization techniques to deliver a list of refactoring suggestions to start working on the refactoring process. We evaluated our approach using two real-life case-studies by using internal validity criteria for the clustering quality.

Keywords: Service-oriented applications · Web services · Unsupervised machine learning · Web service description language · Service understandability · Software visualization

1 Introduction

Nowadays, Web services are in the cutting-edge of the Service-Oriented Computing (SOC) paradigm. Encouraged by the rapid advances in distributed system technologies, most organizations capitalize on SOC by discovering and reusing services already accessible over the Internet. As a consequence, software developers take advantage of platform neutrality and self-descriptiveness of Web services to build distributed applications in heterogeneous contexts [4]. To successfully evolve in distributed applications, refactoring is a commonly used engineering practice viewed as a prerequisite to adding new functionality or features to software systems. In this context, WSDL documents are at the heart of refactoring processes, since these files must be transformed and adapted, preserving the specified API description. Refactoring WSDL documents forces developers to

© Springer International Publishing Switzerland 2016
S. Hartmann and H. Ma (Eds.): DEXA 2016, Part II, LNCS 9828, pp. 335–342, 2016.
DOI: 10.1007/978-3-319-44406-2_27

invest time and effort into discovering refactoring opportunities, while browsing the whole body of WSDL documents, as well as their dependencies. Although software evolution has been widely addressed in object-oriented software development, refactoring has yet to be explored in the web services domain [20].

In this context, we claim that it is necessary to further simplify the process of refactoring service-oriented applications by reducing the set of WSDL documents and the number of operations without neglecting functional and non-functional requirements. This paper proposes *VizSOC*, an unsupervised learning approach to assist developers in detecting opportunities, as an initial step towards the process of refactoring SOC applications. To address this issue, the approach utilizes different clustering algorithms, namely K-Means [13], Partitioning Around Medoids (PAM) [8], X-Means [16] and COBWEB [5], which allow developers to alleviate the task of discovering refactoring opportunities in WSDL documents. Nonetheless, clustering might yield numerous clusters and relationships that could burden developers' understanding. In this context, software visualization might help software developers enhance software comprehension, maintenance and evolution [19].

To evaluate our approach, we have utilized two datasets of WSDL documents. The first dataset belongs to a large Argentinean government agency and contains 32 WSDL documents that represent 261,688 lines of code and 39 operations. The second dataset is a medium-size system taken from the literature and consists of 211 non-refactored WSDL documents, representing 44,627 lines of code and 252 operations. To measure the quality of clustering algorithms internal validation indexes (such as intra-cluster diameters and inter-cluster distances) were calculated.

The paper is organized as follows. Section 2 discusses related works in the research field. Section 3 presents the *VizSOC* approach and provides a detailed account of the assistance process. Section 4 describes the experiments performed to validate our approach. Finally, in Sect. 5 we present our conclusions.

2   Related Work

Several research works have explored clustering to reduce efforts in the development of service-oriented applications. For instance, Sabou et al. proposed Cluster Map as a visualization technique to support analysis, comparison, and search of Web Services [18]. Kuhn et al. used semantic hierarchical clustering to group source artifacts that use similar vocabulary to improve software maintenance [9]. Liu et al. created a search engine that reduces search space for service discovery by using tree-traversing ant algorithm [11]. Along this line, Ma et al. aimed to eliminate Web services irrelevant with respect to a query during the discovery process by using K-Means [12]. Elgazzar et al. proposed to cluster Web services based on function similarity prior to retrieving the relevant Web services for a user query in the context of Web service engines. The approach uses K-Means and Normalized Google Distance as a featureless distance measure between words [3].

Unlike our approach, Fokaefs et al.'s approach used a WSDL documents clustering to study the evolution of the files in the software lifecycle of service-oriented applications. In that case, the approach first recognizes changes in WSDL documents; then, it analyzes changes that occur in subsequent versions of the WSDLs [6]. Dong et al. proposed an approach that uses hierarchical agglomerative clustering, underlying the Woogle search engine for Web services. By using this approach, traditional keyword-based search is outperformed by exploiting the underlying structure and semantics of the Web services [2]. Kumara et al. proposes an approach to help developers search Web services by visualizing the Web service data on a spherical surface [10]. Most of the aforementioned approaches have updated clustering techniques to optimize service search engines; instead, our work proposes to cluster similar operations by comparing Web service interfaces, as a step towards the detection of refactoring opportunities in service-oriented applications.

3   Proposed Approach

In this paper, we propose an approach based on the application of Web mining and machine learning techniques to WSDL documents for the detection of software refactoring opportunities. This approach can aid software developers in refactoring service-oriented applications and increase their quality in terms of maintainability, performance and flexibility, among others. Figure 1 gives a general overview of the proposed approach named *VizSOC*, which consist of three main modules: Web mining module, Clustering module and Visualization module. The first module receives the list of non-refactored WSDL documents and applies a series of operations so as to the clustering techniques can work properly on these files. The operations performed in this module are splitting combined words, stemming, stop words removal and vector space model representation. The second module is in charge of performing clustering techniques selected by the user. By means of the *VizSOC* user interface, the user can select and configure a clustering technique from the following set: PAM, K-Means, COBWEB and X-Means. Finally, the third module displays the clustered Web service operations by applying a suitable software visualization technique named Hierarchical Edge Bundles. Afterwards, the list of clusters (i.e., refactoring opportunities or suggestions) are presented to the software developer who will be responsible for conducting the manual refactoring on the WSDL documents.

3.1   A Motivating Example

The first stage of VizSOC is the Web mining module, in which a set of WSDL documents belonging to a service-oriented application is given as input. VizSOC mines three types of features of a WSDL document, namely operations, messages and input/output parameters. Then, the splitting of compound words is carried out, along with stop-word removal and stemming. These filtering techniques have been selected based on their proven effectiveness in supervised machine learning [1]; moreover, using these techniques allows the approach to detect refactoring opportunities as effectively as expert developers might. The last Web mining step is to use the Vector Space Model (VSM) to represent mined WSDL features as a vector of

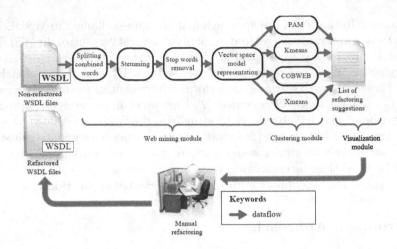

Fig. 1. Overview of the approach

terms. For example, let us suppose there are three operations of a WSDL document represented as follows:

O_1=(*getPersonByID, id, int, nameAndsurname, string, dateOfbirth, string*)

O_2=(*getPersonByWIN, win, string, nameAndsurname, string, dateOfbirth, string*)

O_3=(*getWINByID, id, int, nameAndsurname, string, win, string*)

The first operation obtains a person by her ID (datatype integer); the result is in the form of name and surname, (datatype string) and date of birthday (datatype string). The second operation retrieves a person given a WIN (Work Identification Number, datatype string due to formatting issues); the result is also in the form of name and surname, (datatype string) and date of birthday (datatype string). The third operation obtains a person by her WIN (datatype string); the result is in the form of name and surname, (datatype string), date of birthday (datatype string) and WIN (datatype string). Once the operations vectors are built, the process of building the T vector takes place. T vector represents the list of all the different terms (i.e., features) in the operations vectors. Following our example, T vector is built as follows:

T=(*getPersonByID, id, int, nameAndsurname, string, dateOfbirth, string, getPersonByWIN, win, getWINByID*).

For the sake of simplicity, in the example we disregarded the use of splitting of compound word, stop-word, and stemming. Then, our approach represents each operation as a numerical vector whose size is the length of T and each position means the weight for each term operation. The approach to determine the weight is to assign the number of times that each element of T appears in O_i. Then each operation is represented as follows:

$$O_1 = (1, 1, 1, 1, 2, 1, 0, 0, 0)$$

$$O_2 = (0, 0, 0, 1, 3, 1, 1, 1, 0)$$

$$O_3 = (0, 1, 1, 0, 1, 0, 0, 1, 1)$$

The aforedescribed representations of the WSDL documents will be given as input to different interchangeable clustering algorithms, namely K-Means, Partitioning Around Medoids, X-Means, and COBWEB in the clustering module. These algorithms are utilized to group similar features of the WSDL documents and facilitate the refactoring process.

In the last stage, the visualization module, *VizSOC* displays clustering visualization techniques with a list of refactoring suggestions, which represents similar operations that should be grouped in a single WSDL document. This visualization is suitable for helping software developer to better comprehend the context of the application and identify the refactoring opportunities [19]. Out of the set of clustering visualization techniques, we chose Hierarchical Edge Bundles (HEB) since it is a new and attractive technique and reduces considerably efforts to interpret results and analyze visualized data [7]. HEB is a flexible and generic technique for the visualization of compound (di)graphs, which is based on the principle of visually bundling adjacency edges together. HEB technique has remarkable features that lead us to select it as our visualization approach to visualize clustering of Web Services. Firstly, HEB is a flexible and generic method that can be used in conjunction with existing tree visualization techniques to enable users to choose the tree visualization that they prefer and to facilitate integration into existing tools. Secondly, HEB reduces visual clutter when tackling large numbers of adjacency edges. Thirdly, HEB provides an intuitive and continuous way to control the strength of bundling. For instance, low bundling strength mainly provides low-level, node-to-node connectivity information, whereas high bundling strength provides high-level information determined by implicit visualization of adjacency edges between parent nodes, which are the result of explicit adjacency edges between their respective child nodes.

4   Experimental Evaluation

The first case-study was carried out in a system belonging to a large Argentinean government agency [17]. On average, the system contains 32 non-refactored WSDL documents that represent 261,688 lines of code and 39 operations. The second case-study was carried out in a medium-size system taken from the dataset described in [14]. The system dataset consists of 211 non-refactored WSDL documents, representing 44,627 lines of code and 252 operations. We set the k values for the K-Means and PAM, which are the algorithms that require specification of the k value beforehand. These values were obtained by calculating the *silhouette coefficient* with an iteration value set to 10,000. The selected k value (X axis) is the one that maximizes the *silhouette coefficient* (Y axis). In our context, K is 6 (*silhouette coefficient*=0.31) in case-study 1 and 101 (*silhouette coefficient*=0.25) in case-study 2 in the context of K-Means, whereas k is 11 (*silhouette coefficient*=0.32) in case-study 1 and 51 (*silhouette coefficient*=0.26) in case-study 2 in the context of PAM.

To measure the quality of the clustering algorithms, internal validity criteria were utilized. Internal validity criteria compare different sets of clusters without reference to external knowledge about the given data and/or similarities. For instance, six *intra-cluster distances* and three *inter-cluster diameters* are used to obtain measures of cluster compactness and separation [15]. *Inter-cluster distances* are *Single Linkage Distance* (SLD), *Complete Linkage Distance* (CLD), *Average Linkage Distance* (ALD), *Centroid Linkage Distance* (CeLD), *Average of Centroids Linkage* (ACL) and *Hausdorff Metric* (HM); whereas *intra-cluster diameters* are *Complete Diameter* (CD), *Average Diameter* (AD) and *Centroid Diameter* (CeD). All these metrics were calculated by using the Euclidean distance. The criterion to assess the clustering techniques was defined as follows: the higher the intra-cluster diameter and the lower the inter-cluster distance are, the better the clustering algorithm performs.

After carrying out the experiments, Table 1 summarizes the results in terms of the internal validity metrics. We calculated the inter-cluster distances (rows 1 to 6) and the intra-cluster diameters (rows 8 to 10) of each clustering algorithm for the two case-studies. Each cell value in Table 1 represents the arithmetic mean of each algorithm for each of the aforementioned metrics. In all the experiments, the average scores of the metrics are reported for 1,000 different runs of each algorithm. The results show that COBWEB minimizes most of the intra-cluster diameters and maximizes most of the inter-cluster distances. In case-study 1, COBWEB maximized all the inter-cluster distances and minimized all the intra-cluster diameters; whereas, in case-study 2, COBWEB maximized 83.33 % of the inter-cluster distances and minimized 33.33 % of the intra-cluster diameters. In this case, K-Means minimized 66.66 % the intra-cluster diameters, but maximized 0 % of the inter-cluster distances.

Table 1. Results in terms of internal validity criteria.

Metric	K-Means		PAM		X-Means		COBWEB	
	case 1	case 2	case 1	case 2	case 1	case 2	case 1	case 2
SLD	20.622	11.090	21.814	12.294	18.580	11.342	**24.063**	**12.869**
CLD	25.189	11.920	23.877	**14.592**	22.918	13.283	**25.857**	14.464
ALD	22.695	11.524	22.721	13.510	20.646	12.422	**24.838**	**13.719**
CeLD	17.832	10.540	21.558	12.395	16.494	10.645	**23.853**	**12.891**
ACL	19.861	11.134	22.273	13.057	18.282	11.704	**24.515**	**13.415**
HM	23.525	11.526	22.798	13.382	21.105	12.397	**24.788**	**13.504**
CD	15.370	**4.660**	6.042	6.433	14.197	7.025	**5.359**	4.960
AD	13.405	4.137	4.933	4.858	12.259	5.748	**4.338**	**4.126**
CeD	7.717	**2.364**	3.089	3.036	7.073	3.664	**2.743**	2.450

5 Conclusions

We presented an unsupervised learning approach to assist software developers in detecting refactoring opportunities as an initial step towards the process of refactoring service-oriented applications. The approach constitutes a software aid to improve the maintainability of service-oriented applications by following the list of refactoring suggestions. The software visualization techniques applied in clustering results allowed software developers to ease the identification of refactoring opportunities; moreover, an encouraging scenario to customize views, parameters and inputs of the clustering techniques was also provided.

The approach was evaluated by means of two real world case-studies. We measured the performance of different clustering algorithms by utilizing internal validity criteria. Experimental results showed that COBWEB seems to be the most suitable clustering technique to detect refactoring opportunities. We implemented this approach as part of a framework that includes various clustering techniques, and also, a component to detect Web service discoverability anti-patterns. Furthermore, the flexibility of our approach allows for easily incorporating a new clustering technique to detect refactoring opportunities.

There are some limitations in our approach that should be mentioned. Firstly, the performance of the clustering techniques may be affected by the chosen value of k; therefore, instead of using a single predefined k, we are planning to incorporate an approach to optimally initialize the k value. Secondly, we are planning to improve the detection process by incorporating expert feedback on refactoring suggestions. Finally, more WSDL features will be considered to enrich the assistance by exploiting the structure of these documents.

Acknowledgments. We acknowledge the financial support provided by ANPCyT through grant PICT 2014-1387.

References

1. Crasso, M., Zunino, A., Campo, M.: Awsc: an approach to web service classification based on machine learning techniques. Revista Iberoamericana de Inteligencia Artificial **12**(37), 25–36 (2008)
2. Dong, X., Halevy, A., Madhavan, J., Nemes, E., Zhang, J.: Similarity search for web services. In: 30th International Conference on Very large data bases, pp. 372–383. VLDB Endowment (2004)
3. Elgazzar, K., Hassan, A.E., Martin, P.: Clustering wsdl documents to bootstrap the discovery of web services. In: IEEE International Conference on Web Services, pp. 147–154. IEEE (2010)
4. Erickson, J., Siau, K.: Web services, service-oriented computing, and service-oriented architecture: Separating hype from reality. Principle Advancements in Database Management Technologies: New Applications and Frameworks, p. 176 (2009)
5. Fisher, D.H.: Knowledge acquisition via incremental conceptual clustering. Mach. Learn. **2**(2), 139–172 (1987)

6. Fokaefs, M., Mikhaiel, R., Tsantalis, N., Stroulia, E., Lau, A.: An empirical study on web service evolution. In: IEEE International Conference on Web Services, pp. 49–56. IEEE (2011)
7. Hop, W., de Ridder, S., Frasincar, F., Hogenboom, F.: Using hierarchical edge bundles to visualize complex ontologies in glow. In: Proceedings of the 27th Annual ACM Symposium on Applied Computing, pp. 304–311. ACM (2012)
8. Kaufman, L., Rousseeuw, P.J.: Finding Groups in Data: An Introduction to Cluster Analysis, vol. 344. Wiley, Hoboken (2009)
9. Kuhn, A., Ducasse, S., Gírba, T.: Semantic clustering: identifying topics in source code. Inf. Softw. Technol. **49**(3), 230–243 (2007)
10. Kumara, B.T., Yaguchi, Y., Paik, I., Chen, W.: Clustering and spherical visualization of web services. In: IEEE International Conference on Services Computation, pp. 89–96. IEEE (2013)
11. Liu, W., Wong, W.: Web service clustering using text mining techniques. Int. J. Agent-Oriented Softw. Eng. **3**(1), 6–26 (2009)
12. Ma, J., Zhang, Y., He, J.: Efficiently finding web services using a clustering semantic approach. In: International Workshop on Context Enabled Source and Service Selection, Integration and Adaptation, p. 5. ACM (2008)
13. MacQueen, J., et al.: Some methods for classification and analysis of multivariate observations. In: 5th Berkeley Symposium on Mathematical Statistics and Probability, California, USA, vol. 1, pp. 281–297 (1967)
14. Mateos, C., Crasso, M., Zunino, A., Coscia, J.L.O.: Detecting wsdl bad practices in code-first web services. Int. J. Web Grid Serv. **7**(4), 357–387 (2011)
15. Nieweglowski, L.: clv: cluster validation techniques. R package version 0.3-2. http://cran.r-project.org/web/packages/clv
16. Pelleg, D., Moore, A.W., et al.: X-means: extending k-means with efficient estimation of the number of clusters. In: ICML, pp. 727–734 (2000)
17. Rodriguez, J.M., Crasso, M., Mateos, C., Zunino, A., Campo, M.: Bottom-up and top-down cobol system migration to web services. IEEE Internet Comput. **17**(2), 44–51 (2013)
18. Sabou, M., Pan, J.: Towards semantically enhanced web service repositories. Web Semant. Sci. Serv. Agents WWW **5**(2), 142–150 (2007)
19. Teyseyre, A.R., Campo, M.R.: An overview of 3d software visualization. IEEE Trans. Vis. Comput. Graph. **15**(1), 87–105 (2009)
20. Webster, D., Townend, P., Xu, J.: Interface refactoring in performance-constrained web services. In: 2012 IEEE 15th International Symposium on Object/Component/Service-Oriented Real-Time Distributed Computing (ISORC), pp. 111–118. IEEE (2012)

A Survey on Visual Query Systems in the Web Era

Jorge Lloret-Gazo[✉]

Dpto. de Informática e Ingeniería de Sistemas, Facultad de Ciencias,
Edificio de Matemáticas, Universidad de Zaragoza, 50009 Zaragoza, Spain
jlloret@unizar.es

Abstract. As more and more collections of data are becoming available on the web to everyone, non expert users demand easy ways to retrieve data from these collections. One solution is the so called Visual Query Systems (VQS) where queries are represented visually and users do not have to understand query languages such as SQL or XQuery. In 1996, a paper by Catarci reviewed the Visual Query Systems available until that year. In this paper, we review VQSs from 1997 until now and try to determine whether they have been the solution for non expert users. The short answer is no because very few systems have in fact been used in real environments or as commercial tools. We have also gathered basic features of VQSs such as the visual representation adopted to present the reality of interest or the visual representation adopted to express queries.

1 Introduction

In recent years, and mainly because of the arrival of the web, more and more collections of data are becoming available to everyone in fields ranging from biology to economy or geography. One of the consequences of this fact is that end users, but not experts in Computer Science, demand easy ways to retrieve data from these collections.

Beginning in 1975 with Query By Example (QBE) [39] there have been many proposals in this direction, that is, to facilitate the work of the final user. In [8], the authors reviewed the so-called Visual Query Systems (VQS) from 1975 to 1996 defined as "systems for querying databases that use a visual representation to depict the domain of interest and express related requests".

In this paper, we extend the review from 1997 to date, concentrating our efforts on visual queries to structured information, for example, queries to underlying relational or XML databases. We do not consider the typical search on semistructured documents such as web pages through search engines like Google. Although they are also a good solution for end-users, in this survey we do not take into account natural language interfaces for database query formulation.

The main goal of this survey is to answer the following question: To what extent have the VQS been the solution for novel users for querying databases?

The author would like to thank Rafael Bello for making the initial collection of papers for this review.

S. Hartmann and H. Ma (Eds.): DEXA 2016, Part II, LNCS 9828, pp. 343–351, 2016.
DOI: 10.1007/978-3-319-44406-2_28

To answer this question, we have studied two features: web availability of and validation undergone by the systems. The first feature indicates that the system was designed to be reached easily by novel users simply by means of a web browser, without the burden of installation and with universal availability. The second feature indicates the widespread use of VQSs in practice. Thus, the more systems commercially available, the greater the extension reached by VQSs.

The short answer to the question is that, as far as we know, there is only one system commercially available and designed for the web: Polaris [34].

Moreover, we have included two basic features extracted from the paper [8]: the visual representation adopted to present the reality of interest and the visual representation adopted to express queries. With respect to web features, we have also considered relevant whether the prototype deals with data formatted for the web, that is, XML data or RDF data.

The rest of the paper is organized as follows. In Sect. 2 we state the method followed for elaborating the survey and we briefly describe the values of the relevant features included in the paper. Finally, in Sect. 3, we have drawn several conclusions about the VQSs.

2   Statement of the Method

A survey about a particular object must determine the relevant features of the object with respect to a particular purpose. Once the features have been determined, the next step is to find the possible values of these features. Finally, we have to determine the best combinations of the pairs (feature, value) for the particular purpose.

Usually, we can extract the relevant features and their possible values from published papers about the object, by assuming features in their entirety or by adapting them to new perspectives appearing after the papers have been published. Moreover, we can add features detected by ourselves which were not previously included in any paper.

The survey develops through several steps, which are usually interspersed. In the first step, a complete search of sources determines the candidate papers that deal with the object. In the second step, the relevant features of the object with respect to the particular purpose are determined.

Our object in this survey are the visual query systems with the purpose of facilitating querying databases to non expert in Computer Science users.

The survey [8] reviews up to 80 references from 1975 until 1996 used for querying traditional databases. For this survey, we have searched for papers related with VQS from 1997 to date and we have found 194 candidate papers. Next, we have discarded papers about query languages but without visual part (122) and papers about natural language query languages (8) because they deserve a separate survey. In the remaining 64 works, we have determined sets of 'similar papers' and we have discarded all but one paper in each set. A set of similar papers is composed of several papers built on different aspects of the same idea for a VQS. They also include preliminary versions of the VQS which were later on subsumed by more complete journal publications. We have found 30 similar papers. So, we have discarded 122 + 8 + 30 papers, that is, 160 papers. As a result, the number of papers reviewed in this survey is 34.

As for relevant features, we have extracted the following from the survey of Catarci [8]: Visual representation adopted to present the reality of interest and visual representation adopted to express queries. The values of these features have been determined from the work [8] and from other papers, such as [11], where the faceted option appeared. For answering the question of this paper, we have added the following features: Web orientation and validation.

Let us explain briefly each of the features as well as their values.

2.1   Visual Representation Adopted to Present the Reality of Interest

This feature has been borrowed from the work of Catarci [8]. The reality of interest is modeled by a designer by means of a data metamodel as, for example, the entity/relationship metamodel or a graph data metamodel. As a result of the modelization process, a data model is obtained and it is presented to the user so that (s)he formulates queries on it.

The ways the data model is presented to the user are briefly described next and a more detailed explanation of some of the papers is given in [20].

Diagram-based. Data metamodels come with an associated typical representation for their elements. For example, in the entity/relationship metamodel, there are many representations available and one of them consists of drawing rectangles for the entity types, diamonds for the relationship types and ovals for the attributes. In the diagram-based option, the user has available a diagrammatical representation of the data model elaborated with the typical graphical representation for the elements of the metamodel.

Icon-based. Unlike the diagram-based approach, in this representation there are only iconic representations of some elements of the data model, but the user does not have available the complete data model. According to Catarci [8], 'these VQS are mainly addressed to users who are not familiar with the concepts of data models and may find it difficult to interpret even an E-R diagram'. The aim of the icons is to represent a certain concept by means of its metaphorical power. The problem of these systems is how to construct them in such a way that they express a meaning which is understandable without ambiguity to the users.

Form-based. The typical forms of web pages serve for presenting the extensional database. This occurs in papers such as [34].

Faceted. The data are modeled as faceted classifications which organize a set of items into multiple, independent taxonomies. Each classification is known as a facet and the collection of classification data is faceted metadata. The specific category labels within a facet are facet values. For example, the set of items can be architectural works. For these items, the facets are the architect, the location or the materials. The facet values for materials are stone, steel, etc.

Unknown. As the data model always exists, this option refers to the case where the data model is unknown. For example, the data model may be presented in a paper in textual form but there is no explanation about the way it is presented to the user. For example, paper [26] hides the database and tries to guess the paths for the query from the entities chosen by the user.

2.2 Visual Representation Adopted to Express the Queries

This feature has been borrowed from the work of Catarci [8] and we have adapted it to the object of the survey by adding the Faceted value.

The ways the queries are formulated are briefly described next and a more detailed explanation of some of the papers is given in [20].

Diagram-based. The diagram-based option means that the query is expressed on a diagrammatic representation of the data model.

Icon-based. The icon-based option includes two cases. In the first case, the system offers icons for representing the elements involved in the query. For building a query, the user drags and drops the appropriate icons into a canvas. The second case is the same as in [8], where the icons 'denote both the entities of the real world and the available functions of the system'.

Form-based. Another way to facilitate the query is the form option where the user composes the query by completing options of different elements of a form. The drawback is that the query logic of the end-user does not always fit into a form.

Faceted. We have added as a new value 'Faceted' for describing a system which includes data and metadata in the same page. There, the user specifies the query by clicking on the appropriate links. We have found this situation only in one paper [11].

2.3 Web Orientation

For the web orientation, we have selected two features which are not mutually orthogonal. The first feature is whether the prototype is working on the web or has been conceived to be used in local mode. For the first situation, the value is *Available on the web* and this means that the final user can query the database by means of a prototype which is working on the web. The two values are: There is no web orientation and Available on the web. The second feature indicates whether the user can query data formatted for the web and the values are: Data not formatted for the web, Query XML data, Query RDF data. The values are not orthogonal. So, a paper can have the two values. This is the case, for example, of paper [7].

2.4 Validation

The validation of an idea can be done from several points of view. Regarding query systems, there are, at least, two dimensions: usability and performance.

For example, paper [10] focuses on performance and explains query rewriting techniques that improve the query evaluation performance so that the query execution time is reduced. However, in this paper we concentrate on the usability dimension, that is, the experiments made with users in order to determine the ease of use of the proposed prototype. For this feature, the list of values is: Only prototype, Prototype tested with users, Prototype tested in a real environment, Commercial tool.

Next, we describe briefly each value of this feature. The option *only prototype* means that a prototype has been built but no test has been made with users. The value *prototype tested with users* means that several experiments have been carried out in order to determine the usability of the prototype. The value *prototype tested in a real environment* means that it has been used for real tasks in a particular setting, for example in a department of a university. Finally, the option *commercial tool* means that the VQS has been fully implemented, offered to the public and is in real use in diverse installations.

Table 1. Visual query systems (1997–2003)

Cite	Database	Query	Web	Validation
[2]	Unknown	Icon	No	Only prototype
[31]	Unknown	Form	No	Tested with users
[5]	Diagram	Diagram	No	Only prototype
[9]	Diagram	Diagram	No	Tested with users
[21]	Diagram	Diagram	No	Only prototype
[32]	Diagram	Diagram	No	Only prototype
[24]	Icon	Icon	No	Only prototype
[3]	Diagram	Diagram	No	Tested with users
[12]	Unknown	Form	Query XML data	No
[13]	Icon	Icon	Available on the web	Tested with users
[28]	Diagram	Diagram	No	Only prototype
[33]	Icon	Icon	No	Tested with users
[27]	Diagram	Form	Query XML data	Only prototype
[25]	Icon	Icon	No	Tested with users
[23]	Unknown	Icon	No	Only prototype
[4]	Unknown	Icon	Query XML data	Only prototype
[26]	Form	Form	No	Tested with users
[1]	Unknown	Form	Query XML data	Only prototype
[14]	Unknown	Form	Query XML data	Only prototype

3   Discussion

The arrival of the web brought with it more facilities for users to query databases. As a consequence, users expect to access easily through the web databases situated anywhere in the world.

For expert users, one solution is to express queries in query languages such as SQL or XQuery. However, for novice users whose main concern is to extract data from the database but not the query languages themselves, learning SQL or XQuery is a huge task that is very far from their main concern.

One solution for novice users is to hide the complexity of query languages behind a visual scenery where it is supposed that the complexity is softened with the aid of visual metaphors. This is the idea of Visual Query Systems (VQS) defined in [8] as "systems for querying databases that use a visual representation to depict the domain of interest and express related requests".

In this paper, we have reviewed basic features of Visual Query Systems, such as the representation of databases and the representation of queries. We have also considered the feature of accessing data formatted for the web. Finally, we have reviewed two features we consider relevant to determine whether the VQSs ease querying for novel users: web availability and validation. Next, we discuss the results for each of these features.

Table 2. Visual query systems (2004–2015)

Cite	Database	Query	Web	Validation
[7]	Diagram	Icon	Available on the web; Query XML data	Only prototype
[22]	Diagram	Diagram	Query XML data	Tested with users
[15]	Unknown	Icon	Query RDF data	No
[29]	Diagram	Diagram	No	Tested in a real environment
[16]	Unknown	Form	Query XML data	Tested in a real environment
[36]	Unknown	Form	No	Tested in a real environment
[30]	Diagram	Icon	Query XML data	Only prototype
[34]	Form	Form	Available on the web	Commercial tool
[11]	Diagram	Faceted	Available on the web	Only prototype
[17]	Unknown	Diagram	Query RDF data	Only prototype
[37]	Diagram	Icon	No	No
[6]	Diagram	Diagram	Available on the web	Tested with users
[18]	Unknown	Diagram	No	Tested with users
[35]	Diagram	Icon	No	Tested with users
[38]	Unknown	Icon	No	No
[10]	Diagram	Diagram	Query XML data	Only prototype

The majority of papers offer a diagrammatic representation of the database, only four papers an iconic one [2,13,25,33] and one paper with form representation [34]. For several reasons, there are many papers whose database representation is unknown. For example, paper [26] hides the database and tries to guess the paths for the query from the entities chosen by the user.

With respect to the query representation, the distribution is more balanced between the icon (12 papers), the diagram (11 papers) and the form (8 papers) representation. A special form of query, the faceted one, appears only in one paper [11].

Regarding the data format, there are 9 papers [1,4,7,10,12,16,22,27,30] out of 34 which query XML data and only two papers which query RDF data [15,17]. The rest of the papers do not query web data.

The rest of the features we have identified deal with the main question we have formulated in this paper, that is, to what extent have the VQS been the solution for novel users for querying databases?

For answering this question with respect to the web availability, we can distinguish two periods. From 1997 to 2003 (see Table 1), when the web usage was beginning to spread, there was only one paper oriented to the web [13]. This was very understandable because of the time needed for reorienting the research into the new web setting. In the period 2004 to 2015, only papers [6,7,11,34] propose a web implementation (see Table 2). Although the number of web oriented papers in this period is greater than in the 1997–2003 period, the low number of papers indicates that web orientation has scarcely been taken into account.

For the validation feature, we have found a great number of papers which have only a prototype or have been tested with users in reduced experiments. Only three prototypes have been tested in real environments [16,29,36] and we have found only one commercial tool [34]. So, few papers go beyond testing the prototype with a few users.

As a conclusion of these two features, very few papers are web oriented and also very few papers offer a prototype which has been tested in a real environment. In fact, the combination of both features is only found in paper [34]. Then, although the visual query systems seem to be a great idea for easing the query process for novice users, the reality is that very few papers describe real implementations.

So, the answer to the main question of the paper is that, for the moment, VQSs have not been a widely accepted solution for novel users. From this observation a new, more general question arises: Is there any solution for easing the specification of queries?

If the answer is no, novel users have to learn by themselves query languages or they have to ask computer experts for the specification of queries. In the latter case, no new research would be needed in this field. If the answer is 'we do not know', then new research is required in order to find simple visual query languages which help novice users.

We strongly believe that the idea of VQSs is a good one and that the research should continue in this direction. Recent papers such as [19] also support the idea that a solution for naive users is not available but is necessary in this world in which the use of databases is democratized. The paper proposes as a solution visual systems in which the user writes examples of queries and the system extracts and specifies the desired query in the corresponding query language.

References

1. Abraham, R.: Foxq-xquery by forms. In: Proceedings of the 2003 IEEE Symposium on Human Centric Computing Languages and Environments, pp. 289–290. IEEE (2003)
2. Balkir, N.H., Sükan, E., Özsoyoglu, G., Özsoyoglu, Z.M.: Visual: a graphical icon-based query language. In: Su, S.Y.W. (ed.) ICDE, pp. 524–533. IEEE Computer Society (1996)
3. Benzi, F., Maio, D., Rizzi, S.: Visionary: a viewpoint-based visual language for querying relational databases. J. Vis. Lang. Comput. 10, 117–145 (1999)
4. Berger, S., Bry, F., Schaffert, S., Wieser, C.: Xcerpt and visxcerpt: from pattern-based to visual querying of xml and semistructured data. In: Proceedings of the 29th International Conference on Very Large Data Bases, vol. 29, pp. 1053–1056. VLDB Endowment (2003)
5. Bloesch, A.C., Halpin, T.A.: Conceptual queries using conquer-ii. In: Embley, D.W. (ed.) ER 1997. LNCS, vol. 1331, pp. 113–126. Springer, Heidelberg (1997)
6. Borges, C.R., Macías, J.A.: Feasible database querying using a visual end-user approach. In: Proceedings of the 2nd ACM SIGCHI Symposium on Engineering Interactive Computing Systems, pp. 187–192. ACM (2010)
7. Braga, D., Campi, A., Ceri, S.: XQBE (xquery by example): a visual interface to the standard xml query language. ACM Trans. Database Syst. 30(2), 398–443 (2005)
8. Catarci, T., Costabile, M.F., Levialdi, S., Batini, C.: Visual query systems for databases: a survey. J. Vis. Lang. Comput. 8, 215–260 (1997)
9. Catarci, T., Santucci, G., Cardiff, J.: Graphical interaction with heterogeneous databases. VLDB J. 6, 97–120 (1997)
10. Choi, R.H., Wong, R.K.: VXQ: A visual query language for XML data. Inf. Syst. Front. 17(4), 961–981 (2015)
11. Clarkson, E., Navathe, S.B., Foley, J.D.: Generalized formal models for faceted user interfaces. In: JCDL, pp. 125–134 (2009)
12. Cohen, S., Kanza, Y., Kogan, Y.A., Nutt, W., Sagiv, Y., Serebrenik, A.: Equix easy querying in xml databases. In: WebDB (Informal Proceedings), pp. 43–48 (1999)
13. Cruz, I.F., Leveille, P.S.: As you like it: personalized database visualization using a visual language. J. Vis. Lang. Comput. 12, 525–549 (2001)
14. Erwig, M.: Xing: a visual xml query language. J. Vis. Lang. Comput. 14(1), 5–45 (2003)
15. Harth, A., Kruk, S.R., Decker, S.: Graphical representation of rdf queries. In: WWW, pp. 859–860 (2006)
16. Jagadish, H.V., Chapman, A., Elkiss, A., Jayapandian, M., Li, Y., Nandi, A., Cong, Y.: Making database systems usable. In: SIGMOD Conference, pp. 13–24 (2007)
17. Jarrar, M., Dikaiakos, M.D.: Querying the data web: the mashql approach. IEEE Internet Comput. 14, 58–67 (2010)
18. Jin, C., Bhowmick, S.S., Xiao, X., Cheng, J., Choi, B.: Gblender: towards blending visual query formulation and query processing in graph databases. In: Proceedings of the 2010 ACM SIGMOD International Conference on Management of Data, pp. 111–122. ACM (2010)

19. Li, F., Jagadish, H.V.: Usability, databases, and hci. IEEE Data. Eng. Bull. **35**(3), 37–45 (2012)
20. Lloret-Gazo, J.: A survey on visual query systems in the web era (extended version). http://www.unizar.es/ccia/articulos/VQSCompleto.pdf
21. Madurapperuma, A.P., Gray, W.A., Fiddian, N.J.: A visual query interface for a customisable schema visualisation system. In: IDEAS, pp. 23–32 (1997)
22. Meuss, H., Schulz, K.U., Weigel, F., Leonardi, S., Bry, F.: Visual exploration and retrieval of xml document collections with the generic system x2. Int. J. Digit. Libr. **5**(1), 3–17 (2005)
23. Morris, A.J., Abdelmoty, A.I., El-Geresy, B.A.: A visual query language for large spatial databases. In: Proceedings of the Working Conference on Advanced Visual Interfaces, AVI 2002, pp. 359–360. ACM, New York (2002)
24. Murray, N., Paton, N.W., Goble, C.A.: Kaleidoquery: a visual query language for object databases. In: AVI, pp. 247–257 (1998)
25. Narayanan, A., Shaman, T.: Iconic sql: rractical issues in the querying of databases through structured iconic expressions. J. Vis. Lang. Comput. **13**, 623–647 (2002)
26. Owei, V.: Development of a conceptual query language: adopting the user-centered methodology. Comput. J. **46**(6), 602–624 (2003)
27. Papakonstantinou, Y., Petropoulos, M., Vassalos, V.: Qursed: querying and reporting semistructured data. In: Proceedings of the 2002 ACM SIGMOD International Conference on Management of Data, pp. 192–203. ACM (2002)
28. Poulovassilis, A., Hild, S.G.: Hyperlog: a graph-based system for database browsing, querying, and update. IEEE Trans. Knowl. Data Eng. **13**(2), 316–333 (2001)
29. Rontu, M., Korhonen, A., Malmi, L.: System for enhanced exploration and querying. In: AVI, pp. 508–511 (2006)
30. Sans, V., Laurent, D.: Ifox: interface for ordered xquery an algebraic oriented tool for ordered xquery visualization. In: SAC, pp. 1252–1257 (2008)
31. Sengupta, A., Dillon, A.: Query by templates: a generalized approach for visual query formulation for text dominated databases. In: Proceedings of the IEEE International Forum on Research and Technology Advances in Digital Libraries, ADL 1997, pp. 36–47. IEEE (1997)
32. Shin, D.-G., Grajewski, W., Chu, L.-Y.: An epistemological display query interface. In: AVI, pp. 286–288 (1998)
33. Silva, S.F., Catarci, T., Schiel, U.: Formalizing visual interaction with historical databases. Inf. Syst. **27**, 487–521 (2002)
34. Stolte, C., Tang, D., Hanrahan, P.: Polaris: a system for query, analysis, and visualization of multidimensional databases. Commun. ACM **51**(11), 75–84 (2008)
35. Störrle, H.: Vmql: a visual language for ad-hoc model querying. J. Vis. Lang. Comput. **22**(1), 3–29 (2011)
36. Terwilliger, J.F., Delcambre, L.M.L., Logan, J.: Querying through a user interface. Data Knowl. Eng. **63**, 774–794 (2007)
37. Varga, V., Sacarea, C., Takacs, A.: Conceptual graphs based representation and querying of databases. In: 2010 IEEE International Conference on Automation Quality and Testing Robotics (AQTR), vol. 3, pp. 1–6. IEEE (2010)
38. Zongda, W., Guandong, X., Zhang, Y., Cao, Z., Li, G., Zhiwen, H.: Gmql: a graphical multimedia query language. Knowl.-Based Syst. **26**, 135–143 (2012)
39. Zloof, M.M.: Query by example. In: Proceedings of the May 19–22, 1975, National Computer Conference and Exposition, pp. 431–438. ACM (1975)

Query Answering and Optimization

Query Similarity for Approximate Query Answering

Verena Kantere[✉]

University of Geneva, Geneva, Switzerland
verena.kantere@unige.ch

Abstract. Query rewriting in heterogeneous environments assumes mappings that are complete. In reality and especially in the Big Data era it is rarely the case that such complete sets of mappings exist between sources, and the presence of partial mappings is the norm rather than the exception. So, practically, existing rewriting algorithms fail in the majority of cases. The solution is to approximate original queries with others that can be answered by existing mappings. Approximate queries bear some similarity to original ones in terms of structure and semantics. In this paper we investigate the notion of such query similarity and we introduce the use of query similarity functions to this end. We also present a methodology for the construction of such functions. We employ exemplary similarity functions created with the proposed methodology into recent algorithms for approximate query answering and show experimental results for the influence of the similarity function to the efficiency of the algorithms.

1 Introduction

In data exchange [1], integration [2], and sharing [3], schema mappings (often Local-As-View (LAV) [2]) are used to alleviate heterogeneity across *pairs* of autonomous sources. In data exchange, mappings are used for instance generation of a *target* schema based on those of the *source* schema. In data integration and sharing, mappings are used to *translate* or *rewrite* queries over a source schema to new queries that can be evaluated over the constructs (schemas and attributes) of a target schema.

Here, we focus on the query rewriting problem. Classical rewriting algorithms, like Inverse Rules [4], Bucket [5] and MiniCon [6] have addressed the issue of computing complete or maximally-contained rewritings efficiently. Yet, a common assumption in all is that the available mappings should provide at least one combination that *fully* rewrites the query. If the input query refers to a relation or to an attribute that does not participate in any of the input mappings, then the rewriting *fails* (no output is produced).

This becomes an insurmountable problem in the new era of Big Data management, where data may be sought across many autonomous and heterogeneous

This research is funded from the EU FP7 project ASAP, under Grant Agreement n° 619706.

S. Hartmann and H. Ma (Eds.): DEXA 2016, Part II, LNCS 9828, pp. 355–367, 2016.
DOI: 10.1007/978-3-319-44406-2_29

sources that store huge amounts of data. In many practical settings, there is a necessity to query such heterogeneous data sources that do not hold complete mappings among them. In fact, mappings are inherently incomplete. The size of the data prohibits manual resolution of incomplete mappings. Therefore query answering using classical rewriting is not possible. It is necessary to produce approximate answers to posed queries. ITo achieve this, we need to approximate the original query with a version that can be rewritten on the target data source. The approximated version of the original query can then be rewritten on the target data source schema and answered. Therefore, to perform approximate query answering it is necessary (a) to create one or more approximate query versions, i.e. versions of the original query that are similar, and, furthermore (b) to compare such versions with respect to their similarity to the original query. These are two tasks that are orthogonal to each other. In other work [7] we have proposed solutions for task (a). In this work we deal with task (b), by reflecting on the notion of query similarity, and how the latter can be qualitatively and quantitatively measured.

Motivating Example. Consider a travel web-site iTravel.com offering services similar to the ones found in web-sites like Expedia or Travelocity. Like these web-sites, a user can go to iTravel.com and enter (a) the origin; (b) the destination; and (c) the dates of the vacation; and the system can generate candidate vacation packages that include the flight(s) and hotel room(s) for the duration of the vacation. Conceptually, one can think of this as a single query over iTravel.com database, as illustrated in Fig. 1(a).

VacPackage (<u>vacid</u>, fno, hid, from, until)
Flights (<u>fno</u>, depart, <u>return</u>, carrier, origin, dest, price)
Hotels (<u>hid</u>, <u>chkin</u>, <u>chkout</u>, name, location served, rate)
(a) The iTravel.com schema

iAirline.com: Segment (<u>fno</u>, depart, <u>return</u>, origin, dest, price)
iHotel.com: Locations (<u>location</u>, served, rate)
iInn.com: Inn (<u>name</u>, chkin, chkout, served, rate)
(c) Related travel sources

Segment (fno, dep, ret, org, dst, pr) :- Flights (fno, dep, ret, car, org, dst, pr)
Locations (loc, srv, pr) :- Hotels (hid, chkin, chkout, nm, loc, srv, pr)
Inn (nm, ckin, ckout, srv, pr) :- Hotels (hid, ckin, ckout, nm, loc, srv, pr)
(d) Local-as-view Mappings

```
SELECT *
FROM    VacPackage V, Flights F, Hotels H
WHERE   V.fno = F.fno AND V.from = F.depart AND
        V.until = F.return AND F.depart = "11/01" AND
        F.return = "11/09"
        AND F.origin = "NYC" AND F.depart = H.chkin AND
        F.return = H.chkout AND F.dest = H.served AND
        V.hid = H.hid AND H.location = "Hana"
```

(b) Planning a vacation to Hana, Hawaii

Fig. 1. Querying the iTravel.com

Consider a user looking for vacation packages between the 1st and 9th of November, flying out of New York to the city of Hana, Hawaii (Fig. 1(b)) Web-sites like iTravel.com usually act as brokers. They do not store the latest information for flight and hotel prices (or availability). Instead, they often access the airline and hotel databases (at query-time) to get the latest data. Figures 1(c) and (d) show three such databases along with their corresponding mappings to the iTravel.com schema (expressed as LAV mappings).

```
SELECT *                                          SELECT *
FROM    Segment S, Locations L                    FROM    Segment S, Inn I
WHERE   S.depart = "11/01" AND S.return = "11/09" AND    WHERE   S.depart = "11/01" AND S.return = "11/09" AND
        S.origin = "NYC" AND S.depart = L.chkin AND              S.origin = "NYC" AND S.depart = I.chkin AND
        S.return = L.chkout AND S.dest = L.served AND            S.return = I.chkout AND S.dest = I.served
        L.location = "Hana"
```

Fig. 2. Approximate queries for the iTravel.com

The query in Fig. 1(b) *cannot* be rewritten using the above mappings. Identifiers like vacid and hid are internal to the iTravel.com database and cannot be mapped to the attributes of the individual sources. As well, the join on these identifiers is also not *covered* by any combination of mappings. Therefore, classical rewriting fails. Yet, the individual sources provide us with enough information to *reconstruct* vacation packages since all the *important* attributes, like the dates, origin and destination of flights, and the check-in and check-out dates, are present in the sources. The two queries in Fig. 2 can be computed by approximating (removing the join with VacPackage) the query in Fig. 1(b). The first query identifies flight/hotel combinations where the user gets to stay in a hotel, while in the second the user gets to stay in an inn.

Challenge: Given an input query and a set of mappings there are *multiple* possible approximations and not all approximations are equally *good*. How can we determine which are the *best* query approximations?

2   Query Similarity

Approximate query answering aims to produce a query answer that meets the requirements of the information requested by the original query Q_{orig} in the *best* possible way. The latter can be roughly interpreted in two ways.

- **Structural similarity:** The structure of the returned answers is similar to that returned by Q_{orig}.
- **Semantic similarity:** The content of the returned answers, (i.e. the answer tuples), is similar to the content of the ideal answers to Q_{orig}.

The second approach is very hard to define and impossible to predetermine, whereas we could coarsely define and invent guidelines in order to predetermine the first one.

Essentially, the ideal answers to Q_{orig} are tuples that contain the exact data that the user who poses Q_{orig} has in mind with respect to the answering database. Therefore, it would be necessary to compare the ideal dataset that answers Q_{orig} with the actual dataset that is retrieved from a database. However, in an environment of federated big databases, the autonomy and the size of sources may make this impossible.

Even so, assuming that the ideal answer is accessible, the comparison with the actual answer would require huge human effort in order to be deterministic: In order to decide if an approximation is the most similar to Q_{orig} all the approximations have to be constructed, rewritten and answered; the answers to all the approximations should be compared in order to decide which one is best. Otherwise, a comparison based on statistics and on probabilistic models [8,9] could be possible, but details for such metadata are practically never available.

Oppositely to the similarity of raw data that are returned as query answers, the similarity of the structure of the answers is more approachable, but also of great usefulness. Approximations of Q_{orig} can be employed in order to retrieve data from schemas that are partially mapped on the schema of Q_{orig}; thus, a priori knowledge of which approximate version is most similar to Q_{orig} can lead to answers that are structurally compatible with the schema of Q_{orig}, and, therefore, understandable by the user who posed Q_{orig}, or, even further, easier for her to store locally[1]. Furthermore, the schema itself constitutes, actually, metadata; therefore, structural similarity, which entails schema similarity, leads to metadata similarity. Overall, structural similarity can be a good indication of semantic similarity. Since schemas and mappings are available at query time, it is possible to define the structural similarity and also define guidelines for the comparison of approximations. Therefore, the following discussion is on the definition of structural query similarity. We focus on the structural similarity of SQL and conjunctive queries.

3   Related Work

Query similarity has been explored in several works in the recent past. Some of these works deal with keyword matching in the database environment [10,11] or with the processing of imprecise queries [12–14]. The work in [15] talks about attribute similarity but focuses on numeric data and on conclusions about similarity that can be deduced from the workload. Furthermore, in [16] queries are classified according to their structural similarity; yet, the authors focus on features that differentiate queries with respect to optimization plans. The only work relevant to ours is that of [17], where overall semantic similarity of queries is explored. Yet, our focus is on query versions that are produced through the use of mappings, and we are interested in the effect of the mappings in query similarity.

There has been a great and growing interest in the past few years on how to execute, specifically, a query workload in a way that it is approximate with respect to its actual execution, and, therefore, gain in response time. Some of the work is on approximate query processing. The recent work in [18] as well as the works in [19,20] explore querying large data by accessing only a bounded amount of it, based on formalized access constraints. These works give theoretical results on the classes of queries for which bounded evaluation is possible. Other works focus on how to pre-treat the data in order to create synopses: histograms

---

[1] The problem of storing approximate answers to the database on which Q_{orig} is posed, is related to database versioning and is out of the scope of this paper.

(e.g. [21]), wavelets (e.g. [22] and sampling (e.g. [23]); or to perform execution which terminates based on cost constraints and returns intermediate results (e.g. [24]). Our work is orthogonal to such approaches, in that we do not try to achieve approximation through alteration of the data, but through alteration of the query workload.

Another type of work is on approximate query answering, in which a query that is more *suitable* in some sense is executed in the place of the original one. In [25] a datalog program is approximated with a union of conjunctive queries, and in [18] the same example is followed with the creation of approximate versions of classes of FO queries. In a similar spirit, the works in [26,27] deal with tractable queries for conjunctive queries and the work in [28] deals with subgraph isomorphism for graph queries. Our work is on the same lines of these works, but we focus on the approximation of queries based on specific views that are available in order to perform such approximation, so that the approximated versions are used to query data on heterogeneous sources. Actually, our work proposed here on query similarity is applied in our previous work on relaxing queries that are exchanged in a heterogeneous environment of federated sources that hold large data collections [7].

3.1 Similarity of SQL Queries

Users usually pose their queries in SQL form. The similarity of two SQL queries is confined by the semantic similarity of their elements (namely, 'select' attributes and 'where' conditions). The definition of query similarity should be based on a qualitative study about the semantic relativeness of query versions revealed by their structure. Furthermore, the definition itself should describe a measure that quantifies query similarity. Moreover, such a query similarity measure may be different depending on the application or even the query in hand. In the following we discuss a methodology for the construction of a query similarity measure. Coarsely, the methodology includes two steps: (a) the qualification and (b) the quantification of query similarity. In step (a) the qualitative role of query characteristics in query similarity is assessed, and in (b) the qualitative results of step (a) are quantified and correlated.

A. Query Similarity Qualification. The assessment of the qualitative characteristics of query similarity appoints a role to each query element and prioritizes their importance to the overall query semantics; these can then be used to determine the correlation and interpret similarity of atomic elements to a compound overall query similarity. Let us assume that Q_{apprx} is an approximate version of the original query Q_{orig}.

Definition 1. *A query Q_{orig} is a set of elements $Q_{orig} = \{E_1, \cdots, E_m\}$. An approximate version of it, Q_{apprx}, is another set of elements $Q_{apprx} = \{E'_1, \cdots, E'_n\}$. There is a function $sat(E_i, E'_j) : Q_{orig}xQ_{apprx} \rightarrow \{0,1\}$ that shows if an element $E_i \in Q_{orig}$ is satisfied, i.e. represented by an element*

$E'_j \in Q_{apprx}$. *A query similarity measure M_{sim} of Q_{orig} and Q_{apprx} is a function of sat, i.e. $M_{sim}(Q_{orig}, Q_{apprx}) = f(sat(.))$.*

The above defines a query as a set of elements. Two versions of a query are compared based on their set of elements, and the similarity of this pair of queries is confined by the similarity of their elements. Hence, a query similarity measure should be in the same spirit as such measures in the field of schema matching (e.g. [29]) and matching taxonomies (e.g. [30]). The function *sat* decides for the similarity of two elements; it takes as input two query elements E_i, E'_j, one of each of the two query versions, and returns a boolean (or equivalently, binary) value that shows if E_i is satisfied, i.e. semantically represented by E'_j. The function *sat* can be defined in more detail and have characteristics according to the application in hand. For example, *sat* can be commutative, i.e. $sat(E_i, E'_j) = sat(E'_j, E_i)$, which means that if an element of one query version is represented by an element of the other, the opposite holds, too; this may hold for an application where the semantics of the data in the two databases are considered known, and it may not hold for applications with data semantics that can be considered unknown (e.g. one local database and one web database). Another example is that *sat* can be a function that measures the similarity of two elements, e.g. it outputs values in the range $[0, 1]$; binary output is suitable for applications with a certain knowledge of data semantics, whereas an output within a range is suitable of applications with uncertain data semantics.

Proposition 1. *Adopting a conservative point of view, elements in Q_{orig}, i.e. 'select' attributes or 'where' conditions that are missing from Q_{apprx}, or additional elements, i.e. 'select' attributes or 'where' conditions in Q_{apprx} that do not exist in Q_{orig}, are considered to decrease query similarity.*

The above proposition states, intuitively, that any deviation in the structure of Q_{apprx} from the structure Q_{orig} decreases their similarity. Therefore missing or additional query elements decrease similarity. Key attributes are highly important in a relational schema since their values uniquely prescribe the values of other attributes. The role of keys in queries is as important as in the schema itself, no matter if such an attribute appears in a 'select' or 'where' clause. Thus, deficient approximations of key attributes may result in severe semantic deviations from the original query. Second, 'select' attributes represent the exact information that the user requires. Thus, their lack in the approximate query is decisively irreparable. Third, even though the lack of join conditions is a negative factor for query similarity, it results in an approximate query version that retrieves a superset of the data that would be retrieved by a query with all the original joins. Furthermore, the lack of value constraints has the same effect in the query as the lack of joins. However, the lack of joins probably results in much bigger supersets of retrieved data than the lack of value constraints. Finally, the introduction of new value constraints and joins on non-key attributes is considered a deficiency.

The above are facts that play a role in the estimation of query similarity. These facts, summarised in Table 1, are assessed by the $sat(.)$ function. Such an

Table 1. Summary of facts assessed by the $sat(.)$ function.

Query element	Fact assessed by $sat(.)$
key attributes	key attributes are represented, no matter their position in the query
'select' attributes	'select' attributes are represented
joined attributes	joined attributes are represented
value conditions	value conditions are represented by some attribute constraints
additional elements	existence of new value constraints
	existence of new joined non-key attributes
	existence of new joined key and foreign key attributes

assessment should quantify the similarity of pairs of query elements, based on some qualitative ordering of the importance of these facts. Of course the ordering of the fact importance is application-specific. A generic rationale would indicate that the most important elements of a query are the attributes that are keys or 'select' attributes. Joins are very important; yet their lack results in supersets of answers that the peer might be able to refine. Finally, additional conditions in approximate versions result in rewriting that are classically contained in the rewriting of the original query, therefore, they may be considered as the least important to query similarity. Moreover, the lack of representation of keys or conditions (joins and value constraints) results in answers that are not sound, whereas the lack of representation of 'select' attributes and additional conditions results in answers that are not complete. Again, the importance of answer soundness over completeness or the opposite is application-specific: For example, in a medical application, soundness may be more important than completeness, whereas in an application of web crawling, as in the motivating example, the opposite may hold.

B. Query Similarity Quantification. The qualitative characteristics of query similarity are the guidelines along which a measure for the quantification of query similarity should be constructed. For Q_{orig} and a Q_{apprx}, this measure quantifies (a) the importance of each query element of Q_{orig} being represented by one or more elements of Q_{apprx}, (b) the semantic correlation of the elements of Q_{orig}, and (c) the deviation of Q_{apprx} from Q_{orig} with respect to additional restrictions (if any) of the first that are not present in the latter[2]. We make the following proposition:

Proposition 2. *A query similarity measure should: (a) be monotone to the representation of the elements of Q_{orig} in Q_{apprx}, (b) decrease with the presence of additional elements in Q_{apprx}, and (c) be directional, i.e. the similarity of Q_{orig} with Q_{apprx} can be different from the similarity of Q_{apprx} with Q_{orig}*

---

[2] Intuitively, Q_{apprx} deviates from Q_{orig} only by additional constraints and not by additional 'select' attributes.

The above suggests that the representation of individual elements of Q_{orig} in Q_{apprx} is cumulative and, therefore, the representation of one element cannot cancel or diminish the representation of another. Also, whatever additional in Q_{apprx} can only play a negative role to the similarity of the original and the approximate query; if this was otherwise, it means that an additional element E' in Q_{apprx} would represent some existing element E in Q_{orig}; in such a case E would not be characterised as *additional*. Furthermore, because of the negative role in similarity of additional elements, a similarity measure is directional, since elements are characterised as *additional* with respect to which is the posed query Q_{orig}. Beyond this, the function *sat* may also be directional.

In accordance to the above, Table 2 shows three exemplary similarity measures. In function M_{sim_1} E_i's are elements of Q_{orig}, E_i''s and E_j''s are elements of Q_{apprx}, and $\sum w_i = 1$. The weights w_is and w_js emphasize the importance of individual elements and add to the ordering of the importance of elements represented by the output of *sat*. In functions M_{sim_2} and M_{sim_3} S_i are the 'select' attributes, C_i are any type of constraints (so 'where' conditions), and C_j are additional constraints. The function M_{sim_1} assumes no correlation of similarities of individual elements, expressed as a total summation, whereas M_{sim_2} assumes such a correlation, expressed as the multiplication of the summation of 'select' attributes and the summation of constraints; finally, M_{sim_3} focuses more on the correlation of constraints, rather than any other elements. In Sect. 4 we show experimentally how these three different functions influence the efficiency of algorithms that seek for the best approximate query version and what similarity values the output. In [31] we elaborate more on specific variations of query similarity measures.

Table 2. Exemplary similarity functions.

$$M_{sim_1}(Q_{orig}, Q_{apprx}) = \frac{\sum w_i \cdot sat(E_i, E_i') - \sum w_j \cdot sat(E_j', E_j')}{\sum w_i \cdot sat(E_i, E_i)}$$

$$M_{sim_2}(Q_{orig}, Q_{apprx}) = \frac{(\sum w_i \cdot sat(S_i, S_i')) \cdot (\sum w_i \cdot sat(C_i, C_i')) - \sum w_j \cdot sat(C_j', C_j')}{\sum w_i \cdot sat(S_i, S_i,) \cdot \sum w_i \cdot sat(C_i, C_i,)}$$

$$M_{sim_3}(Q_{orig}, Q_{apprx}) = \frac{(\sum w_i \cdot sat(S_i, S_i')) \cdot (\Pi w_i \cdot sat(C_i, C_i')) - \sum w_j \cdot sat(C_j', C_j')}{\sum w_i \cdot sat(S_i, S_i,) \cdot \Pi w_i \cdot sat(C_i, C_i,)}$$

3.2 Similarity of Conjunctive Queries

Even though the user forms Q_{orig}^{SQL}, the existing rewriting algorithms [4–6] deal with the conjunctive form of this query, Q_{orig}^{conj}, and, therefore, an approximate query version should be in this form in order to be rewritten. In general, Q_{orig}^{conj} contains elements (these are always constraints) that are not apparent in Q_{orig}^{SQL} but are implied by the query elements in the latter.

Example 1. Assume that there are three additional relations: Hotels(<u>name</u>, room#, rate, lowestprice), HotelLocation(<u>locname</u>, lowestprice, <u>location</u>), Season Prices(<u>season</u>, lowestprice)

Q_{orig}^{SQL} requests names and locations for the hotels with the lowest price in spring:

Q_{orig}^{SQL}:   **SELECT** H. name, HL. location

        FROM   Hotels H, HotelLocation HL, SeasonPrices SP

        WHERE H.lowestprice $=$ HL.lowestprice

           AND HL.lowestprice $=$ SP.lowestprice

           AND SP.season $=$ "spring"

Q_{orig}^{conj}(name, location):-Hotels (name, room#, rate, lowestprice) HotelLocation (locname, lowestprice, location) SeasonPrices(season, lowestprice)
Q_{orig}^{conj} implies an additional join H.LowestPrice $=$ SP.LowestPrice that is not apparent in Q_{orig}^{SQL}. The rewriting algorithms considers for elimination both explicit and implicit constraints with respect to Q_{orig}^{SQL}.   □

There are two ways to define the similarity of the conjunctive to the SQL form of Q_{orig} concerning the query elements that are constraints[3]: (a) implicit constraints appearing only in Q_{orig}^{conj} are considered associative to the explicit ones (i.e. those that appear in Q_{orig}^{SQL}); in this case, the role of implicit constraints in similarity should be supportive to that of explicit constraints in a seamless way; (b) both implicit and explicit constraints are considered of equal value to the query similarity; in this case the initial influence to similarity of the explicit constraints has to be disseminated to all of the implicit and explicit constraints, appearing in Q_{orig}^{conj}. The dissemination can be performed either in a absolute way: add similarity value to the implicit constraints, or in a relative way: disperse the value of explicit constraints to the all of the constraints, such that the total is constant. Both ways diminish the influence of the explicit constraints to the overall query similarity.

We propose following the first approach, since, in general, it adheres better to the user preferences: Specifically, the user, who defines the initial role and influence of the query elements, is not aware of the implicit query constraints; therefore, it is better to treat them as auxiliary to the explicit constraints, rather than equal to the latter in ignorance of the user. Practically, this means that whenever the approximate query answering algorithm needs to eliminate an implicit constraint, it has to determine the explicit constraints of Q_{orig} to which this implicit one is auxiliary.

3.3   Lower Bounds for Query Similarity

A similarity measure that is defined along Proposition 2 guarantees a low limit for the similarity of the rewritten approximate query, Q_{apprx_R} to the original query. Formally:

Proposition 3. *If Q_{apprx} is α similar to Q_{orig}, then the respective Q_{apprx_R} is at least α similar to Q_{orig}.*

---

[3] Query elements that correspond to 'select' attributes have a 1-1- correspondence in both Q_{orig}^{SQL} and Q_{orig}^{conj}. In the conjunctive form, these are called *distinguished variables*.

Intuitively, Q_{apprx_R} requests information constrained at least by the conditions of the respective Q_{apprx}[4]. Additional constraints in Q_{apprx_R} are complementary, meaning that they do not annul existing constraints in Q_{apprx} (and therefore in Q_{orig}). Thus, when they are rewritten they (a) do not decrease the similarity of Q_{orig} with Q_{apprx} and, furthermore (b) they may even narrow the answer towards the constraints of Q_{orig} that are missing from Q_{apprx}. Thus, the similarity of Q_{apprx} to Q_{orig} denotes that the approximately requested information is at least as similar to the originally requested information as the similarity measure denotes.

The following example exhibits the above proposition.

Example 2. Assume that there are three views which can be used for the rewriting of Q_{orig}^{conj} of Example 1:

V_1(name, location):- Hotels(name, room#, rate, lowestprice) HotelLocation (name, lowestprice, location)

V_2(name):- Hotels(name, room#, rate, lowestprice) SeasonPrices ('spring', lowestprice)

V_3(location):- HotelLocation (locname, lowestprice, location) SeasonPrices ('spring', lowestprice) V_1, V_2 and V_3 rewrite the following approximate versions:

Q_{apprx1}(name, location):-Hotels (name, room#, rate, lowestprice) HotelLocation (locname, lowestprice, location)

Q_{apprx2}(name, location):-Hotels (name, room#, rate, lowestprice1) HotelLocation (locname, lowestprice2, location) SeasonPrices ('spring', lowestprice1)

Q_{apprx3}(name, location):-Hotels (name, room#, rate, lowestprice1) HotelLocation (locname, lowestprice2, location) SeasonPrices ('spring', lowestprice2)

Q_{apprx1} misses a relation, while Q_{apprx2} and Q_{apprx3} miss a join. The following are the rewritten versions of the three approximate queries using the above views, respectively:

Q_{apprx1_R}(name, location):- V_1(name, location)

Q_{apprx2_R}(name, location):- V_2(name) V_3(location)

Q_{apprx3_R}(name, location):- V_2(name) V_3(location)

Q_{apprx1} is rewritten such that the all requested information is also requested via the rewriting Q_{apprx1_R}. Let us expand Q_{apprx1_R}:

Q_{apprx1_R}(name, location):-Hotels(name, room#, rate, lowestprice) HotelLocation (name, lowestprice, location)

Q_{apprx2_R} and Q_{apprx3_R}, however, are rewritten such that more information is requested and this additional information is part of the information that was requested by Q_{orig} and was lost because of the eliminations of the approximation. The expansion of both rewritten queries is:

$Q_{apprx2/3_R}$(name, location):- Hotels(name, room#, rate, lowestprice1) SeasonPrices('spring', lowestprice1) HotelLocation (locname, lowestprice2, location) SeasonPrices('spring', lowestprice2)

Both Q_{apprx2_R} and Q_{apprx3_R} request names of hotels with the lowest price in spring and locations of hotels with the lowest price in spring. Since the 'spring' constraint is associated with both the name an the location of the hotel,

---

[4] Actually, this is guaranteed by the classical query rewriting methodology, which creates contained rewritten versions.

Q_{apprx2_R} and Q_{apprx3_R} are more similar to Q_{orig} than Q_{apprx2} and Q_{apprx3}, respectively. Moreover, since there is only one lowest price for every season (see key constraints), Q_{apprx2_R} and Q_{apprx3_R} manage to recover the full information requested by Q_{orig}.   □

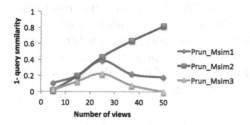
Fig. 3. Quantification of query similarity

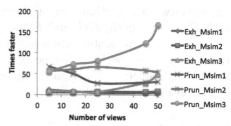
Fig. 4. Influence of Msim on the efficiency of approximate query answering

4   Experiments

We present an experimental evaluation of the discussed query similarity measures, employing the exemplary functions M_{sim_1}, M_{sim_3} and M_{sim_3}, presented in Sect. 3.1. We use these three measures in algorithms that search for approximate query versions, in order to approximately answer original query versions that cannot be classically rewritten on remote data sources. Specifically, we employ two algorithms we have already proposed in [7], which take as an input Q_{orig} and a set of views $\mathcal{V}$ that can serve as mappings with remote data sources. The algorithms create series of Q_{apprx} that are compared with Q_{orig} based on the similarity function $M_{sim_{1,2,3}}$ and pick for rewriting and answering the Q_{apprx} that is most similar to Q_{orig}. One algorithm, Exh, is exhaustive, i.e. it may search the whole search space of solutions, and the other, $Prun$, prunes the search space of solutions according to some heuristics.

The experiments are performed on synthetic data, following the example of works presented in the field of query rewriting using views [6,32,33]. Synthetic data give the opportunity to control the form/size of Q_{orig} and the set of views $\mathcal{V}$; therefore, it enables the case-study of a big variety of query-views combinations, that cannot be found in real datasets. The parameters are plenty. In short, in the presented experiments, the following parameter values are used. Number of: relations = 10, views = 10, view subgoals = 4, query subgoals = 5, repeated query subgoals = 5, query constants = 1, view distinguished variables = 2, view joins = 3, view constants = 3, view comparisons = 3, distinguished variables = 20.

Figure 3 shows that the average dissimilarity of Q_{orig} and Q_{apprx} increases significantly for M_{sim_2} with the increase in the number of available views, whereas for M_{sim_1} and M_{sim_2} the dissimilarity increases for some number of views and decreases as this number becomes bigger. This shows that tightening the correlation of distinguished variables and constraints maybe too strict and the increase in available mappings may not really help to find, or to find early the best Q_{apprx}. Oppositely, it is shown that keeping looser correlations between elements and even focusing only on the correlation of constraints we can find more or find earlier Q_{apprx}s that are quite similar to Q_{orig}.

Figure 4 shows the effect of the similarity function on the efficiency of the algorithms. Algorithms Exh and $Prun$ are compared with a baseline, i.e. the straightforward algorithm that produces all possible Q_{apprx} and compares them with Q_{orig}. For both, Exh and $Prun$, employing M_{sim_3}, which focuses on the correlation of constraints, results in much faster execution. This is natural, since the solutions can be fast rejected or pruned. Also, employing M_{sim_2} is better than M_{sim_1}, since it also requires tighter correlations of query elements, which, again, results in early solution rejection and pruning.

5   Conclusions

In this paper we discuss the notion of query similarity with the prospect of using it in answering approximate versions of originally posed queries in an environment of big heterogeneous data sources. We focus on the similarity of SQL and conjunctive queries, as these forms are necessary for the expression and classical rewriting, respectively, of posed queries. We present a methodology for the creation of query similarity functions, based on qualitative and quantitative characteristics of query similarity. We give examples of query similarity functions and we present experiments with the employment of such functions in approximate query answering.

References

1. Fagin, R., Kolaitis, P.G., Miller, R.J., Popa, L.: Data exchange: semantics and query answering. Th. Comput. Sci. 336 (1)
2. Lenzerini, M.: Data Integration: a theoretical perspective. In: PODS (2002)
3. Rodríguez-Gianolli, P., Kementsietsidis, A., Garzetti, M., Kiringa, I., Jiang, L., Masud, M., Miller, R.J., Mylopoulos, J.: Data sharing in the hyperion peer database system. In: VLDB (2005)
4. Duschka, O.M., Genesereth, M.R.: Answering recursive queries using views. In: PODS (1997)
5. Levy, A.Y., Rajaraman, A., Ordille, J.O.: Query-answering algorithms for information agents. In: 13th International Conference on Artificila Intelligence (1996)
6. Pottinger, R., Levy, A.: A scalable algorithm for answering queries using views. In: VLDB (2000)
7. Kantere, V., Orfanoudakis, G., Kementsietsidis, A., Sellis, T.: Query relaxation across heterogeneous data sources. In: ACM CIKM 2015, pp. 473–482
8. Batista, G., Monard, M.C.: A study of k-nearest neighbour as an imputation method. In: HIS (2002)
9. Poosala, V., Ganti, V.: Fast approximate query answering using precomputed statistics. In: ICDE, p. 252 (1999)
10. Agrawal, S., Chaudhuri, S., Das, G.: DBXplorer: a system for keyword-based search over relational databases. In: ICDE (2002)

11. Cohen, W.: Integration of heterogeneous databases without common domains using queries based on textual similarity. In: SIGMOD (1998)
12. Motro, A.: VAGUE: A user interface to relational databases that permis vague queries. TOIS **6**(3), 187–214 (1988)
13. Fuhr, N.: A probabilistic framework for vague queries and imprecise information in databases. In: VLDB (1990)
14. Kiebling, W., Kostner, G.: Preference SQL - design, implementation, experiences. In: VLDB (2002)
15. Agrawal, S., Chaudhuri, S., Das, G., Gionis, A.: Automated ranking of database query results. In: CIDR (2003)
16. Ghosh, A., Parikh, J., Sengar, V.S., Haritsa, J.R.: Plan selection based on query clustering. In: Intelligent Information Integration (1999)
17. Chu, W.W., Zhang, G.: Associative query answering via query feature similarity. In: IIS (1997)
18. Potti, N., Patel, J.M.: Daq: a new paradigm for approximate query processing. In: VLDB, vol. 8
19. Fan, W., Geerts, F., Libkin, L.: On scale independence for querying big data. In: ACM PODS, pp. 51–62 (2014)
20. Cao, Y., Fan, W., Wo, T., Yu, W.: Bounded conjunctive queries. PVLDB **7**(12), 1231–1242 (2014)
21. Jagadish, H.V., Koudas, N., Muthukrishnan, S., Poosala, V., Sevcik, K.C., Suel, T.: Optimal histograms with quality guarantees. In: VLDB, pp. 275–286 (1998)
22. Garofalakis, M.N., Gibbons, P.B.: Wavelet synopses with error guarantees. In: ACM SIGMOD, pp. 476–487 (2002)
23. Agarwal, S., Milner, H., Kleiner, A., Talwalkar, A., Jordan, M.I., Madden, S., Mozafari, B., Stoica, I.: Knowing when you're wrong: building fast and reliable approximate query processing systems. In: ACM SIGMOD, pp. 481–492 (2014)
24. Agarwal, S., Mozafari, B., Panda, A., Milner, H., Madden, S., Stoica, I.: Blinkdb: queries with bounded errors and bounded response times on very large data. In: EuroSys, pp. 29–42 (2013)
25. Chaudhuri, S., Kolaitis, P.G.: Can datalog be approximated? J. Comput. Syst. Sci. **55**(2), 355–369 (1997)
26. Barceló, P., Libkin, L., Romero, M.: Efficient approximations of conjunctive queries. SIAM J. Comp. **43**(3), 1085–1130 (2014)
27. Fink, R., Olteanu, D.: On the optimal approximation of queries using tractable propositional languages. In: ICDT (2011)
28. Fan, W., Li, J., Ma, S., Tang, N., Wu, Y., Wu, Y.: Graph pattern matching: From intractable to polynomial time. PVLDB **3**(1), 264–275 (2010)
29. Melnik, S., Garcia-Molina, H., Rahm, E.: Similarity flooding: a versatile graph matching algorithm and its application to schema mathcing. In: ICDE (2002)
30. Doan, A., Madhavan, J., Dhamankar, R., Domingos, P., Halevy, A.: Learning to match ontologies on the semantic web. VLDB J. **12**(4), 303–319 (2003)
31. Kantere, V., Tsoumakos, D., Sellis, T., Roussopoulos, N.: GrouPeer: dynamic clustering of P2P databases. In: Information Systems (2008). doi:10.1016/j.is.2008.04.002
32. Lín, V., Vassalos, V., Malakasiotis, P.: Minicount: Efficient rewriting of count-queries using views. In: ICDE, p. 1 (2006)
33. Steinbrunn, M., Moerkotte, G., Kemper, A.: Heuristic and randomized optimization for the join ordering problem. VLDB J. **6**(3), 191–208 (1997)

Generalized Maximal Consistent Answers in P2P Deductive Databases

Luciano Caroprese$^{(\boxtimes)}$ and Ester Zumpano

DIMES, University of Calabria, 87036 Rende, CS, Italy
{lcaroprese,zumpano}@dimes.unical.it

Abstract. The paper provides a contribution in computing consistent answers to logic queries in a P2P environment. Each peer joining a P2P system imports data from its neighbors by using a set of mapping rules, i.e. a set of semantic correspondences to a set of peers belonging to the same environment. By using mapping rules, as soon as it enters the system, a peer can participate and access all data available in its neighborhood, and through its neighborhood it becomes accessible to all the other peers in the system. The declarative semantics of a P2P system is defined in terms of minimal weak models. Under this semantics each peer uses its mapping rules to import minimal sets of mapping atoms allowing to satisfy its local integrity constraints. The contribution of the present paper consists in extending the classical notion of consistent answer by allowing the presence of partially defined atoms, i.e. atoms with "unknown" values due to the presence of tuples in different minimal weak models which disagree on the value of one or more attributes. The basic proposal is the following: in the presence of alternative minimal weak models the choice is to extracts the minimal consistent portion of the information they all hold, i.e. the information on which the minimal weak models agree. Therefore, true information is that "supported"' by all minimal weak models, i.e. the set of atoms which maximizes the information shared by the minimal weak models.

1 Introduction

A flurry of research, in the social, academic and commercial communities, is devoted to the different topics related to the management of Peer-to-peer (P2P) systems. Each peer, joining a P2P systems relies on the peers belonging to the same environment and can both *provide or import data*. More specifically, each peer joining a P2P system exhibits a set of *mapping rules*, i.e. a set of semantic correspondences to a set of peers which are already part of the system (neighbors). Thus, in a P2P system the entry of a new source, *peer*, is extremely simple as it just requires the definition of the mapping rules. By using mapping rules, as soon as it enters the system, a peer can participate and access all data available in its neighborhood, and through its neighborhood it becomes accessible to all the other peers in the system.

S. Hartmann and H. Ma (Eds.): DEXA 2016, Part II, LNCS 9828, pp. 368–376, 2016.
DOI: 10.1007/978-3-319-44406-2_30

This paper aims to provide a contribution to the specific topic related to the integration of information and the computation of queries in an open ended network of distributed peers [1–7, 10].

Many approaches investigate the data integration problem in a P2P system by considering each peer as initially consistent, therefore the introduction of inconsistency is just relied to the operation of importing data from other peers. These approaches assume that for each peer it is preferable to import as much knowledge as possible.

This paper, stems from the work in [9] in which a different perspective is proposed. Intuitively, the basic idea, yet very simple, is the following: a peer can be initially inconsistent. In the case of inconsistent database the information provided by the neighbors can be used in order to restore consistency, that is to only integrate the missing portion of a correct, but incomplete database. Specifically, in [9] the semantics of a P2P system is defined in terms of minimal weak models. Under this semantics an inconsistent peer, in the interaction with different peers, uses its mapping rules to import minimal sets of mapping atoms allowing to satisfy its local integrity constraints, that is *minimal sets of atoms allowing the peer to enrich its knowledge so that restoring inconsistency anomalies* This behavior results to be useful in real world P2P systems in which peers often use the available import mechanisms to extract knowledge from the rest of the system only if this knowledge is strictly needed to repair an inconsistent local database. The proposal in [9] follows the classical approach, that is in the presence of more alternative minimal weak models a deterministic consistent semantic is given by selecting as true information that present in all minimal weak models, i.e. the set of atoms belonging to the intersection of minimal weak models.

Example 1. Consider a P2P system consisting of the following two peers

- The peer P_2 stores information about vendors of devices and contains the facts $vendor(dan, laptop)$, whose meaning is *'Dan is a vendor of laptops'*, and $vendor(bob, laptop)$, whose meaning is *'Bob is a vendor of laptops'*.
- The peer P_1 contains the fact $order(laptop)$, stating that there exists the order of a laptop, the standard rule $available(Y) \leftarrow supplier(X, Y)$, stating that a device Y is available if there is a supplier X of Y, and the constraint $\leftarrow order(X), not\ available(X)$, stating that there cannot exist the order of a device which is not available. Moreover, it also exhibits the mapping rule $supplier(X, Y) \leftarrow vendor(X, Y)$, used to import tuples from the relation $vendor$ of P_2 into the relation $supplier$ of P_1.

The local database of P_1 is inconsistent because the ordered device *laptop* is not available (there is no supplier of laptops). The peer P_1 has to import some supplier of laptops in order to make its database consistent. Then, the mapping rule $supplier(X, Y) \leftarrow vendor(X, Y)$ will be used to import one supplier from the corresponding facts of P_2: $supplier(dan, laptop)$ or $supplier(bob, laptop)$. P_1 will not import both facts because just one of them is sufficient to satisfy the local integrity constraint $\leftarrow order(X), not\ available(X)$.

We observe that if $\mathcal{P}_1$ does not contain any fact its database is consistent and no fact is imported from $\mathcal{P}_2$.   □

In this paper we extend the work in [8,9] by proposing a more flexible semantics that selects as true information that "supported" by all repaired databases, i.e. the set of atoms which maximizes the information shared by the minimal weak models.

Example 2. Consider the P2P system presented in Example 1 and suppose to be interested in *supplier*. As previously stated, in order to restore consistency, either the fact *supplier*(*dan, laptop*) or *supplier*(*bob, laptop*) have to be imported. Considering the standard semantics the set of true atoms does not contain any vendor, whereas, intuitively, it also contains *supplier*($\perp$, *laptop*), stating that it is true that *laptop* is provided even thought the name of the supplier is unknown. In fact *laptop* is provided by *dan* or by *bob*, thus we don't know "exactly" who is the supplier providing *laptop*.   □

The paper extends the classical notion of consistent answer by allowing the presence of partially defined atoms, i.e. atoms with "unknown" values due to the presence of tuples in different minimal weak models which disagree on the value of one or more attributes. In other words, the paper proposes an alternative semantics that, in the presence of alternative minimal weak models, extracts the maximal consistent portion of the information they all hold, i.e. the information on which the minimal weak models agree.

2   Background

We assume that there are finite sets of predicate symbols, constants and variables. A *term* is either a constant or a variable. An *atom* is of the form $p(t_1, \ldots, t_n)$ where p is a predicate symbol and $t_1, \ldots, t_n$ are terms. A *literal* is either an atom A or its negation *not* A. A *rule* is of the form $H \leftarrow \mathcal{B}$, where H is an atom (*head* of the rule) and $\mathcal{B}$ is a conjunction of literals (*body* of the rule). A program $\mathcal{P}$ is a finite set of rules. $\mathcal{P}$ is said to be positive if it is negation free. The definition of a predicate p consists of all rules having p in the head. A ground rule with empty body is a *fact*. A rule with empty head is a *constraint*. It is assumed that programs are *safe*, i.e. variables appearing in the head or in negated body literals are range restricted as they appear in some positive body literal. The ground instantiation of a program $\mathcal{P}$, denoted by $ground(\mathcal{P})$ is built by replacing variables with constants in all possible ways. An interpretation is a set of ground atoms. The truth value of ground atoms, literals and rules with respect to an interpretation M is as follows: $val_M(A) = A \in M$, $val_M(not\ A) = not\ val_M(A)$, $val_M(L_1, \ldots, L_n) = min\{val_M(L_1), \ldots, val_M(L_n)\}$ and $val_M(A \leftarrow L_1, \ldots, L_n) = val_M(A) \geq val_M(L_1, \ldots, L_n)$, where A is an atom, $L_1, \ldots, L_n$ are literals and $true > false$. An interpretation M is a model for a program $\mathcal{P}$, if all rules in $ground(\mathcal{P})$ are *true* w.r.t. M. A model M is said to be minimal if there is no model N such that

$N \subset M$. We denote the set of minimal models of a program $\mathcal{P}$ with $\mathcal{MM}(\mathcal{P})$. Given an interpretation M and a predicate symbol g, $M[g]$ denotes the set of g-tuples in M. The semantics of a positive program $\mathcal{P}$ is given by its unique minimal model which can be computed by applying the *immediate consequence operator* $\mathbf{T}_{\mathcal{P}}$ until the fixpoint is reached ($\mathbf{T}_{\mathcal{P}}^{\infty}(\emptyset)$). The semantics of a program with negation $\mathcal{P}$ is given by the set of its stable models, denoted as $\mathcal{SM}(\mathcal{P})$. An interpretation M is a *stable model* (or *answer set*) of $\mathcal{P}$ if M is the unique minimal model of the positive program $\mathcal{P}^{M}$, where $\mathcal{P}^{M}$ is obtained from $ground(\mathcal{P})$ by (i) removing all rules r such that there exists a negative literal $not\ A$ in the body of r and A is in M and (ii) removing all negative literals from the remaining rules [11]. It is well known that stable models are minimal models (i.e. $\mathcal{SM}(\mathcal{P}) \subseteq \mathcal{MM}(\mathcal{P})$) and that for negation free programs, minimal and stable model semantics coincide (i.e. $\mathcal{SM}(\mathcal{P}) = \mathcal{MM}(\mathcal{P})$).

3   P2P Systems: Syntax and Semantics

3.1   Syntax

A *(peer) predicate symbol* is a pair $i : p$, where i is a *peer identifier* and p is a predicate symbol. A *(peer) atom* is of the form $i : A$, where i is a *peer identifier* and A is a standard atom. A *(peer) literal* is a peer atom $i : A$ or its negation $not\ i : A$. A conjunction $i : A_1, \ldots, i : A_m, not\ i : A_{m+1}, \ldots, not\ i : A_n, \phi$, where ϕ is a conjunction of built-in atoms[1], will be also denoted as $i : \mathcal{B}$, with $\mathcal{B}$ equals to $A_1, \ldots, A_m, not\ A_{m+1}, \ldots, not\ A_n, \phi$.

A *(peer) rule* can be of one of the following three types:

1. STANDARD RULE. It is of the form $i : H \leftarrow i : \mathcal{B}$, where $i : H$ is an atom and $i : \mathcal{B}$ is a conjunction of atoms and built-in atoms.
2. INTEGRITY CONSTRAINT. It is of the form $\leftarrow i : \mathcal{B}$, where $i : \mathcal{B}$ is a conjunction of literals and built-in atoms.
3. MAPPING RULE. It is of the form $i : H \leftarrow j : \mathcal{B}$, where $i : H$ is an atom, $j : \mathcal{B}$ is a conjunction of atoms and built-in atoms and $i \neq j$.

In the previous rules $i : H$ is called *head*, while $i : \mathcal{B}$ (resp. $j : \mathcal{B}$) is called *body*. Negation is allowed just in the body of integrity constraints. The definition of a predicate $i : p$ consists of the set of rules in whose head the predicate symbol $i : p$ occurs. A predicate can be of three different kinds: *base predicate*, *derived predicate* and *mapping predicate*. A base predicate is defined by a set of ground facts; a derived predicate is defined by a set of standard rules and a mapping predicate is defined by a set of mapping rules.

An atom $i : p(X)$ is a *base atom* (resp. *derived atom*, *mapping atom*) if $i : p$ is a base predicate (resp. standard predicate, mapping predicate). Given an interpretation M, $M[\mathcal{D}]$ (resp. $M[\mathcal{LP}]$, $M[\mathcal{MP}]$) denotes the subset of base atoms (resp. derived atoms, mapping atoms) in M.

---

[1] A *built-in atom* is of the form $X\theta Y$, where X and Y are terms and θ is a comparison predicate.

Definition 1. P2P SYSTEM. A *peer* $\mathcal{P}_i$ is a tuple $\langle \mathcal{D}_i, \mathcal{LP}_i, \mathcal{MP}_i, \mathcal{IC}_i \rangle$, where (i) $\mathcal{D}_i$ is a set of facts (*local database*); (ii) $\mathcal{LP}_i$ is a set of standard rules; (iii) $\mathcal{MP}_i$ is a set of mapping rules and (iv) $\mathcal{IC}_i$ is a set of constraints over predicates defined by $\mathcal{D}_i$, $\mathcal{LP}_i$ and $\mathcal{MP}_i$. A *P2P system* $\mathcal{PS}$ is a set of peers $\{\mathcal{P}_1, \ldots, \mathcal{P}_n\}$.   □

Without loss of generality, we assume that every mapping predicate is defined by only one mapping rule of the form $i:p(X) \leftarrow j:q(X)$. The definition of a mapping predicate $i:p$ consisting of n rules of the form $i:p(X) \leftarrow \mathcal{B}_k$, with $k \in [1..n]$, can be rewritten into $2 * n$ rules of the form $i:p_k(X) \leftarrow \mathcal{B}_k$ and $i:p(X) \leftarrow i:p_k(X)$, with $k \in [1..n]$. Given a P2P system $\mathcal{PS} = \{\mathcal{P}_1, \ldots, \mathcal{P}_n\}$, where $\mathcal{P}_i = \langle \mathcal{D}_i, \mathcal{LP}_i, \mathcal{MP}_i, \mathcal{IC}_i \rangle$, $\mathcal{D}, \mathcal{LP}, \mathcal{MP}$ and $\mathcal{IC}$ denote, respectively, the global sets of ground facts, standard rules, mapping rules and integrity constraints, i.e. $\mathcal{D} = \bigcup_{i \in [1..n]} \mathcal{D}_i$, $\mathcal{LP} = \bigcup_{i \in [1..n]} \mathcal{LP}_i$, $\mathcal{MP} = \bigcup_{i \in [1..n]} \mathcal{MP}_i$ and $\mathcal{IC} = \bigcup_{i \in [1..n]} \mathcal{IC}_i$. In the rest of this paper, with a little abuse of notation, $\mathcal{PS}$ will be also denoted both with the tuple $\langle \mathcal{D}, \mathcal{LP}, \mathcal{MP}, \mathcal{IC} \rangle$ and the set $\mathcal{D} \cup \mathcal{LP} \cup \mathcal{MP} \cup \mathcal{IC}$; moreover whenever the peer is understood, the peer identifier will be omitted.

3.2   The Minimal Weak Model Semantics

This section reviews the *Minimal Weak Model* semantics for P2P systems [8] which is based on a special interpretation of mapping rules. The semantics presented in this paper stems from the observations that in real world P2P systems often the peers use the available import mechanisms to extract knowledge from the rest of the system only if this knowledge is strictly needed to repair an inconsistent local database. In more formal terms, each peer uses its mapping rules to import minimal sets of mapping atoms allowing to satisfy local integrity constraints.

In this paper we refer to a particular interpretation of mapping rules. Intuitively, a mapping rule $H \leftarrow \mathcal{B}$ states that if the body conjunction $\mathcal{B}$ is *true* in the source peer the atom H can be imported in the target peer, that is H is *true* in the target peer only if it implies (directly or even indirectly) the satisfaction of some constraints that otherwise would be violated. The following example should make the meaning of mapping rules crystal clear.

Example 3. Consider the P2P system presented in Example 1.

As we observed, the local database of $\mathcal{P}_1$ is inconsistent because the ordered device *laptop* is not available. The peer $\mathcal{P}_1$ has to import some supplier of laptops in order to make its database consistent. Then, the mapping rule $supplier(X,Y) \leftarrow vendor(X,Y)$ will be used to import one supplier from the corresponding facts of $\mathcal{P}_2$: $supplier(dan, laptop)$ or $supplier(bob, laptop)$. $\mathcal{P}_1$ will not import both facts because just one of them is sufficient to satisfy the local integrity constraint $\leftarrow order(X), not\ available(X)$.

We observe that if $\mathcal{P}_1$ does not contain any fact its database is consistent and no fact is imported from $\mathcal{P}_2$.   □

Before formally presenting the minimal weak model semantics, we introduce some notation. Given a mapping rule $r = A \leftarrow \mathcal{B}$, with $St(r)$ we denote the corresponding logic rule $A \leftarrow B$. Analogously, given a set of mapping rules $\mathcal{MP}$, $St(\mathcal{MP}) = \{St(r) \mid r \in \mathcal{MP}\}$ and given a P2P system $\mathcal{PS} = \mathcal{D} \cup \mathcal{LP} \cup \mathcal{MP} \cup \mathcal{IC}$, $St(\mathcal{PS}) = \mathcal{D} \cup \mathcal{LP} \cup St(\mathcal{MP}) \cup \mathcal{IC}$. Informally, the idea is that for a ground mapping rule $A \leftarrow B$, the atom A *could be inferred* only if the body B is *true*. Formally, given an interpretation M, a ground standard rule $C \leftarrow \mathcal{D}$ and a ground mapping rule $A \leftarrow B$, $val_M(C \leftarrow \mathcal{D}) = val_M(C) \geq val_M(\mathcal{D})$, whereas $val_M(A \leftarrow B) = val_M(A) \leq val_M(B)$.

Definition 2. WEAK MODEL. Given a P2P system $\mathcal{PS} = \mathcal{D} \cup \mathcal{LP} \cup \mathcal{MP} \cup \mathcal{IC}$, an interpretation M is a *weak model* for $\mathcal{PS}$ if $\{M\} = \mathcal{MM}(St(\mathcal{PS}^M))$, where $\mathcal{PS}^M$ is the program obtained from $ground(\mathcal{PS})$ by removing all mapping rules whose head is *false* w.r.t. M. □

We shall denote with $M[\mathcal{D}]$ (resp. $M[\mathcal{LP}]$, $M[\mathcal{MP}]$) the set of ground atoms of M which are defined in $\mathcal{D}$ (resp. $\mathcal{LP}$, $\mathcal{MP}$).

Definition 3. MINIMAL WEAK MODEL. Given two weak models M and N, we say that M is *preferable* to N, and we write $M \sqsupseteq N$, if $M[\mathcal{MP}] \subseteq N[\mathcal{MP}]$. Moreover, if $M \sqsupseteq N$ and $N \not\sqsupseteq M$ we write $M \sqsupset N$. A weak model M is said to be *minimal* if there is no weak model N such that $N \sqsubset M$. □

The set of weak models for a P2P system, $\mathcal{PS}$, will be denoted by $\mathcal{WM}(\mathcal{PS})$, whereas the set of minimal weak models will be denoted by $\mathcal{MWM}(\mathcal{PS})$. We say that a P2P system $\mathcal{PS}$ is *consistent* if $\mathcal{MWM}(\mathcal{PS}) \neq \emptyset$; otherwise it is *inconsistent*.

Proposition 1. *For any P2P system $\mathcal{PS}$, $\sqsupseteq$ defines a partial order on the set of weak models of $\mathcal{PS}$.* □

We observe that, if each peer of a P2P system is locally consistent then no mapping atom is inferred. Clearly not always a minimal weak model exists. This happens when there is at least a peer which is locally inconsistent and there is no way to import mapping atoms that could repair its local database so that its consistency can be restored.

Example 4. Consider our running example (Example 1). The weak models of the system are: $M_1 = \{vendor(dan, laptop), vendor(bob, laptop), order(laptop), supplier (dan, laptop), available(laptop)\}$, $M_2 = \{vendor(dan, laptop), vendor(bob, laptop), order(laptop), supplier(bob, laptop), available (laptop)\}$, and $M_3 = \{vendor(dan, laptop), vendor(bob, laptop), order(laptop), supplier(dan, laptop), supplier(bob, laptop), available(laptop)\}$, whereas the minimal weak models are M_1 and M_2 because they contain minimal subsets of mapping atoms (resp. $\{supplier(dan, laptop)\}$ and $\{supplier(bob, laptop)\}$). □

4   Generalized Minimal Weak Model

In this section we extend the semantics of a P2P system by introducing a deterministic model, the Generalized Minimal Weak Model, allowing the presence of partially defined atoms, i.e. atoms having the special constant $\perp$ as a value for one or more attribute, due to the presence of tuples which disagree on the values of those attributes in different minimal weak models. In other words, while in the previous section we have assigned to every ground atom a truth value which can be *true* or *false* in different minimal weak models, in this section we extract the maximal consistent portion of information from the set of minimal weak models that is the portion of the information on which the minimal weak models agree.

We introduce the truth value *undefined* for atoms that are not *true* nor *false* w.r.t. the Generalized Minimal Weak Model. We assume that $true \geq undefined \geq false$.

We first introduce a binary relationship for comparing two ground atoms or two sets of ground atoms.

Definition 4. Given two ground atoms $A = p(t_1, ..., t_n)$ and $B = p(u_1, ..., u_n)$, we say that A *supports* B (written $A \vdash B$) if $\forall i$ either $t_i = u_i$ or $t_i = \perp$. Given two sets of ground atoms S_1 and S_2, S_1 *supports* S_2 (written $S_1 \vdash S_2$) if $\forall B \in S_2$, $\exists A \in S_1$ s.t. $A \vdash B$ and $\forall A \in S_1$, $\exists B \in S_2$ s.t. $A \vdash B$. □

Example 5. The set $\{p(a, \perp), p(\perp, b)\}$ supports the set $\{p(a, d), p(c, b)\}$.   □

Thus, given two ground atoms, A supports B if each term in A is equal or less specific that the correspondent term in B. Note that the relation $\vdash$ is transitive (if $A \vdash B$ and $B \vdash C$, then $A \vdash C$) and antisymmetric (if $A \vdash B$ and $B \vdash A$, then $A = B$). We define the truth value of an atom w.r.t. a set S.

Definition 5. Given a set S and an atom A,

- A is *true* in S iff there is $B \in S$ s.t. $A \vdash B$.
- A is *undefined* in S iff there is $B \in S$ s.t. $B \vdash A$ and it is not *true* in S.
- A is *false* in S iff it is not *true* nor *false* in S.

Definition 6. Given a class of sets of ground atoms S and a set of ground atoms T we say that:

- T *generalizes* S (written $T \vDash S$) if for each $S_i \in S$ is $T \vdash S_i$;
- T is the *minimal generalization* of S if $T \vDash S$ and there is no set U s.t. $U \neq T$, $T \vdash U$ and $U \vDash S$.   □

Example 6. The minimal generalization of the class of sets $\{\{p(a, b)\}, \{p(a, d), p(c, b)\}\}$ is $\{p(a, \perp), p(\perp, b)\}$; whereas the minimal generalization of the class of sets $\{\{p(a, b)\}, \{p(a, b), p(c, b)\}\}$ is $\{p(a, b), p(\perp, b)\}$.   □

Given a P2P system $\mathcal{PS}$ and its set of minimal weak models $\mathcal{MWM}(\mathcal{PS})$, the *Generalized Minimal Weak Model* of $\mathcal{PS}$, denoted as $\mathcal{GMWM}(\mathcal{PS})$, is the minimal generalization of $\mathcal{MWM}(\mathcal{PS})$.

Example 7. Let us consider the P2P system $\mathcal{PS}$ presented in Example 1. Its minimal weak models are: $M_1 = \{vendor(dan, laptop), vendor(bob, laptop), order$ $(laptop), supplier(dan, laptop), available(laptop)\}$ and, $M_2 = \{vendor(dan, laptop), vendor(bob, laptop), order(laptop), supplier(bob, laptop), available(laptop)\}$. The Generalized Minimal Weak Model is: $M = \{vendor(dan, laptop), vendor(bob, laptop), order(laptop), supplier(\bot, laptop), available(laptop)\}$ □

The following proposition shows that the Generalized Minimal Weak Model of a P2P system is a superset of the insersection of its Minimal Weak Models.

Proposition 2. *Given a P2P system* $\mathcal{PS}, \bigcap_{M \in \mathcal{MWM}(\mathcal{PS})} M \subseteq \mathcal{GMWM}(\mathcal{PS})$. □

5 Discussion

Complexity Results. We consider now the computational complexity of calculating minimal weak models and answers to queries.

Proposition 3. *Given a P2P system* $\mathcal{PS}$*, checking if there exists a minimal weak model for* $\mathcal{PS}$ *is a NP-complete problem.*   □

As a P2P system may admit more than one minimal weak model, the answer to a query is given by considering *brave* or *cautious* reasoning (also known as *possible* and *certain* semantics).

Definition 7. Given a P2P system $\mathcal{PS} = \{\mathcal{P}_1, \ldots, \mathcal{P}_n\}$ and a ground peer atom A, then A is *true* under (i) brave reasoning if $A \in \bigcup_{M \in \mathcal{MWM}(\mathcal{PS})} M$, (ii) cautious reasoning if $A \in \bigcap_{M \in \mathcal{MWM}(\mathcal{PS})} M$.   □

Theorem 1. *Let* $\mathcal{PS}$ *be a consistent P2P system, then: i) Deciding whether an atom A is* true *in some minimal weak model of* $\mathcal{PS}$ *is in* Σ_2^p*. (ii) Deciding whether an atom A is* true *in every minimal weak model of* $\mathcal{PS}$ *is in* Π_2^p *and* coNP*-hard.*   □

References

1. Bernstein, P.A., Giunchiglia, F., Kementsietsidis, A., Mylopulos, J., Serafini, L., Zaihrayen, I.: Data management for peer-to-peer computing: a vision. In: WebDB, pp. 89–94 (2002)
2. Calvanese, D., De Giacomo, G., Lenzerini, M., Rosati, R.: Logical foundations of peer-to-peer data integration. In: PODS Conference, pp. 241–251 (2004)
3. Caroprese, L., Greco, S., Zumpano, E.: A logic programming approach to querying and integrating P2P deductive databases. In: FLAIRS, pp. 31–36 (2006)
4. Caroprese, L., Molinaro, C., Zumpano, E.: Integrating and querying P2P deductive databases. In: IDEAS, pp. 285–290 (2006)
5. Caroprese, L., Zumpano, E.: Consistent data integration in P2P deductive. In: SUM, pp. 230–243 (2007)

6. Caroprese, L., Zumpano, E.: Modeling cooperation in P2P data management systems. In: An, A., Matwin, S., Raś, Z.W., Ślęzak, D. (eds.) Foundations of Intelligent Systems. LNCS (LNAI), vol. 4994, pp. 225–235. Springer, Heidelberg (2008)
7. Caroprese, L., Zumpano, E.: Handling preferences in P2P systems. In: Lukasiewicz, T., Sali, A. (eds.) FoIKS 2012. LNCS, vol. 7153, pp. 91–106. Springer, Heidelberg (2012)
8. Caroprese, L., Zumpano, E.: Restoring consistency in P2P deductive databases. In: Hüllermeier, E., Link, S., Fober, T., Seeger, B. (eds.) SUM 2012. LNCS, vol. 7520, pp. 168–179. Springer, Heidelberg (2012)
9. Caroprese, L., Zumpano, E.: A logic based approach for restoring consistency in P2P deductive databases. In: Chen, Q., Hameurlain, A., Toumani, F., Wagner, R., Decker, H. (eds.) DEXA 2015. LNCS, vol. 9262, pp. 3–12. Springer, Heidelberg (2015)
10. Franconi, E., Kuper, G.M., Lopatenko, A., Serafini, L.: A robust logical and computational characterisation of peer-to-peer database systems. In: Aberer, K., Koubarakis, M., Kalogeraki, V. (eds.) DBISP2P 2003. LNCS, vol. 2944, pp. 64–76. Springer, Heidelberg (2004)
11. Gelfond, M., Lifschitz, V.: The stable model semantics for logic programming. In: Proceedings of the Fifth Conference on Logic Programming, pp. 1070–1080 (1998)
12. Sakama, C., Inoue, K.: Prioritized logic programming and its application to commonsense reasoning. Artif. Intell. **123**, 185–222 (2000)

Computing Range Skyline Query
on *Uncertain Dimension*

Nurul Husna Mohd Saad[1](✉), Hamidah Ibrahim[1], Fatimah Sidi[1],
Razali Yaakob[1], and Ali Amer Alwan[2]

[1] Faculty of Computer Science and Information Technology,
Universiti Putra Malaysia, Serdang, Malaysia
nhusna.saad@gmail.com, {hamidah.ibrahim,
fatimah,razaliy}@upm.edu.my
[2] Kulliyyah of Information and Communication Technilogy,
International Islamic University Malaysia, Kuala Lumpur, Malaysia
aliamer@iium.edu.my

Abstract. A user sometimes prefers to not be restricted when querying for information. Querying information within a range of search often provides a different perspective to user as opposed to a rigid search. To compute skyline within a given range would be easy on traditional dataset. The challenge is when the dataset being queried consists of both atomic values as well as continuous range of values. For a set of objects with *uncertain dimension*, a skyline with a range query $[q_j : q'_j]$ on that *uncertain dimension* returns objects which are not dominated by any other objects in the range query. A method is proposed to determine objects and answer skyline query that satisfy the range query. The correctness of the method is proven through comparisons between two naïve methods that strictly reject and loosely accept objects that intersect with the range query.

Keywords: Probabilistic skyline · *Uncertain dimension* · Range query

1 Introduction

Skyline queries retrieved a set of objects that are not dominated by any other objects in a dataset. An object v is said to dominate another object w if and only if v has a lower value than w in at least one dimension, and v has a lower value or equal to w in every other dimensions. Accordingly, a range skyline query is computing skyline on objects that are within a specified range query. To compute skyline objects in a range query, it is preferable to find objects that are not dominated by any other objects. In principle, an object v is preferable if, within the range query, v is better than another object w on at least one dimension and v is not worse than w on every other dimension. It is quite straightforward to report skyline that is within a range query when dealing with a set of objects that are all points. Only points that fall within the range query will be considered for skyline computations. Consider the apartments example in Fig. 1, where a tenant desired to rent apartments within the rent range 5–9. Hence, the skyline in this

© Springer International Publishing Switzerland 2016
S. Hartmann and H. Ma (Eds.): DEXA 2016, Part II, LNCS 9828, pp. 377–388, 2016.
DOI: 10.1007/978-3-319-44406-2_31

scenario would be points *e*, *k*, and *l*, as they are the most desirable apartments in the specified rent range.

Motivation. The apartment rental database that supports *uncertain dimensions* contains listing on apartments for rent. Each apartment is associated with its rental price (can be fixed or within some range) and the commute length. To limit his search, a potential tenant may query for apartments that are within a rent budget of $250 and $440. Thus, can skyline be efficiently reported straightforwardly using existing skyline algorithms when dealing with a set of objects with *uncertain dimension*, which can be points or line segments, and the objects with line segments intersect the range query? It is very obvious that every object that does not lie within the range query can definitely be filtered out, and every object that clearly falls within the range query will be accepted for further skyline computation. Nevertheless, how does one determine to accept or reject objects that intersect with the boundary of a range query? Figure 2 illustrates the above discussion.

To the best of our knowledge, this is the first work that tackles the problem of range skyline query on *uncertain dimension*. A method that implements a threshold to filter objects and computing skyline query based on a range query is proposed. First, objects that do not fall within the range query or objects that intersect with the range query but have the probability to occur within the range query that is less than *t*, where *t* is a user-defined threshold value, will be filtered out. Then, skyline query will be computed on the remaining objects with regards to the range query.

The rest of the paper is organized as follows. Section 2 reviewed the evolution of skylines. Then, the preliminaries of this paper and the problem of range skyline query are formerly defined in Sect. 3. The proposed approach is discussed in Sect. 3 as well, followed by an empirical study in Sect. 4. Finally, Sect. 5 concludes the paper.

2 Related Works

The evolution of skylines in the context of databases can be seen from the first work by Borzsonyi *et al.* [1]. Then, Chomicki *et al.* [2] introduced presorting into the algorithm in [1] to build a more effective algorithm, in which then Godfrey *et al.* [3] have further improved it. Kossmann *et al.* [6] then introduced an algorithm based on the nearest

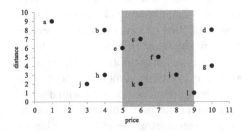
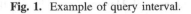

Fig. 1. Example of query interval.

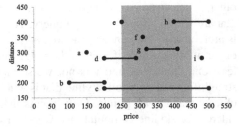

Fig. 2. Example of query interval on *uncertain dimension*.

neighbor search using R-tree. Later on, Papadias *et al.* [8] proposed an algorithm that is based on the sorted R-tree to enhance the algorithm in [6]. Pei *et al.* [9] then pioneered the concept of probabilistic skyline on uncertain data, in which each object is represented by a set of instances and is part of the skyline answer with a certain probability. Inspired by this work, Qi and Atallah [10], Lian and Chen [7], and Zhang *et al.* [14] then proposed a variant of algorithm in the same area. On the other hand, Khalefa *et al.* [5] introduced the concept of probabilistic skyline on uncertain data, where the objects are of continuous range instead of having multiple instances, which was then has inspired the work by Saad *et al.* [12]. Papadias *et al.* [8] coined the term constrained skyline query where the query would returned the most interesting points in the data space defined by the constraints. Jiang and Pei [4], Rahul and Janardan [11], and Wang *et al.* [13] then proposed a variant of algorithm tackling different issues involving range queries as well. The works in [4, 8, 11, 13] focused on implementing a suitable data structure to efficiently search and report skyline points that lie within the range query, and the datasets involved are considered as points only. Following [12], this work focuses on datasets with *uncertain dimensions*, while investigating on how to compute skyline on objects when they intersect with the range query.

3   Preliminaries and Proposed Method

In this section, the concept of *uncertain dimensions* is briefly described, followed by the introduction on the issue of computing range skyline query on *uncertain dimensions*.

Definition 1   (Uncertain Dimension). Given a dataset of n-dimensional space $D = (D_1, D_2, \ldots, D_n)$. A dimension is said to be an *uncertain dimension*, denoted $U'(D_i)$ where $1 \leq i \leq n$, if there exists at least two objects in D with different forms (i.e. points and continuous range) in that dimension.

Let $v = (v.D_1, v.D_2, \ldots, v.D_n)$ and $w = ([w.D_1 : w.D'_1], w.D_2, \ldots, w.D_n)$, where $[w.D_1 : w.D'_1]$ is an interval representing continuous range in dimension D_1 and $w.D_1 < w.D'_1$, be two objects in D, such that $\{v, w\} \in D$. Here, the uncertain dimension would be $U'(D_i)$ since both objects v and w are represented in different forms in D_1.

For ease of description and without loss of generality, this work assumes that the dataset has only one *uncertain dimension*, namely the first dimension, and as such the dataset D has the form $(U'(D_1), D_2, \ldots, D_n)$. Given the nature of the *uncertain dimension*, the results of skyline query executed on this kind of dataset are bound to be probabilistic, since each object with continuous range is now associated with a probability value of it being a query answer. This issue has been solved in [12]. Subsequently, when range query is introduced into skyline query, the aim is to determine and compute skyline on objects that *satisfy* a given range query.

Definition 2 (Range Query). A range query $[q_j : q'_j]$ indicates a range that is being queried on jth dimension, where $q_j < q'_j$.

Definition 3 (*Satisfy*). Given a dataset of n-dimensional space $D = (U'(D_1), D_2, \ldots, D_n)$, with two different sets of objects in it, $\{A, I\} \in D$, where $A = (D_1, D_2, \ldots, D_n)$ and $I = ([D_1 : D'_1], D_2, \ldots, D_n)$, and a range query $[q_1 : q'_1]$. An object $v \in A$ is said to satisfy the given range query if $q_1 \leq v.D_1 \leq q'_1$, while an object $w \in I$ is said to satisfy the given range query if $q_1 \leq w.D_1 < w.D'_1 \leq q'_1$. Object w that intersects with the range query but has endpoints out of the range query is said to satisfy the query range if $P(q_1 < D_1(w) < q'_1) \geq t$, where P is a probability function, $D_1(w)$ represents object w in D_1, and t is a user-defined threshold value.

Note that since object w in D_1 is a continuous range modeled as a uniform probability density function $pdf\, f(x)$ defined on the real range $[w.D_1 : w.D'_1]$, then $P(w.D_1 < D_1(w) < w.D'_1) = \int_{w.D_1}^{w.D'_1} f(x)dx = 1$. Based on these definitions, it can be shown that range queries on *uncertain dimension* can be answered effectively. Having a set of objects that does not involve any *uncertain dimension* would be straightforward to report skyline objects that lie within the range query.

Definition 4 (Range Skyline Query). Given a range query $[q_j : q'_j]$. An object v is a skyline object only if there does not exist an object k that dominates v. Hence, v is said to dominate k with respect to the range query, denoted by $v \prec_{[q_j:q'_j]} k$, if (1) $\exists D_j, v.D_j \in [q_j : q'_j] < k.D_j \in [q_j : q'_j]$ and (2) $\forall D_{i, i \neq j}, v.D_i \leq k.D_i$.

On the contrary, having a range query on *uncertain dimension* would be challenging as there will be objects with continuous range, denoted by $[w.D_j : w.D'_j]$ where $w \in I$, that can intersect the range query, and therefore, the concept of $w.D_j \in [q_j : q'_j]$ as previously defined is not applicable in this case.

Definition 5 (Range Skyline Query on *Uncertain Dimension*). Given a range $[q_j : q'_j]$ queried on an *uncertain dimension* D_j. An object w is a skyline object if there does not exist an object l that dominates w. Hence, w is said to dominate l with respect to the range query, denoted by $w \prec_{[q_j:q'_j]} l$, if (1) $\exists D_j, P(q_j < D_j(w) < q'_j) \geq t$, (2) $P(D_j(w) <_{[q_j : q'_j]} D_j(l)) \geq t$, and (3) $\forall D_{i, i \neq j}, w.D_i \leq l.D_i$.

Problem Definition. Let S be a set of objects in D, where $D = (U'(D_1), D_2, \ldots, D_n)$, and a range $[q_1 : q'_1]$ queried on the *uncertain dimension* $U'(D_1)$. Answer skyline query on S with respect to the range query $[q_1 : q'_1]$ in such a way that the skyline objects satisfy the range query $[q_1 : q'_1]$.

For the purpose of this paper, this work assumes that the range query posed by a user is on a single *uncertain dimension*.

3.1   Range Query on *Uncertain Dimension*

Let S be a set of objects in a n-dimensional space with *uncertain dimension*, $D = (U'(D_1), D_2, \ldots, D_n)$. To answer range query on *uncertain dimension*, an algorithm that filters S is needed, in such a way that all objects reported satisfy the range query $[q_1 : q_1']$. There are several cases where two sets of objects A and I, such that $A \cup I = S$, can lie within or intersect with the range query $[q_1 : q_1']$. The easiest and simplest case is when object $v \in A$ lies entirely inside the range query $[q_1 : q_1']$ (as illustrated in Fig. 3), such that $q_1 \leq v.D_1 \leq q_1'$, and it can definitely be reported as object that satisfies the range query $[q_1 : q_1']$. The same can be said for object $w \in I$ that lies entirely inside the range query $[q_1 : q_1']$ (as illustrated in Fig. 3), in such a way that $q_1 \leq w.D_1 < w.D_1' \leq q_1'$.

The next case is when object w has one endpoint inside the range query $[q_1 : q_1']$ (as illustrated in Fig. 4), where $w.D_1 < q_1 < w.D_1' \leq q_1'$ or $q_1 \leq w.D_1 < q_1' < w.D_1'$. And lastly, when object w intersects the range query $[q_1 : q_1']$ but does not have an endpoint inside the range query $[q_1 : q_1']$ (as illustrated in Fig. 5), in which $w.D_1 < q_1 < q_1' < w.D_1'$. In the latter two cases, it remains to decide whether those objects should be reported as objects that satisfy the range query $[q_1 : q_1']$ and be included in skyline computation at a later stage. Definition 3 can be used to find object w that satisfies the range query $[q_1 : q_1']$ by having the probability of $D_1(w)$ being between the range query $[q_1 : q_1']$ above a threshold t value. Without loss of generality, t is set to 50 %, yet changing t to a higher and lower value would mean a result set that is the most matched and the least match, respectively, to the range query.

Example 1. If $[w.D_1 : w.D_1']$ intersects with the range query $[q_1 : q_1']$ and has an endpoint within the range query $[q_1 : q_1']$, such that $w.D_1 < q_1 < w.D_1' \leq q_1'$, to determine if w will be reported as object that satisfies the range query $[q_1 : q_1']$, then the probability of $D_1(w)$, $\int_{q_1}^{w.D_1'} f(x)dx$, shall be more than t. When $[w.D_1 : w.D_1']$ intersects with the range query $[q_1 : q_1']$ yet with both its endpoints being outside of the range query $[q_1 : q_1']$, then the probability of $D_1(w)$ being within the range query $[q_1 : q_1']$ is computed as $\int_{q_1}^{q_1'} f(x)dx$. The threshold t value is important as it is impossible to determine that for objects with continuous range, they will always satisfy the range query $[q_1 : q_1']$ and the same t value will be used when computing skyline on *uncertain dimension*.

3.2   Range Skyline Query on *Uncertain Dimension*

Let S' be a set of objects in a n-dimensional space with *uncertain dimension*, $D = (U'(D_1), D_2, \ldots, D_n)$, that satisfies the range query $[q_1 : q_1']$. Since S' contains an *uncertain dimension*, S' contains two different sets of objects in it, $A = (D_1, D_2, \cdots, D_n)$ and $I = ([D_1 : D_1'], D_2, \ldots, D_n)$. For two objects $v \in A$ and $w \in I$ that satisfy the range query $[q_1 : q_1']$, four cases may arise when computing skyline. Note that the basic skyline computation on *uncertain dimension* follows the computation proposed in [12].

Case 1. When the interval $[w.D_1 : w.D_1']$ intersects with the range query $[q_1 : q_1']$, $w.D_1 < v.D_1 \leq w.D_1'$, and $v.D_j \prec w.D_j$, $(2 \leq j \leq n)$. Then the probability of w to be a skyline object will be affected by v in D_1 as well as the range query $[q_1 : q_1']$, while v will always be a skyline object.

Example 2. Object w in D_1 intersects with the range query $[q_1 : q_1']$ in such a way $w.D_1 < q_1 < w.D_1' \leq q_1'$. Then the probability of $D_1(w)$ to not be dominated by $D_1(v)$ with respect to the range query $[q_1 : q_1']$ can be represented as $P(D_1(w)) = P\{w \in [q_1 : (v - \varepsilon)]\}$, where ε is the *continuity correction* value [12]. A continuity correction is needed in this case in order for a continuous distribution to be used to approximate a discrete distribution. The continuity correction requires adding or subtracting 0.5 from the value of object v as needed [15].

Fig. 4. Intersecting objects have one endpoint within the query interval.

Fig. 5. Both endpoints of an intersecting object lie outside of the query interval.

Fig. 3. Objects that definitely satisfy the query interval.

Case 2. When the interval $[w.D_1 : w.D_1']$ intersects with range query $[q_1 : q_1']$, $w.D_1 \leq v.D_1 < w.D_1'$, and $w.D_j \prec v.D_j$, $(2 \leq j \leq n)$. Then the probability of w to be a skyline object will be affected by the range query $[q_1 : q_1']$, while the probability of v to be a skyline object will be affected by w in D_1.

Example 3. Object w in D_1 intersects with the range query $[q_1 : q_1']$ in such a way $w.D_1 < q_1 < w.D_1' \leq q_1'$. Then the probability of $D_1(w)$ to be a skyline object with respect to the range query $[q_1 : q_1']$ can be represented as $P(D_1(w)) = P\{w \in [q_1 : w.D_1']\}$, while the probability of $D_1(v)$ to not be dominated by $D_1(w)$ with respect to the range query $[q_1 : q_1']$ can be represented as $P(D_1(v)) = P\{w \in [w.D_1 : q_1]\} + P\{w \in [(v + \varepsilon) : w.D_1']\}$.

Case 3. When the interval $[w.D_1 : w.D_1']$ falls within the range query $[q_1 : q_1']$, $w.D_1 < v.D_1 \leq w.D_1'$, and $v.D_j \prec w.D_j$, $(2 \leq j \leq n)$. Then the probability of w to be a skyline object will be affected by v in D_1, while v will always be a skyline object.

Example 4. Object w in D_1 falls within the range query $[q_1 : q_1']$ in such a way $q_1 \leq w.D_1 < w.D_1' \leq q_1'$. Then the probability of $D_1(w)$ to not be dominated by $D_1(v)$ with respect to the range query $[q_1 : q_1']$ can be represented as $P(D_1(w)) = P\{w \in [w.D_1 : (v - \varepsilon)]\}$.

Case 4. When the interval $[w.D_1 : w.D_1']$ falls within the range query $[q_1 : q_1']$, $w.D_1 \leq v.D_1 < w.D_1$, and $w.D_j \prec v.D_j$, $(2 \leq j \leq n)$. Then the probability of v to be a skyline object will be affected by w in D_1, while w will always be a skyline object.

Example 5. Object w in D_1 falls within the range query $[q_1 : q_1']$ in such a way $q_1 \leq w.D_1 < w.D_1' \leq q_1'$. Then the probability of $D_1(v)$ to not be dominated by $D_1(w)$ with respect to the range query $[q_1 : q_1']$ can be represented as $P(D_1(v)) = P\{w \in [(v + \varepsilon) : w.D_1']\}$.

Hence, according to Definition 5, since in all of these cases it is assumed that both objects v and w satisfied the range query $[q_1 : q_1']$ (which fulfilled the first condition in Definition 5) and either v or w dominates the other object in every D_i, $(2 \leq i \leq n)$ (the third condition in Definition 5), then the second condition in Definition 5 requires $P(D_1(w) \prec_{[q_1:q_1']} D_1(v)) \geq t$, or $P(D_1(v) \prec_{[q_1:q_1']} D_1(w)) \geq t$, where t is a threshold value. Algorithm 1 gives a method to retrieve objects that satisfy the range query $[q_1 : q_1']$ and compute skyline with a probability above a given threshold value.

On the other hand, for two objects $v \in I$ and $w \in I$ that satisfy the range query $[q_1 : q_1']$, several cases may arise when computing skyline. However, due to space limitation, this scenario is briefly discussed and detailed implementations have been omitted from this paper.

Algorithm 1 Range skyline query on *uncertain dimension*

Input: dataset S with $U(D_1)$, query interval $[q_1 : q_1']$, threshold t;

Output: a set of objects *Sky* which is the skyline on $[q_1 : q_1']$;

```
 1:  Initialize S', Sky;
 2:  for each object v ∈ S do
 3:    if q₁ ≤ v.D₁ ≤ q₁' then
 4:      add v to S';
 5:    end if
 6:    else if v.D₁ < q₁ < v.D₁' ≤ q₁' or q₁ ≤ v.D₁ < q₁' < v.D₁' or
        v.D₁ < q₁ < q₁' < v.D₁' then
 7:      compute P(D₁(v)) w.r.t [q₁ : q₁'];
 8:      if P(D₁(v)) ≥ t then
 9:        add v to S';
10:      end if
11:    end if
12:  end for
13:  for each object v ∈ S' do
14:    for each object w ∈ Sky do
15:      if the pair (v,w) qualifies for case 1 then
16:        P(D₁(w)) = P{w ∈ [q₁ : (v - ε)]};
17:      end if
18:      else if the pair (v,w) qualifies for case 2 then
19:        P(D₁(v)) = P{w ∈ [w.D₁ : q₁]} + P{w ∈ [(v + ε) : w.D₁']};
          P(D₁(w)) = P{w ∈ [q₁ : w.D₁']};
20:      end if
21:      else if the pair (v,w) qualifies for case 3 then
22:        P(D₁(w)) = P{w ∈ [w.D₁ : (v - ε)]};
23:      end if
24:      else if the pair (v,w) qualifies for case 4 then
25:        P(D₁(v)) = P{w ∈ [(v + ε) : w.D₁']};
26:      end if
27:      if P(D₁(w)) < t then
28:        remove w from Sky;
29:      end if
30:      if P(D₁(v)) < t then
31:        continue to next object v;
32:      end if
33:    end for
34:    add v to Sky;
35:  end for
36:  return Sky;
```

Example 6. Object w with interval $[w.D_1 : w.D_1']$ and v with interval $[w.D_1 : w.D_1']$ in D_1 falls within the range query $[q_1 : q_1']$ in such a way $w.D_1 < q_1 < q_1' < w.D_1'$ and $q_1 < v.D_1 < q_1' < v.D_1'$, respectively, while w and v overlapped in such a way that $w.D_1 < v.D_1$ and $w.D_1' < v.D_1'$, and $w.D_j \prec v.D_j$, $(2 \leq j \leq n)$. Then the probability of $D_1(w)$ to be a skyline object with respect to the range query $[q_1 : q_1']$ can be represented as $P(D_1(w)) = P\{w \in [q_1 : q_1']\}$, while the probability of $D_1(v)$ to not be dominated by $D_1(w)$ with respect to the range query $[q_1 : q'_1]$ can be represented as $P(D_1(v)) = \frac{1}{2}(P\{v \in [v.D_1 : q'_1]\} * P\{w \in [v.D_1 : q'_1]\}) + P\{w \in [w.D_1 : q_1]\} + P\{w \in [q'_1 : w.D'_1]\}$.

4 Empirical Study

To study the correctness of the proposed method, denoted as SkyQUD-T, which is adopted from SkyQUD algorithm [12], a comparison on the set of skyline objects reported is conducted. Due to the lack of previous work, the following two naïve methods are considered as a basis of comparison: the SkyQUD algorithm, yet instead (1) utilizing the concept of *strictly rejecting* any object that intersects the boundary of range query, denoted as SkyQUD-SR, and (2) utilizing the concept of *loosely accepting* any object that intersects the boundary of range query, denoted as SkyQUD-LA. For both of these two methods, the definition of *satisfy* differs from the term defined in Definition 3. In SkyQUD-SR, only objects that directly fall within the boundaries of a given range query are considered as objects that satisfy the range query. Meanwhile, in SkyQUD-LA all objects that fall within the boundaries of a given range query as well as objects that intersect with the boundaries of the range query are considered as objects that satisfy the range query.

The comparison is performed on a synthetic dataset that has been generated for 100,000 objects on two dimensions, with 50,000 objects generated as points and continuous ranges, respectively, in the *uncertain dimension*. The size of the continuous range is randomly generated. Each dimension represents a uniform random variable from 1 to 10,000, and the first dimension is set as *uncertain dimension*. The threshold t is set to 50 %. For the purpose of this paper, it is assumed that the given range query is on the *uncertain dimension*.

Figure 6(a) and (b) present the result of range skyline query on *uncertain dimension* when the range query $[q_1 : q'_1]$ is set to [0 : 11000]. This means the range query encompasses the whole dataset, which is the same as computing skyline query on the dataset without any range query. Thus, all three methods yielded the same results. Figure 7(a) exhibits objects for all three methods that satisfy the boundaries set by a

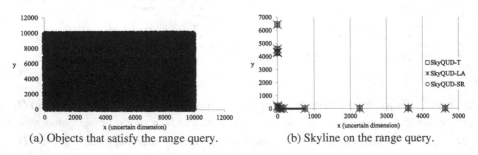

(a) Objects that satisfy the range query.   (b) Skyline on the range query.

Fig. 6. Range query [0 : 11000].

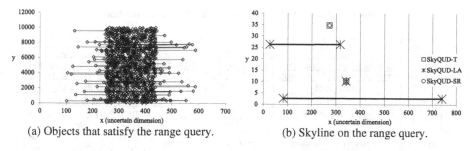

(a) Objects that satisfy the range query.   (b) Skyline on the range query.

Fig. 7. Range query [250 : 440].

given range query [250 : 440], while Fig. 7(b) demonstrates that given a range query $[q_1 : q_1']$, skyline objects reported with respect to the range query by SkyQUD-*T* are reported by SkyQUD-*SR* and SkyQUD-*LA* as well. In Fig. 7(b), the cases discussed in previous section are illustrated, where continuous range objects intersect the range query and their endpoints either lie within or outside of the query boundaries.

Table 1 exhibits a detailed analysis on the behavior of all three methods with several different range queries. The objects that satisfy a given query are divided into two different sets, where A represents a set of objects with atomic values, while I represents a set of objects with continuous range of values. It can be seen that the number of objects reported in I is equal for all three methods since it is quite straightforward to filter out atomic values that do not satisfy the range query. On the other hand, the number of objects reported in A differs for all three methods as each method has a different definition on objects that *satisfy* a given range query. From the analysis, clearly SkyQUD-*SR* would have the lowest number of objects that satisfies a given range query as the method strictly rejects objects that do not fall precisely within the range query, while SkyQUD-*LA* would have the highest number of objects that satisfies a given range query since the method would simply accept all objects that fall within or intersect the range query. However, these facts do not always equate to a less number of skyline objects reported by SkyQUD-*SR* or more skyline objects reported by

Table 1. Behaviour of SkyQUD-*T*, SkyQUD-*LA*, and SkyQUD-*SR* on different range queries.

Query ID	Range Query	Before skyline computation						After skyline computation			Processing time (sec.)			Number of skyline objects reported		
		Number of objects that satisfies the range query						Number of pairwise comparisons								
		SkyQUD-T		SkyQUD-LA		SkyQUD-SR		SkyQUD-T	SkyQUD-LA	SkyQUD-SR	SkyQUD-T	SkyQUD-LA	SkyQUD-SR	SkyQUD-T	SkyQUD-LA	SkyQUD-SR
		A	I	A	I	A	I									
1	[0:11000]	50000	50000	50000	50000	50000	50000	114629	114629	114629	0.264	0.264	0.264	12	12	12
2	[1000:1001]	3	0	3	9071	3	0	2	10791	2	0.662	1.177	0.662	2	1	2
3	[250:440]	1003	57	1003	4226	1003	9	1346	6293	1235	0.667	0.725	0.658	6	3	6
4	[2000:5000]	14984	17221	14984	35645	14984	4547	40634	59894	22639	0.778	6.63	0.632	4	4	10
5	[3186:4233]	5202	2222	5202	28469	5202	544	10452	39250	8098	0.838	3.355	0.648	5	2	10
6	[2:2.5]	2	0	2	18	2	0	1	31	1	0.659	0.667	0.659	1	4	1
7	[98:5418]	26620	29161	26620	39572	26620	14323	66364	77197	50359	0.586	0.799	0.581	11	8	12
8	[888:10545]	45525	49242	45525	49629	45525	41540	138894	139584	122188	2.508	3.131	0.386	11	8	17
9	[4645:4705]	315	2	315	25292	315	0	535	40816	532	0.707	27.227	0.657	9	2	9
10	[87:487]	2095	204	2095	4704	2095	69	2852	8665	2634	0.667	0.931	0.651	9	8	8

Table 2. IDs of skyline objects reported by each method.

Query ID	Range Query	Object ID		
		SkyQUD-*T*	SkyQUD-*LA*	SkyQUD-*SR*
1	[0:11000]	54680, 68292, 69232, 75146, 75917, 77989, 78393, 81111, 82110, 90537, 96081, 34783	54680, 68292, 69232, 75146, 75917, 77989, 78393, 81111, 82110, 90537, 96081, 34783	54680, 68292, 69232, 75146, 75917, 77989, 78393, 81111, 82110, 90537, 96081, 34783
2	[1000:1001]	72201, 79716	21764	72201, 79716
3	[250:440]	71638, 76824, 81465, 87192, 93588, 95580	71638, 34783, 47593	71638, 76824, 81465, 87192, 93588, 95580
4	[2000:5000]	75146, 81111, 92086, 96081	75146, 81111, 92086, 96081	51934, 63907, 75146, 78365, 78499, 81111, 86180, 92086, 92133, 96081
5	[3186:4233]	50881, 61418, 75146, 77895, 83537	75146, 46628	50881, 54533, 61418, 69184, 75146, 77895, 81131, 83537, 88464, 97219
6	[2:2.5]	94992	94992, 16101, 21083, 46746	94992
7	[98:5418]	74737, 75146, 77989, 80037, 81111, 81604, 94445, 96081, 96720, 99674, 34783	75146, 77989, 81111, 81604, 96081, 96720, 34783, 48262	50281, 54482, 74737, 75146, 77989, 80037, 81111, 81604, 94445, 96081, 96720, 99674
8	[888:10545]	55943, 64055, 64399, 75146, 75960, 78212, 80581, 81111, 96081, 21764, 40520	55943, 64399, 75146, 81111, 96081, 7122, 21764, 40520	52940, 55943, 56221, 63598, 63384, 64055, 64399, 75146, 75960, 78212, 80581, 81111, 90187, 93668, 96081, 21764, 40520
9	[4645:4705]	64205, 67006, 68465, 78945, 80670, 81651, 83322, 83819, 89168	19170, 46628	64205, 67006, 68465, 78945, 80670, 81651, 83322, 83819, 89168
10	[87:487]	53070, 77989, 81604, 88025, 94685, 96720, 96932, 34783, 48262	53070, 77989, 81604, 94685, 96720, 96932, 34783, 48262	53070, 77989, 81604, 88025, 94445, 94685, 96720, 96932

SkyQUD-*LA*, since all three methods would have a different set of objects that satisfies a given range query. The IDs of skyline objects reported in previous table is reported in Table 2 to certify that all skyline objects reported by SkyQUD-*T* will always be reported either by SkyQUD-*LA* or SkyQUD-*SR*, or both. SkyQUD-*T* will never report a skyline object that has not been reported by either of the other two methods. The underlined IDs in SkyQUD-*T* indicate that the IDs are reported as well by at least SkyQUD-*LA* or SkyQUD-*SR*. Figure 8(a) and (b) exhibit the effect of increasing the value of threshold *t* in terms of number of pairwise comparisons and processing time, respectively. The figures show that the performance of the algorithm increases with the increase of threshold. With a larger threshold, the algorithm filters more objects earlier, and hence exhibits a better processing time.

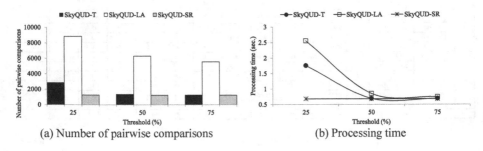

(a) Number of pairwise comparisons   (b) Processing time

Fig. 8. Effects of threshold.

5 Conclusion

The issue of computing skyline on *uncertain dimension* within a given range query is investigated. A method that incorporated a threshold value is proposed in order to filter out objects that intersect with the range query yet having a probability of them being within the range query less than the threshold value. The method is then compared with two naïve methods (1) strictly rejecting, and (2) loosely accepting objects that intersect with the given range query. The skyline objects reported by these two methods are then compared to the skyline objects reported by the proposed method.

References

1. Borzsonyi, S., Kossmann, D., Stocker, K.: The skyline operator. In: Proceedings of International Conference on Data Engineering (ICDE 2001), Heidelberg, Germany, pp. 421–430 (2001)
2. Chomicki, J., Godfrey, P., Gryz, J., Liang, D.: Skyline with presorting. In: Proceedings of International Conference Data Engineering (ICDE 2003), Bangalore, India, pp. 717–719 (2003)
3. Godfrey, P., Shipley, R., Gryz, J.: Maximal vector computation in large data sets. In: Proceedings of the 31th International Conference on Very Large Data Bases (VLDB 2005), Trondheim, Norway, pp. 229–240 (2005)
4. Jiang, B., Pei, J.: Online interval skyline queries on time series. In: Proceedings of the 2009 IEEE International Conference on Data Engineering, pp. 1036–1047 (2009)
5. Khalefa, M.E., Mokbel, M.F., Levandoski, J.J.: Skyline query processing for uncertain data. In: Proceedings of the Conference on Information and Knowledge Management, pp. 1293–1296 (2010)
6. Kossmann, D., Ramsak, F., Rost, S.: Shooting stars in the sky: an online algorithm for skyline queries. In: Proceedings of International Conference on Very Large Data Bases (VLDB 2002), Hong Kong, China, pp. 275–286 (2002)
7. Lian, X., Chen, L.: Monochromatic and bichromatic reverse skyline search over uncertain databases. In: Proceedings of the ACM SIGMOD International Conference on Management of Data, pp. 213–226 (2008)
8. Papadias, D., Tao, Y., Fu, G., Seeger, B.: Progressive skyline computation in database systems. ACM Trans. Database Syst. (TODS). **30**, 41–82 (2005)
9. Pei, J., Jiang, B., Lin, X., Yuan, Y.: Probabilistic skylines on uncertain data. In: Proceedings of the International Conference on Very Large Database, pp. 15–26 (2007)
10. Qi, Y., Atallah, M.: Identifying interesting instances for probabilistic skylines. In: Bringas, P.G., Hameurlain, A., Quirchmayr, G. (eds.) DEXA 2010, Part II. LNCS, vol. 6262, pp. 300–314. Springer, Heidelberg (2010)
11. Rahul, S., Janardan, R.: Algorithms for range-skyline queries. In: Cruz, I.F., Knoblock, C. A., Kröger, P., Tanin, E., Widmayer, P. (eds.) SIGSPATIAL/GIS, pp. 526–529 (2012)
12. Saad, N.H.M., Ibrahim, H., Alwan, A.A., Sidi, F., Yaakob, R.: A framework for evaluating skyline query over uncertain autonomous databases, vol. 29, pp. 1546–1556. Elsevier (2014)
13. Wang, W.-C., Wang, E.T., Chen, A.L.P.: Dynamic skylines considering range queries. In: Yu, J.X., Kim, M.H., Unland, R. (eds.) DASFAA 2011, Part II. LNCS, vol. 6588, pp. 235–250. Springer, Heidelberg (2011)

14. Zhang, W., Lin, X., Zhang, Y., Wang, W., Yu, J.: Probabilistic skyline operator over sliding windows. In: Proceedings of the International Conference on Data Engineering, pp. 1060–1071 (2009)
15. Weisstein, E.W.: Continuity Correction, MathWorld–A Wolfram Web Resource. http://mathworld.wolfram.com/ContinuityCorrection.html

Aging Locality Awareness in Cost Estimation for Database Query Optimization

Chihiro Kato[1]([✉]), Yuto Hayamizu[1], Kazuo Goda[1], and Masaru Kitsuregawa[1,2]

[1] The University of Tokyo, Komaba 4-6-1, Meguro-ku, Tokyo, Japan
kato@tkl.iis.u-tokyo.ac.jp
[2] National Institute of Informatics, Hitotsubashi 2-1-2, Chiyoda-ku, Tokyo, Japan
http://www.u-tokyo.ac.jp/
http://www.nii.ac.jp/

Abstract. A number of insertions, updates and deletions eventually deteriorate the structural efficiency of database storage, and then cause performance degradation. This phenomenon is called "aging." In real-world database systems, aging often exhibits strong locality because of the inherent skewness of data access; specifically speaking, the cost of I/O operations is not uniform throughout the storage space. Potentially query execution cost is influenced by the aging. However, conventional query optimizers do not consider the aging locality; thus they cannot accurately estimate the cost of query execution plans at times. In this paper, we propose a novel method of cost estimation that has the key capability of accurately determining aging phenomena, even though such phenomena are non-uniformly incurred. Our experiment on PostgreSQL and TPC-H data sets showed that the proposed method can accurately estimate the query execution cost even if it is influenced by the aging.

Keywords: Database systems · Query optimizer · Database aging

1 Introduction

The structural efficiency of database storage is fundamental to the query execution performance in database systems. Insertions, updates and deletions can scatter densely packed records and disturb the physical ordering of records. Repeated execution of these operations eventually deteriorates the structural efficiency and then causes performance degradation. This phenomenon is called "aging". Aging can greatly affect the I/O cost of query execution, and its influence is different for each candidate plan owing to the difference of I/O strategies. The progress of aging phenomenon can cause errors in cost estimation. Thus query optimizers may choose non-optimal plans because they are unaware of aging phenomenon.

C. Kato — Currently, Fujitsu Laboratories.

The original version of this chapter was revised: The authors' affiliations were incorrect. An erratum to this chapter can be found at 10.1007/978-3-319-44406-2_40

© Springer International Publishing Switzerland 2016
S. Hartmann and H. Ma (Eds.): DEXA 2016, Part II, LNCS 9828, pp. 389–396, 2016.
DOI: 10.1007/978-3-319-44406-2_32

The difficulty of aging-aware cost estimation is that aging often has strong locality due to the skewness of data access by user activities. Actual cost of query executions can differ even when queries are the same except for access ranges. While the significance of aging on database performance has been recognized from the early history of database systems, awareness of aging locality in query optimization has been remained largely unexplored to the best of our knowledge.

In this paper, we propose a novel method of cost estimation for query optimization that has the key capability of figuring out the aging phenomenon accurately even though this occurs non-uniformly. Our experiments showed that the proposed method yields good cost estimation and helps the choice of optimal query plans. The rest of this paper is organized as follows. We describe the proposed method in Sect. 2. We then present the evaluation of the proposed method in Sect. 3. We summarize related work in Sect. 4 and conclude the paper in Sect. 5.

2   Aging Locality Aware Cost Estimation

2.1   Influence of Aging and Its Locality on Query Optimization

While initially loaded databases can enjoy good efficiency, repeated execution of insertions, updates and deletions eventually disturb the physical ordering of a table, spatially scatter records across a table, and then degrade performance. This phenomenon is called aging. As described in the previous section, databases in production are inherently in aged states for most of their lifetimes.

In terms of query optimization, this performance degradation due to aging means the increase of the I/O cost. This cost increase has two aspects: temporal and spatial variation. Both can lead to wrong choices of query execution plans in different ways. The temporal variation of the I/O cost is caused by the progress of aging and can change the optimal query execution plan for a certain query. The spatial variation of the I/O cost is caused by aging locality. Even if queries are the same except for access ranges, as seen in prepared statements, the optimal query execution plans can be different in the presence of spatial variation of the I/O cost.

In situations with aging locality, a conventional optimizer cannot reflect the spatial variation of the I/O cost in the cost estimation, which can result in choosing non-optimal query execution plans. In order to choose the optimal query execution plan on aged databases, cost models should be aware of aging locality. In the next subsection, we present I/O cost models with aging locality and provide a method to measure the increase of the I/O cost for the presented models.

2.2   I/O Cost Models with Aging Locality

First, we model the I/O access cost for only one table. We define $S(x)$ as a window function of the access range, and $D(x)$ as the distribution density of data, where x can be the value of an indexing key or an address in a table space. If the cost $C(x)$ of accessing a record pointed by x is given, the I/O cost for a single table access can be described as follows:

$$\Gamma = \int S(x)D(x)C(x)dx \tag{1}$$

$S(x)$ is equal to 1 if x is in an accessed range; otherwise, it is 0. $D(x)$ denotes the number of records for x. When the table is initialized or reorganized, $C(x)$ should be nearly a constant; as the table ages, $C(x)$ changes its shape according to the increase of the I/O cost. Note that we do not consider a composite primary key in this paper; this will be left for future work.

For join queries, we combine these functions to estimate the I/O cost. For example, a nested loop join query picks up matching records in table t_1 one by one. For each record in table t_1, scans records in table t_2 that satisfy join conditions. Thus, its I/O cost can be described with the join cardinality $j_{t12}(x)$ between tables t_1 and t_2 as follows:

$$\Gamma_{NLJ} = \int S_{t1}(x)D_{t1}(x)C_{t1}(x)dx + \int j_{t12}(x)\{S_{t1}(x)D_{t1}(x)\}C_{t2}(x)dx \tag{2}$$

On the other hand, the I/O cost of hash join queries are rather simple:

$$\Gamma_{HJ} = \int S_{t1}(x)D_{t1}(x)C_{t1}(x)dx + \int S_{t2}(x)D_{t2}(x)C_{t2}(x)dx \tag{3}$$

In order to calculate a value of the I/O cost of a requested query, $C(x)$ must be available before query requests. In this paper, we focus on two fundamental access methods; full-table scan and index scan. We employed a measurement-based approach with performance test queries to approximate $C(x)$ for each access method. For the full-table scan, regardless of the actual $C(x)$, the average of the I/O cost increase is enough for cost estimation, so its performance test query is just a simple full-table scan. For the index scan, the x-space is divided equally into N parts, and performance test queries are given as index scan queries of each part. By measuring the execution time of these performance test queries beforehand, approximate values of $C(x)$ can be provided for our cost models.

The purpose of this paper is showing that aging locality aware cost modeling can improve the accuracy of cost estimation. This approach requires non-negligible amount of workload. We would like to further investigate efficient calculation of $C(x)$ in future.

3   Experiment

In order to validate the potential benefits of the proposed cost estimation, we performed intensive experiments by using an open-source database system and an industry-standard benchmark data set.

Table 1. Experimental setup

Server model	Dell PowerEdge R720xd
Processor	2x Intel Xeon E5-2690 v2
Main memory	64 GB DRAM
Storage devices	1x 900 GB HDD dedicate for database
	1x 900 GB HDD dedicate for operating system
Operating system	CentOS release 5.8 (64 bit)
Database system	PostgreSQL 9.4.0 (buffer size 128 MB)
Data set and schema	TPC-H, dbgen 2.17.0

3.1   Experimental Setup

Table 1 summarizes the laboratory environment that we built. PostgreSQL was configured with default configuration parameters unless specially noted.

First, we generated an initial data set by executing dbgen with a scale factor 100 and loaded the data set into the PostgreSQL database. After loading the data set, we executed the VACUUM command because this is well-known as a best practice to obtain the maximum performance. Following this, we performed a measurement; we executed a query and measured execution information, such as the taken execution time and deployed query execution plan. Note that, every time we started execution of a query, we cleaned up Linux disk buffer and PostgreSQL database buffer to measure cold-start performance by preventing some data from being cached there.

After we completed a measurement in the initial status, we iterated a bulk update on the database and took another measurement on the updated database. We performed the bulk update by executing refresh functions generated by dbgen. Logically, the database size does not change even as we update the database. To ensure fair measurement, we also ran the VACUUM command every time we completed a refresh function. By iterating database refreshing and performance measurement, we observed how query execution behavior would change as we incrementally updated the database.

```
SELECT SUM(l_extendedprice) FROM lineitem
WHERE l_orderkey < x AND l_orderkey > y
```
$$\cdots \ (1)$$

```
SELECT SUM(l_extendedprice) FROM lineitem
WHERE l_partkey < x AND l_partkey > y
```
$$\cdots \ (2)$$

Fig. 1. Test queries (Example)

```
SELECT SUM(l_extendedprice) FROM lineitem, part
WHERE p_partkey < x AND l_orderkey > a AND l_orderkey < b
AND l_partkey = p_partkey                              ···(A)
```

Fig. 2. Validation queries

3.2 Cost Estimation Accuracy

This section presents the experimental results that validate the benefits of the proposed cost estimation.

We performed a measurement for the initial (non-aged) status and the refreshed (aged) status in the same database for each table and each access method. The refreshed status meant that the refresh function (updating the 10 % of the storage space) had been performed four times. For each measurement, we first performed each test query (example is depicted in Fig. 1) to measure aging degrees throughout the database. For example, regarding the test queries (1)–(2), we performed the query for different combinations of x and y so that the series of query executions would eventually cover the whole database space. Specifically, we divided the key space described by l_orderkey into ten pieces. In the first query trial for the test query (1), we set (x, y) to $(\min(\text{l\_orderkey}), \min(\text{l\_orderkey}) \times 9/10 + \max(\text{l\_orderkey}) \times 1/10)$. As well, in the second query trial, we set (x, y) to $(\min(\text{l\_orderkey}) \times 9/10 + \max(\text{l\_orderkey}) \times 1/10, \min(\text{l\_orderkey}) \times 8/10 + \max(\text{l\_orderkey}) \times 2/10)$. And we execute the same query with different (x, y) until we could cover the whole key space to obtain aging degrees over the space.

Based on the measured aging degrees, we estimated the query cost for the validation query (depicted in Fig. 2) in accordance with the estimation method introduced in Sect. 2. We also actually performed the validation query and compared the estimated cost and actual execution time to investigate how accurately the proposed method could estimate the query execution costs.

For comparison, we also measured the estimated cost reported by the EXPLAIN command in PostgreSQL to execute the validation query. This estimated cost is an internal value that is used for query optimization in PostgreSQL.

Figure 3(a) and (b) present aging degrees that we measured over the key space described by l_orderkey for the initial status and aged status, respectively, of the same database. As is clearly illustrated, access cost were uniformly distributed with the initial status, but in the aged status, access cost in the first 10 % region dramatically increased. In other words, aging phenomena were incurred in this region.

Figure 4(a) shows how query optimization was performed for query (A) with $x = 1000$ in the aged status. Aging was incurred in a limited portion in the database. To investigate the aging locality, we set $a = 60,000,000$ and $b = 120,000,000$ so that the query could fall in the non-aged region and we set $a = 0$ and $b = 60,000,000$ so that it could go into the aged region. The graph summarized the EXPLAIN cost and actual execution time for two different query

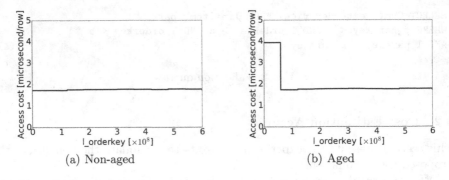

Fig. 3. Aging degrees of l_orderkey

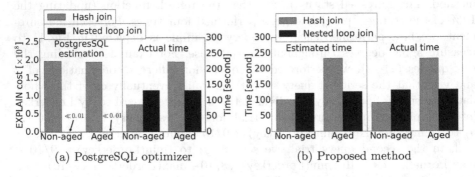

(a) PostgreSQL optimizer   (b) Proposed method

Fig. 4. Compare estimated cost with execution time about query (A) (selectivity of part table is 0.005 %)

execution plans; nested-loop join and hash join. In both regions, PostgreSQL estimated much smaller cost for nested loop join rather than hash join. In terms of the actual execution time, however, hash join outperformed nested loop join for the non-aged region but vice versa for the aged-region. This experiment confirmed that the current implementation in PostgreSQL cannot accurately estimate aging phenomena that were incurred in the database storage.

In contrast, Fig. 4(b) presents the estimated cost with the proposed method for the same query configuration. As is clearly shown, the proposed method successfully obtained a lower cost for hash join in the non-aged region but for nested loop join in the aged region.

4   Related Work

4.1   Aging and Database Reorganization

Performance degradation due to aging phenomenon has been a big headache for database administrators. Besides a mathematical analysis of performance degradation [1], database reorganization has been studied as a practical solution. In the '70s, the size of the database was generally small. Off-line reorganization was an reasonable approach [2], and arbitration between performance degradation and reorganization cost was the main concern at that time [3,4].

As the size of databases in operation grew rapidly, online reorganization became mainstream. Online reorganization technologies are largely placed into two categories: replicating a database and writing back the result afterwards [5], and incrementally reorganizing a database in place avoiding conflict with running queries by users [6]. Starting from the '80s, online reorganization has remained an active field of research [7,8] to the present.

However, despite intensive studies on database reorganization, these approaches still require too many resources to be executed frequently enough for keeping databases from being aged. In realistic situations, a certain level of aging phenomenon is unavoidable. In this paper, we propose a novel method of aging-aware cost estimation for query optimization. The proposed method can accurately estimate the cost of query execution even in the existence of non-uniform aging phenomenon, while conventional methods cannot.

4.2 Query Optimization

A query optimizer is a key component of database systems that converts an incoming query into an optimal query execution plan [9]. It has been studied intensively and extensively [10,11], such as parallelization of query optimization for utilizing the increasing number of CPU cores [12], caching results for future query optimization [13], and so on. In recent years, query optimization for emerging parallel query engine has been actively studied, such as using intermediate results for query optimization [14]. I/O cost modeling is a centerpiece of cost-based optimization. Storage systems were mostly based on magnetic disks before the 2000s [15], but in recent years Flash-based SSDs have increased its adoption rapidly in enterprise systems, and revisiting the I/O cost modeling has gained momentum [16]. In this paper, we focus on I/O cost modeling in the presence of non-uniform aging phenomenon.

5 Conclusion

We proposed a novel technology of query cost estimation that has the key capability of figuring out aging phenomena accurately even though the aging phenomena are non-uniformly incurred on the storage space. Our experiments confirmed that the proposed technology can improve the accuracy of query cost estimation as aging is incurred in the database.

As a first step, this paper has focused on a careful but fundamental investigation of our purposed approach. Many open problems still remain. First, we would like to extend our technical investigation toward database queries of higher complexity. Second, we would like to extend our experiments by using different real-world data sets and queries in order to validate the benefits for a wide spectrum of applications. Finally, we also plan to work on implementation of the proposed framework into PostgreSQL so as to share our knowledge among the community.

References

1. Heyman, D.P.: Mathematical models of database degradation. ACM Trans. Database Syst. (TODS) **7**(4), 615–631 (1982)
2. Sockut, G.H., Goldberg, R.P.: Database reorganization-principles and practice. ACM Comput. Surv. (CSUR) **11**(4), 371–395 (1979)
3. Shneiderman, B.: Optimum data base reorganization points. Commun. ACM **16**(6), 362–365 (1973)
4. Bing Yao, S., Sundar Das, K., Teorey, T.J.: A dynamic database reorganization algorithm. ACM Trans. Database Syst. (TODS) **1**(2), 159–174 (1976)
5. Sockut, G.H., Beavin, T.A., Chang, C.-C.: A method for on-line reorganization of a database. IBM Syst. J. **36**(3), 411–436 (1997)
6. Omiecinski, E., Scheuermann, P.: A global approach to record clustering and file reorganization. In: Proceedings of the Seventh Annual International ACM SIGIR Conference on Research and Development in Information Retrieval, pp. 201–219. British Computer Society (1984)
7. Kitsuregawa, M., Goda, K., Hoshino, T.: Storage fusion. In: Proceedings of the 2nd International Conference on Ubiquitous Information Management and Communication (ICUIMC2008), pp. 270–277. ACM (2008)
8. Ghandeharizadeh, S., Gao, S., Gahagan, C., Krauss, R.: An on-line reorganization framework for SAN file systems. In: Manolopoulos, Y., Pokorný, J., Sellis, T.K. (eds.) ADBIS 2006. LNCS, vol. 4152, pp. 399–414. Springer, Heidelberg (2006)
9. Chaudhuri, S.: An overview of query optimization in relational systems. In: Proceedings of the 17th ACM SIGACT-SIGMOD-SIGART Symposium on Principles of Database Systems, pp. 34–43. ACM (1998)
10. Jarke, M., Koch, J.: Query optimization in database systems. ACM Comput. Surv. (CsUR) **16**(2), 111–152 (1984)
11. Graefe, G.: The cascades framework for query optimization. Data Eng. Bull. **18**(3), 19–29 (1995)
12. Waas, F.M., Hellerstein, J.M.: Parallelizing extensible query optimizers. In: Proceedings of the 2009 ACM SIGMOD International Conference on Management of Data, pp. 871–878. ACM (2009)
13. Chen, C.M., Roussopoulos, N.: The implementation and performance evaluation of the ADMS query optimizer: Integrating query result caching and matching. Springer, Heidelberg (1994)
14. Perez, L.L., Jermaine, C.M.: History-aware query optimization with materialized intermediate views. In: 2014 IEEE 30th International Conference on Data Engineering (ICDE), pp. 520–531. IEEE (2014)
15. Haas, L.M., Carey, M.J., Livny, M., Shukla, A.: Seeking the truth about ad hoc join costs. The VLDB J. **6**(3), 241–256 (1997)
16. Ghodsnia, P., Bowman, I.T., Nica, A.: Parallel I/O aware query optimization. In: Proceedings of the 2014 ACM SIGMOD International Conference on Management of Data, pp. 349–360. ACM (2014)

Information Retrieval, and Keyword Search

Constructing Data Graphs for Keyword Search

Konstantin Golenberg[(✉)] and Yehoshua Sagiv

The Hebrew University, 91094 Jerusalem, Israel
{konstg01,sagiv}@cs.huji.ac.il

Abstract. Data graphs are convenient for supporting keyword search that takes into account available semantic structure and not just textual relevance. However, the problem of constructing data graphs that facilitate both efficiency and effectiveness of the underlying system has hardly been addressed. A conceptual model for this task is proposed. Principles for constructing good data graphs are explained. A transformation for generating data graphs from XML is developed.

Keywords: Data-graph construction · Keyword search · XML

1 Introduction

Considerable research has been done on effective algorithms for keyword search over data graphs (e.g., [3,4,7,10–12,14,17]). Usually, a data graph is obtained from RDB, XML or RDF by a rather simplistic transformation. In the case of RDB [3,6,12], tuples are nodes and foreign keys are edges. When the source is XML [11,13], elements are nodes, and the edges reflect the document hierarchy and IDREF(S) attributes.

In many cases, the source data suffers from certain anomalies and some papers (e.g., [13,15]) take necessary steps to fix those problems. For example, when citations are represented by XML elements, they should be converted to IDREF(S) attributes. As another example, instead of repeating the details of an author in each one of her papers, there should be a single element representing all the information about that author and all of her papers should reference that element. These are examples of necessary transformations on the source data. If they are not done, existing algorithms for keyword search over data graphs will not be able to generate meaningful answers.

Once a source data is ameliorated, it should be transformed into a graph. The literature hardly discusses how it should be done. In [3,14], the source is an RDB and the naive approach mentioned earlier is used (i.e., tuples are nodes and foreign keys are edges). In [11,20], the source data is XML and the simplistic transformation described at the beginning of this section is applied. In [2,5,12,16,18], they do not mention any details about the construction of data graphs. The lack of a thoughtful discussion in any of those papers is rather surprising, because the

---

This work was supported by the Israel Science Foundation (Grant No. 1632/12). The full version of this paper appears in [9].

S. Hartmann and H. Ma (Eds.): DEXA 2016, Part II, LNCS 9828, pp. 399–409, 2016.
DOI: 10.1007/978-3-319-44406-2_33

actual details of constructing a data graph have a profound effect on both the efficiency and the quality of keyword search, regardless of the specific algorithms and techniques that are used for generating answers and ranking them.

Construction of effective data graphs is not a simple task, since the following considerations should be taken into account. For efficiency, a data graph should be as small as possible. It does not matter much if nodes have large textual contents, but the number of nodes and edges is an important factor. However, lumping together various entities into a single node is not a good strategy for increasing efficiency, because answers to queries would lose their coherence.

The structure of a data graph should reflect succinctly the semantics of the data, or else answers (which are subtrees) would tend to be large, implying that finding them would take longer and grasping their meaning quickly would not be easy.

An effective engine for keyword search over data graphs must also use information-retrieval techniques. Those tend to perform better on large chunks of text, which is another reason against nodes with little content.

In this paper, we address the problem of how to construct data graphs in light of the above considerations. In Sect. 4, we develop a transformation for constructing data graphs from XML. In the full version of this paper [9], we also present a transformation for RDB. In addition, we show there that the format of the source data (i.e., RDB or XML) has a significant impact on the quality of the generated data graph. Moreover, XML is a better starting point than RDB. This is somewhat surprising given the extensive research that was done on designing relational database schemes.

As a conceptual guideline for constructing a good data graph, we use the OCP model [1], which was developed for supporting a graphical display of answers so that their meaning is easily understood. In Sect. 3, we explain why the OCP model is also useful as a general-purpose basis for constructing data graphs in a way that takes into account all the issues mentioned earlier.

In summary, our contributions are as follows. First, we enunciate the principles that should guide the construction of data graphs. Second, we develop transformations for doing so when the source data is XML or RDB (the latter is done in the full version [9]). These transformations are more elaborate than the simplistic approach that is usually applied. Third, the full version [9] shows how the format of the source data impacts the quality of the generated graphs. Moreover, it also explains why XML is a better starting point than RDB.

Our contributions are valid independently of a wide range of issues that are not addressed in this paper, such as the algorithm for generating answers and the method for ranking them. We only assume that an answer is a non-redundant subtree that includes all the keywords of the query. However, our results still hold even if answers are subgraphs, as sometimes done.

2   Preliminaries

2.1   The OCP Model

The *object-connector-property* (OCP) model for data graphs was developed in [1] to facilitate an effective GUI for presenting subtrees. (As explained in the next section, those subtrees are answers to keyword search over data graphs.) In the OCP model, objects are entities and connectors are relationships. We distinguish between two kinds of connectors: *explicit* and *implicit*. Objects and explicit connectors can have any number of properties. Two special properties are *type* and *name*.

Parts (a) and (b) of Fig. 1 show an object and a snippet of a data graph, respectively. An object is depicted as a rectangle with straight corners. The top line of the rectangle shows the name and type of the object. The former appears first (e.g., Ukraine) and the latter is inside parentheses (e.g., country). The other properties appear as pairs consisting of the property's name and value, as shown in Fig. 1(a). Observe that properties can be nested; for example, the property percentage is nested inside ethnicgroup. Nesting is indicated in the figure by indentation.

An implicit connector is shown as a directed edge between two objects. Its meaning should be clear from the context. In Fig. 1(b), the implicit connector from Ukraine to Odeska means that the latter is a province in the former.

An explicit connector is depicted as a rectangle with rounded corners. It has at most one incoming edge from an object and any positive number of outgoing edges to some objects. An explicit connector has a type, but no name, and may also possess other properties. Figure 1(b) shows an explicit connector of type border from Ukraine to Russia that has the property length whose value is 1576 km.

2.2   Answers to Keyword Search

We consider keyword search over a directed data graph G. (A data graph must be directed, because relationships among entities are not always symmetric.) A *directed subtree* t of G has a unique node r, called the *root*, such that there is exactly one directed path from r to each node of t.

A query Q over a data graph G is a set of keywords, namely, $Q = \{k_1, \ldots, k_n\}$. An *answer* to Q is a directed subtree t of G that contains all the keywords of Q and is nonredundant, in the sense that no proper subtree of t also contains all of them.

For example, consider Fig. 2, which shows a snippet of the data graph created from the XML version of the Mondial dataset,[1] according to the transformation of Sect. 4. To save space, only the name (but not the type) of each object is shown. The dashed edges should be ignored for the moment. The subtree in Fig. 3(a) is an answer to the query {Dnepr, Russia, Ukraine}. There are additional

---

[1] http://www.dbis.informatik.uni-goettingen.de/Mondial/.

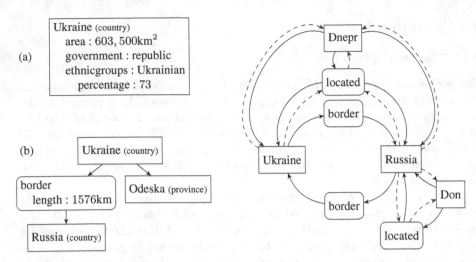

Fig. 1. An object and a tiny snippet of a data graph (not all properties are shown)

Fig. 2. A tiny portion of Mondial

answers to this query, but all of them have more than three nodes and at least one explicit connector.

For the query {Dnepr, Don}, there is no answer (with only solid edges) saying that Dnepr and Don are rivers in Russia, although the data graph stores this fact. The reason is that the connectors (in the data graph of Fig. 2) have a symmetric semantics, but the solid edges representing them are in only one direction. The only exception is the connector border, which is already built into the graph in both directions (between Russia and Ukraine). In order not to miss answers, we add *opposite edges* when symmetric connectors do not already exist in both directions. Those are shown as dashed arrows. Now, there are quite a few answers to the query {Dnepr, Don} and Fig. 3(b)–(d) shows three of them. The first two of those say that Dnepr and Don are rivers in Russia. These two answers have the same meaning, because the relationship between a river and a country is represented twice: by an implicit connector and by the explicit connector located. The answer in Fig. 3(d) has a different meaning, namely, Dnepr and Don are rivers in Ukraine and Russia, respectively, and there is a border between these two countries.

To generate relevant answers early on, weights are assigned to the nodes and edges of a data graph. Existing algorithms (e.g., [3,7,8,10] enumerate answers in an order that is likely to be correlated with the desired one. Developing an effective weighting scheme is highly important, but beyond the scope of this paper.

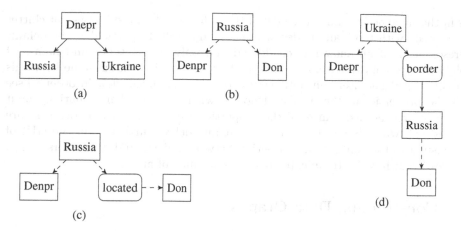

Fig. 3. Answers to queries

3 Advantages of the OCP Model

In this section, we discuss some of the advantages of the OCP model. In a naive approach of building a data graph, there is only one type of nodes (i.e., no distinction between objects and connectors). Moreover, sometimes there is even a separate node for each property. This approach suffers from three drawbacks. First, from the implementation's point of view, this is inefficient in both time and space. That is, even if there is not much data, the number of nodes and edges is likely to be large. As a result, searching a data graph for answers would take longer (than the alternative described later in this section). In addition, if all the processing is done in main memory, the size of the data graph is more likely to become a limiting factor.

The second drawback of the naive approach is from the user's point of view. A meaningful answer is likely to have quite a few nodes; hence, displaying it graphically in an easily understood manner is rather hard. Another problem is the following. The definition of an answer is intended to avoid redundant parts in order to cut down the search space. However, sometimes an answer must be augmented to make it clear to the user. For example, an answer cannot consist of just some property that contains the keywords of the query, without showing the context.

The third drawback pertains to ranking, which must take into account textual relevance (as well as some other factors). In the naive approach, many nodes have only a small amount of text, making it hard to determine their relevance to a given query.

In comparison to the naive approach, the OCP model dictates *fat* nodes. That is, an object or an explicit connector is represented by a node that contains all of its properties. Consequently, we get the following advantages. First, a data graph is not unduly large, which improves efficiency. Second, relevance is easier to determine, because all the text pertaining to an object or an explicit connector is in the same node. Third, the GUI of [1] is effective, because it does not clutter the screen with too many nodes or unnecessary stuff. In particular, the default presentation of an answer is condensed and only shows: types and names of objects; types of explicit connectors; and properties that match some keywords of the query. The user can optionally choose an expanded view in order to see all the properties of the displayed nodes, when additional information about the answer is needed. Since all the properties are stored in the nodes that are already shown, this can be done without any delay. Furthermore, the GUI of [1] visualizes the conceptual distinction between objects and connectors, which makes it much easier to quickly grasp the meaning of an answer.

4   Constructing Data Graphs

An XML document is a rooted hierarchy of *elements*. Each element can have any number of *attributes*. Three special types of attributes are ID, IDREF and IDREFS. An attribute of the first type has a value that uniquely identifies its element. The last two types serve as references to other elements. For an attribute defined (in the DTD) as IDREF, the value is a single ID (of the referenced element); and if an attribute is defined as IDREFS, its value is a set of IDs. In our terminology, a *reference* attribute is one defined as either IDREF or IDREFS. An attribute is *plain* if it is neither ID, IDREF nor IDREFS.

In XML lingo, an element has a *name* that appears in its tag (e.g., <city>). To avoid confusion, we call it the *type* of the element, because it corresponds to the notion of a type in the OCP model

In this section, we describe how to transform an XML document to a data graph. We assume that the document has a DTD and use it in the transformation. As we shall see, the DTD provides information that is essential to constructing the data graph. Conceivably, this information can also be gleaned from the document itself. However, if the document does not conform to a reasonable DTD, the resulting data graph (similarly to the document itself) is likely to be poorly designed. By only assuming that there is a DTD (as opposed to an XML schema), we make our transformation much more applicable to real-world XML documents.

In XML documents, redundancies commonly occur due to reference attributes. For example, a course may have an IDREF attribute `teacher` that points to an element of type `teacher`. Converting the attribute `teacher` to an explicit (rather than implicit) connector is redundant, because the type of the referenced element makes it clear what is the semantic meaning of the connector.

Formally, consider an attribute A that is defined as IDREF. A DTD does not impose any restriction on the type E of an element that can be referenced by the value of A. In a given XML document, A (i.e., its name) and E could be the same (e.g., `teacher`). If so, we say that A is an *insignificantly named* reference attribute. In the constructed data graph, the reference described by A can be represented by an implicit connector. If the opposite holds, namely, A and E are different, then we say that A is a *significantly named* reference attribute.

In this case, the constructed data graph should retain A as the type of an explicit connector.

If attribute A is defined as IDREFS, then it is insignificantly named if all the IDs (in the value of A) are to elements of a type that has the same name as A; otherwise, it is significantly named.

Whether a reference attribute is significantly named depends on the given XML document (and not just on the DTD). It may change after some future updates. As a general rule, we propose the following. It is safe to assume that a reference attribute A is significantly named if there is no element of the DTD, such that its type is the same as A. In any other case, it is best to get some human confirmation before deciding that a reference attribute is insignificantly named.

Let E_1 and E_2 be element types. We say that E_2 is a *child element type* of E_1 if the DTD has a rule for E_1 with E_2 on its right side. In this case, E_1 is a *parent element type* of E_2.

Rudimentary rules for transforming an XML document to a data graph were given in [19]. However, they are applicable only to simple cases. Next, we describe a complete transformation that consists of two stages. We assume that prior to these two stages, both the DTD and the XML document are examined to determine for each reference attribute whether it is significantly named or not.

In the first stage, we analyze the DTD and classify element types as either objects, connectors or properties. This also induces a classification over the elements themselves. That is, when a type E is classified as an object, then so is every element of type E (and similarly when E is classified as a connector or a property). In the second stage, the classification is used to construct the data graph from the given XML document. The first stage starts by classifying all the element types E that satisfy one of the following *base rules*.

1. If E does not have any child element type and all of its attributes are plain, then E is a property.
2. If E has an ID attribute or a significantly named reference attribute, then it is an object.
3. If E has neither any child element type nor an ID attribute, but it does have some reference attributes and all of them are insignificantly named, then E is a connector.

As an example, consider the DTD of Fig. 4. Base Rule 2 implies that the element types `country`, `province`, `river` and `confluence` are objects, because the first three have an ID attribute and the fourth has a significantly named IDREFS attribute (i.e., `rivers`). No base rule applies to `economy`. By Base Rule 1, all the other element types are properties.

Next, we find all the element types that should be classified as properties by applying the following recursive rule. If (according to the DTD rules) element type E only has plain attributes and all of its child element types are already classified as properties, then so is E. It is easy to show that a repeated application of this recursive rule terminates with a unique result.

```
<!ELEMENT country (name,population,
        economy,province)>
<!ATTLIST country (code ID #REQUIRED
        area CDATA #IMPLIED)>
<!ELEMENT economy (gdp,inflation)>
<!ELEMENT province (name,area)>
<!ATTLIST province (id ID #REQUIRED)>
<!ELEMENT river (name,length)>
<!ATTLIST river (id ID #REQUIRED)>
<!ELEMENT confluence (lng,lat)>
<!ATTLIST confluence
        rivers IDREFS #REQUIRED)
        province IDREF #REQUIRED)>
<!ELEMENT name (#PCDATA)>
<!ELEMENT population (#PCDATA)>
<!ELEMENT gdp (#PCDATA)>
<!ELEMENT inflation (#PCDATA)>
<!ELEMENT area (#PCDATA)>
<!ELEMENT length (#PCDATA)>
<!ELEMENT lng (#PCDATA)>
<!ELEMENT lat (#PCDATA)>
```

```
<country code="F" area="547030">
  <name>France</name>
  <population>58M</population>
  <economy>
    <gdp>$37,728</gdp>
    <inflation>1.7%</inflation>
  </economy>
  <province id="prov-France-25">
    <name>Rhône Alpes</name>
    <area>43698</area>
  </province>
</country>
<river id="riv-Saone">
  <name>Saône</name>
  <length>473</length>
</river>
<river id="riv-Rhone">
  <name>Rhône</name>
  <length>813</length>
</river>
<confluence
    province="prov-France-25"
    rivers="riv-Saone riv-Rhone">
  <lng>45°43'N</lng>
  <lat>4°49'E</lat>
</confluence>
```

Fig. 4. DTD snippet of Mondial   **Fig. 5.** XML snippet of Mondial

Continuing with the above example, a single application of the recursive rule shows that economy is a property, because all of its child elements have already been classified as such by Base Rule 1.

Now, we apply the following generalization of Base Rule 3. If E does not have an ID attribute, all of its child element types are classified as properties, and it has some reference attributes and all of them are insignificantly named, then E is a connector.

We end the first stage by classifying all the remaining element types as objects, and then the following observations hold. First, if an element type is classified as a property, then so are all of its descendants. Second, the classification (when combined with the construction of the data graph that is described below) ensures that a connector is always between two objects. Third, if an element type is classified as a connector, then it has some reference attributes and all of them are insignificantly named.

In the second stage, we transform the XML document to a data graph. At first, we handle PCDATA as follows. If an element e (of the document) includes PCDATA as well as either sub-elements or attributes, then we should create a new attribute having an appropriate name (e.g., text) and make the PCDATA its value. This is not needed if e has neither sub-elements nor attributes, because in this case, e becomes (in the data graph constructed below) a non-nested property, such that the element type of e is the name of that property and the PCDATA is its value.

Now we construct the data graph as follows. For each element e, such that e is not classified as a property, we generate a node n_e. This node is either an object or a connector (and hence an explicit one) according to the classification of e. The type of n_e is the same as that of e. If n_e is an object, we should choose one of its properties (which will be created by the rules below) as its name. As usual, we prefer a property (e.g., `title`) that describes the meaning of n_e, even if it is not a unique identifier. For each n_e, we create properties and add additional edges and nodes by applying the following six *construction rules*.

1. Every plain attribute of e is a property of n_e.
2. For each child p of e, such that p is classified as a property, the subtree (of the given document) that is rooted at p becomes a property of n_e. Note that this property is nested if p has either plain attributes or descendants of its own. Also observe that element types and attribute names appearing in p become names of properties nested in n_e.
3. For each child o of e, such that o is classified as an object (hence, so is e), we add an edge from n_e to n_o (which is the node created for o).
4. For each child c of e, such that c is classified as a connector, we add an edge from n_e to n_c. Observe that if such a c exists, then e is classified as an object and n_c is the node of the explicit connector corresponding to c.
5. For each reference attribute R of e, we create new connectors or add edges to existing ones, according to the following two cases. First, if R is insignificantly named, then for each object o that (the value of) R refers to, we add an edge from n_e to o. Note that this edge is an implicit connector if n_e is an object; otherwise, it is part of the explicit connector n_e.

   The second case applies when R is significantly named. In this case, the classification rules imply that n_e is an object. We first create a node n_r, such that its only incoming edge is from n_e. This node represents an explicit connector that gets the name of attribute R as its type and has no properties. In addition, for each object o that (the value of) R refers to, we add an edge from n_r to o.

Figure 6 shows the data graph created from the XML document of Fig. 5 with the DTD of Fig. 4.

We divide the original edges (i.e., those created by the above transformation) into two kinds. The *hierarchical edges* are those created by Construction Rule 3. They are implicit connectors that reflect the parent-child relationship between XML elements. The *reference edges* are the ones introduced by Construction

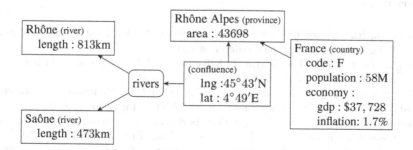

Fig. 6. A data graph constructed from the Mondial XML

Rule 5 (i.e., due to reference attributes). Construction Rule 4 creates edges due to the element hierarchy, but they enter nodes of explicit connectors; hence, we also refer to them as reference edges.

As explained in Sect. 2.2, we add opposite edges. However, our experience indicates that even if it is done just for the reference edges (i.e., no opposite edges are added for the hierarchical ones), we generally do not miss meaningful answers to queries. Furthermore, as we argue in the full version [9], a strategy that works well is to assign higher weights to opposite edges than to original ones. In this way, relevant answers are likely to be generated first without having too many duplicates early on.

5   Conclusions

We showed that the OCP model is an effective conceptual basis for constructing data graphs. Using it, we developed transformations for generating data graphs from XML and RDB (the latter is done in the full version of this paper [9]). These transformations are quite elaborate and provide much better results than the ad hoc methods that have been used in the literature thus far. In particular, the produced data graphs are better in terms of both efficiency (i.e., answers are generated more quickly) and effectiveness (i.e., the most relevant answers are produced early on).

It should be emphasized that the presented transformations are based on the principle of creating fat nodes (as explained in Sect. 3) and avoiding redundancies (e.g., due to insignificantly named references). Thus, they are applicable and useful (in most if not) all cases, regardless of how answers are generated or ranked.

In the full version of this paper [9], we show that XML is the preferred starting point for constructing data graphs. However, we need to better understand how to create XML documents that yield the best possible data graphs. Toward this end, we plan to develop appropriate design rules for XML documents.

An interesting topic for future work is to how to construct data graphs from XML documents without DTDs.

References

1. Achiezra, H., Golenberg, K., Kimelfeld, B., Sagiv, Y.: Exploratory keyword search on data graphs. In: SIGMOD Conference (2010)
2. Bao, Z., Ling, T.W., Chen, B., Lu, J.: Effective XML keyword search with relevance oriented ranking. In: ICDE (2009)
3. Bhalotia, G., Hulgeri, A., Nakhe, C., Chakrabarti, S., Sudarshan, S.: Keyword searching and browsing in databases using BANKS. In: ICDE (2002)
4. Coffman, J., Weaver, A.C.: An empirical performance evaluation of relational keyword search techniques. IEEE Trans. Knowl. Data Eng. **26**(1), 30–42 (2014)
5. Dalvi, B.B., Kshirsagar, M., Sudarshan, S.: Keyword search on external memory data graphs. PVLDB **1**(1), 1189–1204 (2008)

6. Ding, B., Yu, J.X., Wang, S., Qin, L., Zhang, X., Lin, X.: Finding top-k min-cost connected trees in databases. In: ICDE (2007)
7. Golenberg, K., Kimelfeld, B., Sagiv, Y.: Keyword proximity search in complex data graphs. In: SIGMOD Conference (2008)
8. Golenberg, K., Kimelfeld, B., Sagiv, Y.: Optimizing and parallelizing ranked enumeration. PVLDB **4**(11), 1028–1039 (2011)
9. Golenberg, K., Sagiv, Y.: Constructing data graphs for keyword search. arXiv: https://arxiv.org/abs/1605.07865 (2016)
10. Golenberg, K., Sagiv, Y.: A practically efficient algorithm for generating answers to keyword search over data graphs. In: ICDT (2016)
11. Guo, L., Shao, F., Botev, C., Shanmugasundaram, J.: XRANK: ranked keyword search over XML documents. In: SIGMOD Conference (2003)
12. He, H., Wang, H., Yang, J., Yu, P.S.: BLINKS: ranked keyword searches on graphs. In: SIGMOD Conference (2007)
13. Hristidis, V., Papakonstantinou, Y., Balmin, A.: Keyword proximity search on XML graphs. In: ICDE (2003)
14. Kacholia, V., Pandit, S., Chakrabarti, S., Sudarshan, S., Desai, R., Karambelkar, H.: Bidirectional expansion for keyword search on graph databases. In: VLDB (2005)
15. Kasneci, G., Ramanath, M., Sozio, M., Suchanek, F.M., Weikum, G.: STAR: steiner-tree approximation in relationship graphs. In: ICDE (2009)
16. Li, G., Ooi, B.C., Feng, J., Wang, J., Zhou, L.: EASE: an effective 3-in-1 keyword search method for unstructured, semi-structured and structured data. In: SIGMOD Conference (2008)
17. Mass, Y., Sagiv, Y.: Virtual documents and answer priors in keyword search over data graphs. In: Proceedings of the Workshops of the EDBT/ICDT 2016 Joint Conference (2016)
18. Park, C., Lim, S.: Efficient processing of keyword queries over graph databases for finding effective answers. Inf. Process. Manage. **51**(1), 42–57 (2015)
19. Sagiv, Y.: A personal perspective on keyword search over data graphs. In: ICDT, pp. 21–32 (2013)
20. Xu, Y., Papakonstantinou, Y.: Efficient LCA based keyword search in XML data. In: EDBT (2008)

Generating Pseudo Search History Data in the Absence of Real Search History

Ashraf Bah[✉] and Ben Carterette

Department of Computer Sciences, University of Delaware, Newark, DE, USA
{ashraf,carteret}@udel.edu

Abstract. Previous studies in Information Retrieval literature have shown that users' search history can be leveraged to improve current search results. However sometimes we have little to no search history available. In such cases, it would be helpful to obtain data *similar* to search history data. One way of doing this is by simulating previous search interactions. In the present study, we focus on generating simulated "related queries" that can serve as an additional source of information about the current search [1]. Assuming that users reformulate their queries by leveraging some of the terms and key phrases they find in ranked documents during their search, we proposed simple models for generating such related queries.

Keywords: Query reformulation · Simulation · Session history · Relevance feedback · Data fusion · Session search

1 Introduction

In this paper, we consider and address the problem of generating data similar to search history data in the absence of actual search history. This is an important problem to address as some recent studies have shown that we can leverage users' session search history to improve their current search results. One way of obtaining data similar to search history data is by simulating previous search interactions. Our current study focuses solely on "related queries" such as the ones leveraged by Bah and Carterette [1]. For the sake of succinctness, we intentionally leave out other possible exploitable resources such as clicks and dwell times for future work. We hypothesize that users reformulate their queries by leveraging some of the terms and key phrases they find in ranked documents during their search process. Our study is thus focused on generating "related queries" by leveraging the most significant key-phrases from documents in our simulated interactions.

More specifically, our problem formulation is as follows: suppose we have a real user who provides one single query and nothing else. Can we generate data that can be considered to be similar to search history data, and that leads to results similar to the ones we obtain when we leverage real users' search history?

Our contributions consist in addressing the following: Can we improve search effectiveness by leveraging simulated queries, and how does such a method compare to leveraging real search history? What are the effects of concatenating the simulated

© Springer International Publishing Switzerland 2016
S. Hartmann and H. Ma (Eds.): DEXA 2016, Part II, LNCS 9828, pp. 410–417, 2016.
DOI: 10.1007/978-3-319-44406-2_34

queries with the original user query and/or aggregating the resulting rankings with the ranking of the user's original query? And should we explore deeper layers in our model?

2   Related Work

The present work is partially motivated by the assertion of Keskustalo et al. [13] that IR test collections should "model processes where the searcher may try out several queries for one topic." In subsequent studies, Baskaya et al. proposed a simulation of query reformulation technique based on query modification, with the aim of modeling how words are selected to form an initial simulated query or subsequent queries [2, 4]. In a different study, Baskaya et al. model scenarios in which the user involved in a search process based on relevance feedback can make mistakes by providing erroneous feedback [3]. Other work (such as that by Verberne et al. [17]) sets the goal of a simulated user to collect as much gain (Cumulative Gain [10]) as possible in a five minute search session. Cartette et al. [5] proposed to generate query reformulations by leveraging individual terms from previous queries' rankings.

Our work is also motivated by recent studies that leverage users' search history. Researchers proposed to tackle the problem using a relevance feedback model that takes advantage of query changes in a session [19], or by modeling sessions as Markov Decision Processes [9], or by diversifying results while maintaining cohesion with the current query [15], or using query aggregation [8], or using anchor texts for query expansion proposed by Kruschwitz [14]. Other approaches include the work of Jiang et al. [11], who combine Sequential Dependence Model features in both current queries and previous queries in the session for one system.

Our ultimate objective is to obtain data similar to, and as effective as search history data, in the absence of real search history data, and unlike the work of Baskaya et al., it is based on the assumption that users base their reformulations on terms and key phrases they find in ranked documents. Our goal is to obtain, through simulation, (partial) search history data similar to the ones that lead to results as good as the results we can obtain when we leverage real users' search history.

3   Methodology

3.1   A Simple Model for Generating Search History Data

Our task is to generate simulated search history data that can be utilized as a substitute for real search history. Our assumption is that a user's next query reformulations are inspired and informed by the information she gets from reading the top-ranked documents from the current ranking. This implies that at each phase of our query reformulation simulation, there is a document retrieval step first, followed by the proper generation of simulated queries. There are two phases in our model, as depicted in Fig. 1: "layer 1" and "layer 2". Layer 1 begins with a real user query and a ranked list of results, which are used to generate simulated possible "next" queries. In layer 2, each of these simulated queries are used to retrieve documents, which in turn are used to generate a second set of simulated possible next queries.

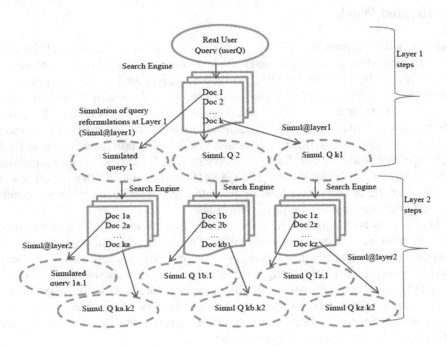

Fig. 1. A somewhat simple model for generating search history data: The elliptical shapes with large dashes represent the generated simulated queries

3.2   A Somewhat More Complex Model

Our second model is more complex. It starts the same as the previous model, using a user query and ranked results to generate simulated queries. In addition, the user query is submitted to the general web to obtain a ranking of URLs with snippets. Rather than use the simulated queries to retrieve documents at layer 2, we extract key phrases from the snippets of the web results to use for document retrieval. Then, as in the original model, a second set of simulated queries are generated from these retrieved documents. Thus the models differ only in the source of queries used to rank documents at layer 2 (Fig. 2).

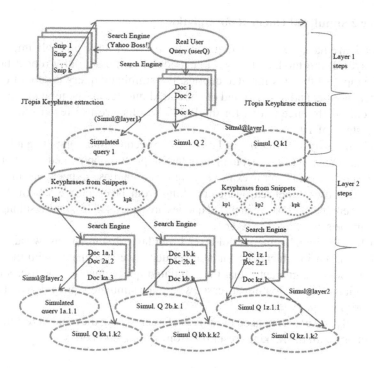

Fig. 2. A somewhat more complex model for simulating session search data: The elliptical shapes with large dashes represent the generated simulated queries. The elliptical shapes with dashes represent the key phrases generated in the new steps (they can also be used as simulated queries)

4   Implementations of the Methods

To implement our models, we need a retrieval engine and methods for generating simulated queries from ranked results. Below we describe three different implementations of each of the two models.

4.1   Layer 1 Simulated Query Reformulations

At layer 1, models 1 and 2 are identical. For the search engine, we use either Indri [16], which uses Dirichlet-smoothed language model scoring to rank full-text documents, or Yahoo! BOSS [18], which returns a SERP with URLs, titles, and snippets. When we use Indri, we extract key phrases from each of the top-10 full-text documents using JTopia [12], then concatenate the top key phrases from each document to form a simulated query. When we use BOSS, we select either titles or snippets from the top-10 ranked URLs to be used as simulated queries.

4.2 Layer 2 Simulated Query Reformulations

After layer 1, we have 10 simulated queries from one of three possible implementations. At layer 2, the models diverge. For model 1, we essentially repeat layer for
each of the top-4 of the 10 simulated queries: the simulated query is submitted to the
same search engine, and a new round of simulated queries are generated in the same
way. The only difference is that for the BOSS results, we use fewer ranked documents (5 instead of 10).

Model 2 differs by using the original user query a second time, submitting it to Yahoo!
BOSS (and only Yahoo! BOSS, not Indri) to obtain snippets of top-ranked documents from
the general web. We use JTopia to extract key phrases from those snippets, and then, unlike
model 1 (which uses simulated queries resulting from layer 1), we submit those key
phrases to our engine of choice. Simulated queries are generated from the resulting ranked
documents in the same way as in layer 1.

For each of the implementations of layer 1 and layer 2 simulations, we also experiment with a variant in which each simulated query is concatenated with the original
query. In this way we guarantee that the simulated query contains the user's original
query, potentially helping to mitigate cases where the simulated query does not contain
any of the original query terms.

Additionally, we experiment on the effects of aggregating the ranking resulting from
the user's query with the rankings resulting from simulated queries.

5 Experiment and Results

5.1 Dataset and Evaluation Measure

We use the Session track 2013 dataset [6]. It contains several user sessions which contain
one or more interactions. Each interaction consists of a query related to a given information need, a ranked list of results from a search engine, user clicks on the results, and
the time spent by the user reading the clicked document. Finally, there is a "current
query", the last query in the search session. The 2013 data consists of 87 sessions.

For the effectiveness measure, we adopted the official primary measure used by the
TREC Session track organizers, namely nDCG@10. nDCG is a graded relevance
measure that rewards documents with high relevance grades and discounts the gains of
documents that are ranked at lower positions [10].

5.2 Effectiveness of Simulated Queries: Leveraging Simulated Queries

We leverage our generated simulated queries by applying the CombCAT rank fusion
method introduced by Bah and Carterette [1]. For each query formulation, each top-k
ranked document is placed into different bins such that documents that appeared in n
different rankings are put in the same bin, labeled category$_n$. Documents are then
reranked in decreasing number of rankings they appeared in. Each simulated query was
submitted to Indri for document retrieval. We compare to the baseline of simply submitting the original user query to the Indri retrieval engine.

5.3 Results

Can We Improve Effectiveness at All by Simulating Queries and Leveraging Them? The results in Table 1 show that by applying layer 1 simulations alone, we are able to improve the results over the baseline. The highest improvements occur when we leverage Q1X + QNSnip (41.24 % improvement over the baseline) and QNSnip (40.71 % improvement over the baseline). It is to be noted that although using JTopia alone leads to a significant decrease in effectiveness, using it in addition to user query leads to relatively large improvement (19.53 %, 19.79 % and 19.97 % improvements respectively for Q2x + Jtopia, Q3x + Jtopia and Q4x + Jtopia).

Table 1. Results for layer 1 of both models on Session track 2013 dataset. Q0x, Q1x, Q2x, Q3x and Q4x respectively denote incorporating the real user query ranking 0, 1, 2, 3 and 4 times in the set of rankings that are being aggregated. QN denotes the concatenation of the real user query to the simulated query.

Resources	Q0X nDCG	%Δ	Q1X nDCG	%Δ	Q2X nDCG	%Δ	Q3X nDCG	%Δ	Q4X nDCG	%Δ
Baseline	0.1147	0.00 %								
Jtopia	0.0746	−34.96 %	0.0810	−29.38 %	0.1371	19.53 %	0.1374	19.79 %	0.1376	19.97 %
QNJTopia	0.1007	−12.21 %	0.1017	−11.33 %	0.1207	5.23 %	0.1216	6.02 %	0.1204	4.97 %
Titles	0.1465	27.72 %	0.1520	32.52 %	0.1504	31.12 %	0.1474	28.51 %	0.1468	27.99 %
QNTitles	0.1420	23.80 %	0.1331	16.04 %	0.1349	17.61 %	0.1330	15.95 %	0.1338	16.65 %
Snip	0.1407	22.67 %	0.1596	39.15 %	0.1533	33.65 %	0.1473	28.42 %	0.1459	27.20 %
QNSnip	0.1614	40.71 %	0.1620	41.24 %	0.1573	37.14 %	0.1565	36.44 %	0.1577	37.49 %

We would also like to compare to a stronger baseline that uses real search history. These results are given in Table 2. We can see that using layer 1 alone is not competitive with using real session history, nor is the first model with both layers. However, using the second model with both layers gives a substantial improvement over using real session data in all cases but the QNSnip method. This suggests that the second model is more than good enough to substitute real session history in the absence of no/little real session history data.

Table 2. Comparing layer1 to "simpler L1 + L2" as well as "complex L1 + L2"

Resources	L1 nDCG	Simpler L1 + L2 nDCG %Δ		Complex L1 + L2 nDCG %Δ		Real search history nDCG %Δ	
Titles	0.1465	0.1085	−25.94 %	**0.1893**	29.22 %	0.1598	9.08 %
QNTitles	0.1420	0.1265	−10.92 %	**0.1907**	34.30 %	0.1722	21.27 %
Snip	0.1407	0.1323	−5.97 %	**0.1850**	31.49 %	0.1715	21.89 %
QNSnip	0.1614	0.1326	−17.84 %	0.1914	18.59 %	**0.1963**	21.62 %

When We Aggregate the Ranking Resulting From the User's Query with the Rankings Resulting From Simulated Queries, Does It Affect the Results? The results in Table 1 suggest that there is generally a positive impact when we aggregate the ranking resulting from the user's query (userQ) with the rankings resulting from leveraging layer 1 simulated queries, as can be seen by comparing results across rows. However the results start degrading when we start over-representing the userQ rankings.

For instance, in the case of Snip, leveraging Snip + Q1x leads to a performance increase of 13.43 % over leveraging Snip only (from 0.1407 to 0.1596). This means that including the ranking resulting from the actual user query (only once) helps improve the result by 13.43 % over the effectiveness of simply leveraging Snip. But, including those results two times (Snip + Q2x), three times (Snip + Q3x), or four times as much voting rights as the simulated query (Snip) leads to 8.96 %, 4.69 %, or 3.70 % increases over using Snip only.

We conclude that we obtain better results by aggregating the ranking resulting from the user's query with the rankings resulting from leveraging layer 1 simulated queries once, but in most cases only once (except notably for JTopia and QNJTopia).

Does Concatenating the Simulated Queries with the Original Query Impact the Results? Comparing the QN variants in Table 1 clearly suggests that, in general, concatenating the original query to the simulated query improves the results. For instance, when we go from using Snip to using QNSnip, the results improve from 0.1407 to 0.1614 (14.71 % improvement). Results improve by 34.99 % from JTopia to QNJTopia. It is worth noting that when leveraging Titles, the nDCG went down from 0.1465 to 0.1420. But that negative change is negligible (3.07 % decrease) compared to the 14.71 % and 34.99 % increase.

Is There Any Added Value in Going Down to Layer 2 and Deeper? Table 2 shows that, for our first model, layer 2 provides no benefit and in fact hurts effectiveness. This was somewhat foreseeable, in that the queries generated in layer 2 are drifting further away from the original intent.

The second model, however, benefits greatly from the addition of the second layer. Using layer 1 results as strong baselines for the purpose of comparison, the increases in effectiveness from layer 1 to "Complex L1 + L2" are in fact 29.22 %, 34.30 %, 31.49 % and 18.59 % respectively for Titles, QNTitles, Snip, QNSnip.

6   Conclusions

In this paper, we address the problem of simulating a user who is reformulating queries based on terms and key phrases s/he encountered during the search process, in order to obtain data similar to search history data that studies leverage for improved effectiveness. In the current study, we assumed a real user provides one single query and nothing else prior to that event, and proposed ways to simulate and generate such data that can be considered to be similar to search history data given that they provide results similar to the ones we obtain when we leverage real users' search history.

References

1. Bah, A., Carterette, B.: Aggregating results from multiple related queries to improve web search over sessions. In: Jaafar, A., et al. (eds.) AIRS 2014. LNCS, vol. 8870, pp. 172–183. Springer, Heidelberg (2014)
2. Baskaya, F.: Simulating Search Sessions in Interactive Information Retrieval Evaluation. Tampere University, Tampere (2014)
3. Baskaya, F., Keskustalo, H., Järvelin, K.: Simulating simple and fallible relevance feedback. In: Clough, P., Foley, C., Gurrin, C., Jones, G.J., Kraaij, W., Lee, H., Mudoch, V. (eds.) ECIR 2011. LNCS, vol. 6611, pp. 593–604. Springer, Heidelberg (2011)
4. Baskaya, F., Keskustalo, H., Järvelin, K.: Time drives interaction: simulating sessions in diverse searching environments. In: Proceedings of SIGIR, August 2012
5. Carterette, B., Bah, A., Zengin, M.: Dynamic test collections for retrieval evaluation. In: Proceedings of the 2015 International Conference on the Theory of Information Retrieval, pp. 91–100. ACM, September 2015
6. Carterette, B., Kanoulas, E., Hall, M.M., Clough, P.D.: Overview of the TREC 2013 Session Track. In: TREC (2013)
7. Cormack, G.V., Smucker, M.D., Clarke, C.L.: Efficient and effective spam filtering and re-ranking for large web datasets. Inf. Retr. **14**(5), 441–465 (2011)
8. Guan, D.: Structured Query Formulation and Result Organization for Session Search (Doctoral dissertation, Georgetown University) (2013)
9. Guan, D., Zhang, S., Yang, H.: Utilizing query change for session search. In: Proceedings of SIGIR, July 2013
10. Järvelin, K., Kekäläinen, J.: Cumulated gain-based evaluation of IR techniques. ACM Trans. Inf. Syst. (TOIS) **20**(4), 422–446 (2002)
11. Jiang, J., He, D., Han, S.: On duplicate results in a search session. In: Proceedings of the 21st TREC (2012)
12. JTopia. https://github.com/srijiths/jtopia
13. Keskustalo, H., Järvelin, K., Pirkola, A., Sharma, T., Lykke, M.: Test collection-based IR evaluation needs extension toward sessions – a case of extremely short queries. In: Lee, G.G., Song, D., Lin, C.-Y., Aizawa, A., Kuriyama, K., Yoshioka, M., Sakai, T. (eds.) AIRS 2009. LNCS, vol. 5839, pp. 63–74. Springer, Heidelberg (2009)
14. Kruschwitz, U.: University of essex at the TREC 2012 session track. In: Proceedings of the 21st TREC (2012)
15. Raman, K., Bennett, P.N., Collins-Thompson, K.: Toward whole-session relevance: exploring intrinsic diversity in web search. In: Proceedings of the 36th International ACM SIGIR, pp. 463–472. ACM, July 2013
16. Strohman, T., Metzler, D., Turtle, H., Croft, W.B.: Indri: a language model-based search engine for complex queries. In: Proceedings of the International Conference on Intelligent Analysis, vol. 2, no. 6, pp. 2–6, May 2005
17. Verberne, S., Sappelli, M., Järvelin, K., Kraaij, W.: User simulations for interactive search: evaluating personalized query suggestion. In: Hanbury, A., Kazai, G., Rauber, A., Fuhr, N. (eds.) ECIR 2015. LNCS, vol. 9022, pp. 678–690. Springer, Heidelberg (2015)
18. Yahoo! BOSS. https://developer.yahoo.com/search/boss/
19. Zhang, S., Guan, D., Yang, H.: Query change as relevance feedback in session search. In: Proceedings of the 36th International ACM SIGIR Conference, pp. 821–824. ACM, July 2013

Variable-Chromosome-Length Genetic Algorithm for Time Series Discretization

Muhammad Marwan Muhammad Fuad[(✉)]

Aarhus University, MOMA, Palle Juul-Jensens Boulevard 99,
8200 Aarhus N, Denmark
marwan.fuad@clin.au.dk

Abstract. The symbolic aggregate approximation method (SAX) of time series is a widely-known dimensionality reduction technique of time series data. SAX assumes that normalized time series have a high-Gaussian distribution. Based on this assumption SAX uses statistical lookup tables to determine the locations of the breakpoints on which SAX is based. In a previous work, we showed how this assumption oversimplifies the problem, which may result in high classification errors. We proposed an alternative approach, based on the genetic algorithms, to determine the locations of the breakpoints. We also showed how this alternative approach boosts the performance of the original SAX. However, the method we presented has the same drawback that existed in the original SAX; it was only able to determine the locations of the breakpoints but not the corresponding alphabet size, which had to be input by the user in the original SAX. In the method we previously presented we had to run the optimization process as many times as the range of the alphabet size. Besides, performing the optimization process in two steps can cause overfitting. The novelty of the present work is twofold; first, we extend a version of the genetic algorithms that uses chromosomes of different lengths. Second, we apply this new version of variable-chromosome-length genetic algorithm to the problem at hand to simultaneously determine the number of the breakpoints, together with their locations, so that the optimization process is run only once. This speeds up the training stage and also avoids overfitting. The experiments we conducted on a variety of datasets give promising results.

Keywords: Discretization · Time series · Variable-chromosome-length genetic algorithm

1 Introduction

A *time series* $S = \langle s_1 = \langle v_1, t_1 \rangle, s_2 = \langle v_2, t_2 \rangle, \ldots, s_n = \langle v_n, t_n \rangle \rangle$ of length n is a chronological collection of observations v_n measured at timestamps t_n. Time series data mining handles several tasks, the most important of which are query-by-content, clustering, and classification. Executing these tasks requires performing another fundamental task in data mining which is the *similarity search*. A similarity search problem consists of a database D, a query or a pattern q, and a tolerance ε that determines the proximity of the data objects to qualify as answers to that query.

© Springer International Publishing Switzerland 2016
S. Hartmann and H. Ma (Eds.): DEXA 2016, Part II, LNCS 9828, pp. 418–425, 2016.
DOI: 10.1007/978-3-319-44406-2_35

Sequential scanning compares every single time series in D against q to answer the similarity search problem. This is not an efficient approach given that time series databases can be very large.

Data transformation techniques transform the time series from the original high-dimension space into a low-dimension space so that they can be managed more efficiently. *Representation Methods* apply appropriate transformations to the time series to reduce their dimension. The query is then processed in those low-dimension spaces.

There are several representation methods in the literature, the most popular are: *Piecewise Aggregate Approximation* (PAA) [1, 2] and *Adaptive Piecewise Constant Approximation* (APCA) [3].

The *Symbolic Aggregate approXimation* method (SAX) [4] stands out as probably the most powerful representation method for time series discretization. The main advantage of SAX is that the similarity measure it utilizes, called MINDIST, uses statistical lookup tables. SAX is based on an assumption that normalized time series have "highly Gaussian distribution" (quoting from [4]), so by determining the locations of the breakpoints that correspond to a particular alphabet size, one can obtain equal-sized areas under the Gaussian curve. SAX is applied in four steps: in the first step the time series are normalized. In the second step the dimensionality of the normalized time series is reduced using PAA [1, 2]. In the third step the PAA representation resulting from the second step is discretized by determining the number and locations of the breakpoints. The number of the breakpoints *nrBreakPoints* is related to the alphabet size *aphabetSize* (chosen by the user); i.e. *nrBreakPoints* = *aphabetSize* − 1. As for their locations, they are determined, as mentioned above, by using Gaussian lookup tables. The interval between two successive breakpoints is assigned to a symbol of the alphabet, and each segment of PAA that lies within that interval is discretized by that symbol. The last step of SAX is using the following similarity measure:

$$MINDIST\left(\hat{S}, \hat{R}\right) \equiv \sqrt{\frac{n}{N}} \sqrt{\sum_{i=1}^{N} (dist(\hat{s}_i, \hat{r}_i))^2} \tag{1}$$

Where n is the length of the original time series, N is the number of segments, $\hat{S}$ and $\hat{R}$ are the symbolic representations of the two time series S and R, respectively, and where the function $dist()$ is implemented by using the appropriate lookup table.

There are other versions and extensions of SAX [5, 6]. These versions use it to index massive datasets, or they compute MINDIST differently. However, the version of SAX that we presented earlier is the basis of all these versions and extensions and it is actually the most widely-known one.

In this paper we determine the locations of the breakpoints by using a version of the genetic algorithms that uses chromosomes of variable lengths. This enables us to simultaneously determine the number of the breakpoints, together with their location, so that the optimization process is run only once, and the side effects resulting from overfitting, which happens when optimization is processed in two steps, can be avoided.

The paper is organized as follows; in Sect. 2 we present the new method to discretize the time series, we test it in Sect. 3. We conclude with Sect. 4.

2 Discretizing Time Series Using Variable-Chromosome-Length Genetic Algorithms

At the very heart of SAX, as we saw in Sect. 1, is the assumption that normalized time series have a highly Gaussian distribution. This is an intrinsic part of SAX on which the locations of the breakpoints are determined. This, in turn, allows SAX to use pre-computed distances, which is the main advantage of SAX over other methods.

However, the assumption that normalized time series follow a Gaussian distribution oversimplifies the problem as it does not take into account the dataset to which SAX is applied. The direct result of this assumption is the poor performance of SAX on certain datasets as we showed in [7]. That was the motivation behind the alternative method we presented in [7], which does not assume any particular distribution of the time series. Instead, the method we presented formulates the problem of determining the locations of the breakpoints as an optimization problem. This approach, as we showed in [7], substantially boosts the performance of the original SAX.

However, the method we presented in [7] has a drawback that also exists in the original SAX; it can only optimize the locations of the breakpoints for a given value of the alphabet size, but it cannot determine the optimal alphabet size for a given dataset. In other words, during the training stage the optimization process should be run for each value of the alphabet size for a given dataset to determine the optimal value of the objective function for all these runs, which is then used in the testing stage. As we can easily see, this approach is time consuming. Another adverse consequence is that such an approach – finding the optimal alphabet size first and then determining the locations of the breakpoint – may result, as we showed for a similar problem in [8], in overfitting.

The optimization process should handle the above mentioned problem in one step. In other words, its outcome should yield the optimal alphabet size for a particular dataset together with the locations of the breakpoints that correspond to that alphabet size.

To solve this problem we propose a variant of the genetic algorithms called variable-chromosome-length genetic algorithm (VCL-GA). But before we present VCL-GA we start by giving a brief outline of the genetic algorithm.

2.1 The Genetic Algorithm (GA)

GA is the most popular bio-inspired optimization algorithm. GA belongs to a larger family of bio-inspired optimization algorithms which is the *Evolutionary Algorithms*. In the following we present a description of the simple, classical GA. GA starts with a collection of individuals, also called *chromosomes*. Each chromosome represents a possible solution to the problem at hand. This collection of randomly chosen chromosomes constitutes a population whose size *popSize* is chosen by the algorithm designer. This step is called *initialization*. A candidate solution is represented as a vector whose length is equal to the number of parameters of the problem. This dimension is denoted by *nbp*. The fitness function of each chromosome is evaluated in order to determine the chromosomes that are fit enough to survive and possibly produce offspring. This step is called *selection*. The percentage of chromosomes selected for mating is denoted by *sRate*. *Crossover* is the next step in which the offspring of two parents are produced to enrich the population with fitter chromosomes. *Mutation*, which is a random alteration of a certain percentage *mRate* of chromosomes, is the other mechanism that enables GA to explore the search space. Now that a new generation is formed, the fitness function of the offspring is calculated and the above procedures repeat for a number of generations *nGen* or until a *stopping criterion* terminates the algorithm.   □

2.2   Variable-Chromosome-Length Genetic Algorithm (VCL-GA)

Whereas a large number of optimization problems can be modeled by a definite number of parameters, and consequently apply a genetic algorithm with a predefined chromosome-length, there is a category of applications where the number of parameters is not known *a priori*. These problems require a representation which is not based on fixed length of chromosomes, and also a fitness function that is independent of the number of parameters in each chromosome.

Chromosomes with variable length were introduced in [9] as a variant of classifier systems. Later, this concept was used to solve different optimization problems where the number of parameters is not fixed. In [10] the authors apply genetic algorithms with variable chromosome lengths to structural topology optimization. Their approach was based on a progressive refinement strategy, where GA starts with a short chromosome and first finds an optimum solution in the simple design space. The optimum solutions are then transferred to the next stages with longer chromosomes. This is the main difference between this method and ours, where there is no possibility of a "gradual" refinement by adding more complexity as, in our problem, the optimal solution for each alphabet size is independent of that for another alphabet size.

In [11] the authors presented a genetic planner method that uses chromosomes of variable length. The method they presented applies a particular genetic scheme (complex fitness function, multi-population, population reset, weak memetism, tournament selection and elitist genetic operators).

2.3   VCL-GA for Discretizing Time Series

In this section we present our version of VCL-GA which is designed to solve the problem of determining the locations of the breakpoints, together with the corresponding alphabet size, which give the minimum classification error according to first nearest-neighbor (1NN) rule using leaving-one-out cross validation. This means that every data object is compared to the other data objects in the dataset. If the 1NN does not belong to the same class, the error counter is incremented by 1.

In order for the optimization process to converge, the value of the alphabet size should be constrained by two values: *upperAlphaSize* and *lowerAlphaSize*. Also, the value of the breakpoints is constrained by *upperVal* and *lowerVal*.

We implemented the method based on the locations of the breakpoints, which implicitly encodes for the alphabet size, taking into account that *nrBreakPoints = aphabetSize* − 1 (see Sect. 1).

The algorithm starts by initializing a population whose size is *popSize*. Each chromosome is a vector of *chromLength* real different values par_i, $i \in \{1, 2, \ldots,$ *chromLength*$\}$ and where *chromLength* is an integer chosen randomly between *upperAlphaSize* $- 1$ and *lowerAlphaSize* $- 1$. The values par_i encode the locations of the breakpoints. Although par_i are, theoretically, not constrained, but given that the locations of the breakpoints using the original SAX for *aphabetSize* $= 20$ (the maximum value of the alphabet size in the original SAX) are constrained between -1.64 and $+1.64$, we constrained par_i in our experiments between -2 and $+2$.

Another feature of our VCL-GA that is different from the classical (fixed-chromosome length) GA is crossover (recombination). Classical GA applies different recombination schemes. In the *single-point crossover* (SPX) scheme (which we adopt in this paper, for its simplicity), the two chromosomes are split at one common locus, or crossover point, and the segments at that crossover point are swapped.

In VCL-GA the split locus is not necessarily the same for the two chromosomes. As a result, the two resulting offspring chromosomes may have different chromosome-length from the parent chromosomes. One of the consequences of this is that the algorithm should check that the length of the offspring chromosomes is always larger or equal to *lowerVal* and smaller or equal to *upperVal*.

Formally, let $chrom^i = \langle par_1^i, par_2^i, \ldots, par_m^i \rangle$, $chrom^j = \langle par_1^j, par_2^j, \ldots, par_n^j \rangle$, where $m \neq n$ in the general case and where *lowerAlphaSize* $- 1 \leq m, n \leq$ *upperAlphaSize* $- 1$, be the two mating parent chromosomes. The crossover operation uses two crossover points: cp_1, cp_2; two real numbers sampled from a uniform distribution, which split the first parent chromosome into two segments: $chrom_{left}^i = \langle par_1^i, par_2^i, \ldots, par_p^i \rangle$ and $chrom_{right}^i = \langle par_{p+1}^i, par_{p+2}^i, \ldots, par_m^i \rangle$, where $p \leq cp_1 \leq p + 1$, and the second parent chromosome into: $chrom_{left}^j = \langle par_1^j, par_2^j, \ldots, par_q^j \rangle$ and $chrom_{right}^j = \langle par_{q+1}^j, par_{q+2}^j, \ldots, par_n^j \rangle$, where $q \leq cp_2 \leq q + 1$. The resulting offspring are: $offspring^1 = \langle par_1^i, par_2^i, \ldots, par_p^i, par_{q+1}^j, par_{q+2}^j, \ldots, par_n^j \rangle$. The second offspring is: $offspring^2 = \langle par_1^j, par_2^j, \ldots, par_q^j, par_{p+1}^i, par_{p+2}^i, \ldots, par_m^i \rangle$. As we can see, the first possible consequence of this crossover scheme is that the length of the resulting offspring may be smaller than *lowerAlphaSize* $- 1$ or larger than *upperAlphaSize* $- 1$. There are several scenarios that can applied to guarantee that the lengths of the resulting offspring satisfy this constraint, but we opted for a very simple scenario, which is to choose other crossover points cp_1, cp_2 if the ones chosen result in offspring lengths that violate this constraint.

Our problem also has another constraint; for any chromosome $chrom = \langle par_1, par_2, \ldots, par_r \rangle$ we have: $k < l \Rightarrow par_k < par_l, \forall 0 \leq k, l \leq r$. Given that the parameters *par* all are of the same nature, we simply sort the components of the offspring chromosomes to satisfy this latter condition.

3 Experiments

We conducted experiments on 20 datasets chosen at random from the UCR time series archive [12]. Each dataset consists of a training set and a testing set.

The length of the time series on which we conducted our experiments varies between 24 (ItalyPowerDemand) and 1024 (MALLAT). The size of the training sets varied between 16 (DiatomSizeReduction) and 300 (synthetic_control). The size of the testing sets varied between 28 (Coffee) and 2345 (MALLAT). The number of classes varied between 2 (ItalyPowerDemand), (Coffee), (ECG200), (SonyAIBORobotSurfaceII) (TwoLeadECG), (ToeSegmentation2), (SonyAIBORobotSurface), (ECGFive-Days), (Wine), and 8 (MALLAT).

The purpose of our experiments is to compare our method (that we refer to from now on as VCL-GA-SAX), which uses VCL-GA to obtain the locations of the breakpoints, together with the corresponding alphabet size, which yield the minimum classification error, with the classical SAX which, as indicated in previous sections, determines the locations of the breakpoints from lookup tables.

In fact, VCL-GA, because it does not presume any distribution of the time series, does not require normalization of the time series to be applied, and can be applied to normalized as well as non- normalized time series. This is another advantage VCL-GA-SAX has over classical SAX. However, in our experiments we normalize the time series so that SAX can be applied to them.

The range of the alphabet size on which the two methods were tested is $\{3, 4, \ldots, 20\}$, because SAX is defined on this range. However, because VCL-GA-SAX does not require predefined lookup tables, it can practically be applied to any value of the alphabet size.

The experimental protocol was as follows: during the training stage VCL-GA-SAX is trained on the training set by performing an optimization process to obtain the locations of the breakpoints and the corresponding alphabet size, which yield the minimum classification error. In the testing stage the locations of the breakpoints and the corresponding alphabet size are used to perform a classification task.

As for SAX, its application also includes two stages; in the training stage we obtain the alphabet size that yields the minimum classification error. Then in the testing stage we apply SAX to the corresponding dataset using the alphabet size obtained in the training stage to obtain the classification error of the testing dataset.

VCL-GA uses the following control parameters: the number of generations $nGen$ is set to 100. The population size $popSize$ is set to 24. The mutation rate $mRate$ is set to 0.2 and the selection rate $sRate$ is set to 0.5. As for the number of parameters nbp it is variable, which is the main feature of our algorithm.

In addition to $nGen$, we also used another stopping criterion, which is the classification error, which is set to 0. VCL-GA terminates and exists as soon as one of these stopping criteria is met. Table 1 summarizes the control parameters we used in the experiments.

Table 1. The control parameters of VCL-GA

popSize	Population size	24
nrGen	Number of generations	100
mRate	Mutation rate	0.2
sRate	Selection rate	0.5
nbp	Number of parameters	variable

Table 2. The classification errors of SAX and VCL-GA-SAX

Datasets	SAX		VCL-GA-SAX	
	classification error	alphabet size	classification error	alphabet size
CBF	0.076	17	**0.026**	10
synthetic_control	0.023	15	**0.007**	13
Beef	0.433	18	**0.333**	13
Symbols	**0.103**	18	0.109	7
Coffee	0.286	19	**0.000**	18,19
SonyAIBORobotSurfaceII	**0.144**	11	0.175	14
DiatomSizeReduction	0.082	20	**0.036**	17
ECGFiveDays	0.150	14	**0.075**	9
Gun_Point	0.147	18	**0.060**	20
ItalyPowerDemand	0.192	19	**0.066**	20
ECG200	**0.120**	12	0.130	13
OliveOil	0.833	3→20	**0.367**	17
SonyAIBORobotSurface	0.298	14,17	**0.186**	5
TwoLeadECG	0.309	20	**0.225**	3
Trace	0.370	18	**0.120**	13
FaceFour	0.144	11	**0.125**	3
MALLAT	0.143	18	**0.078**	16
ArrowHead	0.246	18	**0.229**	17
ToeSegmentation2	0.146	19,20	**0.138**	3
Wine	0.500	3→20	**0.389**	15

In Table 2 we present a comparison of the classification errors between SAX and VCL-GA-SAX for the 20 datasets tested. The best result (the minimum classification error) for each dataset is shown in bold-underlined printing in yellow-shaded cells.

As we can see from the results, of all the 20 datasets tested VCL-GA-SAX outperformed SAX 17 times, whereas SAX outperformed VCL-GA-SAX for 3 datasets only (SonyAIBORobotSurfaceII), (Symbols), and (ECG200).

For some datasets (Coffee) and (OliveOil) the difference in performance was spectacular. We believe the reason for this is that the assumption of Gaussianity is completely erroneous for these datasets.

4 Conclusion

In this work we applied a version of the genetic algorithms that uses chromosomes of variable length to determine the locations of the breakpoints and the corresponding alphabet size of the SAX representation method of time series discretization. The main advantage of using chromosomes of variable lengths is that the locations of the breakpoints and the corresponding alphabet size are all determined in one optimization process. This avoids overfitting problems and speeds up the training stage because we do not need to train the algorithm for each value of the alphabet size. Comparing our new method to SAX shows how the new method outperforms SAX for the great majority of datasets.

In the future we intend to apply VCL-GA to several problems in bioinformatics where the number of parameters is variable, yet the solutions presented in the literature attempt to circumvent this fact in different ways. We believe these problems are particularly adapted to VCL-GA.

References

1. Keogh, E., Chakrabarti, K., Pazzani, M., Mehrotra, S.: Dimensionality reduction for fast similarity search in large time series databases. J. Knowl. Inf. Syst. 3(3), 263–286 (2000)
2. Yi, B.K., Faloutsos, C.: Fast time sequence indexing for arbitrary Lp norms. In: Proceedings of the 26th International Conference on Very Large Databases, Cairo, Egypt (2000)
3. Keogh, E., Chakrabarti, K., Pazzani, M., Mehrotra, S.: Locally adaptive dimensionality reduction for similarity search in large time series databases. In: SIGMOD (2001)
4. Lin, J., Keogh, E., Lonardi, S., Chiu, B.Y.: A symbolic representation of time series, with implications for streaming algorithms. DMKD 2003, 2–11 (2003)
5. Muhammad Fuad, M.M., Marteau, P.-F.: Enhancing the symbolic aggregate approximation method using updated lookup tables. In: Setchi, R., Jordanov, I., Howlett, R.J., Jain, L.C. (eds.) KES 2010, Part I. LNCS, vol. 6276, pp. 420–431. Springer, Heidelberg (2010)
6. Shieh, J., Keogh, E.: iSAX: disk-aware mining and indexing of massive time series datasets. Data Min. Knowl. Discov. 19(1), 24–57 (2009)
7. Muhammad Fuad, M.M.: Genetic algorithms-based symbolic aggregate approximation. In: Cuzzocrea, A., Dayal, U. (eds.) DaWaK 2012. LNCS, vol. 7448, pp. 105–116. Springer, Heidelberg (2012)
8. Muhammad Fuad, M.M.: One-step or two-step optimization and the overfitting phenomenon: a case study on time series classification. In: The 6th International Conference on Agents and Artificial Intelligence- ICAART 2014, 6–8 March 2014, Angers, France. SCITEPRESS Digital Library (2014)
9. Smith, S.F.: A Learning System Based on Genetic Adaptive Algorithms. Doctoral dissertation, Department of Computer Science, University of Pittsburgh, PA (1980)
10. Kim, L.Y., Weck, O.L.: Variable chromosome length genetic algorithm for progressive refinement in topology optimization. Struct. Multidisciplinary Optim. 29(6), 445–456 (2005)
11. Brié, A.H., Morignot, P.: Genetic planning using variable length chromosomes. In: Proceedings of the 15th International Conference on Automated Planning and Scheduling (2005)
12. Chen, Y., Keogh, E., Hu, B., Begum, N., Bagnall, A., Mueen, A., Batista, G.: The UCR Time Series Classification Archive (2015). www.cs.ucr.edu/~eamonn/time_series_data

Approximate Temporal Aggregation
with Nearby Coalescing

Kai Cheng$^{(\boxtimes)}$

Faculty of Information Science, Kyushu Sangyo University, 2-3-1, Mtsukadai,
Higashi-ku, Fukuoka 813-8503, Japan
chengk@is.kyusan-u.ac.jp
http://www.is.kyusan-u.ac.jp/~chengk

Abstract. Temporal aggregation is an important query operation in temporal databases. Although the general forms of temporal aggregation have been well researched, some new applications such as online calendaring systems call for new temporal aggregation. In this paper, we study the issue of approximate temporal aggregation with nearby coalescing, which we call NSTA. NSTA improves instant temporal aggregation by coalescing nearby (not necessarily adjacent) intervals to produce more compact and concise aggregate results. We introduce the term of coalescibility and based on it we develop efficient algorithms to compute coalesced aggregates. We evaluate the proposed methods experimentally and verify the feasibility.

Keywords: Temporal aggregation · Temporal coalescing · Interval-valued timestamp · Coalescibility

1   Introduction

Temporal aggregation is an important query operation in temporal databases. In temporal databases, tuples are typically stamped with time intervals that capture the valid time of the information or facts. When aggregating temporal relations, tuples are grouped according to their timestamp values. There are basically two types of temporal aggregation: instant temporal aggregation and span temporal aggregation [2,5]. *Instant temporal aggregation (ITA)* computes aggregates on each time instant and consecutive time instants with identical aggregate values are coalesced into so-called *constant intervals*, i.e., tuples over maximal time intervals during which the aggregate results are constant. ITA works at the smallest time granularity and produces a result tuple whenever an argument tuple starts or ends. Thus the result relation is often larger than the argument relation, up to $2n - 1$ tuples, where n is the size of the argument relation [6]. *Span temporal aggregation (STA)* on the other hand allows an application to control the result size by specifying the time intervals, such as year, month, or day, for which to report a result tuple. For each of these intervals a result tuple is produced by aggregating over all argument tuples that overlap that interval.

© Springer International Publishing Switzerland 2016
S. Hartmann and H. Ma (Eds.): DEXA 2016, Part II, LNCS 9828, pp. 426–433, 2016.
DOI: 10.1007/978-3-319-44406-2_36

Table 1. A sample temporal relation and its aggregates

(a) Activities Relation

	Name	Content	Time
r_1	Jim	A	$[1, 9]$
r_2	Wang	A	$[14, 17]$
r_3	Tom	F	$[7, 12]$
r_4	Susan	G	$[19, 21]$
r_5	Abe	A	$[15, 19]$
r_6	Steve	D	$[3, 5]$

(b) ITA

Time	COUNT
$[1, 3]$	1
$[3, 5]$	2
$[5, 7]$	1
$[7, 9]$	2
$[9, 12]$	1
$[14, 16]$	1
$[16, 17]$	2
$[17, 19]$	2
$[19, 21]$	1

(c) NSTA

Time	COUNT
$[1, 3]$	1
$[3, 5]$	2
$[5, 7]$	1
$[7, 9]$	2
$[9, 16]$	1
$[16, 19]$	2
$[19, 21]$	1

Nowadays a handful of new applications motivate more flexible aggregation operation. Consider an online calendaring system such as Google Calendar[1], where a temporal relation stores scheduled activities for individuals or groups. The information about an activity includes name, content, and the scheduled period of time. Table 1(a) shows a sample temporal relation of six activities. Suppose we want to create a new activity for a group of people. We must find a time interval so that all members can participate. We first compute the count aggregate for each occupied timespan as shown in Table 1(b). The result relation contains all information about occupied time intervals, for example, 1 people in $[1, 3]$ and 2 in $[3, 5]$ are occupied. Based on the count aggregate, we then derive free time intervals from outside of the occupied parts. For example, $[12, 14]$ is free at this time.

It is often important to take into account more constraints and/or preferences when we a new activity is scheduled. First, the length of free time is crucial. For instance there must be at least 60 min left for the new activity. In addition, some people may prefer morning to afternoon, or think Friday is better than Monday. In practice, when a completely free time interval is not available, a time interval with a few occupied members should be considered as a feasible choice. For example, a query for free time intervals of 10 members may accept results with just 1 or 2 members not completely free.

All these entail a new form of approximate temporal aggregation that returns more compact results. In [9] the authors introduced parsimonious temporal aggregation (PTA) that aimed to reduce the ITA result by merging similar and temporally adjacent tuples until a user-specified size or error bound is satisfied. Tuples are adjacent only if they are not separated by a temporal gap.

In the calendaring application, however, the required free time must meet length constraint, which implies a temporal gap can be ignored when it is shorter than the length constraint. This is the case of Table 1(c), where $[9, 16]$ is the coalesced result from $[9, 12]$ and $[14, 16]$ in Table 1(b) although there is a gap

---

[1] https://calendar.google.com/.

between them. This relaxation is reasonable also because timestamps in real-world are not always exact.

In this paper, we study a new form of temporal aggregation, called *NSTA* (NS stands for the magnetic poles), where nearby time intervals (not necessarily adjacent or overlapping) are coalesced to obtain more compact and concise result whenever possible. We formally define the term of *coalescibility* and based on it we develop algorithms for efficient query precessing.

The rest of paper is organized as follows. In Sect. 2, we define the problem and proposes the main techniques. Section 3 introduces the experimental results. Section 4 concludes the paper and points out some future directions.

2   Nearby Coalescing

Conventionally, two intervals are candidates for temporal coalescing if they are adjacent to or overlapping with each other. In Allen's term [1], two intervals can be coalesced only if one interval *meets* or *extends* another one. For example, in Fig. 1, since interval **b** extends **a**, and **c** meets **a**, both pairs can be coalesced. However neither **a,d** nor **a,e** are coalescible because **a** is before **d** and **e**.

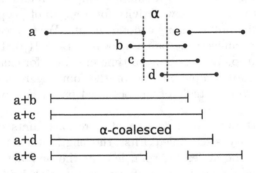

Fig. 1. α-Nearby coalescing

2.1 α-Coalescibility

In this work, we relax the constraint by allowing a user specified threshold to control the coalescibility. Consider a set of N real-valued time intervals $\mathcal{I}$. Each interval is associated with a weight w_i $(i = 1, 2, \cdots, N)$, which can be any numeric attribute of a time interval, such as revenue or number of overlapped intervals.

Definition 1 (α-nearby, α-coalescible, α-coalesced). *Given $\alpha \geq 0$ and two intervals $\mathbf{s} = [s^-, s^+] \in \mathcal{I}$, $\mathbf{t} = [t^-, t^+] \in \mathcal{I}$ where $s^- < t^-, s^+ < t^+$. We say $\mathbf{s}$ and $\mathbf{t}$ are α-nearby if $t^- - s^+ \leq \alpha$. If the weights associated with α-nearby intervals are identical, the intervals are α-coalescible. $[s^-, t^+]$ is called α-coalesced from $\mathbf{s}$ and $\mathbf{t}$. α is called nearby threshold.*

In Fig. 1, **a** and **d** are α-nearby but **a** and **e** are not. α-coalescible **a** and **d** are coalesced to **a** + **d** as shown in Fig. 1. Notice that the adjacent or overlapping intervals, such as **a** and **b** or **a** and **c** also α-nearby. If $\alpha = 0$, it becomes the exact case where only adjacent or overlapping intervals are considered near enough to coalesce.

In this work, two nearby intervals can be coalesced even when there is a small gap between them, just like the N/S magnetic poles. For this reason, *temporal aggregation with nearby coalescing* is named as NSTA.

2.2   Nearby Coalescing

Coalescing is a fundamental operation in many temporal databases. The basic strategies for coalescing are run-time (lazy) coalescing and update (eager) coalescing. The lazy strategy defers coalescing to query evaluation. The eager strategy performs coalescing whenever data update occurs. When new data is inserted, or data is modified or deleted, the tuples are coalesced. Note that update coalescing does not completely obviate the need to coalesce during query evaluation. Value-equivalent intermediate and temporary results may still need to be coalesced.

In [8], a third strategy, called partial coalescing where each temporal relation is split into two parts: an uncoalesced base relation, and a derived relation that records the covered endpoints. An covered endpoint is a time that starts (ends) an interval and is met by (meets) or is contained within the interval of some value-equivalent tuple. Which endpoints are covered or uncovered depends on some query-time information such as the reference time (which is bound to now in the evaluation of the query), the granularity at which the interval is evaluated, and the interpretation of the incomplete information.

In our work, coalescing depends on the user-specified nearby threshold, a covered relation is not helpful so much. To implement nearby coalescing, we adopt the run-time strategy. The input to our algorithm is a (sorted) list of intervals, returned from a temporal aggregate query. An interval is a triple $< B, E, W >$ with a lower bound B, upper bound E and an associated weight W. If x is an interval, then x, $x.B$, $x.E$ and $x.W$ are the lower, upper bounds and weight respectively.

The algorithm uses a working variable t to track the intermediate result in the process of nearby coalescing. A new interval is coalesced by updating $t.E$, the upper bound of t. The algorithm works as follows. The input is a list of uncoalesced intervals sorted in ascending order of the lower bound. Each input interval is checked if it is the α-coalescible. The algorithm checks if it is near enough to coalesced part. If so and if its weights equals to $t.W$, it is coalesced. Otherwise, current coalescing finishes and a new coalescing begins.

2.3   Segment B+ Tree for NSTA

A segment B+ tree uses a B+ tree as a base tree for elementary intervals. All endpoints form a ordered list stored at leaves. Intervals are indexed in this way.

Algorithm 1. Nearby coalescing

Input: A nearby threshold α; a sorted list of intervals $S = \{s_1, s_2, \cdots, s_m\}$
Output: A list of coalesced intervals T

```
 1: T ← ∅
 2: t ← s₁              ♯ t is a working variable for intermediate coalescing result
 3: i ← 2
 4: while i <= |S| do
 5:     if  sᵢ.B > t.E + α ∨ sᵢ.W ≠ t.W then      ♯ not α-coalescible
 6:         T ← T ∪ {t}
 7:         t ← sᵢ
 8:     else if sᵢ.E > t.E ∧ sᵢ.W = t.W then      ♯ coalesce
 9:         t.E ← sᵢ.E
10:     end if
11:     i ← i + 1
12: end while
13: return T
```

(1) If the interval is identical to an elementary interval, it is recorded in a leaf node, with a key-pointer pair where the key equals to the interval's start point; (2) If the interval contains a few adjacent elementary intervals but these elementary intervals belong to a single leaf node, we record each part in the leaf node in different key-point pairs. (3) If the interval contain more elementary intervals that belong to different leaf nodes, one or more parent nodes will record joint part from several leaf nodes. A even larger interval will need more parent nodes or even grandparent nodes and so on. For each elementary interval r, a reference count is used to record the number of intervals overlapping r.

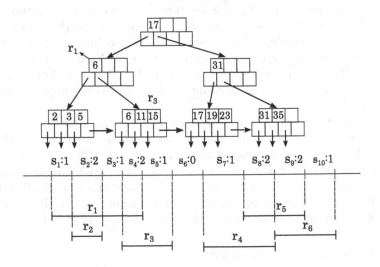

Fig. 2. Structure of a segment B+ tree

Let $p_1, p_2, \cdots, p_m$ be the sorted list of distinct interval endpoints. The elementary intervals are, from left to right, $[p_1 : p_2], [p_2 : p_3], \cdots, [p_{m-1} : p_m]$. In our segment B+ tree, an interval has a record of the following form:

$$< p_i, ref_i >$$

where p_i is the lower bound of elementary interval $[p_i, p_{i+1}]$ and ref_i is its reference of it, i.e. how many indexed intervals contain this elementary interval. When $ref = 0$, we call it a *free* interval. Figure 2 illustrates the segment B+ tree structure where a set of 6 intervals: $\{r_1, r_2, \cdots, r_6\}$ are indexed. The endpoints induce all elementary intervals: $\{s_1, s_2, \cdots, s_{10}\}$ each element is associated with a reference count. r_1 consists of $\{s_1, s_2, s_3, s_4\}$ that are recorded in two leaf nodes and a parent node is needed.

Let $\mathcal{T}$ be a segment B+ tree built for a set of intervals. A range query that reports all intersecting intervals the can be processed as follows. Suppose $[x^-, x^+]$ is the query range. We begin by searching with x^- and x^+ in $\mathcal{T}$ and stopping at a node where two search paths will split. This node is called *splitNode*. We then traverse the subtree rooted at the *splitNode* and report intervals recorded at the visited nodes. In this process, traversing a subtree is most costly, in worst case the whole tree should be read.

3   Experimental Evaluation

To evaluate the performance of our approach, we implement the following techniques in addition to our **Segment B+ Tree (SG-Tree)**. **Interval-Spatial Transformation (IST)**. Using D-order index to support spatial range query. For integer interval bounds $[lower, upper]$, the is equivalent to a composite index on the attributes $(upper, lower)$. **IST with MAX aggregate (IST-MAX)**. For max query (Problem 2), we make use of the DBMS's aggregation capability to reduce computation cost. Intervals with identical lower bound are grouped together. For each group only the maximal upper bound is reported. **Relational Interval Tree (RI-Tree)**. An external memory dynamic interval management technique using relational storage structure [4]. The basic idea is to manage the data objects by common relational indexes rather than to access raw disk blocks directly.

We generate time intervals from the domain of $[0, 2^{20} - 1]$. First, we preserve a set of *free intervals*. Every 100 consecutive time instants, with a probability of 0.25 we decide if free intervals will be generated. If so, an interval of random length is inserted to the free interval table. With the free interval table, we then generate *activity intervals* without intersecting any free interval. Similar to the process of free interval generation, for each 100 consecutive time instants, we randomly generate 10 activity intervals. The synthesized dataset includes 5,592 free intervals and 64,651 activity intervals.

To evaluate the performance of the proposed method, we perform a series of range queries. The query experiments have been performed with query intervals following a uniform distribution with selectivity $\sigma = \{0.01, 0.02, 0.03, \cdots, 0.50\}$.

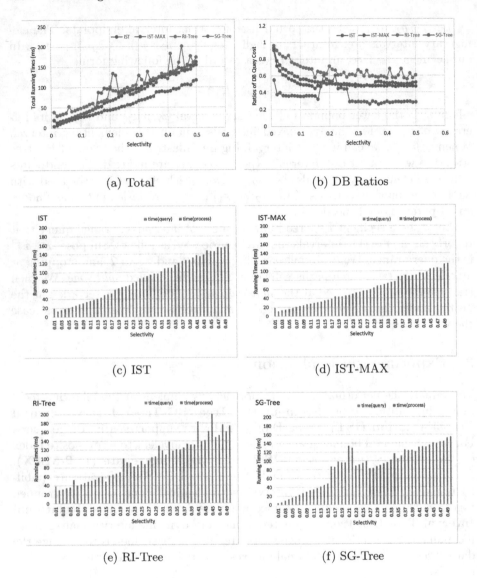

Fig. 3. Approximate temporal count queries

For each σ, a query interval $[B, E]$ is generated randomly as follows: $B \in [2, 300]$ and $E = B + \sigma N$ where $N = 2^{20} - 1$.

The running cost includes two parts: query processing (T_q) and coalescing (T_c). Figure 3(a)–(b) show the running time results. In terms of overall running time ($T_q + T_c$), the result in Fig. 3(a) tells us that IST-MAX outperforms other approaches. To understand the cost result, in Fig. 3(b) we give another result $T_q/(T_q + T_c)$, which tells us the ratio of query processing by database system. From this viewpoint, our segment B+ tree is the most efficient. The details of time cost for each approach under evaluation are given in Fig. 3(c)–(f). The lower part is query processing cost. TI-Tree most heavily utilizes DBMS processing capability but its total cost is highest. SG-Tree on the contrary is the most efficient in terms of query processing. However, SG-Tree is based on elementary intervals, which is nearly twice of the original intervals. This increases the coalescing cost.

4   Conclusion

Temporal aggregation is a fundamental query in temporal databases. Instant temporal aggregation is one basic form of temporal aggregation but the main problem is that the result size is much larger than base relation. In this paper, we improved the ITA temporal aggregation by introducing near coalescing so that nearby intervals, not necessarily being adjacent, have chance to be coalesced. We developed segment B+ tree to implement the proposed scheme. Experimental results showed the performance improvement to some extent although coalescing cost is still high. Some details and related work have been omitted due to the space limit. More details will be presented in a separate paper.

References

1. Allen, J.F.: Maintaining knowledge about temporal intervals. Commun. ACM **26**(11), 832–843 (1983)
2. Kline, N., Snodgrass, R.T.: Computing temporal aggregates. In: Proceedings of the Eleventh International Conference on Data Engineering, pp. 222–231, 06–10 March 1995
3. Böhlen, M.H., Snodgrass, R.T., Soo, M.D.: Coalescing in temporal databases. In: Proceedings of the 22th International Conference on Very Large Data Bases (VLDB 1996), San Francisco, CA, USA, pp. 180–191 (1996)
4. Kriegel, H.-P., Ptke, M., Seidl, T.: Managing intervals efficiently in object-relational databases. In: Proceedings of the 26th International Conference on Very Large Data Bases, pp. 407–418. Morgan Kaufmann Publishers Inc. (2000)
5. Lopez, I.F.V., Snodgrass, R.T., Moon, B.: Spatiotemporal aggregate computation: a survey. IEEE Trans. Knowl. Data Eng. **17**(2), 271–286 (2005)
6. Böhlen, M.H., Gamper, J., Jensen, C.S.: Multi-dimensional aggregation for temporal data. In: Ioannidis, Y., Scholl, M.H., Schmidt, J.W., Matthes, F., Hatzopoulos, M., Böhm, K., Kemper, A., Grust, T., Böhm, C. (eds.) EDBT 2006. LNCS, vol. 3896, pp. 257–275. Springer, Heidelberg (2006)
7. de Berg, M., Cheong, O., van Kreveld, M., Overmars, M.: Computational Geometry. Springer, Heidelberg (2008). (3rd revised edn.)
8. Dyreson, C.E.: Temporal coalescing with now granularity, and incomplete information. In: Proceedings of the 2003 ACM SIGMOD International Conference on Management of Data (SIGMOD 2003), NY, USA, pp. 169–180 (2003)
9. Gordevičius, J., Gamper, J., Böhlen, M.: Parsimonious temporal aggregation. VLDB J. Int. J. Very Large Data Bases **21**(3), 309–332 (2012)

Data Modelling, and Uncertainty

A Data Model for Determining Weather's Impact on Travel Time

Ove Andersen[1,2]($\boxtimes$) and Kristian Torp[1]

[1] Department of Computer Science, Aalborg University, Aalborg, Denmark
{xcalibur,torp}@cs.aau.dk
[2] FlexDanmark, Aalborg, Denmark
oan@flexdanmark.dk

Abstract. Accurate estimating travel times in road networks is a complex task because travel times depends on factors such as the weather. In this paper, we present a generic model for integrating weather data with GPS data to improve the accuracy of the estimated travel times. First, we present a data model for storing and map-matching GPS data, and integrating this data with detailed weather data. The model is generic in the sense that it can be used anywhere GPS data and weather data is available. Next, we analyze the correlation between travel time and the weather classes *dry*, *fog*, *rain*, and *snow* along with winds impact on travel time. Using a data set of 1.6 billion GPS records collected from 10,560 vehicles, over a 5 year period from all of Denmark, we show that *snow* can increase the travel time up to 27 % and strong headwind can increase the travel time with up to 19 % (compared to *dry* calm weather). This clearly shows that accurate travel time estimation requires knowledge about the weather.

Keywords: Data model · Data integration · Spatiotemporal · GPS · Travel time

1 Introduction

Estimating travel times in road networks is of great importance for a wide range of applications such as road-network monitoring, driving directions, and traffic planning. When a user requests the travel time from A to B, it is expected that the duration of the trip to be as accurate as possible. Travel time is complex to estimate because it is affected by several factors such as rush hours, road construction, accidents and weather conditions.

Until now the work of determining weather's impact on travel time has mainly been focusing on analyzing single or few selected road segments and the data foundation is often limited to few months of data. In this paper, we determine the weather's impact on a country-size road network using 1.6 billion GPS positions collected from 10,560 vehicles over a 5 year period. We present a generic model for integrating large scale GPS data with weather information for country-size road networks. We present a model for storing and preparing data for performing

© Springer International Publishing Switzerland 2016
S. Hartmann and H. Ma (Eds.): DEXA 2016, Part II, LNCS 9828, pp. 437–444, 2016.
DOI: 10.1007/978-3-319-44406-2_37

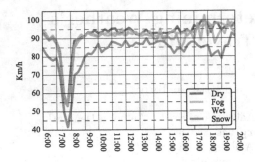
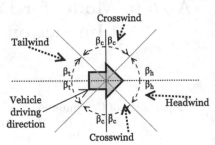

Fig. 1. Average speed by weather

Fig. 2. Wind direction identifiers (Color figure online)

a broad variation of analysis with regards to weather impact. The GPS data is map-matched to the road-network of Denmark (~1.8 million edges). Using this data model we analyze in details how the weather conditions *dry*, *fog*, *rain*, and *snow* impacts the travel-time on the entire road network. The analysis includes (a) determining the correlation between weather conditions and travel time Fig. 1 and (b) the impact of head-, tail-, and crosswind on travel time, show by Fig. 2

The contribution of this paper is twofold. First, we present a generic data model for storing large GPS data sets and integrate this with detailed weather information. Second, to the best of our knowledge, we present the first detailed, nation-size study on how weather impacts travel times. One such analysis can be seen by Fig. 1, showing the average speed on a motorway on Monday through Friday depending on weather conditions. Here it can be seen that the speed varies across they day and *snow* is in general ~10 km/h slower than *dry*, *fog*, and *rain*.

The paper is organized as follows. Section 2 describes the data foundation and Sect. 3 presents in details the model integrating GPS and weather data. A thorough analysis of weather's impact on travel time is presented in Sect. 4. Section 5 lists related work and Sect. 6 concludes the paper.

2   Data Foundation

This section describes the GPS, map, and weather data sources integrated to be able to analyze weather's impact on travel-time. First the data model is presented, next, the concrete data sources used are introduced.

The positions of vehicles are tracked using GPS data. A GPS record, r, is a 6-tuple defined as follows.

$$r = \langle vid, lat, lon, time, speed, course \rangle$$

The tuple contains a unique vehicle id, *vid*, the position as latitude, *lat*, and longitude, *lon*, a timestamp, *time*, a vehicle speed, *speed*, and a compass direction, *course*. The set R denotes all GPS records.

The map foundation is a directed, weighted graph $G = \langle V, E, W \rangle$ where V is a set of vertices and $E \subseteq V \times V$ is a set of edges. Each $v \in V$ is defined by two attributes $v = \langle lat, lon \rangle$ that denote the latitude and longitude. For each edge $e \in E$ we define two attribute $e = \langle course, road\text{-}category \rangle$ where course is the compass direction defined by the straight-line between the two vertices that defines e. The road-category is the road category, e.g., motorway. The weight $w \in W$ is an array of four speed values describing average speed for four time intervals Monday through Friday. These are free-flow (20:00–06:00), morning peak (7:30–8:15), afternoon peak (15:00–16:30), and non-peak (6:00–7:30, 8:15–15:00, and 16:30–20:00).

A set of weather observations O are reported from a set of stationary weather stations s. A weather station is defined by a three tuple $s = \langle sid, lat, lon \rangle$ where sid is a unique station ID, and lat and lon are the latitude and longitude of the weather station. A weather observation is defined as $o = \langle weather\text{-}class, time, speed, course, temperature, sid \rangle$ where weather-class is the type of weather, e.g., rain or snow, time is the timestamp when the weather observation is recorded, speed is the mean wind speed in m/s, course is the wind direction, temperature is the temperature, and finally sid is the weather station ID.

Each GPS record is map-matched to an edge in the road-network G and a weather observation in O, see Sect. 3. A matched GPS record is called a point $p = \langle r, e, o \rangle$ where $r \in R$, $e \in E$, and $o \in O$. The map foundation is OpenStreetMap (OSM) [12], from Geofabrik [9]. Four road categories, extracted from the OpenStreetMap Highway tag [13] and four categories, motorway, secondary, tertiary, and residential are selected for analysis. Historic weather data is integrated from National Oceanic and Atmospheric Administration (NOAA) [2,5].

3   Method

In this section, we describe how data is prepared to produce the results in Sect. 4. The data foundation presented in Sect. 2 is referenced in this section.

To determine the weather's impact on travel time we match each GPS record to the weather class at the nearest weather station at the time the GPS record was recorded. The work presented here is a generalization and an extension to existing work [4].

Each point p is matched against all weather stations S within 200 km radius. If a weather observation is present for a station, ordered and processed by the distance between p and S, the weather observation o at the station S is assigned to the observation attribute of the p point. To study the effects of the wind, we define three wind attack classes, that is tailwind, crosswind, and headwind. The three classes are defined by an angle β describing the accepted offset from direct tail-, cross-, or head-, illustrated by Fig. 2, showing the angles β_t, β_c, and β_h for tail-, cross, and headwind respectively. The yellow arrow illustrates a vehicle and its driving direction. The mean wind speed will be classified into four groups of 1–5 m/s, 6–10 m/s, 11–15 m/s, and 16- m/s, describing calm, light, moderate, and heavy wind conditions.

4 Results

We first examine the weather impact on the entire road network and then on an individual street level. Next, we study the effect of the wind speed and direction.

4.1 Weather Class Analysis

Figure 3 shows the distribution of GPS reports from vehicles within eight weather classes. Due to the uneven distribution of weather only the top four classes as selected for analysis, that is *dry*, *fog*, *rain*, and *snow*. From Fig. 4 it can be seen that *snow* is typically present from November through March, while *fog* is fairly even distributed over the year. *Dry* and *rain* is varying across the seasons.

Road Categories and Weather. The average speeds on all roads in an entire road network is a good indicator of the weather's impact in general. Figure 5a shows the average speed on the four road categories in non-peak intervals, depending on different weather classes. It can be seen that *dry*, *fog*, and *rain* are very comparable and the speed only varies approximately 2 % on all road categories. *Snow* has an impact of up to 8 % on *motorway*, *secondary*, and *tertiary*. On *residential* roads there is no measurable impact of *snow*. The effects are similar when looking at morning peak, Fig. 5b, and afternoon peak (Fig. 5c) where *dry*, *fog*, and *rain* are comparable, and *snow* leads to lower speeds in morning traffic.

Road Stretch Analysis. While aggregated analysis are good at providing an overview of the weather's impact, more detailed analysis can give a deeper insight on individual roads. We study the weather's impact on four different motorway stretches, labeled $M1$ through $M4$, two rural stretches, labeled $R1$ to $R2$ and four urban stretches, labeled $U1$ through $U4$.

Figure 6 shows heat maps of the routes in the morning traffic, where D is *Dry*, F is *Fog*, R is *Rain*, and S is *Snow*. *Dry* speeds are the baseline speeds and the cells are colored by their relative difference to *dry* speed, that is the percentage

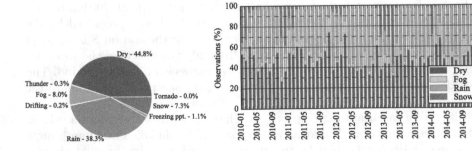

Fig. 3. Weather distribution   **Fig. 4.** Monthly weather distribution

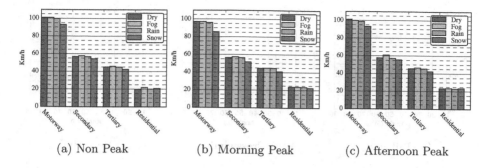

(a) Non Peak   (b) Morning Peak   (c) Afternoon Peak

Fig. 5. Average speed on road categories in different intervals

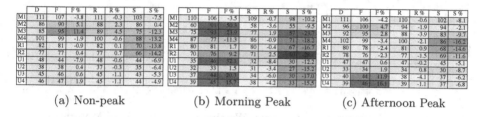

	D	F	F%	R	R%	S	S%
M1	111	107	-3.8	111	-0.3	103	-7.5
M2	86	90	5.1	88	2.3	86	0.4
M3	85	95	11.4	89	4.5	75	-12.3
M4	101	99	-1.9	100	-0.6	88	-13.2
R1	82	81	-0.9	82	0.1	70	-13.8
R2	77	77	0.4	77	0.7	66	-14.2
U1	48	44	-7.9	48	-0.6	44	-6.9
U2	38	38	0.4	37	-0.3	35	-6.4
U3	45	46	0.6	45	-1.1	43	-5.3
U4	46	47	1.9	45	-1.1	44	-4.9

	D	F	F%	R	R%	S	S%
M1	110	106	-3.5	109	-0.7	98	-10.2
M2	60	91	50.3	58	-3.6	55	-9.5
M3	75	93	23.8	77	1.9	57	-23.7
M4	87	77	-11.3	86	-0.9	71	-18.2
R1	80	81	1.7	80	-0.4	67	-16.7
R2	70	76	9.2	71	2.5		
U1	35	46	32.1	32	-8.4	30	-12.2
U2	32	33	1.5	31	-3.4	27	-15.2
U3	37	44	20.3	34	-6.0	30	-17.0
U4	39	45	15.7	38	-4.2	33	-15.5

	D	F	F%	R	R%	S	S%
M1	111	106	-4.2	110	-0.6	102	-8.1
M2	96	100	4.7	94	-1.9	94	-2.1
M3	92	95	2.8	88	-3.9	83	-9.7
M4	102	99	-3.4	100	-2.1	86	-16.2
R1	80	78	-2.4	81	0.9	68	-14.6
R2	78	76	-2.3	77	-1.5	69	-11.6
U1	47	47	0.6	47	-0.2	45	-5.1
U2	33	34	1.9	34	0.8	30	-8.7
U3	40	44	11.9	38	-4.1	37	-6.2
U4	39	46	16.1	39	-1.1	37	-6.8

(a) Non-peak   (b) Morning Peak   (c) Afternoon Peak

Fig. 6. Weather impact on road stretches (Color figure online)

for each weather class. Yellow/red indicates slower speeds and dark green/blue indicates faster speeds than *dry* weather. Figure 6a shows only limited impact by *rain* for all routes. *Fog* shows a significant impact for three routes, $M2$, $M3$, and $U1$, while only limited impact for the remaining seven routes. Only *snow* causes a significant reduction in speed by up to 13.8 % for the $M1$, $M3$, $M4$, $R1$, and $R2$. The urban roads are only slightly affected by *snow*. Morning peak speeds, Fig. 6b, shows that *fog* is often faster than *dry* weather. For the weather condition snow the speed is reduced with up to 27 %. Afternoon peak speeds, Fig. 6c, also shows tendencies to faster speed at *fog* similar to non-peak, with relative speed differences of up to 16 %.

When comparing road stretches it is interesting that afternoon peak is more closely related to non-peak intervals than morning peak. This is likely due to that morning peak traffic is a shorter and more compressed period compared to the afternoon peak traffic.

4.2   Wind Analysis

To analyze the impact of wind we will study the wind impact on *motorway* segments as vehicles tend to have relative stable speeds on these segments.

Figure 7a shows the effect of tailwind on motorway stretches. The figure shows that the speed is slightly affected by the angle of accepting winds used. A narrow angle means only very direct tailwind is accepted, while a broad angle means accepting more crosswind. Accepting a wider angle only yields a decreased speed

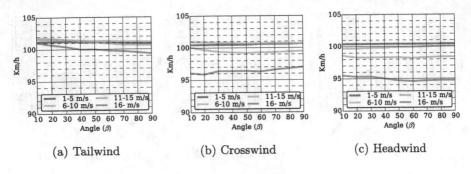

Fig. 7. Wind direction impact

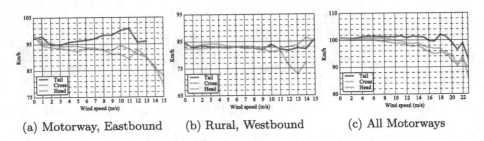

Fig. 8. Wind speed and direction impact

of 1 km/h, except for very strong winds where speeds decreases by 2 km/h going from a β of 10 to 90°. Figure 7b shows the impact of increasing the angle for accepting crosswinds. It can be seen that for winds of 11–15 and 16- m/s there is an impact when increasing β, thus accepting evenly more tail- and headwind. It can be seen that a wider angle yields faster speeds for 16- m/s winds, which indicates tailwind has a stronger effect than headwind. Figure 7c shows accepting more crosswind has a little impact in speed, mainly at 16- m/s, though speed is only varying 1 km/h.

Based on the analysis of the wind attack angle, we decide on an angle (β) of 45°. Comparing Fig. 7a through Fig. 7c it can be seen, that for wind speeds $\leq$10 m/s there is no significant difference between tail-, cross-, and headwind, while stronger winds of 11–15 and 16- m/s indicates faster speeds of tailwind than crosswind and faster speeds of crosswind than headwind.

Two road segments have been selected for performing detail analysis of the wind impact, along with an aggregated analysis on all *motorway* segments, Fig. 8. In general it can be seen, that vehicle speeds decreases at cross- and headwind when wind speed increases. Figure 8b shows though that this road stretch is not impacted by low wind speeds while wind speeds of 11 m/s or stronger. Tailwind results in slightly increased speeds.

5 Related Work

The field of analyzing weather impact on vehicle speeds have been studied for years. Most existing work, [1,3,6–8,10,11,14–17,19], utilizes induction loop detectors to obtain traffic data. The works study the impact of weather on travel time, traffic flow, and traffic levels. In general, they find that rain has a varying impact on travel time while snow can have a larger impact. As the studies are limited to induction loop detectors, the studies are mainly on single or few road segments. Most of the work utilizes data for shorter periods, weeks or months, while some has data for multiple years.

In contradiction to using loop detectors, [18] uses an Automated Number Plate Recognition (ANPR) system for obtaining similar results for London, showing that temperatures below 0°C implies delays and the intensity of rain and snow can impact speeds.

GPS data has been utilized by [21] where 8,000 taxis provide 800,000 records over a 4 months period in Hongzhou, China. They propose a prediction framework and while doing so they analyze weather impact. Another work, [20], utilizes 10M GPS records over 2 months from 1,570 taxis in Nagoya City, Japan.

Existing work for analyzing weather impact on road networks often suffer from at least one of two factors. Firstly, most related work only utilizes data for shorter periods, e.g., few months, making the analysis suffering from seasonal variations. Secondly, existing studies only performs analysis on reduced samples of a road networks, either due to fixed measuring stations (loop detectors, ANPR) or spatially limited extent of GPS data.

6 Conclusion

This paper presents a large-scale nation-wide study of how weather impacts the speed in road networks. 1.6 billion GPS data is collected from 10,560 vehicles over five years from 2010 through 2014 across all of Denmark. The data is integrated with OpenStreetMap and detailed weather information from NOAA.

A generic data model is presented which has global scope and is applicable if a set of GPS data and a road network graph is present.

Using the weather classes *dry*, *fog*, *rain*, and *snow* we show that *snow* has the greatest impact, primarily on *motorway*, *secondary*, and *tertiary* roads with a reduction in speed of up to 27 %. *Residential* roads show only little to no impact on *snow*. For the other weather classes (*dry*, *fog*, and *rain*) there are only smaller differences across all road categories. Similarly we show that wind can reduce speeds with up to 19 %. Wind direction only impacts the vehicle speed at strong wind speeds.

In conclusion, to compute or predict the average speed accurately it is necessary to take into consideration, the factors weather conditions and wind speed.

References

1. Agarwal, M., Maze, T.H., Souleyrette, R.: Impacts of weather on urban freeway traffic flow characteristics and facility capacity. Technical report (2005)

 2. NOAA Agency: http://www.noaa.gov/
 3. Akin, D., Sisiopiku, V.P., Skabardonis, A.: Impacts of weather on traffic flow characteristics of urban freeways in istanbul. Procedia-Soc. Behav. Sci. **16**, 89–99 (2011)
 4. Andersen, O., Krogh, B.B., Thomsen, C., Torp, K.: An advanced data warehouse for integrating large sets of gps data. In: ACM DOLAP, pp. 13–22 (2014)
 5. National Climate Data Center: Federal climate complex datadocumentation for integratedsurface data. Technical report (2015). ftp://ftp.ncdc.noaa.gov/pub/data/noaa/ish-format-document.pdf
 6. Chung, E., Ohtani, O., Warita, H., Kuwahara, M., Morita, H.: Does weather affect highway capacity. In: 5th International Symposium on Highway Capacity and Quality of Service, Yakoma, Japan (2006)
 7. Datla, S., Sahu, P., Roh, H.J., Sharma, S.: A comprehensive analysis of the association of highway traffic with winter weather conditions. Procedia-Soc. Behav. Sci. **104**, 497–506 (2013)
 8. Edwards, J.B.: Speed adjustment of motorway commuter traffic to inclement weather. Transp. Res. Part F **2**(1), 1–14 (1999)
 9. GeoFabrik: Openstreetmap data extracts. http://download.geofabrik.de/
10. Mashros, N., Ben-Edigbe, J., Hassan, S.A., Hassan, N.A., Yunus, N.Z.M.: Impact of rainfall condition on traffic flow and speed: a case study in johor and terengganu. Jurnal Teknologi **70**(4), 65–69 (2014)
11. Maze, T., Agarwai, M., Burchett, G.: Whether weather matters to traffic demand, traffic safety, and traffic operations and flow. TRB J. **1948**, 170–176 (2006)
12. OpenStreetMap: http://www.openstreetmap.org
13. OpenStreetMap: Key:highway - openstreetmap. http://wiki.openstreetmap.org/wiki/Key:highway
14. Rakha, H., Farzaneh, M., Arafeh, M., Hranac, R., Sterzin, E., Krechmer, D.: Empirical studies on traffic flow in inclement weather. Virginia Tech Transportation Institute (2007)
15. Saberi, K.M., Bertini, R.L.: Empirical analysis of the effects of rain on measured freeway traffic parameters. TRB 89th Annual Meeting (2010)
16. Smith, B.L., Byrne, K.G., Copperman, R.B., Hennessy, S.M., Goodall, N.J.: An investigation into the impact of rainfall on freeway traffic flow. TRB 83rd Annual Meeting (2006)
17. Thakuriah, P., Tilahun, N.: Incorporating weather information into real-time speed estimates: comparison of alternative models. J. Transp. Eng. **139**(4), 379–389 (2013)
18. Tsapakis, I., Cheng, T., Bolbol, A.: Impact of weather conditions on macroscopic urban travel times. J. Transp. Geogr. **28**, 204–211 (2013)
19. Tu, H., van Lint, H.W., van Zuylen, H.J.: Impact of adverse weather on travel time variability of freeway corridors. In: TRB 86th Annual Meeting (2007)
20. Wang, L., Yamamoto, T., Miwa, T., Morikawa, T.: An analysis of effects of rainfall on travel speed at signalized surface road network based on probe vehicle data. In: ICTTS, Xian, China, pp. 2–4 (2006)
21. Zhang, R., Shu, Y., Yang, Z., Cheng, P., Chen, J.: Hybrid traffic speed modeling and prediction using real-world data. In: IEEE Big Data, pp. 230–237 (2015)

Simplify the Design of XML Schemas by Type Dependencies

Jia Liu[✉] and Husheng Liao[✉]

College of Computer Science, Beijing University of Technology, Beijing, China
jeromeliu2006@sina.com, liaohs@bjut.edu.cn

Abstract. In XML Schema, the type definition mechanism is responsible for defining types as well as passing contextual information, which may cause some design problems such as artificial types. This paper proposes a new kind of XML schema called Type Dependencies (TD) schema to realize the separation of those two tasks. A TD schema includes two parts, a set of type dependencies which is responsible for passing contextual information and a complete competition grammar which is only responsible for defining types. It can help users to design better schemas more easily, since there are no problems related to Element Declarations Consistent (EDC) rule and artificial types in TD schemas. Furthermore, the expressiveness of TD schemas is more powerful than XML Schema and it also satisfies the semantic concept of 1-pass preorder typing, which make it more suitable for streaming data.

Keywords: XML · Schema language · Type · Streaming data

XML Schema specifies how to formally describe the constraints on the structure and content of an XML document, above and beyond the basic syntax imposed by XML itself. Although XML Schema is successful in that it has been widely adopted and largely achieves what it set out to, it has been the subject of a great deal of criticism due to its complexity. There are two problems which make the design of schemas more difficult and are hard to be settled by design skills. One is the Element Declarations Consistent (EDC) rule in the XML Schema specification. Simply put, this semantic constraint requires that elements with the same name in the same content model must have the same type. The purpose of this restriction is to facilitate a simple one-pass top-down validation algorithm [1]. But it is difficult to understand the effect of this constraint for non-expert users. More importantly, this constraint limits the expressiveness of XSD, making it cannot fully meet the needs of streaming data. The other one is called artificial types [5], which purpose is to pass the contextual information to other elements. However, it makes the elements with the same children may have the different types, and it is also difficult to understand for normal users.

In fact, all the above two problems can be traced back to its theoretical model, that is the regular tree grammar(RTG), or more precisely, single-type tree grammar (STTG)–a subset of RTG. The above two problems are caused by the same reason, which is the two tasks, defining types and passing contextual information, are undertaken by the one type definition mechanism. A good idea for avoiding those problems and making design of schema easier is to separate the task of passing contextual

© Springer International Publishing Switzerland 2016
S. Hartmann and H. Ma (Eds.): DEXA 2016, Part II, LNCS 9828, pp. 445–453, 2016.
DOI: 10.1007/978-3-319-44406-2_38

information from the type definition mechanism. To solve it, we propose a new kind of semantic constraints for XML, namely Type Dependencies (TD), to express the relationships between types. Based on the concept of type dependencies, we propose a new kind of XML schema called Type Dependencies schema to realize the separation of the two tasks. There are two major benefits: Firstly, there are no problems related to EDC rule and artificial types in TD schemas, so it is convenient for users to design better schemas. Secondly, the TD schema has more powerful expressiveness than XML Schema and it also satisfies the 1-pass preorder typing [2], which make it more suitable for streaming data.

1   Motivation

In this paper we will use the XML tree shown in Fig. 1 as the running example.

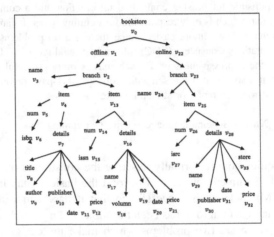

Fig. 1. An XML tree for the bookstore

The considered XML tree describes the information of items related to a bookstore. This example is a microcosm of the complexity in practical applications. The elements *item*, *num* and *details* all has more than one types. Non-expert users may design an XSD schema for this case, however, this kind of XML trees cannot be handled by Schema. The schema represented by regular tree grammar for this example is shown in Fig. 2. Let us consider the following problems involved in this example:

1. The grammar is in conflict with the EDC rule since the element *item* in the content models of OfflineBranch and OnlineBranch has different types. For example, the content model of OfflineBranch has three types related to *item*: OfflineBook, OfflineMagazine, OfflineDVD. Therefore, XML Schema cannot be used in this case.
2. The elements *branch* or *item* have the same children structures but may have different types, therefore, the types of elements *branch* or *item* are all artificial types.

3. Verification process will generate backtrackings for the *num* elements, therefore, this XML tree cannot be verified by using the streaming fashion since there must be no backtrackings in the verification process for streaming data, that is the constraint of 1-pass preorder typing.

1.	BookStore	→ bookstore[Offline Online]
2.	Offline	→ offline[OfflineBranch*]
3.	OfflineBranch	→ branch[Name (OfflineBook|OfflineMagazine|OfflineDVD)*]
4.	OfflineBook	→ item[IsbnNum OfflineBookDetails]
5.	OfflineMagazine	→ item[IssnNum OfflineMagazineDetails]
6.	OfflineDVD	→ item[IsrcNum OfflineDVDDetails]
7.	IsbnNum	→ num[Isbn]
8.	IssnNum	→ num[Issn]
9.	IsrcNum	→ num[Isrc]
10.	OfflineBookDetails	→ details[Title Author Publisher Date Price]
11.	OfflineMagazineDetails	→ details[Name Volumn No Date Price]
12.	OfflineDVDDetails	→details[Name Publisher Date Price]
13.	Online	→ Online[OnlineBranch*]
14.	OnlineBranch	→ branch[Name (OnlineBook|OnlineMagazine|OnlineDVD)*]
15.	OnlineBook	→ item[IsbnNum OnlineBookDetails]
16.	OnlineMagazine	→ item[IssnNum OnlineMagazineDetails]
17.	OnlineDVD	→ item[IsrcNum OnlineDVDDetails]
18.	OnlineBookDetails	→ details[Title Author Publisher Date Price Store]
19.	OnlineMagazineDetails	→ details[Name Volumn No Date Price Store]
20.	OnlineDVDDetails	→details[Name Publisher Date Price Store]

Fig. 2. The schema for bookstore

2 Type Dependencies

We first present the definitions of the XML trees and the regular tree grammars. Let E denote a finite set of labels of element nodes, A denote a finite set of labels of attribute nodes and $text denote the label of text nodes, where $E \cap A = \{\emptyset\}$ and $text \notin E \cup A$.

Definition 1 XML Trees. Formally, we define an XML tree as a 5-tuple $< V, label, parent, value, num >$, denoted by T, where V denotes the set of all nodes in T, *label* is a mapping from V to $E \cup A \cup \{$ text\}$, *parent* is a mapping from V to $V \cup \perp$, which returns the parent of a node in T. *value* is a mapping from V to *Strings*, which returns an attribute value for attribute nodes, a text for text nodes and an empty string for element nodes. *num* is a mapping from V to *NUM*, which assigns a unique positive integer to each node in the order of preorder traversal.

Definition 2 Regular Tree Grammars. A regular tree grammar, denoted by $\mathcal{G}$, can be defined as a 4-tuple $< \mathcal{N}, \mathcal{T}, \mathcal{S}, \mathcal{P} >$, where $\mathcal{N}$ is a finite set of non-terminals appeared in $\mathcal{P}$, $\mathcal{T} \subseteq $ text \cup E \cup A$ is a finite set of terminals appeared in $\mathcal{P}$, $\mathcal{S} \in \mathcal{N}$ is the start symbol and the production with symbol $\mathcal{S}$ in its left part is called start production, $\mathcal{P}$ is a finite set of productions and each production in $\mathcal{P}$ is an expression of form $n \to t[c]$, where $n \in \mathcal{N}$, $t \in \mathcal{T}$ and c is a regular expression defined on $\mathcal{N} \cup \epsilon$, called content model.

Definition 3 Valid XML Trees. Let $T = <V, label, parent, value, num>$ be an XML tree and $\mathcal{G} = <\mathcal{N}, \mathcal{T}, \mathcal{S}, \mathcal{P}>$ be a regular tree grammar. If there is a mapping I from each node v in T to a certain non-terminal in $\mathcal{G}$, such that: When v is the root of T, $I(v)$ is the start symbol $\mathcal{S}$ of $\mathcal{G}$; For each node v in T and the children (if any) of v, denoted by $v_1, v_2, \cdots, v_n$ and let $v_i \prec v_{i+1}(1 \leq i < n)$, then there always exist a production $n \to \ell[c]$, such that $I(v) = n$, $label(v) = \ell$ and the $I(v_1)I(v_2)\cdots I(v_n) \in \mathcal{L}(c)$. Especially, if v has no any children, then $\epsilon \in \mathcal{L}(c)$; then we say that T is valid to $\mathcal{G}$, mapping I is an interpretation of T to $\mathcal{G}$ and the set of all valid XML document trees to $\mathcal{G}$ is called the regular tree language defined by $\mathcal{G}$, which is denoted by $\mathcal{L}(\mathcal{G})$.

Definition 4 Types. Let $\mathcal{G} = <\mathcal{N}, \mathcal{T}, \mathcal{P}, s>$ be a regular tree grammar, let $T = <V, label, parent, value, num>$ be an XML document tree in $\mathcal{L}(\mathcal{G})$, and let I be an interpretation of T on $\mathcal{G}$. Given an element node on T, denoted by v, the type of node v is defined as $I(v)$. Given an element label $l \in E$, if there is a production p in $\mathcal{P}$, such that $p.\ell = l$, then the non-terminal $p.n$ is said to be a type of label l. Let $types(l)$ denote the set of all types of l on $\mathcal{G}$, and let $\parallel types(l) \parallel$ be the number of types in $types(l)$.

Definition 5 Prefix Path Patterns. The syntax of the prefix path pattern is defined as follows:

d: The label of root node, n: E, l: $\$text \cup E \cup A$

$\phi ::= d/\phi' | d//\phi' | d, \phi' ::= \phi'/\phi' | \phi'//\phi' | \phi' [r] | n | *, r ::= r \cdot r | (r|r) | r^* | l | * | \epsilon$

All prefix path patterns ϕ are begin with the label of root node. / and // are called children axis and descendant axis respectively. The expressions before and after axes and called step, r is a regular expression defined on $\$ text \cup E \cup A \cup \epsilon$, which is called predicate. Let $V' \subseteq V$ is a set of nodes in XML tree T, if the string which is composed by the labels of nodes in V' according to the document order is matched to r, then $r(V')$ returns $TRUE$, else it returns $FALSE$. The formal semantics for prefix path pattern is defined as follows:

$S: Pattern \to Node \to Set(Node)$
$S[d]v = root(v), if\ d = label(root(v))$
$S[d/\phi']v = \{v' | v' \in S[\phi']root(v), d = label(root(v))\}$
$S[d//\phi']v = \{v'' | v'' \in S[\phi']v', v' \in descendants(root(v)) \cup root(v), d = label(root(v))\}$
$S[\phi'_1/\phi'_2]v = \{v'' | v' \in S[\phi'_1]v, v'' \in S[\phi'_2]v'\}$
$S[\phi'_1//\phi'_2]v = \{v''' | v' \in S[\phi'_1]v, v'' \in descendants(v') \cup v', v''' \in S[\phi'_2]v''\}$
$S[\phi'[r]]v = \{v' \in S[\phi], r(leftsiblings(v')) = TRUE\}$
$S[l]v = \{v' \in children(v), label(v') = l\}, S[*]v = \{v' \in children(v)\}$

The predicate in a prefix path pattern is used to filter nodes. Especially, if a predicate r is ϵ, then for any element node, $r(leftsiblings(v)) = TRUE$ if and only if $leftsiblings(v) = \emptyset$. For example, the prefix path pattern of nodes named *details* and under the *online* node is:

bookstore/online/branch[branch*]/item[name item*]/details[num] ϕ_d

Definition 6 Reachable Nodes and Reachable Types. Let T be an XML document tree, ϕ be a prefix path pattern and let v be a node on T. The nodes in $S[\phi]v$ are said to be the reachable nodes of ϕ on T, where S is the semantic function for prefix path pattern defined above, and the set of all reachable nodes of ϕ on T is said to be the reachable node set of ϕ on T, denoted by $\phi(T)$. Let $\mathcal{G}$ be a regular tree grammar, T be an XML document tree in $\mathcal{L}(\mathcal{G})$, I be an interpretation of T on $\mathcal{G}$ and let v be a node in T. $I(v)$ is said to be a reachable type of ϕ on $\mathcal{G}$ if $v \in \phi(T)$, and the set of all reachable types of ϕ on $\mathcal{G}$ is said to be the reachable type set of ϕ on $\mathcal{G}$, denoted by $\phi(\mathcal{G})$.

Definition 7 Nearest Nodes. Given a prefix path pattern ϕ and an XML document tree T, v' is the nearest node of v in $\phi(T)$ if there is no other node v'' in $\phi(T)$ such that $v' \prec v'' \prec v$, where v is a node on T and v' is a node in $\phi(T)$. Especially, if there is no node which satisfies the above definition in $\phi(T)$, then we let $\perp$ denote the nearest node of v in $\phi(T)$. Given a set of prefix path patterns $\Phi = \{\phi_1, \phi_2, \cdots, \phi_n\}$, the nodes $v_{\phi_1}, v_{\phi_2}, \cdots, v_{\phi_n}$ are said to be the nearest node set of v on Φ if v_{ϕ_n} is the nearest node of v in $\phi_n(T)$, and $v_{\phi_{i-1}}$ is the nearest node of v_{ϕ_i} in $\phi_{i-1}(T)$, where $1 < i \leq n$. In particular, if $v_{\phi_i} = \perp$, then $v_{\phi_{i-1}} = \perp$.

Definition 8 Type Dependencies. Let $\mathcal{G} = \langle \mathcal{N}, \mathcal{T}, \mathcal{P}, \delta \rangle$ be a regular tree grammar, and let Φ_1 and Φ_2 denote two sets of prefix path patterns. A type dependency on $\mathcal{G}$ is an expression defined as $\Phi_1 \rightarrow \Phi_2$.

Let $\mathcal{D}$ be a set of type dependencies on $\mathcal{G}$, $\Phi_1 \rightarrow \Phi_2$ is an XML type dependency in $\mathcal{D}$. An XML document tree, denoted by T, satisfies $\Phi_1 \rightarrow \Phi_2$ if $T \in \mathcal{L}(\mathcal{G})$ and for any two nodes in $\phi_j(T)$, denoted by v_j and $v'_j (n+1 \leq j \leq n+m)$, if there are two node sets $V_j = \{v_{\phi_1}, v_{\phi_2}, \cdots, v_{\phi_n}\}$ and $V'_j = \{v'_{\phi_1}, v'_{\phi_2}, \cdots, v'_{\phi_n}\}$, such that V_j and V'_j are the nearest node sets of v_j and v'_j on Φ_1 respectively and any node in V_j or V'_j is not $\perp$, then if $I(v_j) \neq I(v'_j)$, then for any v_{ϕ_i} and v'_{ϕ_i}, we have $I(v_{\phi_i}) \neq I(v'_{\phi_i})$, where $1 \leq i \leq n$. Furthermore, we say that $\Phi_1 \rightarrow \Phi_2$ is a n-ary type dependency if the number of prefix path patterns in Φ_1 is n. For example, in the schema of Fig. 2, there is a type dependency $\phi_1, \phi_2 \rightarrow \phi_3$, where ϕ_1 : bookstore/ $*$ [offline$*$],

ϕ_2 : bookstore/ $*$ [offline$*$]/branch[branch$*$]/item[name item$*$]/num/$*$
ϕ_3 : bookstore/ $*$ [offline$*$]/branch[branch$*$]/item[name item$*$]/details[num].

Definition 9 Instance Constraints. Let $\mathcal{G} = \langle \mathcal{N}, \mathcal{T}, \mathcal{S}, \mathcal{P} \rangle$ be a regular tree grammar, $\Phi_1 = \{\phi_1, \phi_2, \cdots, \phi_n\}$ and $\Phi_2 = \{\phi_{n+1}, \phi_{n+2}, \cdots, \phi_{n+m}\}$ are two prefix path pattern sets, $\Phi_1 \rightarrow \Phi_2$ is a type dependency on $\mathcal{G}$, an instance constraint of $\mathcal{G}$ is an expression as follows: $n_1, n_2, \cdots, n_n \rightarrow n_{n+1}, n_{n+2}, \cdots, n_m$, where $n_1, n_2, \cdots, n_{n'} n_{n+1}, n_{n+1}, \cdots, n_m$ are the non-terminals in $\mathcal{N}$. An XML document tree T satisfy the instance constraint if and only if for any node v_i in $\phi_i(T)(n+1 \leq i \leq m)$, if types of the nearest nodes $v_1, v_2, \cdots, v_n$ of v_i on $\phi_1(T), \phi_2(T), \cdots, \phi_n(T)$ are $n_1, n_2, \cdots, n_{n'}$, then the type of v is n_i. For example, the type dependency showed in Definition 8 has the following instance constraints:

Offline, Isbn $\rightarrow$ OfflineBookDetails, Offline, Issn $\rightarrow$ OfflineMagazineDetails
Offline, Isrn $\rightarrow$ OfflineDVDDetails, Online, Isbn $\rightarrow$ OnlineBookDetails
Online, Issn $\rightarrow$ OnlineMagazineDetails, Online, Isrn $\rightarrow$ OnlineDVDDetails

3 Type Dependencies Schema

In this section, we will introduce the type dependencies schema. Firstly, we will give the definition of complete competition grammars, which are only response for type definitions and there is no type dependencies in complete competition grammars.

Definition 10 Single-types and Multi-types. Given a regular tree grammar $\mathcal{G} = <\mathcal{N}, \mathcal{T}, \mathcal{P}, \mathfrak{d}>$, the label l in E is said to be a single-type label on $\mathcal{G}$ if $\| types(l) \| = 1$ or l is said to be a multi-type label on $\mathcal{G}$ if $\| types(l) \| > 1$.

Definition 11 Prefixes. Let $\mathcal{G} = <\mathcal{N}, \mathcal{T}, \mathcal{P}, \mathfrak{d}>$ be a regular tree grammar, let c be the content model of a production p in $\mathcal{P}$, and let n be a non-terminal in $\mathcal{N}(p.c)$. For any non-terminal string, denoted by $xn'y$, if $xn'y \in \mathcal{L}(c)$ and $n = n$, then x is said to be a prefix of n on c, where x and y are non-terminal strings on $\mathcal{N} \cup \epsilon$ and n' is a non-terminal in $\mathcal{N}$. The prefix set of n on c is the set of all prefixes of n on c, and the prefix expression of n on c is a regular expression on $\mathcal{N} \cup \epsilon$ which defines the prefix set of n on c.

Definition 12 Competition. Let $\mathcal{G} = <\mathcal{N}, \mathcal{T}, \mathcal{P}, \mathfrak{d}>$ be a regular tree grammar, let c be a content model of a production p in $\mathcal{P}$, let $\mathcal{J} = \{n_1, n_2, \cdots, n_n (n \geq 2)\}$ be a set of non-terminals such that $\mathcal{J} \subseteq \mathcal{N}(p.c) \wedge \mathcal{J} \subseteq types(l)$, where l is a multi-type label, and let $q_1, q_2, \cdots, q_n$ be the prefix expressions of $n_1, n_2, \cdots, n_n$. If $\cap_{i=1}^{n} \mathcal{L}(tot(q_i)) \neq \emptyset$, then non-terminals $n_1, n_2, \cdots, n_n$ are said to be competitive non-terminals on c, the set $\mathcal{J}$ is said to be a competition set on c, the regular expression r such that $\mathcal{L}(r) = \cap_{i=1}^{n} \mathcal{L}(tot(q_i))$ is said to be the competition prefix expression of $\mathcal{J}$, and the terminal strings in $\mathcal{L}(r)$ are said to be the competition prefixes of $\mathcal{J}$ on c, where tot is a mapping from non-terminals to terminals.

Definition 13 Complete Competition. Let $\mathcal{G} = <\mathcal{N}, \mathcal{T}, \mathcal{P}, \mathfrak{d}>$ be a regular tree grammar, c be a content model of a production p in $\mathcal{P}$, let $\mathcal{J} = \{n_1, n_2, \cdots, n_n (n \geq 2)\}$ be a set of non-terminals such that $\mathcal{J} \subseteq \mathcal{N}(p.c) \wedge \mathcal{J} \subseteq types(l)$, where l is a multi-type label, and let $q_1, q_2, \cdots, q_n$ be the prefix expressions of $n_1, n_2, \cdots, n_n$. If $\cap_{i=1}^{n} \mathcal{L}(q_i) \neq \emptyset$, then the non-terminals $n_1, n_2, \cdots, n_n$ are said to be complete competition non-terminals on c, the set $\mathcal{J}$ is said to be a complete competition set on c, the regular expression r' such that $\mathcal{L}(r') = \cap_{i=1}^{n} \mathcal{L}(q_i)$ is said to be the complete competition prefix expression of $\mathcal{J}$, and the terminal strings in $\mathcal{L}(r')$ are said to be the complete competition prefixes of $\mathcal{J}$ on c. For example, let $\mathcal{J} = \{OfflineBook, OfflineMagazine, OfflineDVD\}$, then $\mathcal{J}$ is a complete competition set and the complete competition prefix expression of $\mathcal{J}$ is *name item**. Similarly, let $\mathcal{J}' = \{OnlineBook, OnlineMagazine, OnlineDVD\}$, then $\mathcal{J}'$ is a complete competition set and the complete competition prefix expression of $\mathcal{J}'$ is *name item**.

Definition 14 Complete Competition Grammars. Given a competition grammar, denoted by $\mathcal{G} = <\mathcal{N}, \mathcal{T}, \mathcal{P}, \mathfrak{d}>$, $\mathcal{G}$ is said to be a complete competition grammar if for any competition set on $\mathcal{G}$, denoted by $\mathcal{J}$, $\mathcal{J}$ is a complete competition set, and the complete competition prefix of $\mathcal{J}$ is also the prefix of $\mathcal{J}$, and if non-terminal $n \in \mathcal{J}$,

```
1.   BookStore → bookstore[Offline Online]
2.   Offline → offline[Branch*]
3.   Online → online[Branch*]
4.   Branch → branch[Name Item*]
5.   Item → item[Num (OnlineBookDetails|OnlineMagazineDetails|OnlineDVDDetails|
      OfflineBookDetails|OfflineMagazineDetails|OfflineDVDDetails)*]
6.   Num → num[Isbn| Issn| Isrc]
7.   OfflineBookDetails → details[Title Author Publisher Date Price]
8.   OfflineMagazineDetails → details[Name Volumn No Date Price]
9.   OfflineDVDDetails →details[Name Publisher Date Price]
10.  OnlineBookDetails → details[Title Author Publisher Date Price Store]
11.  OnlineMagazineDetails → details[Name Volumn No Date Price Store]
12.  OnlineDVDDetails →details[Name Publisher Date Price Store]
```

Fig. 3. A complete competition grammar for bookstore

then every non-terminal $n' \in \mathcal{N}$ such that $p'.t = p.t$ is also in set $\mathcal{J}$, where p, p' are productions in $\mathcal{P}$ and, $p'.n = n'$. The schema in Fig. 2 is not a complete competition grammar since the complete competition sets {*OfflineBook*, *OfflineMagazine*, *OfflineDVD*} and {*OnlineBook*, *OnlineMagazine*, *OnlineDVD*} are not equal. The following schema is a complete competition grammar.

Definition 15 Type dependencies schemas. A type dependencies schema, denoted by $\mathcal{F}$, is a triple $< \mathcal{G}, \mathcal{DS}, inst >$, where $\mathcal{G}$ is a complete competition grammar, $\mathcal{DS}$ is a set of 1-ary XML type dependencies, and *inst* is a mapping from $\mathcal{DS}$ to instance constraints, which returns the instance constraint set for each XML type dependency in $\mathcal{DS}$, and for any XML tree T in $\mathcal{L}(\mathcal{G})$, if v is node with multi-type label on T, then $\mathcal{DS}$ implies a type dependency, denoted by $\Phi \to \phi$, such that v is a reachable node on T, V is the nearest node set of v on Φ and $\perp \notin V$. Given an XML tree T, T is said to be valid to $\mathcal{F}$ if T is valid to $\mathcal{G}$ and for each XML type dependency in $\mathcal{DS}$, denoted by $\Phi \to \phi$, T satisfies all instance constraint set in $inst(\Phi \to \phi)$. For example, a TD schema is showed in Fig. 4. There are following features in this schema:

```
1.   the complete competition grammar showed in Fig.3
2.   type dependency:  φ₁, φ₂ → φ₃, where:
      φ₁: bookstore/*[offline*]
      φ₂: bookstore/*[offline*]/branch[branch*]/item[name item*]/num/*
      φ₃: bookstore/*[offline*]/branch[branch*]/item[name item*]/details[num].
3.   instance constraints:
      Offline, Isbn → OfflineBookDetails
      Offline, Issn → OfflineMagazineDetails
      Offline, Isrn → OfflineDVDDetails
      Online, Isbn → OnlineBookDetails
      Online, Issn → OnlineMagazineDetails
      Online, Isrn → OnlineDVDDetails
```

Fig. 4. A type dependency schema for bookstore

1. This schema separates the task of passing contextual information from the type definition mechanism. Type dependencies are response for passing contextual information and a complete competition grammar is response for type definitions.
2. There are no EDC rule and artificial types in TD schemas, so users can pay more attention to the design of the schema itself and avoid mistakes
3. There are no backtrackings in the verification process, therefore this schema is suitable for streaming data.

The TD schema is more powerful than XML Schema. In fact, the expressiveness of TD schemas is beyond the restrained competition grammar which is the maximal subset of regular tree grammar for streaming data.

Theorem 1. A forward determined grammar is equivalent to a constrained type dependencies schema.

4   Related Works

Murata et al. [1] investigate the problem of the expressiveness for XML schema languages by using the tree grammars. Murata also point that XML Schema only capture the class of single-type tree grammars. In paper [2], the author gives two conditions for processing streaming data: a single pass scan and constant memory. Martens et al. [4, 5] shows that restrained competition tree grammar is the largest class of XML schemas for streaming data. Martens turns out that a regular tree grammar admits 1-pass preorder typing if and only if it is restrained competition. Gelade et al. [6] defined a pattern-based specification language equivalent in expressive power to the XML Schema.

A number of XML functional dependencies have been proposed to enrich the semantics of XML for developing a normalization theory for XML. Arenas [7] investigates the problem of the normalization of XML documents. In a contextual of DTD, XML functional dependencies are defined based on paths, and tree tuples are used to simulate the tuples in relational databases. Vincent [8] investigates the path-based XML functional dependencies with closest attribute value. Unlike the definitions based on paths, sub-graph is used in paper [9] for locating the data item in XFDs and increase the flexibility of the XFD's definition with the help of the graph structure of XML schema.

5   Conclusion

This paper proposes a new XML schema language called Type Dependencies schema to realize the separation of the tasks of the defining types and passing contextual information which are mixed in XML Schema. It is convenient for users to design better schema, since there are no problems related to EDC rule and artificial types in a TD schema. Furthermore, TD schema has powerful expressiveness, making it more suitable for streaming data than XML Schema.

References

1. Murata, M., Lee, D., Mani, M., et al.: Taxonomy of XML schema languages using formal language theory. ACM Trans. Internet Techn. **5**(4), 660–704 (2005)
2. Segoufin, L., Vianu, V.: validating streaming XML documents. In: PODS, pp. 53–64 (2002)
3. Bex, G.J., Martens, W., Neven, F., Schwentick, T.: Expressiveness of XSDs: from practice to theory, there and back again. In: WWW, pp. 712–721 (2005)
4. Martens, W., Neven, F., Schwentick, T.: Which XML schemas admit 1-pass preorder typing? In: Eiter, T., Libkin, L. (eds.) ICDT 2005. LNCS, vol. 3363, pp. 68–82. Springer, Heidelberg (2005)
5. Martens, W., Neven, F., Schwentick, T., Bex, G.J.: Expressiveness and complexity of XML Schema. ACM Trans. Database Syst. **31**(3), 770–813 (2006)
6. Gelade, W., Neven, F.: Succinctness of pattern-based schema languages for XML. J. Comput. Syst. Sci. **77**(3), 505–519 (2011)
7. Arenas, M., Libkin, L.: A normal form for XML documents. ACM Trans. Database Syst. **29**, 195–232 (2004)
8. Vincent, M.W., Liu, J., Liu, C.: Strong functional dependencies and their application to normal forms in XML. ACM Trans. Database Syst. **29**(3), 445–462 (2004)
9. Hartmann, S., Link, S.: More functional dependencies for XML. In: Kalinichenko, L.A., Manthey, R., Thalheim, B., Wloka, U. (eds.) ADBIS 2003. LNCS, vol. 2798, pp. 355–369. Springer, Heidelberg (2003)

An Efficient Initialization Method for Probabilistic Relational Databases

Hong Zhu, Caicai Zhang, and Zhongsheng Cao[✉]

School of Computer Science and Technology,
Huazhong University of Science and Technology,
1037 Luoyu Road, Wuhan 430074, China
caozhongsheng@163.com

Abstract. Probabilistic relational databases play an important role on uncertain data management. Informally, a probabilistic database is a probability distribution over a set of deterministic databases (namely, possible worlds). The existing initialization methods that transform the possible worlds representation into our chosen representation, make the formulae of tuples very long. An efficient initialization method is proposed by providing an equation that can generate simplified formulae of tuples. The experimental study shows that the proposed method greatly simplifies the formulae of tuples without additional time cost. The subsequent queries benefit from the simplified formulae of tuples.

Keywords: Probabilistic relational databases · Formulae · Simplification

1 Introduction

Modern applications need to process uncertain data that are retrieved from diverse and autonomous sources [3], such as data cleaning [4], sensor networks [12]. Informally, a probabilistic database is a probability distribution over a set of deterministic databases (namely, possible worlds) [11].

Uncertain data in real world are usually represented as possible worlds. However, multiple data sets can lead to considerable large number of possible worlds of probabilistic relational databases. Therefore, probabilistic relational databases need some more concise uncertain data representation formalism [5]. The variable-based formalism [9] is a very general and very powerful representation mechanism [13] that can represent rich correlations between probabilistic data. In this formalism, each tuple is annotated with a propositional formula composed of independent variables.

In this paper, we consider the initialization problem for probabilistic relational databases that is to transform the probability distribution of possible worlds of uncertain data into the variable-based representation formalism. The existing initialization methods make the formulae of tuples very long and complicated. The lengthy formulae lead to the waste of storage and significant overhead in subsequent query processing. Therefore, the formulae simplification is

© Springer International Publishing Switzerland 2016
S. Hartmann and H. Ma (Eds.): DEXA 2016, Part II, LNCS 9828, pp. 454–462, 2016.
DOI: 10.1007/978-3-319-44406-2_39

very important for efficient uncertain data management. However, the formulae simplification is a NP problem. This paper presents an efficient initialization method for the variable-based representation. Our main contributions in this article are summarized as follows.

(1) We propose a new initialization method to generate simplified formulae of tuples without additional time cost.
(2) We prove the correctness of the proposed initialization method by showing that it generates an equivalent formula as the existing initialization method.
(3) We conduct an extensive experimental study to evaluate our initialization method in different configurations, showing its efficiency and scalability.

The remainder of this article is organized as follows. Section 2 discusses the related work. We present in Sect. 3 the necessary preliminaries on the probabilistic databases. Section 4 describes our proposed initialization method for probabilistic relational databases. In Sect. 5, we report the performance evaluation of our initialization method. Section 6 concludes this paper.

2 Related Work

There are significant amounts of work on representation formalism for uncertain data, e.g., [9]. In general, these studies can be divided into two categories, one is based on simple correlation model [5], which associates existence probabilities with individual tuples and assumes that the tuples are mutually independent or exclusive; the other is based on a richer representation formalism which can express complex correlations between tuples [9]. Many application domains naturally produce correlated data. Furthermore, dependencies among tuples arises naturally during query evaluation even when one assumes that the base data tuples are independent [10]. Several formalisms can express complex correlations between tuples, lineage-based [2], possible worlds decompositions [6], U-relations [6], the variable-based representation [8,9]. The lineage-based representation formalism can represent the rich correlations among answering tuples of queries, however, the base tuples are assumed to be independent. MayBMS system [1] successively adopted possible worlds decompositions [6] and U-relations [6] to store probabilistic data. Only several operations can be processed efficiently in the possible worlds decompositions representation. U-relations is a succinct and complete representation system for large sets of possible worlds [1]. However, besides the succinctness and completeness, the variable-based representation can also accelerate the processing of query processing by lots of optimization strategies for Boolean logic expressions processing [7]. Therefore, the variable-based representation is a very general and very powerful representation mechanism [13]. To the best of our knowledge, this paper is the first to study the initialization problem for the variable-based representation.

3 Preliminary

3.1 Data Model

Definition 1. *A probabilistic database in the variable-based representation is a quadruple $\mathbb{D} = \langle D, E, P, f \rangle$ such that*

(i) D: D is a traditional relational database.
(ii) E: $E = \{e_1, \ldots, e_m\}$ is a finite set of independent Boolean variables.
(iii) f: $\forall t \in D$, $f(t)$ is a propositional formula of variables from E.
(iv) P: P is a function that defines a discrete probability distribution for each random variable e_i in E, in other words, $P(e_i) + P(\neg e_i) = 1$. [9].

A possible world is a pair of form $\langle E^j, w_i \rangle$, where E^j is an evaluation of E, and w_i is a set of tuples from D such that $t \in w_i$ iff $E^j \vDash f(t)$.[1] $P(w_i) = \sum_{E^j \Rightarrow w_i} P(E^j)$ and $\sum_{w_i \in W} P(w_i) = 1$, where $E^j \Rightarrow w_i$ represents that E^j leads to the possible world w_i. The truth value of $f(t)$ determines the presence of t in the actual world, and its probability is defined by the probabilities of the composed variables.

We briefly introduce the only existing initialization method for variable-based representation (referred to as *Naive* [14]) by the following example.

Example 1. Assume a set of tuples $\{t_1, t_2, t_3, t_4\}$ are mutually correlated, and their possible worlds are listed in Table 1. By the Naive initialization method, first, a set of tuples $T = \{t_1, t_2, t_3, t_4\}$ are inserted into D, then a new set of variables $\{e_1, e_2, e_3, e_4\}$ is generated to express the set of possible worlds, as shown in Column f in Table 1. The formula of each tuple is the disjunction of formulae of possible worlds where this tuple exist, as shown in Table 2.

Table 1. The set of tuples

w_i	T	$P(w_i)$	f
w_1	$\{t_1, t_2, t_4\}$	0.2	e_1
w_2	$\{t_2, t_3\}$	0.2	$\neg e_1 \wedge e_2$
w_3	$\{t_2, t_3, t_4\}$	0.2	$\neg e_1 \wedge \neg e_2 \wedge e_3$
w_4	$\{t_1\}$	0.2	$\neg e_1 \wedge \neg e_2 \wedge \neg e_3 \wedge e_4$
w_5	$\{t_1, t_4\}$	0.2	$\neg e_1 \wedge \neg e_2 \wedge \neg e_3 \wedge \neg e_4$

The formulae of tuples generated by Naive method are very long and complicated. On one hand, the long and complicated formulae lead to waste of storage, on the other hand, large time cost of processing formulae will happen when evaluating the subsequent queries. Therefore, formulae simplification is a very

---

[1] In this paper, $A \vDash B$ means that A makes B true.

Table 2. The formulae of tuples

RID	f
t_1	$f(w_1) \vee f(w_4) \vee f(w_5) = e_1 \vee (\neg e_1 \wedge \neg e_2 \wedge \neg e_3 \wedge e_4) \vee (\neg e_1 \wedge \neg e_2 \wedge \neg e_3 \wedge \neg e_4)$
t_2	$f(w_1) \vee f(w_2) \vee f(w_3) = e_1 \vee (\neg e_1 \wedge e_2) \vee (\neg e_1 \wedge \neg e_2 \wedge e_3)$
t_3	$f(w_2) \vee f(w_3) = (\neg e_1 \wedge e_2) \vee (\neg e_1 \wedge \neg e_2 \wedge e_3)$
t_4	$f(w_1) \vee f(w_3) \vee f(w_5) = e_1 \vee (\neg e_1 \wedge \neg e_2 \wedge e_3) \vee (\neg e_1 \wedge \neg e_2 \wedge \neg e_3 \wedge \neg e_4)$

important problem here. Since the formulae of possible worlds are very regular, and the formulae of tuples are the disjunction of formulae of possible worlds that contain the tuple, therefore, in fact, formulae of tuples are also regular. Although simplifying an arbitrary formula is a NP-hard problem, simplification of the regular formulae is tractable.

4   The Optimal Initialization Method

This section introduces a theorem to generate simplified formulae of tuples.

If $f = \neg e_{i_1} \wedge \neg e_{i_2} \ldots \wedge \neg e_{(i_n-1)} \wedge e_{i_n}$, then $f^{\neg} = \neg e_{i_1} \wedge \neg e_{i_2} \ldots \wedge \neg e_{(i_n-1)}$.

If $f(i_m) = \neg e_{i_1} \wedge \neg e_{i_2} \ldots \wedge \neg e_{(i_m-1)} \wedge e_{i_m}$, and $m < n$, then $f^{-i_m}(i_n) = \neg e_{i_{(m+1)}} \wedge \ldots \wedge \neg e_{(i_n-1)} \wedge e_{i_n}$.

Assume that $F(start, I = \{i_1, \ldots, i_m\}, n) = \vee_{j \in [1,m]} f(i_j)$, where $start$, i_j ($j \in [1, m]$) and n are natural numbers, and $m < n$, when $i_j = start$, $f(i_j) = e_{start}$; when $start < i_j < n$, $f(i_j) = \neg e_{start} \wedge \ldots \wedge \neg e_{(i_j-1)} \wedge e_{i_j}$; when $i_j = n$, $f(i_j) = \neg e_{start} \wedge \ldots \wedge \neg e_{(i_j-1)}$. Then the formulae of tuples can be represented as $F(start, I = \{i_1, \ldots, i_m\}, n)$, where I can be considered as the set of order number of possible worlds that contain the tuple, $start$ can be considered as the start number of newly generated variable, and n can be considered as the number of possible worlds.

For example, in Example 1, $f(t_1) = f(w_1) \vee f(w_4) \vee f(w_5)$, then $f(t_1) = F(1, \{1, 4, 5\}, 5)$.

Theorem 1. *When $m \geq 2$, we have the following simplification equation for $F(start, I = \{i_1, \ldots, i_m\}, n)$:*

(1) When $m = 2$, and $i_1 \neq (n-1)$, $F(start, I, n) = f^{\neg}(i_1) \wedge (e_{i_1} \vee f^{-i_1}(i_2))$

(2) When $m = 2$, and $i_1 = (n-1)$, $F(start, I, n) = f^{\neg}(i_1)$

(3) When $m > 2$, $F(start, I = \{i_1, \ldots, i_m\}, n) = f^{\neg}(i_1) \wedge (e_{i_1} \vee F((i_1+1), I = \{i_2, \ldots, i_m\}, n))$

Proof. When $|I| = 2$, and $i_1 \neq (n-1)$, $F(start, \{i_1, \ldots, i_m\}, n) = f(i_1) \vee f(i_2)$.

$f(i_1) = \neg e_{start} \wedge \ldots \wedge \neg e_{(i_1-1)} \wedge e_{i_1} = f^{\neg}(i_1) \wedge e_{i_1}$

$f(i_2) = f^{\neg}(i_1) \wedge \neg e_{i_1} \wedge f^{-i_1}(i_2)$

$f(i_1) \vee f(i_2) = f^{\neg}(i_1) \wedge (e_{i_1} \vee (\neg e_{i_1} \wedge f^{-i_1}(i_2))) = f^{\neg}(i_1) \wedge (e_{i_1} \vee f^{-i_1}(i_2))$

When $m = 2$, and $i_1 = (n-1)$, $F(start, \{i_1, \ldots, i_m\}, n) = f(i_1) \vee f(i_2)$.

$f(i_1) = \neg e_{start} \wedge \ldots \wedge \neg e_{(n-2)} \wedge e_{(n-1)}$, $f(i_2) = \neg e_{start} \wedge \ldots \wedge \neg e_{(n-1)}$

$f(i_1) \vee f(i_2) = f^{\neg}(i_1) \wedge (e_{(n-1)} \vee \neg e_{(n-1)}) = f^{\neg}(i_1)$

When $m > 2$, $F(start, \{i_1, \ldots, i_m\}, n) = f(i_1) \vee \ldots \vee f(i_m)$.

$f(w_{i_j}) = \neg e_{start} \wedge \ldots \wedge \neg e_{(i_j-1)} \wedge e_{i_j}$

$f^{\neg}(i_1) = \neg e_{start} \wedge \ldots \wedge \neg e_{(i_1-1)}$ is the common part of the formulae of $f(i_1)$ and $f(i_2)$, and also is the common part of $f(i_1)$ with all other f in $F(start, \{i_1, \ldots, i_m\}, n)$. Thus, $f^{\neg}(i_1)$ can be extracted. Then, the formula transformed to be: $F(start, \{i_1, \ldots, i_m\}, n) = f^{\neg}(i_1) \wedge (e_{i_1} \vee (\neg e_{i_1} \wedge f^{-i_1}(i_2)) \vee \ldots \vee (\neg e_{i_1} \wedge f^{-i_1}(i_m)))$

where $f^{-i_1}(i_j)$ is the formula of $f(i_j)$ with eliminating the common part with $f(i_1)$, that is, $f^{-i_1}(i_j) = \neg e_{(i_1+1)} \wedge \ldots \neg e_{i_j-1} \wedge e_{i_j}$

After extract $\neg e_{i1}$ from each $(\neg e_{i1} \wedge f^{-i_1}(i_j))$, $e_{i_1} \vee (\neg e_{i_1} \wedge f^{-i_1}(i_j)) = e_{i_1} \vee f^{-i_1}(i_j)$. Obviously,

$F(start, \{i_1, \ldots, i_m\}, n) = f^{\neg}(i_1) \wedge (e_{i_1} \vee f^{-i_1}(i_2) \vee \ldots \vee f^{-i_1}(i_m))$

where $f^{-i_1}(i_2) \vee \ldots \vee f^{-i_1}(i_n))$ can be considered as a new $F((i_1 + 1), \{i_2, \ldots, i_m\}, n) = \vee_{j \in [2,m]} f'(i_j)$, where $start$ is $e_{(i_1+1)}$, and $f'(i_j) = \neg e_{(i_1+1)} \wedge \ldots \wedge \neg e_{(i_j-1)} \wedge e_{i_j}$

Therefore, there is rule (3),

$F(start, \{i_1, \ldots, i_m\}, n) = f^{\neg}(i_1) \wedge (e_{i_1} \vee F((i_1 + 1), \{i_2, \ldots, i_m\}, n))$

Based on Theorem 1, in the simplified formula of $F(start, \{i_1, \ldots, i_m\}, n)$, each variable appears at most once.

4.1   The Algorithm of the Optimal Initialization Method

Algorithm 1 describes the procedure of transforming the input uncertain data into the variable-based representation introduced in Definition 1. For each set of possible worlds, the formula of each tuple is generated by the simplification equation introduce in Theorem 1, after obtaining the set of order numbers of possible worlds that contain the tuple.

The complexity of generating the simplified formula of each tuple is $O(|W|)$, which is much smaller than that of the existing formulae simplification method,

---

Input: (W,P) the set of possible worlds and probability distribution
1 insert a set of tuples $T = \{t | t \in w, w \in W\}$
2 $start \leftarrow$ the max no. of variable in database
3 create a new set of $(|W| - 1)$ variables $\{e_{start}, \ldots, e_{(start+|W|-2)}\}$
4 $P(e_{start}) = P(w_1)$
5 **foreach** $i = start, \ldots, (start + |W| - 2)$ **do**
6 $\quad | \quad P(e_i) = P(w_i)/(P(\neg e_{start}) \cdot \ldots \cdot P(\neg e_{(start+|W|-2)}))$
7 **end**
8 **foreach** $t \in T$ **do**
9 $\quad | \quad I = \{i | t \in w_i\}$
10 $\quad | \quad f(t) = F(start, I, |W|)$
11 **end**

Algorithm 1. Data(W,P)

which is $O(2^{|W|})$ (the number of variables in formulae of tuples is $(|W| - 1)$). Thus, for a data set with $|T|$ tuples and $|W|$ possible worlds, the time complexity of Algorithm 1 is $O(|T| \cdot |W|)$.

Example 2. The formulae of tuples in Table 2 are simplified as shown in Table 3. The formula of each tuple can be generated by the simplification equation for $F(start, \{i_1, \ldots, i_m\}, n)$ in Theorem 1 as follows:

$$f(t_1) = F(1, \{2, 4, 5\}, 5) = f^-(w_2) \wedge (e_2 \vee F(3, \{4, 5\}, 5)) = \neg e_1 \wedge (e_2 \vee F(3, \{4, 5\}, 5)) = \neg e_1 \wedge (e_2 \vee \neg e_3)$$

Table 3. The simplified formulae of tuples

t	f
t_1	$e_1 \vee (\neg e_2 \wedge \neg e_3)$
t_2	$e_1 \vee e_2 \vee e_3$
t_3	$\neg e_1 \wedge (e_2 \vee e_3)$
t_4	$e_1 \vee \neg e_2 \wedge (e_3 \vee \neg e_4)$

5   Experiments

In this section, we compare the time cost, variable redundancy reduction ratio of formulae and time cost of computing tuple existence probability of the initialization methods. Let *Naive* be the existing method, and *SimpleTrans* be the method proposed in this paper. Let *n(Naive)* be the number of variables appear in the formulae of tuples generated by naive method, where duplicate variables are counted as different variables. The duplicate reduce ratio is computed as $(n(Naive)-n(SimpleTrans))/n(naive)$. The goal of the experiment is to show *SimpleTrans* can considerably reduce the duplicate of the formulae without additional time cost. The time cost of tuple existence probability computation is also evaluated to show its effect on the performance of subsequent queries.

All of the algorithms are implemented in Java, and all of our experiments are conducted on a Pentium 2.5 GHz PC with 3G memory, on Windows XP.

5.1   Correlated Parameters and Data Sets

Since the specific data set can affect the result. Therefore, we generate 1000 samples for each data set and report the average result of the 1000 samples. The more the number of all possible worlds, the more common variables exist among formulae of possible worlds, then the more redundancy variables exist in the formulae of tuples by Naive method. Thus, we generate the following data sets by varying $|W|$ (the number of possible worlds).

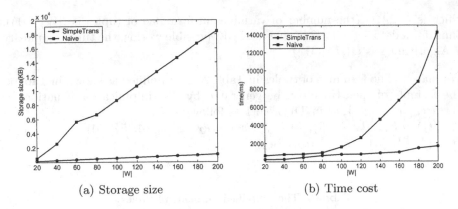

(a) Storage size   (b) Time cost

Fig. 1. Comparison of Naive and SimpleTrans methods when varying $|W|$

Synthetic dataset 1. First, generate 1000 tuples, afterwards, generate $|W|$ possible worlds over these tuples. We must ensure that there are no empty or duplicate possible worlds. $|W|$ varies from 20 to 200.

Synthetic dataset 2. This data set is used for comparing the time cost of probability computation of formulae of tuples generated by Naive and SimpleTrans methods. Since in the formulae of tuples generated by SimpleTrans method, there are no duplicate variables, we choose the tuples whose formulae have duplicate variables generated by Naive method. The number of duplicate variables varies from 10 to 20.

5.2 Result and Analysis

Figure 1a shows that the storage space of Naive method increases from 528 K to 18448 K rapidly as $|W|$ increases, while that of SimpleTrans method increases from 160 K to 1088 K slowly. Figure 1b shows that the time cost of Naive method is larger than that of SimpleTrans, and the difference between them increases as $|W|$ increases. Since the formulae of tuples are the disjunction of formulae of possible worlds that contain the tuple in Naive method, the number of variables in formulae of tuples increases as $|W|$ increases in Naive method. While the number of variable in formulae of tuples by SimpleTrans method is the maximum order number of possible worlds that contain the tuple, thus, it increases slowly as $|W|$ increases. The larger storage size leads to more time cost during data insert operation. Furthermore, the difference between the storage size of Naive and SimpleTrans increases, thus, the difference between their time cost increases.

Figure 2 shows that when the number of duplicate variables in formulae of tuples increases from 11 to 20, the time cost of computing probability after Naive initialization method almost increases exponentially, while the time cost after SimpleTrans nearly stays still. If each variable appears only once in its formula, then the time complexity of computing probability of the formula is $O(n)$, where n is the number of variables in the formula. However, for each duplicate

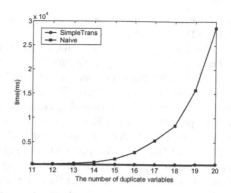

Fig. 2. Time cost of tuple existential probability computation

variable in the formula, two possible assignments need to be enumerated. There-fore, the more the duplicate variables, the more time needed in the probability computation. Since there are no duplicate variables in the formulae of tuples after SimpleTrans method, its time cost of probability computation nearly stays still.

6   Conclusion

This paper proposed an optimal initialization method that transforms uncertain data represented by the set of possible worlds into the variable-based represen-tation. The optimal initialization method generates formulae of tuples without duplicate variables without additional time cost. Subsequent queries can bene-fit from the simplified formulae of tuples by reducing time cost of probability computation of formulae of tuples. The experiments show the efficiency of the optimal initialization method.

References

1. Aggarwal, C.C.: Maybms a system for managing large probabilistic databases. Adv. Database Syst. **35**, 1–34 (2009)
2. Aggarwal, C.C.: Trio a system for data uncertainty and lineage. Managing Min. Uncertain Data **2006**, 1151–1154 (2009)
3. Agreste, S., De Meo, P., Ferrara, E., Ursino, D.: Xml matchers: approaches and challenges. Knowl. Based Syst. **66**, 190–209 (2014)
4. Ayat, N., Akbarinia, R., Afsarmanesh, H., Valduriez, P.: Entity resolution for prob-abilistic data. Inf. Sci. **277**, 492–511 (2014)
5. Dalvi, N., Ré, C., Suciu, D.: Probabilistic databases: diamonds in the dirt. Com-mun. ACM **52**(7), 86–94 (2009)
6. Dan, O., Koch, C., Antova, L.: World-set decompositions: expressiveness and efficient algorithms. Theor. Comput. Sci. **403**(2), 265–284 (2008)

7. Fink, R., Huang, J., Olteanu, D.: Anytime approximation in probabilistic databases. VLDB J. **22**(6), 823–848 (2013)
8. Fink, R., Olteanu, D.: Dichotomies for queries with negation in probabilistic databases. ACM Trans. Database Syst. **41**(1), 4–47 (2016)
9. Fink, R., Olteanu, D., Rath, S.: Providing support for full relational algebra in probabilistic databases. In: ICDE, pp. 315–326. IEEE, Hannover, Germany (2011)
10. Sen, P., Deshpande, A.: Representing and querying correlated tuples in probabilistic databases. In: ICDE, pp. 596–605. IEEE, Istanbul, Turkey (2007)
11. Sen, P., Deshpande, A., Getoor, L.: Prdb: managing and exploiting rich correlations in probabilistic databases. VLDB J. **18**(5), 1065–1090 (2009)
12. Škrbić, S., Racković, M., Takači, A.: Prioritized fuzzy logic based information processing in relational databases. Knowl. Based Syst. **38**, 62–73 (2013)
13. Suciu, D., Olteanu, D., Ré, C., Koch, C.: Probabilistic databases. Synth. Lect. Data Manage. **3**(2), 1–180 (2011)
14. Tang, R., Cheng, R., Wu, H., Bressan, S.: A framework for conditioning uncertain relational data. In: Liddle, S.W., Schewe, K.-D., Tjoa, A.M., Zhou, X. (eds.) DEXA 2012, Part II. LNCS, vol. 7447, pp. 71–87. Springer, Heidelberg (2012)

Erratum to: Aging Locality Awareness in Cost Estimation for Database Query Optimization

Chihiro Kato[1](✉), Yuto Hayamizu[1], Kazuo Goda[1],
and Masaru Kitsuregawa[1,2]

[1] The University of Tokyo, Komaba 4-6-1, Meguro-ku, Tokyo, Japan
kato@tkl.iis.u-tokyo.ac.jp
[2] National Institute of Informatics, Hitotsubashi 2-1-2, Chiyoda-ku, Tokyo, Japan
http://www.u-tokyo.ac.jp/
http://www.nii.ac.jp/

Erratum to:
Chapter 32: S. Hartmann and H. Ma (Eds.)
Database and Expert Systems Applications
DOI: 10.1007/978-3-319-44406-2_32

In an older version of the paper starting on p. 389 of the DEXA proceedings (LNCS 9828), the authors' affiliations were incorrect. This has been corrected.

The updated original online version for this chapter can be found at 10.1007/978-3-319-44406-2_32

S. Hartmann and H. Ma (Eds.): DEXA 2016, Part II, LNCS 9828, p. E1, 2016.
DOI: 10.1007/978-3-319-44406-2_40

Author Index

Printed in the United States
By Bookmasters